Foundations of Stochastic Analysis

This is a volume in
PROBABILITY AND MATHEMATICAL STATISTICS

A Series of Monographs and Textbooks

A complete list of titles in this series appears at the end of this volume.

Foundations of Stochastic Analysis

M. M. RAO

Department of Mathematics
University of California
Riverside, California

1981

ACADEMIC PRESS

A Subsidiary of Harcourt Brace Jovanovich, Publishers

New York London Toronto Sydney San Francisco

ACADEMIC PRESS, INC.
111 Fifth Avenue, New York, New York 10003

United Kingdom Edition published by
ACADEMIC PRESS, INC. (LONDON) LTD.
24/28 Oval Road, London NW1 7DX

Library of Congress Cataloging in Publication Data

Rao, Malempati Madhusudana, Date.
Foundations of stochastic analysis.

(Probability and mathematical statistics)
Bibliography: p.
Includes index.
1. Stochastic analysis. I. Title. II. Series.
QA274.2.R36 519.2 81-10831
ISBN 0-12-580850-X AACR2

PRINTED IN THE UNITED STATES OF AMERICA

81 82 83 84 9 8 7 6 5 4 3 2 1

To my sister, Jamantam,
for her help and encouragement
throughout my education

Contents

Chapter IV **Martingales and Likelihood Ratios**

Chapter V **Abstract Martingales and Applications**

Preface

Stochastic analysis consists of a study of different types of stochastic processes and of their transformations, arising from diverse applications. A basic problem in such studies is the existence of probability spaces supporting these processes when only their finite-dimensional distributions can be specified by the experimenter. The first solution to this problem is provided by the fundamental existence theorem of Kolmogorov (1933), according to which such a process, or equivalently a probability space, exists if and only if the set of all finite-dimensional distributions forms a compatible family. This result has been analyzed and abstracted by Bochner (1955), who showed it to be a problem on projective systems of probability spaces and who then presented sufficient conditions for such a system to admit a limit. The latter becomes the desired probability space, and this abstraction has greatly enlarged the scope of Kolmogorov's idea. One of the purposes of this book is to present the foundations of this theory of Kolmogorov and Bochner and to indicate its impact on the growth of the subject.

An elementary but important observation is that a projective system uniquely associates with itself a set martingale. In many cases the latter can be represented by a (point) martingale. On the other hand, a (point) martingale trivially defines a projective system of (signed) measure spaces. Thus the Kolmogorov–Bochner theory naturally leads to the study of martingales in terms of the basic (and independent) work due to Doob and Andersen–Jessen. However, to analyze and study the latter subject in detail, it is necessary to turn to the theory of conditional expectations and probabilities, which also appears in the desired generality in Kolmogorov's *Foundations* (1933) for the first time. This concept seems simple on the surface, but it is actually a functional operation and is nontrivial. To facilitate dealing with conditional expectations, which are immensely important in stochastic analyses, a detailed structural study of these operators is desirable. But such a general and comprehensive treatment has not yet appeared in book form. Consequently, after presenting the basic Kolmogorov–Bochner theorem in Chapter I, I devote Chapter II to this subject. The rest of the book treats aspects of martingales, certain extensions of projective limits, and applications to ergodic theory, to harmonic analysis, as well as to (Gaussian) likelihood ratios. The topics considered here are well suited for showing the natural interplay between real and abstract methods in stochastic analysis. I have tried to make this explicit. In so doing, I attempted to motivate the ideas at each turn so that one can see the appropriateness of a given method.

As the above description implies, a prerequisite for this book is a standard measure theory course such as that given in the Hewitt–Stromberg or Royden textbooks. No prior knowledge of probability (other than that it is a normed measure) is assumed. Therefore most of the results are proved in detail (at the risk of some repetitions), and certain elementary facts from probability are included. Actually, the present account may be regarded as an updating of Kolmogorov's *Foundations* (English translation, Chelsea, 1950, 74 pp.) referred to above, and thus a perusal of its first 56 pages will be useful. The treatment and the point of view of the present book are better explained by the brief outline that follows. A more detailed summary appears at the beginning of each chapter.

After introducing the subject, the main result proved in Chapter I is the basic Kolmogorov–Bochner existence theorem referred to above. To facilitate later work and to fix some notation and terminology, a résumé of real and abstract analysis is included here. Occasionally, some needed results that are not readily found in textbooks are presented in full detail. Most of these (particularly Section 4) can be omitted, and the reader may refer to them only when they are invoked. Chapter II is devoted entirely to conditional expectations and probabilities containing several characterizations of these operators and measures. The general viewpoint emphasizes that the Kolmogorov foundations are adequate for all the known applications. This is contrasted with (and is shown to include) the new foundations proposed by Rényi (1955). Then the integral representation of Reynolds operators is given as an application of these ideas, to be used later for a unified study of ergodic-martingale theories. Chapter III contains extensions of the Kolmogorov–Bochner theorem. The existence theorem of Prokhorov and certain other results of Choksi are also proved here. A treatment of direct limits of measures is necessary. This topic and infinite product conditional probabilities (Tulcea's theorem) are discussed. The work in this chapter is somewhat technical, and the reader might postpone the study of it until later. Chapters IV and V contain several aspects of (discrete) martingale theory. These include both scalar- and vector-valued martingales, their basic convergence, and many applications. The latter deal with ergodic theory, likelihood ratios, the Gaussian dichotomy theorem, and some results on the convergence of "partial sums" in harmonic analysis on a locally compact group. At the end of each chapter there is a problem section containing several facts, including important results in information theory, and many additions to the text. Most of these are provided with copious hints.

References to the literature are interspersed in the text with (I hope) due credits to various authors, backed up by an extensive bibliography. However, I have not always given the earliest reference of a given result. For instance, all the early work by Doob is referenced to his well-known treatise, and similarly, certain others with references to the monumental work of Dunford–Schwartz, from which an interested reader can trace the original source.

The arrangement of the material is such that this book can be used as a textbook for study following a standard real variable course. For this purpose, the following selections, based on my experience, are suggested: A solid semester's course can be given using Sections 1–3 of Chapter I, Chapter II (minus Section 6), Sections 1 and 2 of Chapter III, and most of Chapter IV. Then one can use any of the omitted sections with a view to covering Chapter V for the second semester. (This may be appropriately divided for a quarter system.) There is a sufficient amount of material for a year's treatment, and several possible extensions and open problems are pointed out, both in the text and in the Complements sections of the book. For ease of reference, theorems, lemmas, definitions, and the like are all consecutively numbered. Thus II.4.2 refers to the second item in Section 4 of Chapter II. In a given chapter (or section) the corresponding chapter (and section) number is omitted.

Several colleagues and students made helpful suggestions while the book was in progress. For reading parts of an earlier draft and giving me their comments and corrections, I am grateful to George Chi, Nicolae Dinculeanu, Jerome Goldstein, William Hudson, Tom S. Pitcher, J. Jerry Uhl, Jr., and Grant V. Welland. This work is part of a project that was started in 1968 with a sabbatical leave from Carnegie-Mellon University, continued at the Institute for Advanced Study during 1970-1972, and completed at the University of California at Riverside. This research was in part supported by the Grants AFOSR-69-1647, ARO-D-31-124-70-G100, and by the National Science Foundation. I wish to express my gratitude to these institutions and agencies as well as to the UCR research fund toward the preparation of the final version. I should like to thank Mrs. Joyce Kepler for typing the final and earlier drafts of the manuscript with diligence and speed. Also D. M. Rao assisted me in checking the proofs and preparing the Index. Finally, I appreciate the cooperation of the staff of Academic Press in the publication of this volume.

CHAPTER

I

Introduction and Generalities

This chapter is devoted to a motivational introduction and to preliminaries on real and abstract analysis to be used in the rest of the book. The main probabilistic result is the Kolmogorov–Bochner theorem on the existence of general, not necessarily scalar valued stochastic processes. Also included is a result on the existence of suprema for sets of measurable functions. Several useful complements are included as problems.

1.1 INTRODUCING A STOCHASTIC PROCESS

Stochastic analysis, in a general sense, is a study of the structural and inferential properties of stochastic processes. The latter object may be described as an indexed family of random variables $\{X_t, t \in T\}$ on a probability space. This brief statement implies much more and contains certain hidden conditions on the family. To explain this point clearly and precisely, we use the axiomatic theory of probability, due to Kolmogorov, and show how the basic probability space may be constructed, with the available initial information, in order that a stochastic process may be defined on it. Other axiomatic approaches, notably Rényi's, are also available, but the methods developed for the Kolmogorov model are adequate for all our purposes. This will become more evident in Chapter II, which elaborates on conditional probabilities, where Rényi's model is discussed and compared.

Thus, if (Ω, Σ, P) is a probability space, a mapping $X_t\colon \Omega \to \mathbb{R}$ (real line) is a (real) *random variable* if X_t is a measurable function. To fix the notation and for precision, we shall present a *résumé* of the main results from real analysis in Section 2, which will then be freely used in the book. Let T be an index set and $\{X_t, t \in T\}$ be a family of random variables on (Ω, Σ, P). If $t_1, \ldots, t_n$ are n points from T and $x_1, \ldots, x_n$ are in $\mathbb{R}$ or are $\pm\infty$, define the function $F_{t_1, \ldots, t_n}$, called the *n-dimensional (joint) distribution function* of $(X_{t_1}, \ldots, X_{t_n})$, by the equation

$$F_{t_1,\ldots,t_n}(x_1, \ldots, x_n) = P\left[\bigcap_{i=1}^{n} \{\omega : X_{t_i}(\omega) < x_i\}\right]. \tag{1}$$

As n and the t points vary, we get a family of multidimensional distribution functions $\{F_{t_1,\ldots,t_n}, t_i \in T, n \geq 1\}$. Since $\{\omega : X_t(\omega) < \infty\} = \Omega$, from (1) we get at once the following pair of relations:

$$F_{t_1,\ldots,t_n}(x_1, \ldots, x_{n-1}, \infty) = F_{t_1,\ldots,t_{n-1}}(x_1, \ldots, x_{n-1}), \tag{2}$$

$$F_{t_{i_1},\ldots,t_{i_n}}(x_{i_1}, \ldots, x_{i_n}) = F_{t_1,\ldots,t_n}(x_1, \ldots, x_n), \tag{3}$$

where $(i_1, \ldots, i_n)$ is any permutation of $(1, \ldots, n)$. The functions $\{F_t, t \in T\}$ are monotone, nondecreasing, nonnegative, and left continuous. Moreover, $F_{t_1,\ldots,t_n}(+\infty, \ldots, +\infty) = 1$ and $F_{t_1,\ldots,t_n}(x_1, x_2, \ldots, x_{n-1}, -\infty) = 0$. The relations (2) and (3) are called the *Kolmogorov compatibility conditions* of the family $\{F_{t_1,\ldots,t_n} : t_i \in T, n \geq 1\}$. Thus any indexed family of (real) random variables on a probability space (or equivalently a stochastic process) determines a compatible collection of finite-dimensional distribution functions whose cardinality is that of D, the directed set (by inclusion) of all finite subsets of T.

The preceding description shows that even if the question of existence of a probability space (Ω, Σ, P) is not settled, it is simple to exhibit compatible families of distribution functions. It will then be natural to inquire into their relation to some (or any) probability space. To see that such families exist, let $f_1, \ldots, f_n$ be positive, measurable functions on the line each of which has integral equal to 1. Define $F_{1,2,\ldots,n}$ $(= F_n$, say):

$$F_n(x_1, \ldots, x_n) = \int_{-\infty}^{x_1} \cdots \int_{-\infty}^{x_n} f_1(t_1) \cdots f_n(t_n)\, dt_n \cdots dt_1. \tag{4}$$

It is clear that $\{F_n, n \geq 1\}$ is a family of distribution functions satisfying (2) and (3) with $T = \mathbb{N}$ there. A less simple collection is the Gaussian family of distribution functions given by

$$G_n(x_1, \ldots, x_n) = C_n \int_{-\infty}^{x_1} \cdots \int_{-\infty}^{x_n} \exp[-\tfrac{1}{2}(t - \alpha)K^{-1}(t - \alpha)']\, dt_n \cdots dt_1, \tag{5}$$

where $K = (k_{ij})$ is a real symmetric positive definite matrix, $\alpha = (\alpha_1, \ldots, \alpha_n)$ is a point of $\mathbb{R}^n$, $C_n = [(2\pi)^n \det(K)]^{-1/2}$, $\det(K) = \text{determinant}(K)$, and a prime denotes the transpose. An easy computation, which we omit, shows that the family $\{G_n, n \geq 1\}$ of (5) satisfies (2) and (3). Thus one can find many compatible families of distribution functions on $\{\mathbb{R}^n, n \geq 1\}$. A fundamental theorem of Kolmogorov states that every such compatible family of distribution functions yields a probability space and a stochastic process on it such that the (joint) finite-dimensional distributions on the process are precisely the given distributions. We shall prove this (in a slightly more general form) in Section 3. Thus the existence of a probability space is equivalent to the selection of a compatible family of distributions. Depending on the type of this family (i.e., *Gaussian*, *Poisson*, etc.), the probability space (Ω, Σ, P), or the stochastic process, is referred to by the same name. Let us first recall some measure theoretical results for convenient reference.

1.2 RÉSUMÉ OF REAL ANALYSIS

In this section we present an account of certain results from measure theory, mostly without proofs. Our purpose is to fix some notation and to make certain concepts precise since the reader is expected to have this background. (The omitted proofs may be found in Halmos [1], Hewitt–Stromberg [1], Royden [1], Sion [1], or Zaanen [1].)

Most of the references to measure will be to the abstract theory set forth by Càrathéodory as follows. Let $\mathscr{A}$ be a collection of subsets of a point set Ω for which $\emptyset \in \mathscr{A}$ and let $\tau: \mathscr{A} \to \bar{\mathbb{R}}^+$ be a function such that $\tau(\emptyset) = 0$. We define the set function μ on Ω by

$$\mu(A) = \inf\left\{ \sum_{i=1}^{\infty} \tau(B_i) : B_i \in \mathscr{A},\ A \subset \bigcup_{i=1}^{\infty} B_i \right\}, \qquad A \subset \Omega, \tag{1}$$

where $\inf(\emptyset) = +\infty$. We say that μ is *generated* by the pair $(\tau, \mathscr{A})$. Then μ is an outer measure. Let $\mathscr{M}_\mu = \{A \subset \Omega : \mu(T) = \mu(A \cap T) + \mu(A^c \cap T)$ for all $T \subset \Omega\}$. The following results holds.

1. Theorem (a) *The restriction of μ to $\mathscr{M}_\mu$, denoted by $\mu | \mathscr{M}_\mu$, is σ-additive, and $\mathscr{M}_\mu$ is a σ-algebra, containing the class of its μ-null sets (i.e., $\mathscr{M}_\mu$ is complete)*;

(b) *if $\mathscr{A}$ is a semi-ring and τ is additive, then $\mathscr{A} \subset \mathscr{M}_\mu$ and μ is an $\mathscr{A}_\sigma$-outer measure, i.e., for any $A \subset \Omega$, $\mu(A) = \inf\{\mu(B) : B \in \mathscr{A}_\sigma,\ B \supset A\}$, where $\mathscr{A}_\sigma$ is the closure of $\mathscr{A}$ under countable unions*;

(c) *under the hypothesis of* (b), *$\mu | \mathscr{A} = \tau$ iff τ is σ-additive*; *and*

(d) *if $\mu(\Omega) < \infty$, for each $A \subset \Omega$ there exists a $B \in \mathscr{A}_{\sigma\delta}$ (the closure of $\mathscr{A}_\sigma$ under countable intersections), $B \supset A$, such that $\mu(B) = \mu(A)$.*

[*Hence each A has a measurable cover B if μ is finite and the hypothesis of* (b) *holds.*]

Note that if $\bar{\mu}$ is defined by $\bar{\mu}(A) = \inf\{\mu(B): B \in \mathscr{M}_\mu, B \supset A\}$ for $A \subset \Omega$, then $\bar{\mu}$ is an outer measure and $\bar{\mu} | \mathscr{M}_\mu = \mu$ even when μ is an arbitrary (not necessarily generated) outer measure on Ω. If $\mathscr{M}_{\bar{\mu}} = \mathscr{M}_\mu$, we say that μ is *Carathéodory regular*, so that μ admits no more extensions by this procedure. (It may be verified that $\bar{\bar{\mu}} = \bar{\mu}$ for all μ, although $\mu \neq \bar{\mu}$ may occur.) Thus μ is Carathéodory regular iff it is an $\mathscr{M}_\mu$-outer measure. In this case, if $A_n \uparrow A$, then $\mu(A_n) \uparrow \mu(A)$. We present later the topological regularity of μ when Ω is a topological space. It should also be remarked that both (b) and (c) of the theorem may be false if $\mathscr{A}$ is only a lattice.

On extension of measures from a smaller ring to a larger one, a positive answer is provided by part (b) of the above theorem. A more precise result is as follows:

2. Theorem (Hahn) *Let Σ_0 be an algebra of sets of Ω and let μ on Σ_0 be a countably additive real, complex, or (more generally) $\mathscr{X}$-valued function, where $\mathscr{X}$ is a reflexive Banach space (here $\sum_{i=1}^{\infty} \mu(A_i) \in \mathscr{X}$ means that the series converges unconditionally in the norm of $\mathscr{X}$), then μ has a unique countably additive extension to $\Sigma = \sigma(\Sigma_0)$, the σ-algebra generated by Σ_0. The same result holds if μ is positive, and extended real valued if it is σ-finite in addition.*

The vector case (i.e., μ is $\mathscr{X}$-valued) uses a theorem of Pettis stating that weak and strong σ-additivities are equivalent in a Banach space. The σ-finite case is the key part of this result; others can then be reduced to this case and can be further generalized.

The following result on the structure of certain measurable functions is particularly useful in probability and is due to Doob [1] and (in the form stated) to Dynkin [1]. We include its proof.

3. Theorem (Doob–Dynkin lemma) *Let (Ω, Σ) and $(S, \mathscr{S})$ be measurable spaces and $f: \Omega \to S$ be measurable for $(\Sigma, \mathscr{S})$. Let $\mathscr{A} = f^{-1}(\mathscr{S}) \subset \Sigma$ and $g: \Omega \to \mathbb{R}$ be a mapping. Then g is $\mathscr{A}$-measurable (relative to the Borel algebra $\mathscr{B}$ of $\mathbb{R}$) iff there exists a measurable $h: S \to \mathbb{R}$ such that $g = h \circ f$.*

Proof Since $g^{-1}(\mathscr{B}) = f^{-1}(h^{-1}(\mathscr{B})) \subset f^{-1}(\mathscr{S}) = \mathscr{A}$, only the converse is nontrivial.

It suffices to prove the relation with the additional assumption that g is a step function.† Indeed, if this is known and g is any $\mathscr{A}$-measurable function, then there exists a sequence of $\mathscr{A}$-measurable step functions g_n such that $g_n(\omega) \to g(\omega)$, $\omega \in \Omega$, and by the special case $g_n = h_n \circ f$ for some

† A step function takes finitely many finite values on disjoint measurable sets. A simple function is a step function with support of finite measure.

measurable $h_n: S \to \mathbb{R}$, $n \geq 1$. Let $\bar{S} = \{s \in S : h_n(s) \to \text{a limit}\}$. Then $\bar{S} \in \mathcal{S}$ and $f(\Omega) \subset \bar{S}$. Define $h: S \to \mathbb{R}$ as $h(s) = \lim_n h_n(s)$, $s \in \bar{S}$, and $=0$ for $s \in S - \bar{S}$. It follows that $g = h \circ f$; so we need to establish the special case.

Thus let $g = \sum_{i=1}^n a_i \chi_{A_i}$, $A_i \in \mathcal{A}$, disjoint. Then there exist $S_i \in \mathcal{S}$ such that $A_i = f^{-1}(S_i)$. We disjunctify $\{S_i\}_1^n$. Let $T_1 = S_1$, and for $j > 1$, $T_j = S_j - \bigcup_{i=1}^{j-1} S_i$. Then $f^{-1}(T_j) = f^{-1}(S_j) - \bigcup_{i=1}^{j-1} f^{-1}(S_i) = A_j - \bigcup_{i=1}^{j-1} A_i = A_j$ by disjointness of A_i. If $h = \sum_{i=1}^n a_i \chi_{T_i}: S \to \mathbb{R}$, then h is measurable,

$$(h \circ f) = \sum_{i=1}^{n} a_i \chi_{T_i} \circ f = \sum_{i=1}^{n} a_i \chi_{f^{-1}(T_i)} = g, \tag{2}$$

and the result follows.

The importance of this becomes clear from a specialization:

4. Corollary *Let $(S, \mathcal{S}) = (\mathbb{R}^n, \mathcal{B}^n)$, the Borelian n-space. If $f = (f_1, \ldots, f_n): \Omega \to \mathbb{R}^n$ is $(\Sigma, \mathcal{S})$-measurable, and $\mathcal{A} = f^{-1}(\mathcal{S})$, then $g: \Omega \to \mathbb{R}$ is $\mathcal{A}$-measurable iff there is a measurable $h: \mathbb{R}^n \to \mathbb{R}$ such that $g = h(f_1, \ldots, f_n)$.*

We shall see later that the case $S = \mathbb{R}^T$, $T \subset \mathbb{R}$, gives an interesting application, related to the Kolmogorov existence theorem noted in Section 1. The Lebesgue limit theorems are deducible from the following monotone convergence criterion:

5. Theorem *Let μ on (Ω, Σ) be a measure (or an $\mathcal{M}_\mu$-outer measure). If $0 \leq f_n \leq f_{n+1} \leq \cdots$ are μ-measurable (or arbitrary) functions on Ω, then*

$$\lim_n \int_\Omega f_n \, d\mu = \int_\Omega \lim_n f_n \, d\mu, \tag{3}$$

where the integral is defined in the usual manner for the measurable case, or more generally (for both cases) if one sets $\int_\Omega f \, d\mu = \mu(\text{region under } f^+) - \mu(\text{region under } f^-)$ when this makes sense for $f: \Omega \to \mathbb{R}$. [This integral is evidently subadditive. The region under f^+ is the set $\{(\omega, y): 0 \leq y < f(\omega)\} \subset \Omega \times \mathbb{R}^+$.]

The dominated convergence theorem and Fatou's lemma are deduced from this result immediately.

We next present the (Lebesgue–)Radon–Nikodým and the (M. H. Stone extension of) Fubini and Tonelli theorems, which are used often in our work.

6. Definition (a) Let (Ω, Σ, μ) be a complete measure space such that μ has the *finite subset property*, i.e., for each $A \in \Sigma$, $\mu(A) > 0$, there is $B \in \Sigma$, $B \subset A$, $0 < \mu(B) < \infty$. Then μ is said to be a *localizable* measure if each nonempty collection $\mathscr{C} \subset \Sigma$ has a supremum $C_0 \in \Sigma$, in the sense that for each $B \in \mathscr{C}$, (i) $\mu(B - C_0) = 0$ and (ii) if $C_1 \in \Sigma$ and $\mu(B - C_1) = 0$, $B \in \mathscr{C}$, then $\mu(C_0 - C_1) = 0$.

(b) The measure μ has the *direct sum property* (or is *strictly localizable*) if there exists a collection $\{A_\alpha, \alpha \in I\} \subset \Sigma$, $0 < \mu(A_\alpha) < \infty$, $\Omega = \bigcup_\alpha A_\alpha$, such that each $B \in \Sigma$, $\mu(B) < \infty$, satisfies $B \subset (\bigcup_{\alpha \in J_B} A_\alpha \cup N)$ where $\mu(N) = 0$ and $J_B \subset I$ is countable.

It can be shown that if μ has the direct sum property, then μ is localizable (cf. Zaanen [1, p. 263]). [Thus each σ-finite measure (special case of the direct sum property) is localizable.] However, it is not known whether or not these two concepts are equivalent.† In case the cardinality of the algebra Σ is at most of the continuum, McShane [1, p. 333] shows that the two notions are equivalent. The concept of localizability was introduced by Segal [1], to whom the precise form of the Radon–Nikodým theorem (given as part (i)) below is due.

7. Theorem (i) (Radon–Nikodým) *Let (Ω, Σ) be a measurable space and ν, μ two measures on Σ with finite subset properties. Let ν vanish on each μ null set ($\nu \ll \mu$ or ν is absolutely continuous for μ). Then there exists a μ-unique μ-measurable function $f: \Omega \to \bar{\mathbb{R}}^+$ such that $\nu(A) = \int_A f\,d\mu$, $A \in \Sigma$ iff μ is localizable. Moreover, $f < \infty$ a.e., if ν is σ-finite (or only has the direct sum property) and is integrable if $\nu(\Omega) < \infty$.*

(ii) (Lebesgue–Radon–Nikodým) *Let μ, ν be arbitrary finite measures on Σ. Then $\nu = \nu_1 + \nu_2$ uniquely, where $\nu_1 \ll \mu$ and ν_2 is μ-singular, in the sense that there exists $A_0 \in \Sigma$, $\mu(A_0) = 0$, $\nu_2(A_0^c) = 0$, and a μ-unique integrable $f: \Omega \to \mathbb{R}^+$ such that*

$$\nu(A) = \int_A f\,d\mu + \nu_2(A_0 \cap A), \qquad A \in \Sigma. \tag{4}$$

For a detailed treatment of the Radon–Nikodým theorem, the reader may consult the textbook by Zaanen [1].

8. Theorem (i) (Fubini–Stone) *Let $(\Omega_i, \Sigma_i, \mu_i)$, $i = 1, 2$, be two measure spaces and $(\Omega, \Sigma, \mu) = (\Omega_1 \times \Omega_2, \Sigma_1 \otimes \Sigma_2, \mu_1 \otimes \mu_2)$, their product. Let $f: \Omega \to \bar{\mathbb{R}}$ be a μ-measurable function. If $\int_\Omega |f|\,d\mu < \infty$, then the functions $\int_{\Omega_1} f(\omega_1, \cdot)\,d\mu_1$ and $\int_{\Omega_2} f(\cdot, \omega_2)\,d\mu_2$ are measurable relative to μ_2 and μ_1, respectively, and moreover,*

$$\int_\Omega f\,d\mu = \int_{\Omega_1} \int_{\Omega_2} f\,d\mu_2\,d\mu_1 = \int_{\Omega_2} \int_{\Omega_1} f\,d\mu_1\,d\mu_2. \tag{5}$$

† It seems that the inequivalence of these concepts has recently been established (oral communication from S. D. Chatterji).

(ii) (Tonelli) *Let μ_1, μ_2 be σ-finite and $f: \Omega \to \bar{\mathbb{R}}^+$ be just μ $(=\mu_1 \otimes \mu_2)$-measurable. Then the conclusions of* (i) *hold. If μ_1, μ_2 are not restricted but if there exist μ-measurable f_n, $0 \leq f_n \uparrow f$ a.e., such that $\int_\Omega f_n \, d\mu < \infty$ for each n, then again* (5) *holds.*

This is the most general result known. For instance, if (5) holds for all nonnegative μ-measurable f, should μ_1, μ_2 be localizable? What can we say about these measures? The answers to these questions have interest in real analysis.

In case the basic space Ω is topological, some refinements are possible. We state the regularity of μ in that situation, as this is needed. To motivate the general definition, it is convenient to recall the Lebesgue–Stieltjes measure μ on $\mathbb{R}^n$. From the standard theory such a μ has the following properties: (a) open sets are μ-measurable, (b) compact sets (are measurable and) have finite measure, and (c) the outer measure of every set can be approximated from above by the measures of open sets, i.e., $\mu^*(A) = \inf\{\mu(G): G \supset A, G \text{ open}\}$. Then μ also has an inner approximation property: (d) $\mu(G) = \sup\{\mu(C): C \subset G, \ C \text{ compact}\}$ for all open G. Conversely, these properties characterize Lebesgue–Stieltjes measures on $\mathbb{R}^n$. (See, e.g., Sion [1].) This last property is called *inner regularity* (and (c) is *outer regularity*) and is the crucial requirement in the general study. It is taken as a definition. Let us state this precisely.

9. Definition (i) Let Ω be a Hausdorff topological space and $\mathscr{B}$ its Borel algebra (i.e., the σ-algebra generated by the open or closed sets of Ω). Then a measure μ on $\mathscr{B}$ is *inner regular*, or a *Radon measure*, if it is locally finite (i.e., every point of Ω is in some open set G of finite μ-measure, so that each compact set has finite μ-measure) and for each $A \in \mathscr{B}$ we have

$$\mu(A) = \sup\{\mu(C): C \subset A, \ C \text{ compact}\} \qquad \text{(inner regularity)}. \tag{6}$$

[It can then be shown that μ must also be outer regular on $\mathscr{B}$.]

(ii) Let Ω be a topological space and Σ_0 be an algebra of subsets of Ω. Then an additive set function μ on $\Sigma_0 \to \mathbb{C}$ is called *regular* (in the sense of Dunford–Schwartz, or *D–S sense*) if for each $A \in \Sigma_0$ and $\varepsilon > 0$ there exists a pair $\{E, F\} \subset \Sigma_0$ such that $\bar{E} \subset A \subset \text{int}(F)$, where $\bar{E}$ is the closure of E and $\text{int}(F)$ is the interior of F, and for each $C \subset F - E$, $C \in \Sigma_0$, we have $|\mu(C)| < \varepsilon$.

If in (ii) μ is a σ-additive measure and Σ_0 is the Borel algebra, then it may be shown that this definition and that of (i) agree. (One needs the work of Schwartz [1, pp. 17–18].) Moreover, the variation measure of μ in (ii) is also regular. If in (i) Ω is a completely regular space and $\mathscr{B}$ is the *Baire σ-algebra* (i.e., the smallest σ-algebra with respect to which each real continuous function on Ω is measurable), then μ of (i) is a regular Baire measure when all sets in the definition are restricted to the Baire compact and open sets.

The following result provides basic information on these (regular) measures and some interrelations.

10. Theorem (i) (Alexandroff) *Let Ω be a compact space and μ be an additive bounded (real or complex) regular (D–S sense) set function on an algebra Σ_0 of Ω. Then μ has a unique σ-additive extension to $\Sigma = \sigma(\Sigma_0)$ and the extended function, also denoted μ, is regular (D–S sense). [The σ-additivity part holds even if the boundedness hypothesis is suppressed—a result due to R. P. Langlands.]*

(ii) *Let Ω be a sigma compact Hausdorff space and μ a regular Baire measure on Ω. Then there exists a unique Radon measure $\bar{\mu}$ on Ω such that its restriction to the Baire σ-algebra coincides with μ. Moreover, for each Borel set B of finite measure, there is a Baire set D and a Borel null set N such that $B = D \Delta N$ (symmetric difference).*

The result (i) is proved in Dunford–Schwartz [1, p. 138] and (ii) is in Royden [1, p. 314]. An extended discussion of regularity and of Radon measures is given in Schwartz [1]. We find an application of regularity in Chapter II when conditional probabilities are treated.

1.3 THE BASIC EXISTENCE THEOREM

It was noted in Section 1 that the first problem for stochastic processes, and then for an anlysis on them, is to establish their existence when the basic information can be put in terms of a (compatible) family of distribution functions. The solution, due to Kolmogorov [1], will be precisely stated here. To make the structure more explicit, we then establish a slightly more general version of this fundamental result. It will be sufficient for many applications.

1. Theorem (Kolmogorov) *Let T be a subset of the real line $\mathbb{R}$ and $t_1 < \cdots < t_n$ be n points from it. With each such n-tuple let there be given a distribution function $F_{t_1,\ldots,t_n}$. Suppose that the family $\{F_{t_1,\ldots,t_n};\ t_i \in T,\ n \geq 1\}$ thus given is compatible, i.e., equations (2) and (3) of Section 1 are satisfied. Let $\Omega = \mathbb{R}^T$. Let $\mathcal{B}$ be the σ-algebra generated by all the sets*

$\{\{\omega \in \Omega : -\infty < \omega(t) < a\} : t \in T,\ a \in \mathbb{R}\}$. Then there exists a unique probability P on $\mathscr{B}$ and a stochastic process $\{X_t, t \in T\}$ on $(\Omega, \mathscr{B}, P)$ such that

$$P\left[\bigcap_{i=1}^{n} \{\omega : X_{t_i}(\omega) < x_i\}\right] = F_{t_1,\ldots,t_n}(x_1, \ldots, x_n), \qquad x_i \in \mathbb{R}, \quad n \geq 1. \tag{1}$$

The process is formed by the coordinate functions defined by the equation, $X_t(\omega) = \omega(t)$, for $\omega \in \Omega$, $t \in T$.

The compatibility conditions may be expressed more symmetrically using the following abstration due to Bochner [1]. If $\mathscr{B}_\alpha$ is the Borel algebra of $\mathbb{R}^\alpha$, $\alpha = (t_1, \ldots, t_n)$, and P_α is the Lebesgue–Stieltjes probability determined by the distribution $F_{t_1,\ldots,t_n}$, so that

$$P_\alpha(A) = \int \cdots \int_A dF_{t_1,\ldots,t_n}, \qquad A \in \mathscr{B}_\alpha, \tag{2}$$

then the family of distributions is equivalent to the set $\{P_\alpha, \alpha \in D\}$ of probabilities where D is the set of all finite subsets of T. If α, β are a pair of elements of D, let $\alpha < \beta$ stand for $\alpha \subset \beta$ so that D is *directed*, i.e., $(D, <)$ is partially ordered and for any two elements there is a third (namely, their union) in D, which dominates both. If $\alpha < \beta$, let $\pi_{\alpha\beta}$ be the coordinate projection of $\mathbb{R}^\beta$ onto $\mathbb{R}^\alpha$. Thus if $\alpha = (t_1, \ldots, t_m)$, $\beta = (t_1, \ldots, t_m, t_{m+1}, \ldots, t_n)$, and $x_\beta = (x_{t_1}, \ldots, x_{t_n})$ in $\mathbb{R}^\beta$, we have $\pi_{\alpha\beta} x_\beta = x_\alpha = (x_{t_1}, \ldots, x_{t_m})$. Then the compatibility conditions (2) and (3) of Section 1 take the form: for $\alpha < \beta$, $P_\alpha = P_\beta \circ \pi_{\alpha\beta}^{-1}$. The conclusion of the above theorem thus becomes the existence of a probability P on $\mathscr{B}$ such that $P_\alpha = P \circ \pi_\alpha^{-1}$, $\alpha \in D$, where $\pi_\alpha : \Omega \to \mathbb{R}^\alpha$ is the coordinate projection. Note that each P_α, being a Lebesgue–Stieltjes probability, is a Radon measure. This formulation admits extensions.

The following result is a generalization of Theorem 1 in that $\mathbb{R}$ is replaced by a more general space, though the proof still uses the basic ideas of Kolmogorov. On the other hand, it is a specialization of a more inclusive theorem due to Bochner [1], which will be given in Chapter III along with other generalizations. The present intermediate version fits in here.

2. Theorem (Kolmogorov–Bochner) *Let T be an index set and D be the directed (by inclusion) family of all finite subsets of T. Let $(\Omega_t, \mathscr{B}_t)_{t \in T}$ be a family of measurable spaces where Ω_t is a Hausdorff space and $\mathscr{B}_t$ is the Borel algebra of Ω_t, $t \in T$. If $\Omega_\alpha = \times_{t \in \alpha} \Omega_t$, $\mathscr{B}_\alpha = \bigotimes_{t \in \alpha} \mathscr{B}_t$, let P_α be a Radon probability on $\mathscr{B}_\alpha$, $\alpha \in D$, where all product spaces are endowed with product topologies. Let $\Omega = \times_{t \in T} \Omega_t$, $\mathscr{B} = \bigotimes_{t \in T} \mathscr{B}_t$; then there exists a probability P $(= P_T)$ on $\mathscr{B}$ such that $P \circ \pi_\alpha^{-1} = P_\alpha$, $\alpha \in D$, where $\pi_\alpha : \Omega \to \Omega_\alpha$ is the coordinate projection (and $\pi_{\alpha\beta} : \Omega_\beta \to \Omega_\alpha$ are also such projections), iff $P_\alpha = P_\beta \circ \pi_{\alpha\beta}^{-1}$ for* *each* α, β *in* D *with* $\alpha < \beta$. *When this holds,* P *is "restrictedly regular" in that for each* $A \in \Sigma_0 = \bigcup_{\alpha \in D} \pi_\alpha^{-1}(\mathscr{B}_\alpha)$, *we have*

$$P(A) = \sup\{P(C) : C \in \mathscr{C}\}, \tag{3}$$

where $\mathscr{C} = \{C = \pi_\alpha^{-1}(K) : K \in \mathscr{B}_\alpha, K \text{ compact}, \alpha \in D\}$, *the cylinders with compact bases.* (Again, $X(\omega): t \mapsto \omega(t) \in \Omega_t$, $\omega \in \Omega$, $t \in T$ gives the process.)

Remark If $\Omega_t = \mathbb{R}$, $t \in T$, and P_α is given by (2), this result becomes Theorem 1. We also must note that P is not necessarily a Radon measure. Its restricted regularity can be extended slightly, as shown in the corollary that follows, but nothing more can be asserted without further hypothesis and (deeper) analysis. The latter is therefore postponed to Chapter III.

Proof The necessity is simple. In fact, the coordinate projections $\pi_{\alpha\beta}: \Omega_\beta \to \Omega_\alpha$ clearly satisfy the composition rules, $\pi_{\alpha\beta} \circ \pi_{\beta\gamma} = \pi_{\alpha\gamma}$ for $\alpha < \beta < \gamma$, $\pi_{\alpha\alpha} = \text{identity}$. Moreover, for each open set $G \subset \Omega_\alpha$, $\pi_{\alpha\beta}^{-1}(G)$ is an open set in Ω_β since $\pi_{\alpha\beta}$ are coordinate projections. Hence each $\pi_{\alpha\beta}$ is continuous and $(\mathscr{B}_\beta, \mathscr{B}_\alpha)$-measurable. If there is a probability P on $\mathscr{B}$ such that $P_\alpha = P \circ \pi_\alpha^{-1}$, then for any $A \in \mathscr{B}_\alpha$, $\alpha < \beta$, we have, since $\pi_\alpha = \pi_{\alpha\beta} \circ \pi_\beta$,

$$\begin{aligned} P_\alpha(A) = (P \circ \pi_\alpha^{-1})(A) &= P((\pi_{\alpha\beta} \circ \pi_\beta)^{-1}(A)) \\ &= P(\pi_\beta^{-1}(\pi_{\alpha\beta}^{-1}(A))) = P_\beta(\pi_{\alpha\beta}^{-1}(A)). \end{aligned} \tag{4}$$

So $P_\alpha = P_\beta \circ \pi_{\alpha\beta}^{-1}$, and $\{(P_\alpha, \mathscr{B}_\alpha), \alpha \in D\}$ is a compatible family of probabilities, proving this part. The converse is nontrivial.

Let $\mathscr{C}$ be the class of cylinders with compact bases, as in the statement. Denote by $\mathscr{A}_\alpha$ the class of all cylinders of Ω whose bases are in $\mathscr{B}_\alpha$, $\alpha \in D$. Then $\mathscr{A}_\alpha$ is an algebra since it is closed under unions and complements. Also for $\alpha < \beta$, if $A \in \mathscr{A}_\alpha$, so that $A = \pi_\alpha^{-1}(A_\alpha)$ for an $A_\alpha \in \mathscr{B}_\alpha$, we have

$$A = \pi_\alpha^{-1}(A_\alpha) = (\pi_{\alpha\beta} \circ \pi_\beta)^{-1}(A_\alpha) = \pi_\beta^{-1}(\pi_{\alpha\beta}^{-1}(A_\alpha)) \in \mathscr{A}_\beta \tag{5}$$

since $\pi_{\alpha\beta}^{-1}(A_\alpha) \in \mathscr{B}_\beta$. Thus $\mathscr{A}_\alpha \subset \mathscr{A}_\beta$ and $\mathscr{A} = \bigcup_{\alpha \in D} \mathscr{A}_\alpha$ is an algebra. Moreover, $\mathscr{B} = \sigma(\mathscr{A})$, by the definition of an infinite product σ-algebra, which in fact is generated by all the cylinders. To prove the sufficiency, and the theorem, it is only required to show that (i) an additive function P_T can be unambiguously defined on $\mathscr{A}$ in terms of the P_α, (ii) the P_T is restrictedly regular on $\mathscr{A}$ (i.e., (3) is true), and (iii) P_T is σ-additive. Then by Theorem 2.2, P_T has a unique extension P to $\mathscr{B}$ satisfying all the conditions of the theorem. Let us prove these three assertions.

(i) To define P_T on $\mathscr{A}$, let $A \in \mathscr{A}_\alpha \cap \mathscr{A}_\beta$. Then there exist $B_1 \in \mathscr{B}_\alpha$, $B_2 \in \mathscr{B}_\beta$ such that $A = \pi_\alpha^{-1}(B_1) = \pi_\beta^{-1}(B_2)$. By directedness of D, there is a $\gamma \in D$ such that $\gamma > \alpha$, $\gamma > \beta$. Since $\pi_\alpha = \pi_{\alpha\gamma} \circ \pi_\gamma$, $\pi_\beta = \pi_{\beta\gamma} \circ \pi_\gamma$, we have

$$\pi_\gamma^{-1} \circ \pi_{\alpha\gamma}^{-1}(B_1) = \pi_\alpha^{-1}(B_1) = A = \pi_\beta^{-1}(B_1) = \pi_\gamma^{-1} \circ \pi_{\beta\gamma}^{-1}(B_1). \tag{6}$$

But $\pi_\gamma(\Omega) = \Omega_\gamma$, so π_γ^{-1} is one-to-one. Hence (6) yields the relation $\pi_{\alpha\gamma}^{-1}(B_1) = \pi_{\beta\gamma}^{-1}(B_2)$. The compatibility of P_αs then gives

$$P_\alpha(B_1) = P_\gamma(\pi_{\alpha\gamma}^{-1}(B_1)) = P_\gamma(\pi_{\beta\gamma}^{-1}(B_2)) = P_\beta(B_2). \tag{7}$$

If we now set $P_T(A) = P_\alpha(B_1)$ $(= P_\beta(B_2))$, where $A = \pi_\alpha^{-1}(B_1)$ $(= \pi_\beta^{-1}(B_2))$, the function P_T on $\mathscr{A}$ is unambiguously defined and is nonnegative. Also if A, B are disjoint, there exists (by directedness of D) a $\gamma_0 \in D$ such that $A = \pi_{\gamma_0}^{-1}(A_{\gamma_0})$, $B = \pi_{\gamma_0}^{-1}(B_{\gamma_0})$, $A_{\gamma_0} \cap B_{\gamma_0} = \varnothing$, $\{A_{\gamma_0}, B_{\gamma_0}\} \subset \mathscr{B}_{\gamma_0}$. Then $P_T(A \cup B) = P_{\gamma_0}(A_{\gamma_0} \cup B_{\gamma_0}) = P_{\gamma_0}(A_{\gamma_0}) + P_{\gamma_0}(B_{\gamma_0}) = P_T(A) + P_T(B)$. This implies the first statement. Note that $P_T(\Omega) = P_T(\pi_\alpha^{-1}(\Omega_\alpha)) = 1$, $\alpha \in D$.

(ii) For the restricted regularity, let $A \in \mathscr{A}$, so $A = \pi_\alpha^{-1}(B)$, $B \in \mathscr{B}_\alpha$. Then by definition of P_T, and the regularity of P_α, one has

$$\begin{aligned} P_T(A) = P_\alpha(B) &= \sup\{P_\alpha(K) : K \subset B,\ K \text{ compact}\} \\ &= \sup\{P_T(\pi_\alpha^{-1}(K)) : \pi_\alpha^{-1}(K) \subset A,\ K \text{ compact}\} \\ &\le \sup\{P_T(C) : C \in \mathscr{C},\ C \subset A\}. \end{aligned} \tag{8}$$

Since $P_T(\cdot)$ is increasing by its (sub-)additivity, we also have $P_T(A) \ge P_T(C)$ for all $C \in \mathscr{C}$, $C \subset A$. Hence, with (8), one has

$$P_T(A) = \sup\{P_T(C) : C \in \mathscr{C},\ C \subset A\}.$$

(iii) To prove the σ-additivity of P_T on $\mathscr{A}$, we use the following property of $\mathscr{C}$: If

$$C_n \in \mathscr{C}, \qquad \bigcap_{k=1}^{n} C_k \ne \varnothing \tag{$*$}$$

for each $n \ge 1$, then $\bigcap_{k=1}^{\infty} C_k \ne \varnothing$. This is a consequence of the topological lemma below, and it will be used now.

For (iii) it suffices to establish that $P_T(A_n) \to 0$ for any $A_n \in \mathscr{A}$, $A_n \downarrow \varnothing$. Thus let $\varepsilon > 0$ be given and $\{A_n, n \ge 1\} \subset \mathscr{A}$ be such a sequence. We may and do assume that $A_n \in \pi_{\alpha_n}^{-1}(\mathscr{B}_{\alpha_n})$ and $\alpha_n < \alpha_{n+1}$ (by directedness of D). Then by (ii) there exists a $C_n \in \mathscr{C}$, $C_n \subset A_n$, and

$$P_T(A_n) \le P_T(C_n) + \varepsilon/2^n. \tag{9}$$

Since C_n need not be monotone, let $B_n = \bigcap_{k=1}^{n} C_k \subset A_n$. Clearly, $B_n \in \mathscr{A}$, and we assert that $\{B_n, n \ge 1\}$ is also an approximating sequence. In fact, the additivity of P_T on $\mathscr{A}$ implies

$$\begin{aligned} P_T(C_1 \cup C_2) + P_T(C_1 \cap C_2) &= P_T(C_1) + P_T(C_2) \\ &\ge P_T(A_1) + P_T(A_2) - (\varepsilon/2 + \varepsilon/2^2). \end{aligned}$$

But $C_1 \cup C_2 = C_1 \subset A_1$, and so $P_T(A_1) - P_T(C_1 \cup C_2) \ge 0$. Thus

$$P_T(B_2) = P_T(C_1 \cap C_2) \ge P_T(A_2) - (\varepsilon/2 + \varepsilon/2^2).$$

Similarly, using B_2 and C_3 and noting that $C_3 \cap B_2 = B_3$, we get

$$P_T(B_3) \geq P_T(A_3) - (\varepsilon/2 + \varepsilon/2^2 + \varepsilon/2^3),$$

and by induction,

$$P_T(B_n) \geq P_T(A_n) - \sum_{i=1}^{n} (\varepsilon/2^i). \tag{10}$$

However, $\bigcap_{n=1}^{\infty} B_n = \bigcap_{k=1}^{\infty} C_k \subset \bigcap_{n=1}^{\infty} A_n = \varnothing$, and by (*), $\bigcap_{k=1}^{m_0} C_k = \varnothing$ for some m_0. Hence $B_{m_0} = \varnothing$, and by (10) for $n \geq m_0$, one has

$$0 \geq P_T(A_n) - \varepsilon \sum_{i=1}^{\infty} (1/2^i). \tag{11}$$

Thus $\lim_n P_T(A_n) \leq \varepsilon$. Since $\varepsilon > 0$ is arbitrary, this proves the σ-additivity of P_T on $\mathscr{A}$ and hence has a unique extension (by Theorem 2.2) to a probability P on $\mathscr{B}$, as asserted.

Remark It is of interest to note that not only $B_n \in \mathscr{A}$, but $B_n \in \mathscr{C}$ itself for each n. In fact, $C_n = \pi_{\alpha_n}^{-1}(K_n)$ for a compact $K_n \in \mathscr{B}_{\alpha_n}$ by definition. Since K_n is compact and π_{α_n} is continuous, C_n is closed in Ω. Let $B_n = \pi_{\alpha_n}^{-1}(J_n)$ for some $J_n \in \mathscr{B}_\alpha$. Since $B_n = \bigcap_{i=1}^{n} C_i$ is a cylinder, we have to show that J_n is compact in Ω_{α_n}. If $\tilde{K}_{in} = \pi_{\alpha_i \alpha_n}^{-1}(K_i)$, where $K_i \subset \Omega_{\alpha_i}$, then $\tilde{K}_{in} \subset \Omega_{\alpha_n}$ and is closed because $\pi_{\alpha_i \alpha_n}$ is continuous. Also $\tilde{K}_{nn} = K_n$. Then $K_n = \bigcap_{i=1}^{n} \tilde{K}_{in}$ by the compatibility of the mappings $\{\pi_{\alpha_i \alpha_{i+1}}, i \geq 1\}$. But $B_n = \bigcap_{k=1}^{n} C_k = \bigcap_{i=1}^{n} \pi_{\alpha_n}^{-1}(\tilde{K}_{in}) = \pi_{\alpha_n}^{-1}(K_n) = \pi_{\alpha_n}^{-1}(J_n)$. Hence $J_n = K_n$ is compact. Thus $B_n \in \mathscr{C}$.

Let us now prove the topological lemma used in (*) of (iii) above. The argument follows Harpain and Sion [1].

3. Lemma *If* $\mathscr{C}_0 = \{C_n, n \geq 1\} \subset \mathscr{C}$, $\bigcap_{i=1}^{n} C_i \neq \varnothing$ *for each* $1 \leq n < \infty$, *then* $\bigcap_{i=1}^{\infty} C_i \neq \varnothing$.

Proof For each n, there is a compact $K_{\alpha_n} \subset \Omega_{\alpha_n}$ such that $C_n = \pi_{\alpha_n}^{-1}(K_{\alpha_n})$. Let $T' = \bigcup_n \alpha_n$. For each $t \in T'$, choose a C_{α_t} in $\mathscr{C}_0$ such that $t \in \alpha_t$ and define $K_{\{t\}} = K_t = \pi_{t\alpha_t}(\pi_{\alpha_t}(C_{\alpha_t})) = \pi_{t\alpha_t}(K_{\alpha_t}) \subset \Omega_t$ (again $\pi_{t\alpha_t} = \pi_{\{t\}\alpha_t}$). Since K_{α_t} has a compact base and $\pi_{t\alpha_t}$ is continuous, K_t is a compact set and is nonempty since each member of $\mathscr{C}_0$ is such. Let $\omega^0 = \{\omega_t^0, t \in T\} \in \Omega$ be chosen such that $\omega_t^0 \in K_t$ for $t \in T'$. Since the C_α in $\mathscr{C}_0$ need not have bases that are rectangles, ω^0 is not necessarily in C_{α_n} for $n \geq 1$. Since Ω_t is not compact, none of the elements of $\mathscr{C}$ need be compact. So we have a nontrivial argument to fill in here.

Define the set $K = \{\omega \in \Omega : \omega_t \in K_t \text{ for } t \in T' \text{ and } \omega_t = \omega_t^0 \text{ for } t \in T - T'\}$. Thus $K = (\times_{t \in T'} K_t) \times (\times_{t \in T - T'} \{\omega_t^0\})$ so that K is compact and nonvoid in

Ω. We now produce a net in K with an element from each $C_\alpha \in \mathscr{C}_0$, so that its limit (by compactness of K) will be in each member of $\mathscr{C}_0$ to complete the argument.

Let $\mathfrak{F}$ be the class of all nonvoid finite subcollections of $\mathscr{C}_0$, directed by inclusion. For each $\mathscr{E} \in \mathfrak{F}$, let $T_{\mathscr{E}} = \bigcup_{C_\alpha \in \mathscr{E}} \alpha$, which is a finite subset of T'. By the (hypothesis of) finite intersection property, $\bigcap_{G \in \mathscr{E}} G \cap \bigcap_{t \in T_{\mathscr{E}}} C_{\alpha_t} \neq \varnothing$. Let $\omega^{\mathscr{E}}$ be a point in this intersection. Then for each $t \in T_{\mathscr{E}}$, $\pi_t(\omega^{\mathscr{E}}) = \omega_t^{\mathscr{E}} \in K_t$ since $T_{\mathscr{E}} \subset T'$. If we define a point $\tilde{\omega}^{\mathscr{E}} \in \Omega$ by the rule

$$\tilde{\omega}_t^{\mathscr{E}} = \pi_t(\tilde{\omega}^{\mathscr{E}}) = \begin{cases} \omega_t^{\mathscr{E}}, & t \in T_{\mathscr{E}} \\ \omega_t^0, & t \in T - T_{\mathscr{E}}, \end{cases}$$

then $\tilde{\omega}^{\mathscr{E}} \in \bigcap_{G \in \mathscr{E}} G$ and $\tilde{\omega}^{\mathscr{E}} \in K$ since $\omega_t^0 \in K_t$ for $t \in T'$ and $T_{\mathscr{E}} \subset T'$. Thus $\{\tilde{\omega}, \mathscr{E} \in \mathfrak{F}\}$ is a net in K. By the compactness of K (and its closure by the fact that the spaces are Hausdorff) this net has a cluster point ω^* in K. If $\mathscr{E} \in \mathfrak{F}$ and $\{C_\alpha\} \subset \mathscr{E}$, then $\tilde{\omega}^{\mathscr{E}} \in C_\alpha$ so that the net is eventually in C_α for each $C_\alpha \in \mathscr{C}_0$. Hence $\omega^* \in C_\alpha$ for each $C_\alpha \in \mathscr{C}_0$, and then $\omega^* \in \bigcap_{C \in \mathscr{C}_0} C$. This completes the proof.

Remark If $\mathscr{C}$ above is replaced by $\mathscr{C}_1 = \{C_\alpha, \alpha \in D_1\}$, $D_1 \subset D$, which is not necessarily countable, and if each C_α has a compact base, then $\bigcap_{C \in \mathscr{E}} C \neq \varnothing$ for each $\mathscr{E} \in \mathfrak{F}$ (the class of finite nonvoid subsets of $\mathscr{C}_1$) implies $\bigcap_{C \in \mathscr{C}_1} C \neq \varnothing$. This follows from the above proof, which did not use the countability hypothesis of $\mathscr{C}$.

4. Corollary (to Theorem 2) *The measure P of Theorem 2 is "restrictedly regular on $\mathscr{B}$" relative to $\mathscr{C}_\delta = \{A = \bigcap_{i=1}^\infty B_i : B_i \in \mathscr{C}\}$ in the sense that* (3) *holds with $\mathscr{C}_\delta$ in place of $\mathscr{C}$ there.*

Proof Let $\mathscr{B}^*$ be the Carathéodory generated algebra by the pair $(P_T, \mathscr{A})$, as in Theorem 2.1. Then $\mathscr{B} \subset \mathscr{B}^*$. By Eq. (1) of Section 3, for any $A \in \mathscr{B}$ and $\varepsilon > 0$ there exists a sequence $\{B_n, n \geq 1\} \subset \mathscr{A}$, such that $A \subset \bigcup_{n=1}^\infty B_n$, and

$$\varepsilon/2 > P\left(\bigcup_{n=1}^{\infty} B_n - A\right) = P\left(A^c - \bigcap_{n=1}^{\infty} B_n^c\right). \tag{12}$$

Since $\mathscr{A}$ and $\mathscr{B}$ are algebras, and $A^c \supset \bigcap_{n=1}^\infty B_n^c$, we deduce from the above that for each $A \in \mathscr{B}$ and $\varepsilon > 0$ there are $\tilde{B}_n \in \mathscr{A}$ such that $A \supset \bigcap_{n=1}^\infty \tilde{B}_n$ and $P(A - \bigcap_{n=1}^\infty \tilde{B}_n) < \varepsilon/2$. By the theorem, P is restrictedly regular on $\mathscr{A}$, so there exist $C_n \in \mathscr{C}$, $C_n \subset \tilde{B}_n$, such that $P(\tilde{B}_n - C_n) < \varepsilon/2^{n+1}$. Let $C^* = \bigcap_{n=1}^\infty C_n$. Then $C^* \in \mathscr{C}_\delta$, $C^* \subset \bigcap_{n=1}^\infty \tilde{B}_n = \tilde{B} \subset A$, and we have

$$P(A - C^*) \leq P(A - \tilde{B}) + \sum_{n=1}^{\infty} P(\tilde{B}_n - C_n) < \varepsilon/2 + \varepsilon/2 = \varepsilon. \tag{13}$$

This completes the proof.

Under the present (general) hypotheses of Theorems 1 and 2, which are now completely demonstrated, we have extracted as much information on the space $(\Omega, \mathscr{B}, P)$ as possible. This gives us the first fundamental solution to the existence problem of a stochastic process. However, it will be necessary for many applications that one find some suitable conditions in order that P (on Ω) be a Radon probability. This involves more detailed analysis. It will be treated in Chapter III. One needs some results on abstract analysis both for this study and for the theory of conditional expectations and probabilities, which are basic to stochastic analysis. In order to emphasize this need, we illustrate by the following result the inadequacy of the basic Theorems 1 and 2 in their general form. We then show in Theorem 5 that a modification of these results takes care of the difficulty.

Let the compatible family of distribution functions given in Theorem 1 be absolutely continuous relative to the Lebesgue measure in $\mathbb{R}^n$. An example of such a class is the Gaussian distributions given by Eq. (5) of Section 1. Let $\{X_t, t \in T\}$ be the resulting process on $(\Omega, \mathscr{B}, P)$ guaranteed by that theorem. The functions $\omega \in \Omega$, where $X_{(\cdot)}(\omega) = \omega(\cdot)$, are called the *sample paths* (*or functions*) of the process. Suppose that the process has almost all continuous sample paths. This means

$$P\{\omega : \lim_{t \to t_0}(X_t(\omega) - X_{t_0}(\omega)) = 0\} = 1, \qquad t_0 \in T = [a, b], \quad a < b. \tag{14}$$

If we complete the P-measure, then there will be no measurability problem here. (Cf. Section 5.) A class of Gaussian processes, called the Brownian motions, can be shown to have the property expressed in (14). (Cf. Problem 3.) Thus if C $(\subset \mathbb{R}^T = \Omega)$ is the set of all continuous functions, then clearly this part of Ω is of primary interest for analysis on such a process. We now show that $C \notin \mathscr{B}$ so that P is not even defined on it!

To prove this, first observe that each element of $\mathscr{B}$ is obtained by a countable number of operations of unions, intersections, and complements of the cylinder sets of the form $\{\omega : a_i \leq \omega(t_i) < b_i,\ i = 1, \ldots, n\}$, and hence *each set of* $\mathscr{B}$ *is determined by a countable set* of t-points from T. (Compare with Corollary 2.4.) Consequently, if C belongs to $\mathscr{B}$, there must exist a countable set $J \subset T$ determining C, and we may as well assume that J is also dense in T. Therefore an ω from Ω belongs to C if $\omega(t) = \omega'(t)$, $t \in J$, where $\omega' \in C$. However, if $t_0 \in T - J$, for each $\omega' \in C$, $\omega'(t_0) = \lim_{t_i \to t_0} \omega'(t_i)$, $t_i \in J$ by continuity. Choose a function that is discontinuous at t_0 $(\in T - J)$ from $\Omega = \mathbb{R}^T$ and that agrees with an element ω' from C at the countable set J. (Such functions clearly exist.) Then by the above we must have $\omega_0 \in C$, which is impossible. Thus $C \notin \mathscr{B}$. In fact this nonmeasurable C is a *thick* set, i.e., $P^*(C) = 1$ (C has outer measure one). This will follow from Theorem 2.1 if it is shown that for each $A \in \mathscr{B}$, $A \cap C = \varnothing$ implies $P(A) = 0$

(so $P^*(C) = \inf\{P(A^c): A \cap C = \emptyset,\ A \in \mathscr{B}\} = 1$). Let A be as above. Then there is a countable set $J_0 \subset T$ such that both A and A^c are determined by J_0. Since almost all sample paths of the process are continuous, by hypothesis, for each $t \in J_0$, there is a P-null set $N_t \in \mathscr{B}$ such that each $\omega \in N_t^c$ is continuous at t. If $N = \bigcup_{t \in J_0} N_t$, then $N \in \mathscr{B}$ and $P(N) = 0$. Moreover, each $\omega \in N^c$ agrees with some $\omega' \in C$ on J_0. Since $A \cap C = \emptyset$, i.e., $A^c \supset C$ for each $\omega \in N^c$, there is an $\omega' \in C \subset A^c$ satisfying $\omega(t) = \omega'(t)$, $t \in J_0$, so that $\omega \in A^c$, i.e., $N^c \subset A^c$ or $N \supset A$ and $P(A) = 0$. This shows $P^*(C) = 1$.

Let us now present a method, due to Doob, to show how one may slightly modify Theorem 1 and replace $(\Omega, \mathscr{B}, P)$ by another $(C, \tilde{\mathscr{B}}, \tilde{P})$, derived from it. This (purely measure theoretical) procedure is useful for most of the problems of this kind and reinforces the primacy of Theorems 1 or 2.

5. Theorem *Let $(\Omega, \mathscr{B}, P)$ be the probability space constructed in Theorem* 1 *(or* 2*). If C $(\notin \mathscr{B})$ is a thick set, let $\tilde{\mathscr{B}} = \mathscr{B}(C) = \{A \cap C : A \in \mathscr{B}\}$, the trace σ-algebra of $\mathscr{B}$ on C, and $\tilde{P}(A \cap C) = P(A)$. Then $\tilde{P}$ is well defined and $(C, \tilde{\mathscr{B}}, \tilde{P})$ is a probability space and $P_\alpha = \tilde{P} \circ \pi_\alpha^{-1}$ for each $\alpha \in D$. If $\tilde{X} = X \mid C$ or $\tilde{X}_t(\omega) = \omega(t)$, $t \in T$ and $\omega \in C$, then $\{\tilde{X}_t, t \in T\}$ is a stochastic process on $(C, \tilde{\mathscr{B}}, \tilde{P})$ with its finite dimensional distributions as the given family.*

Proof Since $(C, \tilde{\mathscr{B}})$ is clearly a measurable space, it is only necessary to show that $\tilde{P}$ is well defined and is σ-additive. The rest follows from the earlier theorem.

Suppose that A_1, A_2 from $\mathscr{B}$ satisfy $A_1 \cap C = A_2 \cap C$. We assert that A_1, A_2 differ by a P-null set so that $\tilde{P}$ is unambiguously defined on $\tilde{\mathscr{B}}$. Thus let $N = A_1 - A_1 \cap A_2 \in \mathscr{B}$. Then $N \cap C = A_1 \cap C - A_1 \cap A_2 \cap C$ so $N \cap C = \emptyset$. Hence $C \subset N^c$, and if P^* is the outer measure generated by $(P, \mathscr{B})$, we get, with the thickness of C.

$$1 = P^*(C) \le P(N^c) = 1 - P(N) \le 1. \tag{15}$$

Hence $P(N) = 0$, and then $P(A_1) = P(A_1 \cap A_2)$. By symmetry this also implies $P(A_2) = P(A_1 \cap A_2) = P(A_1)$, and $\tilde{P}$ is well defined. For the σ-additivity of $\tilde{P}$, let $\{\tilde{A}_n, n \ge 1\} \subset \tilde{\mathscr{B}}$ be a disjoint sequence. If $\tilde{A}_n = A_n \cap C$, $A_n \in \mathscr{B}$, let $\{B_n, n \ge 1\}$ be a disjunctification of $\{A_n, n \ge 1\}$. Thus $B_1 = A_1$, $B_n = A_n - \bigcup_{i=1}^{n-1} A_i$ for $n > 1$, and $(A_n - B_n) \cap C = \emptyset$. Hence $P(A_n - B_n) = \tilde{P}((A_n - B_n) \cap C) = \tilde{P}(\emptyset) = 0$. Consequently,

$$\begin{aligned}\sum_{n=1}^{\infty} \tilde{P}(\tilde{A}_n) &= \sum_{n=1}^{\infty} P(A_n) = \sum_{n=1}^{\infty} P(B_n) = P\left(\bigcup_{n=1}^{\infty} B_n\right) \\ &= P\left(\bigcup_{n=1}^{\infty} A_n\right) = \tilde{P}\left(\bigcup_{n=1}^{\infty} \tilde{A}_n\right).\end{aligned} \tag{16}$$

Thus $(C, \tilde{\mathscr{B}}, \tilde{P})$ is a probability space.

If $\{\tilde{X}_t, t \in T\}$ is as defined in the statement, then for any interval $I_i \subset \mathbb{R}$,

$$\begin{aligned}\tilde{P}\{\omega : \tilde{X}_{t_i}(\omega) \in I_i,\ i = 1, \ldots, n\} &= \tilde{P}[C \cap \pi_\alpha^{-1}(I)], \qquad \alpha = (t_1, \ldots, t_n),\\ &\qquad\qquad I = \mathop{\times}_{i=1}^{n} I_i \in \mathscr{B}_n\\ &= P(\pi_\alpha^{-1}(I)),\\ &= P\{\omega : X_{t_i}(\omega) \in I_i,\ i = 1, \ldots, n\},\end{aligned} \tag{17}$$

where $\pi_\alpha : \Omega \to \mathbb{R}^\alpha$ is the coordinate projection and X_ts are defined on Ω as in Theorem 1. This completes the proof.

Remark If the process $\{X_t, t \in T\}$ is not real valued but takes values in a general topological space Λ, then the distribution function is replaced by a Radon measure, and Theorem 2 is invoked. If there is a nonmeasurable set of outer measure one, we can use the same procedure as in the above theorem even in this case. Thus abstract results are intrinsically useful in stochastic analyses of various problems.

1.4 SOME RESULTS FROM ABSTRACT ANALYSIS AND VECTOR MEASURES

In this section a few results from Linear Analysis and a definition of an integral of scalar functions relative to vector measures, together with some of its properties, are presented. The main references are Dunford–Schwartz [1, Sections IV.8 and IV.10] and Dinculeanu [1].

1. Theorem *If (Ω, Σ, μ) is a measure space, let $L^p = L^p(\Omega, \Sigma, \mu)$, $p \geq 1$, be the space of equivalence classes of μ-measurable scalar functions f such that*

$$\|f\|_p = \left[\int_\Omega |f|^p \, d\mu\right]^{1/p} < \infty.$$

If $p = \infty$, let $\|f\|_\infty = \inf\{\alpha > 0 : \mu\{\omega : |f(\omega)| > \alpha\} = 0\}$. Then $(L^p, \|\cdot\|_p)$, $1 \leq p \leq \infty$, is a Banach space (i.e., a complete normed linear space); and if either $1 < p < \infty$ or $p = 1$ and μ is localizable (cf. Definition 2.6), then the adjoint space $(L^p)^*$ of L^p ($(L^p)^*$ is the space of all bounded linear functionals on L^p) is isometrically isomorphic to $(L^q, \|\cdot\|_q)$ where $q = p/(p-1)$ with $q = \infty$ for $p = 1$. Thus if $x^* \in (L^p)^*$, then there exists a μ-unique $g_{x^*} \in L^q$ such that

$$x^*(f) = \int_\Omega f g_{x^*} \, d\mu, \qquad f \in L^p, \quad \|x^*\| = \|g_{x^*}\|_q. \tag{1}$$

The following result, called an inverse Hölder inequality, is useful in some computations.

2. Proposition *Let L^p be the Banach space on (Ω, Σ, μ) as in the above theorem. If g is μ-measurable such that $fg \in L^1$ for each $f \in L^p$, $1 < p \leq \infty$, then $g \in L^q$ where $q = p/(p-1)$. The same result holds for $p = 1$ (then $q = \infty$) if μ is localizable in addition.*

In a Banach space $\mathscr{X}$, with adjoint $\mathscr{X}^*$, a sequence $\{x_n, n \geq 1\}$ is said to *converge weakly* to $x \in \mathscr{X}$ if $x^*(x_n - x) \to 0$ for each $x^* \in \mathscr{X}^*$, and $\mathscr{X}$ is *weakly sequentially complete* if every weak Cauchy sequence $\{x_n, n \geq 1\}$ converges to some (necessarily unique) element x in $\mathscr{X}$. [If the sequence is replaced by a net, then we have *weak completeness*.]

3. Theorem *Let L^p be the Banach space on (Ω, Σ, μ) as in Theorem 1. Then for each $1 \leq p < \infty$, L^p is weakly sequentially complete. [In general, L^p is not weakly complete and L^∞ is not even weakly sequentially complete.] But if μ is localizable, then L^∞ is weakly* sequentially complete, i.e., for every sequence $\{f_n, n \geq 1\} \subset L^\infty$ such that $\int_\Omega f_n g \, d\mu \to a(g)$ for each $g \in L^1$, one can find a μ-unique $\tilde{f} \in L^\infty$ such that $\int_\Omega (f_n - \tilde{f}) g \, d\mu \to 0$ for each $g \in L^1$.*

For L^p, $1 < p < \infty$, weak* sequential convergence (completeness) and weak sequential convergence are equivalent because of Theorem 1, since by (1) we have $(L^p)^* = (L^q)$ and $(L^q)^* = L^p$ so that $(L^p)^{**} = L^p$ where equalities stand for isometric equivalence. Such a space is called *reflexive*. [In general, a Banach space $\mathscr{X}$ is *reflexive* if $\mathscr{X}^{**} = (\mathscr{X}^*)^* = \mathscr{X}$.] Clearly, weak* convergence is weaker than weak convergence if the space involved is nonreflexive.

Uniform integrability of sets in L^1 plays an important role in probability theory. Recall that a set $\mathscr{A} \subset L^1$ is *uniformly integrable* if it is bounded (i.e., contained in a ball) and if $\lim_{\mu(A) \to 0} \int_A |f| \, d\mu = 0$ uniformly in $f \in \mathscr{A}$. The following result, due to de la Vallée Poussin (1915), gives a convenient characterization of uniformly integrable sets in L^1, which is often used in applications.

4. Theorem *Consider L^1 $(= L^1(\Omega, \Sigma, \mu))$ with $\mu(\Omega) < \infty$. A set $\mathscr{A} \subset L^1$ is uniformly integrable iff there exists a nonnegative convex function φ such that $\varphi(0) = 0$, $\varphi(t)/t \uparrow \infty$, and $\sup\{\int_\Omega \varphi(|f|) \, d\mu : f \in \mathscr{A}\} < \infty$. This may be stated alternatively as : A set $\mathscr{A} \subset L^1$ is uniformly integrable iff $\mathscr{A}$ is contained in a ball of the Orlicz space L^φ on (Ω, Σ, μ) with φ as above.*

The alternative version is more elegant than the original one and is convenient for applications. We therefore prove the result in this form after recalling the definition of an Orlicz space, which also has independent interest. We start with the classical result on the structure of a convex function (cf., e.g., Hardy–Littlewood–Pólya [1, p. 130]). If $I = [a, b] \subset \mathbb{R}$, and $\varphi: I \to \bar{\mathbb{R}}$ is a measurable function, then φ is convex iff there exists a nondecreasing function g on I such that

$$\varphi(x) = \varphi(a) + \int_a^x g(t)\,dt, \qquad x \in I. \qquad (\text{So } \varphi' = g \quad \text{a.e.}) \tag{2}$$

Let φ, ψ be a pair of nonnegative symmetric convex functions on $\mathbb{R}$ such that $\varphi(0) = 0 = \psi(0)$. Thus (2) implies

$$\varphi(x) = \int_0^x g(t)\,dt, \qquad \psi(y) = \int_0^y h(t)\,dt, \qquad x \geq 0, \quad y \geq 0. \tag{3}$$

Then these φ, ψ are called *complementary Young functions* if g and h are mutually inverse to each other and φ, ψ do not both take just $\{0, \infty\}$ values. One may (and does) demand g and h to be left continuous here. It is not hard to show that such a pair (φ, ψ) satisfies the inequality (due to Young)

$$|xy| \leq \varphi(x) + \psi(y), \qquad \{x, y\} \subset \mathbb{R}, \tag{4}$$

with equality iff $y = \varphi'(x)$ or $x = \psi'(y)$ a.e. (Lebesgue); it will be convenient to modify (φ, ψ) by the requirement (φ', ψ' denote derivatives of φ, ψ)

$$\varphi(1) + \psi(1) = 1. \tag{5}$$

Any Young pair can always be so modified. If (Ω, Σ, μ) is a measure space and φ is a Young function (i.e., one of the pair (φ, ψ)), then the space $L^\varphi = L^\varphi(\Omega, \Sigma, \mu)$ is defined as the set of equivalence classes of scalar measurable functions f on Ω for which $N_\varphi(f) < \infty$, where

$$N_\varphi(f) = \inf\left\{k > 0: \int_\Omega \varphi\left(\frac{|f|}{k}\right) d\mu \leq \varphi(1)\right\}. \tag{6}$$

Then $(L^\varphi, N_\varphi(\cdot))$ is a complete normed linear space, called an *Orlicz space*. Note that if $\varphi(x) = |x|^p/p$ and $\psi(x) = |x|^q/q$, $p^{-1} + q^{-1} = 1$, then $L^\varphi = L^p$ and $L^\psi = L^q$. Thus Orlicz spaces are a natural generalization of the classical Lebesgue spaces. A great deal of the theory of the latter spaces extends to Orlicz spaces. For instance, if $f \in L^\varphi$, $g \in L^\psi$, then the *Hölder inequality* holds,

$$\int_\Omega |fg|\,d\mu \leq N_\varphi(f) N_\psi(g), \tag{7}$$

and if $fg \in L^1$ for all $f \in L^\varphi$, then $g \in L^\psi$ whenever μ is localizable (the inverse Hölder inequality). If, moreover, φ is *moderated* (i.e., φ satisfies a growth condition: $\varphi(2x) \leq C\varphi(x)$ for $x \geq x_0$, where $x_0 \geq 0$ if μ is finite and $x_0 = 0$

when $\mu(\Omega) = \infty$), then $(L^\varphi)^* = L^\psi$. Thus Theorem 1 extends. But this identification is false if φ is not moderated, in contrast to the Lebesgue theory. (For the basic facts, we refer to Zygmund [1]. The above-stated representation theory is discussed by the author [1].) There is also a result similar to that of Theorem 3 if φ is moderated (and in the general case), but these are not needed here. We are now ready to prove de la Vallée Poussin's theorem, which was brought back by Meyer [1] for probabilistic applications.

Proof of Theorem 4 Since $\mu(\Omega) < \infty$, $L^\varphi \subset L^1$. In fact, by the support line property of a convex function, $\varphi(x) \geq ax + b$; $a, b, x \in \mathbb{R}$. So $f \in L^\varphi$ implies $(a/k)\int_\Omega |f|\,d\mu + b\mu(\Omega) \leq \int_\Omega \varphi(|f|/k)\,d\mu < \infty$ for a $k > 0$. Hence $f \in L^1$. To prove the necessity, the conditions on φ imply that its (right) derivative g in (3) satisfies $g(t) < \infty$ a.e. (Lebesgue) and $g(t) \nearrow \infty$ as $t \nearrow \infty$. Hence its complementary function ψ (and its (right) derivative h) will have the same properties. By (7),

$$\int_A |f|\,d\mu \leq N_\varphi(f)\cdot N_\psi(\chi_A) \leq K \cdot N_\psi(\chi_A) < \infty, \qquad A \in \Sigma, \tag{8}$$

where one uses the fact that $\mathscr{A}$ is in a ball so that $N_\varphi(f) \leq K$ for all $f \in \mathscr{A}$, and some $K < \infty$. So $\mathscr{A}$ is also contained in a ball of L^1 if we take $A = \Omega$ in (8). On the other hand, if $\alpha_A = N_\psi(\chi_A)$, then by definition of $N_\psi(\cdot)$:

$$\int_A \psi\left(\frac{1}{\alpha_A}\right) d\mu = \int_\Omega \psi\left(\frac{\chi_A}{\alpha_A}\right) d\mu \leq \psi(1). \tag{9}$$

Clearly, (9) implies $\alpha_A = [\psi^{-1}(\psi(1)/\mu(A))]^{-1}$ and $\alpha_A \to 0$ as $\mu(A) \to 0$ since $\psi^{-1}(x) \to \infty$ as $x \to \infty$. Hence (8) yields for all $f \in \mathscr{A}$,

$$\lim_{\mu(A)\to 0} \int_A |f|\,d\mu \leq K \lim_{\mu(A)\to 0} \alpha_A = 0.$$

So $\mathscr{A}$ is uniformly integrable in L^1.

For the converse, let $f \in \mathscr{A}$ and set $\alpha_n(f) = \mu[|f| > n]$ where $\mathscr{A}$ is uniformly integrable. So $\alpha_n(f) = \int_{[|f|>n]} d\mu \leq (1/n)\|f\|_1 \leq (K_0/n) \to 0$ as $n \to \infty$ uniformly in $f \in \mathscr{A}$, where K_0 (≥ 1) is the bound on $\mathscr{A}$. For each n, choose an $\alpha_n > 0$ such that

$$\sup_{f\in\mathscr{A}} \int_{[|f| \geq \alpha_n]} |f|\,d\mu \leq 1/2^n. \tag{10}$$

To see that this is possible, note that for any $1 > \varepsilon > 0$ there is a $\delta_\varepsilon > 0$ such that $\mu(A) < \delta_\varepsilon$ implies $\int_A |f|\,d\mu < \varepsilon/2$ for all $f \in \mathscr{A}$. If $\alpha = K_0/\varepsilon$, $A = [|f| > \alpha]$, then from the fact that $\mu(A) \le (1/\alpha)\int_A |f|\,d\mu \le K_0/\alpha = \delta_\varepsilon$ for all $f \in \mathscr{A}$, we have (actually $\delta_\varepsilon = \varepsilon$ here) with $K_0 \ge 1$,

$$\int_B |f|\,d\mu = \int_{B\cap[|f|>\alpha]} |f|\,d\mu + \int_{B\cap[|f|\le\alpha]} |f|\,d\mu \le \int_{B\cap[|f|>\alpha]} |f|\,d\mu + \alpha\mu(B). \tag{11}$$

Choose $\delta_1 = (\varepsilon/2\alpha) < \delta_\varepsilon$ and $\mu(B) < \delta_1$. Then $\int_{B_f} |f|\,d\mu < \varepsilon/2$ for all $f \in \mathscr{A}$, with $B_f = B \cap [|f| > \alpha]$. So (11) yields

$$\int_B |f|\,d\mu < \frac{\varepsilon}{2} + \alpha\cdot\delta_1 = \varepsilon, \qquad \text{uniformly in} \quad f \in \mathscr{A}. \tag{12}$$

Evidently, (12) implies (10). Hence

$$\begin{aligned} 2^{-n} &\ge \int_{[|f|\ge\alpha_n]} |f|\,d\mu = \int_{\bigcup_{j\ge\alpha_n}[j<|f|\le j+1]} |f|\,d\mu \\ &\ge \sum_{j=\alpha_n}^{\infty} j\mu[j<|f|\le j+1] \\ &= \sum_{j=\alpha_n}^{\infty} j\mu[|f|>j] - \sum_{j=\alpha_n}^{\infty} (j+1)\mu[|f|>j+1] \\ &\quad + \sum_{j=\alpha_n}^{\infty} \mu[|f|>j+1] \\ &\ge \sum_{j=\alpha_n}^{\infty} \mu[|f|>j] = \sum_{j=\alpha_n}^{\infty} \alpha_j(f), \text{ (say)}. \end{aligned} \tag{13}$$

It follows from (13) that $\sum_{n=1}^{\infty}\sum_{j=\alpha_n}^{\infty}\alpha_j(f) < \sum_{n=1}^{\infty} 2^{-n} = 1$ for all $f \in \mathscr{A}$. Let ξ_n be the number of integers j such that $\alpha_j(f) \le n$. Then $\xi_n \uparrow \infty$ and $\sum_{n=1}^{\infty}\xi_n\alpha_n(f) \le 1$. Taking $\xi_0 = 0$ and setting $\varphi(t) = \int_0^{|t|}\xi(u)\,du$, where $\xi(u) = \xi_n$ on $[n, n+1)$, we get φ to be a Young function, and since $\varphi(n+1) - \varphi(n) = \xi_n$,

$$\begin{aligned} \int_\Omega \varphi(|f|)\,d\mu &= \int_{\bigcup_{j\ge0}[j<|f|\le j+1]} \varphi(|f|)\,d\mu \le \sum_{j\ge0}\left(\sum_{i=0}^{j}\xi_i\right)\mu[j<|f|\le j+1] \\ &= \sum_{n\ge1}\xi_n\alpha_n(f) \le 1. \end{aligned}$$

Hence $\mathscr{A} \subset L^\varphi$ and is in its unit ball. This completes the proof.

The following consequence of the theorem is of some interest.

5. Corollary *If $\mu(\Omega) < \infty$, then $L^1 = \bigcup\{L^\varphi : (\varphi(t)/t)\uparrow\infty,\ \varphi$ is moderated$\}$. Thus L^1 is the union of all Orlicz spaces on a finite measure space, while the same is not, in general, true if only Lebesgue spaces are allowed.*

Proof If φ is any Young function, then $L^\varphi \subset L^1$, as noted at the beginning of the proof of Theorem 4. Hence $L^1 \supset \bigcup\{L^\varphi : \varphi \text{ any Young function}\}$. We shall show that there is equality for (a subclass of) moderated φs.

By linearity (and the fact that all spaces are lattices) it suffices to show that any $0 \le f \in L^1$ belongs to L^φ for a moderated φ. Thus let $\alpha = \int_\Omega f\,d\mu$. If $a_1 > 0$, choose by induction $a_n \ge 3a_{n-1}$, $n \ge 2$, such that $\int_{[f \ge a_n]} f\,d\mu < 2^{-n}$. This is possible by the integrability of f. If $A_n = [a_{n-1}, a_n)$, $a_0 = 0$, let $g = \sum_{n=1}^\infty n\chi_{A_n}$. Then $g(t) < \infty$ for $t < \infty$ and $g(t) \nearrow \infty$ as $t \nearrow \infty$. Let $\varphi(x) = \int_0^{|x|} g(t)\,dt$. Then by (2), $\varphi(\cdot)$ is convex, $(\varphi(t)/t) \uparrow \infty$ as $t \uparrow \infty$. If $t \in A_n$ so $2t < 2a_n < a_{n+1}$ and $g(2t) \le n+1$, we have $g(2t) \le 2g(t)$. Hence if $a_n > x > 0$,

$$\varphi(2x) = \int_0^{2x} g(t)\,dt = 2\int_0^x g(2t)\,dt \le 4\int_0^x g(t)\,dt = 4\varphi(x). \tag{14}$$

So with $C = 4$, we get $\varphi(2x) \le C\varphi(x)$, and φ is moderated. Since $\varphi(x) \le nx$ for $x < a_n$, one has

$$\begin{aligned}\int_\Omega \varphi(f)\,d\mu &\le \sum_{n=1}^\infty n\int_{[n-1 \le f < n)} f\,d\mu = \sum_{n=1}^\infty\left[\int_{[f \ge a_{n-1}]} nf\,d\mu - \int_{[f \ge a_n]} nf\,d\mu\right]\\ &= \int_\Omega f\,d\mu + \sum_{n=1}^\infty \int_{[f \ge a_n]} (n+1)f\,d\mu - \sum_{n=1}^\infty \int_{[f \ge a_n]} nf\,d\mu\\ &= \int_\Omega f\,d\mu + \sum_{n=1}^\infty \int_{[f \ge a_n]} f\,d\mu < \int_\Omega f\,d\mu + \sum_{n=1}^\infty 2^{-n} = \alpha + 1.\end{aligned}$$

So $f \in L^\varphi$, and this completes the proof.

Remark This result shows that the Orlicz spaces are more extensive than the Lebesgue spaces. In fact, one can assert that between any two Lebesgue spaces there exists an Orlicz space L^φ with and without moderated functions φ. (To construct such examples, see Krasnoselskiĭ and Rutickiĭ [1, p. 28]. This book contains a systematic exposition of Orlicz spaces.)

The following result, called Vitali's theorem, shows the use of uniform integrability in convergence theory.

6. Theorem *Let $\{X_n, n \ge 1\}$ be a sequence in L^1 on a finite measure space (Ω, Σ, μ). Suppose $X_n \to X$ in measure (or a.e.). Then $\{X_n, n \ge 1\}$ is a Cauchy sequence (hence converges in L^1 to X) iff it is uniformly integrable. If this set is in the positive part of L^1, then the latter integrability condition is satisfied iff $\lim_n \int_\Omega X_n\,d\mu = \int_\Omega X\,d\mu < \infty$.*

In this connection we state one more result, which is a consequence of the classical Dunford–Pettis theorem. Its proof uses a generalization of the preceding, called the Vitali–Hahn–Saks theorem, and another, called the Eberlein–Šmulian theorem. (These are given in Dunford–Schwartz [1].) It is included here because of its importance and its relation with Theorem 4.

7. Theorem *A subset $\mathscr{A} \subset L^1$ (on a finite measure space (Ω, Σ, μ)) is relatively weakly compact (i.e., the closure $\bar{\mathscr{A}}$ is compact in the weak topology of L^1) iff $\mathscr{A}$ is contained in a ball of L^φ (on (Ω, Σ, μ)) where φ is a Young function such that $\varphi(x)/x \uparrow \infty$ as $x \uparrow \infty$.*

It will also be useful to state an isomorphism result on abstract additive set functions and their images on a suitable topological measure space. The precise statement is as follows.

8. Theorem *Let (Ω, Σ) be a measurable space and let $ba(\Omega, \Sigma)$ denote the set of all bounded additive scalar set functions on Σ (which becomes a Banach space under the total variation norm). Then there exists a totally disconnected compact Hausdorff space S with $\mathscr{B}$ denoting its Baire σ-algebra (cf. Section 2) such that $ba(\Omega, \Sigma)$ and the space of all regular bounded scalar σ-additive set functions on $\mathscr{B}$, denoted $rca(S, \mathscr{B})$, are isometrically isomorphic. This mapping $U: ba(\Omega, \Sigma) \to rca(S, \mathscr{B})$ satisfies, in addition to $|v|(\Omega) = |Uv|(S)$ for $v \in ba(\Omega, \Sigma)$, (i) it is onto, and (ii) it is order preserving.*

The space S above is called a Stone space (after M. H. Stone). It has importance in topology as well as in functional analysis. A proof of the theorem may be found in Dunford–Schwartz [1, pp. 312–313]. This result helps to understand the relations between an arbitrary conditional probability measure and regular conditional probabilities in Chapter II and shows how these abstract results enter naturally and inevitably into the foundations of the subject.

In our future analysis and its applications, two situations arise in a standard and important way. These are when vector valued mappings have to be integrated on a measure space (e.g., in applications of martingales) and when scalar functions must be integrated relative to a vector measure (e.g., with conditional probabilities). We include a brief account of the needed theory in order not to interrupt the discussion in Chapters II and V. Both these types of integration can be treated together (using, for instance, a formulation due to Bartle [1]), but then a more abstract framework (with more preliminaries) is desired for it. So the simpler route will be followed here as this is sufficient for our purposes, and the two cases are treated separately.

If (Ω, Σ, μ) is a measurable space and $\mathscr{X}$ is a Banach space, let $\{A_i, i \geq 1\} \subset \Sigma$ be a collection of disjoint sets. Then the function

$$f = \sum_{i=1}^{n} a_i \chi_{A_i}, \qquad a_i \in \mathscr{X}, \quad 0 < \mu(A_i) < \infty, \tag{15}$$

is called, as usual, a *simple function*, and it is elementary or *countably valued* if $n = \infty$. A mapping $f: \Omega \to \mathscr{X}$ is *strongly measurable* if there is a sequence $\{f_n, n \geq 1\}$ of countably valued (simple) functions such that $\|f - f_n\|(\omega) \to 0$, or $f_n(\omega) \to f(\omega)$, for $\omega \in \Omega - \Omega_0$ with $\mu(\Omega_0) = 0$ $(\mu(\Omega) < \infty)$. Thus $f(\Omega - \Omega_0) \subset \mathscr{X}$ is separable. It is *weakly measurable* if for each $x^* \in \mathscr{X}^*$ the numerical function $x^* \circ f$ defined as $(x^* \circ f)(\omega) = x^*(f(\omega))$, $\omega \in \Omega$, is μ-measurable. The following theorem is proved as in the scalar case. Hereafter $\mathscr{X}$ is a Banach space unless otherwise stated.

9. Theorem *If (Ω, Σ, μ) is a measure space, $f, f_n: \Omega \to \mathscr{X}$ are strongly measurable, consider the following statements:*

(i) *$f_n \to f$ almost uniformly, i.e., for each $\varepsilon > 0$ there is a $\delta_\varepsilon > 0$ and $E_\varepsilon \in \Sigma$ with $\mu(E_\varepsilon) < \delta_\varepsilon$ such that $f_n(\omega) \to f(\omega)$ uniformly in $\omega \in \Omega - E_\varepsilon$.*

(ii) *$f_n \to f$ almost everywhere, i.e., $f_n(\omega) \to f(\omega)$ for all $\omega \in \Omega - \Omega_0$, with $\mu(\Omega_0) = 0$.*

(iii) *$f_n \to f$ in μ-measure, i.e., for each $\varepsilon > 0$, if $E_{n,\varepsilon} = \{\omega : \|f - f_n\|(\omega) > \varepsilon\}$, then $\mu(E_{n,\varepsilon}) \to 0$, as $n \to \infty$.*

Then (i) $\Rightarrow$ (ii) *and* (iii) *(even if $\mu(\Omega) = \infty$), and in the case that $\mu(\Omega) < \infty$,* (ii) $\Rightarrow$ (i) *and* (iii). *Moreover, (iii) always implies the existence of a subsequence f_{n_i} which converges to f almost uniformly.*

It is easy to extend the earlier definition of vector valued f to the matrix or operator valued case. If $\mathscr{X}, \mathscr{Y}$ are two Banach spaces, $f: \Omega \to B(\mathscr{X}, \mathscr{Y})$, where $B(\mathscr{X}, \mathscr{Y})$ is the Banach space under the uniform (or operator) norm of continuous linear mappings from $\mathscr{X}$ to $\mathscr{Y}$, then f is *uniformly measurable* if $f: \Omega \to \mathscr{Z} = B(\mathscr{X}, \mathscr{Y})$ is (strongly) measurable in the earlier sense, relative to (Ω, Σ, μ) and $\mathscr{Z}$. It is *strongly operator measurable* if for each $x \in \mathscr{X}$ the mapping $fx: \Omega \to \mathscr{Y}$ is (strongly) measurable as before, and it is *weakly operator measurable* if for each $x \in \mathscr{X}$, $y^* \in \mathscr{Y}^*$ the mapping $y^* \circ f \circ x: \Omega \to \mathbb{R}$ is μ-measurable (i.e., fx is weakly measurable in the earlier sense). The following result gives some connections between these concepts, all of which clearly coincide if $\mathscr{X} = \mathscr{Y} = \mathbb{R}$.

10. Theorem (a) *A function $f: \Omega \to \mathscr{X}$ is strongly measurable iff it is weakly measurable and almost separably valued (i.e., there is an $\Omega_0 \in \Sigma$, with $\mu(\Omega_0) = 0$, and $f(\Omega - \Omega_0)$ is a separable subset of $\mathscr{X}$).*

(b) *A function $f: \Omega \to B(\mathcal{X}, \mathcal{Y})$ is strongly operator measurable iff f is weakly operator measurable and fx is almost separably valued in $\mathcal{Y}$. Further, f is uniformly measurable iff it is weakly operator measurable and almost separably valued in $B(\mathcal{X}, \mathcal{Y})$.*

With these results the set of measurable functions in each sense is a vector space (over the scalars). We now define an integral. If $f: \Omega \to \mathcal{X}$ is elementary or countably valued, then it is said to be *Bochner integrable* whenever $\|f(\cdot)\|: \Omega \to \mathbb{R}$ is μ-integrable. Thus with the representation (15) we have

$$\int_E f\,d\mu = \sum_{i=1}^{\infty} a_i \mu(E \cap A_i), \qquad E \in \Sigma, \tag{16}$$

and the series on the right is absolutely convergent since $\{\sum_{i=1}^{n} a_i \mu(E \cap E_i)\}_1^\infty$ is a Cauchy sequence because

$$\left\| \sum_{i=1}^{n} a_i \mu(E \cap E_i) \right\| \le \sum_{i=1}^{\infty} \|a_i\| \mu(E \cap E_i) = \int_E \|f\|\,d\mu < \infty, \qquad n \ge 1. \tag{17}$$

Let $f: \Omega \to \mathcal{X}$ be strongly measurable. Then f is *Bochner integrable* iff there exists a sequence $\{f_n, n \ge 1\}$ of elementary functions such that $f_n \to f$ a.e., f_n is Bochner integrable, and $\int_\Omega \|f - f_n\|\,d\mu \to 0$. Hence we define the unique limit, which is seen to be independent of the sequence used in getting it, as the integral:

$$\int_E f\,d\mu = \lim_{n\to\infty} \int_E f_n\,d\mu, \qquad E \in \Sigma. \tag{18}$$

We have the inequality $\|\int_E f\,d\mu\| \le \int_E \|f\|\,d\mu$ and also the dominated convergence theorem for this integral.

Extending the proofs in the scalar case and using the results of the preceding two theorems, we can establish:

11. Theorem *Let $B_p = \{f: \Omega \to \mathcal{X}, \|f\|_p < \infty\}$ where one defines*

$$\|f\|_p = \begin{cases} \left[\int_\Omega \|f\|^p\,d\mu\right]^{1/p} & \text{if} \quad 1 \le p < \infty \\ \text{ess. sup of } \|f\| & \text{if} \quad p = \infty \end{cases} \tag{19}$$

*for strongly measurable f. If the equivalence classes of functions are considered, then $(B_p, \|\cdot\|_p)$, $1 \le p \le \infty$, is a Banach space of (equivalence classes of) p*th *power Bochner integrable functions.*

The same result holds in the context of Orlicz spaces if $B_\varphi = \{f: \Omega \to \mathscr{X}, \|f\| \in L^\varphi\}$ whose norm is (cf. (6) above):

$$N_\varphi(f) = \inf\left\{k > 0: \int_\Omega \varphi\left(\frac{\|f\|}{k}\right) d\mu \leq \varphi(1)\right\}. \tag{20}$$

The following result, due to F. Riesz when $\mathscr{X} = \mathbb{R}$ and $\varphi(x) = |x|^p$, on mean convergence in these spaces gives a feeling in handling these concepts.

12. Theorem *Let φ be either a moderated (continuous) Young function or else an increasing (continuous) concave function with $\varphi(0) = 0$. Let $f_n: \Omega \to \mathscr{X}$ be strongly measurable functions such that $\sup_n \int_\Omega \varphi(\|f_n\|)\, d\mu < \infty$. If $f_n \to f$ a.e., and $\int_\Omega \varphi(\|f_n\|)\, d\mu \to \int_\Omega \varphi(\|f\|)\, d\mu$, as $n \to \infty$, then $\int_\Omega \varphi(\|f_n - f\|)\, d\mu \to 0$ and that in any case $\{\varphi(\|f_n\|), n \geq 1\}$ is uniformly integrable.*

Proof Since every convex or concave function φ is absolutely continuous, one has the representation (cf. (2)) for any x, y in $\mathbb{R}^+$,

$$\varphi(x + y) = \int_0^{x+y} \varphi'(t)\, dt = \varphi(x) + \int_0^y \varphi'(t + x)\, dt, \tag{21}$$

where $\varphi': \mathbb{R}^+ \to \mathbb{R}^+$ is increasing or decreasing according to whether φ is convex or concave. If φ is a moderated Young function, then $\varphi(2x) \leq C\varphi(x)$, $x \geq 0$ for some constant $C > 0$ (as defined after (7)). Then (by convexity)

$$\varphi(x + y) \leq \tfrac{1}{2}[\varphi(2x) + \varphi(2y)] \leq (C/2)(\varphi(x) + \varphi(y)), \tag{22}$$

and the same inequality (with $C = 2$) holds also in the concave case. Thus if $\tilde{C} = \max(1, C/2)$, then in either case we have

$$\varphi(x + y) \leq \tilde{C}[\varphi(x) + \varphi(y)], \qquad x \geq 0, \quad y \geq 0. \tag{23}$$

Let us extend the definition of φ to all of $\mathbb{R}$ by setting $\varphi(-x) = \varphi(x)$, $x \in \mathbb{R}^+$. Then using $|x + y| \leq |x| + |y|$ and the increasing property of φ on $\mathbb{R}^+$, one sees that (23) holds for all real x, y.

Now let $f_n \to f$ a.e. and $x^* \in \mathscr{X}^*$ be of unit norm. If $x = x^*(f_n(\omega))$, $y = x^*(f(\omega))$ in (23), and noting that $|x^*(f_n(\omega))| \leq \|f_n(\omega)\|$, one has

$$\begin{aligned} 0 &\leq \tilde{C}[\varphi(|x^* \circ f_n|) + \varphi(|x^* \circ f|)](\omega) - \varphi(|x^*(f_n - f)|)(\omega) \\ &\leq \tilde{C}[\varphi(\|f_n\|) + \varphi(\|f\|)](\omega) - \varphi(|x^*(f_n - f)|)(\omega). \end{aligned} \tag{24}$$

However, by the Hahn–Banach theorem, for each ω there exists $x_\omega^* \in \mathscr{X}^*$ of unit norm such that $x^*(f_n - f)(\omega) = \|f_n - f\|(\omega)$. Using this x_ω^* in (24), we get

$$0 \leq \tilde{C}[\varphi(\|f_n\|) + \varphi(\|f\|)](\omega) - \varphi(\|f_n - f\|)(\omega), \qquad \omega \in \Omega. \tag{25}$$

Since $\|f_n - f\|(\omega) \to 0$ for almost all (ω), so that $\|f_n\|(\omega) \to \|f\|(\omega)$, a.a. (ω), and φ is continuous, the right-hand side of (25) tends (as $n \to \infty$) to $2\tilde{C}\varphi(\|f\|)(\omega)$ for a.a. (ω). So by the Fatou inequality,

$$2\tilde{C}\int_\Omega \varphi(\|f\|)\,d\mu \le \liminf_n \int_\Omega \{\tilde{C}[\varphi(\|f_n\|) + \varphi(\|f\|)] - \varphi(\|f_n - f\|)\}\,d\mu$$

$$= \lim_n \tilde{C}\int_\Omega \varphi(\|f_n\|)\,d\mu + \tilde{C}\int_\Omega \varphi(\|f\|)\,d\mu$$

$$- \limsup_n \int_\Omega \varphi(\|f_n - f\|)\,d\mu$$

$$= 2\tilde{C}\int_\Omega \varphi(\|f\|)\,d\mu - \limsup_n \int_\Omega \varphi(\|f_n - f\|)\,d\mu, \tag{26}$$

where we used the hypothesis in the last line. Hence in either case of the convex or the concave φ, (26) implies $\lim_n \int_\Omega \varphi(\|f_n - f\|)\,d\mu = 0$, as asserted.

In the case of the moderated Young function φ, this statement implies (as in the L^p-theory) that $N_\varphi(f_n - f) \to 0$ and then it is a simple and standard computation to show that $\{\varphi(\|f_n\|), n \ge 1\}$ is uniformly integrable. We omit the details. Actually the uniform integrability follows from Theorem 6 in all cases.

Finally, let us consider the second problem, that of integrating scalar functions relative to a vector valued set function. If (Ω, Σ) is a measurable space and $\mathscr{X}$ is a Banach space, then a mapping $\mu: \Sigma \to \mathscr{X}$ is called a *vector measure* if (i) $\mu(\varnothing) = 0$ and (ii) for any disjoint sequence $\{A_n, n \ge 1\} \subset \Sigma$ we have (σ-*additivity*):

$$\mu\left(\bigcup_{i=1}^{\infty} A_i\right) = \sum_{i=1}^{\infty} \mu(A_i), \tag{27}$$

where the series on the right is unconditionally convergent in $\mathscr{X}$. If instead of (27) one has the scalar function $x^* \circ \mu: \Sigma \to \mathbb{R}$ to be a signed measure for each $x^* \in \mathscr{X}^*$, then μ is *weakly* σ-*additive*. However, the following important theorem of Pettis shows that these are equivalent concepts.

13. Theorem *If $\mu: \Sigma \to \mathscr{X}$ is additive, then the σ-additivity and weak σ-additivity are equivalent. Moreover, every such σ-additive μ is bounded, i.e., $\sup\{\|\mu(A)\| : A \in \Sigma\} < \infty$, Σ being a σ-algebra, or a σ-ring (but not a δ-ring).*

The preceding result implies that for any Bochner integrable $f: \Omega \to \mathscr{X}$ the set function $v: A \mapsto \int_A f\,d\mu$ is well defined on Σ and is σ-additive, i.e., v is a vector measure and, moreover, v is absolutely continuous relative to μ (i.e., given $\varepsilon > 0$, there is a $\delta_\varepsilon > 0$ such that $A \in \Sigma$, $\mu(A) < \delta_\varepsilon$ implies $\|v(A)\| < \varepsilon$). Conversely, if $v: \Sigma \to \mathscr{X}$ is a vector measure that is absolutely continuous relative to μ, $\mu(\Omega) < \infty$, (Ω, Σ, μ) being a measure space (μ can be localizable), under what conditions can we write $v(A) = \int_A f\, d\mu$, $A \in \Sigma$, for a Bochner integrable $f: \Omega \to \mathscr{X}$? This is the (vector) Radon–Nikodým theorem and is a nontrivial extension of Theorem 2.7. We present a solution, as an application of abstract martingale theory, in Chapter V. Here some elementary properties of integrals relative to a vector measure v will be included.

If $v: \Sigma \to \mathscr{X}$ is a vector measure, then there exists a finite positive measure λ on Σ (a σ-algebra of Ω) such that v is absolutely continuous relative to λ and for each $A \in \Sigma$, $\lambda(A) \leq \|v\|(A)$ where, for $a_i \in \mathbb{R}$,

$$\|v\|(A) = \sup\left\{\left\|\sum_{i=1}^{n} a_i v(A_i)\right\| : A_i \in \Sigma, \text{ disjoint}, A_i \subset A, |a_i| \leq 1\right\}, \tag{28}$$

and $\|v\|(\cdot)$ is called the *semivariation* of v. The *variation* $|v|(\cdot)$ of v is given by

$$|v|(A) = \sup\left\{\sum_{i=1}^{n} \|v(A_i)\| : A_i \in \Sigma, \text{ disjoint}, A_i \subset A\right\}. \tag{29}$$

The existence of such a λ, sometimes called a *control measure*, was proved by Bartle–Dunford–Schwartz for the purpose of integration. It may be noted that while $|v|(\cdot)$ is a measure, $\|v\|(\cdot)$ is only an outer measure and that $\|v\|(A) \leq |v|(A)$, $A \in \Sigma$. Also as a consequence of Theorem 13, one deduces that $\|v\|(\Omega) < \infty$ (even though $|v|(\Omega) = \infty$ in many cases of interest). Thus a function $f: \Omega \to \mathbb{R}$ which is measurable on $(\Omega, \Sigma, \lambda)$ is also called *v-measurable*. For step $f = \sum_{i=1}^{n} \alpha_i \chi_{A_i}$, $A_i \in \Sigma$ disjoint, define (as usual) the integral:

$$\int_A f\, dv = \sum_{i=1}^{n} \alpha_i v(A_i \cap A), \qquad A \in \Sigma. \tag{30}$$

If $f: \Omega \to \mathbb{R}$ is any measurable function for (Ω, Σ, μ), then it is said to be integrable in the sense of *Dunford and Schwartz* (or *D–S integrable*) if there exists a sequence of simple λ-measurable $f_n: \Omega \to \mathbb{R}$ such that $f_n \to f$ a.e. $[\lambda]$, and $\{\int_A f_n\, dv, n \geq 1\} \subset \mathscr{X}$ is a Cauchy sequence for each $A \in \Sigma$. The unique limit, shown to be independent of the sequence used, is defined as the *D–S integral* of f for v, denoted $\int_A f\, dv$. The theory of these integrals can be found in Dunford–Schwartz [1, IV.10]. We need the following result for some of the later applications:

14. Theorem *The class of v-integrable functions $L(v)$ is a vector space (over the scalar field) and the following properties hold:*

(i) *the mapping $f \mapsto \int_E f\, dv$ is linear for $f \in L(v)$, $E \in \Sigma$;*

(ii) *the set function $\zeta: A \mapsto \int_A f\, dv$ is a vector measure for $A \in \Sigma$ and $f \in L(v)$; moreover, $\lim_{\lambda(A) \to 0} \zeta(A) = 0$;*

(iii) *if* $|f| \leq k_0$, *then* $\|\int_A f\,dv\| \leq k_0 \|v\|(A)$, $A \in \Sigma$;
(iv) *if* $\{f_n, g\} \subset L(v)$, $f_n \to f$ *a.e.* $[\lambda]$, *and for each* $A \in \Sigma$, $n \geq 1$,

$$\left\| \int_A f_n\,dv \right\| \leq \left\| \int_A g\,dv \right\|,$$

then $f \in L(v)$ *and* $\lim_n \int_A f_n\,dv = \int_A f\,dv$, $A \in \Sigma$. *In particular, if* $|f_n| \leq g$ *a.e.* $[\lambda]$ *and* $g \in L(v)$, *then the same conclusion holds.*

It is the last part that corresponds to the dominated convergence theorem, and part (ii) shows that the indefinite D–S integral is absolutely continuous relative to λ (or equivalently to v). When $\mathscr{X} = L^p$ so that $\mathscr{X}$ is a (Banach) lattice, we may specialize this theory for "positive" vector measures and add the monotone convergence theorem. This is possible, and one can define the integral for a slightly more general vector valued set functions, which agrees with the D–S integral when the corresponding conditions are met. Since conditional probability measures belong to this category, the following additional result is appropriate here, and the argument is based on that of Wright [1].

The main point of the special case where $\mathscr{X}$ is a Banach lattice is to have $v: \Sigma \to \mathscr{X}$ as a σ-additive function in the sense of *order*, instead of norm convergence. Considering the case $\mathscr{X} = L^\infty$ (or $= L^\varphi$, where φ is not moderated), one sees that the order convergence requirement is weaker than the norm convergence of (27). We therefore call the present one a *vector valued measure* (instead of a vector measure), and (27) is replaced by ($\mathscr{X}$ is a Banach function lattice on (Ω, Σ)):

$$v: \Sigma \to \mathscr{X} \quad \text{with (i) } v(\varnothing) = 0, \quad \text{(ii) } v(A) \geq 0, \text{ and for } \{A_n, n \geq 1\} \subset \Sigma, \text{ disjoint, implies}$$
$$v\left(\bigcup_{n=1}^{\infty} A_n\right) = \sum_{n=1}^{\infty} v(A_n) = \sup_n \sum_{i=1}^{n} v(A_i). \tag{31}$$

With this definition one can follow the scalar theory and show that if $A_n \downarrow \varnothing$, $A_n \in \Sigma$, $v(A_k) < \infty$ for some $1 \leq k < \infty$, then $\lim_n v(A_n) = \inf_n v(A_n) = 0$. If f is a step function relative to Σ, then $\int_A f\,dv \in \mathscr{X}$ is defined by (30), and we observe that the thus defined integral (which is seen to be independent of the representation of f) has the following continuity property. Namely, if $f_n \downarrow 0$ pointwise (each f_n a step function of Σ), then $\lim_n \int_\Omega f_n\,dv = 0$. Indeed, let $\alpha = \max\{f_1(\omega): \omega \in \Omega\}$, and for any $\varepsilon > 0$ let $A_n = \{\omega: f_n(\omega) \geq \varepsilon\}$. Then $v(A_n) < \infty$ and $\lim_n A_n = \bigcap_{n=1}^{\infty} A_n = \varnothing$ so that $\lim_n v(A_n) = \inf_n v(A_n) = 0$. Since $0 \leq f_n \leq \alpha\chi_{A_n} + \varepsilon\chi_{A_n^c}$, we have

$$0 \leq \lim_n \int_\Omega f_n\,dv \leq \alpha \lim_n v(A_n) + \varepsilon v(\Omega) \leq \varepsilon v(\Omega), \tag{32}$$

proving the assertion. Hence the result also holds for increasing sequences.

The above remark enables one to define the integral for any measurable $f \geq 0$ and then for general functions. If f is a positive measurable (for Σ) real function, then by the elementary structure theorem there are increasing measurable step functions f_n such that $0 \leq f_n \uparrow f$ pointwise. We set $\int_\Omega f\,dv = \lim_n \int_\Omega f_n\,dv = \sup_n \int_\Omega f_n\,dv$; both sides are defined as $+\infty$ if $v\{\omega : f(\omega) > n\} = \infty$ for some $n \geq 0$. If f is arbitrary (measurable), let $f = f^+ - f^-$ be the decomposition into positive parts and f be called *(order) integrable* iff both f^+ and f^- are. Then define

$$\int_\Omega f\,dv = \int_\Omega f^+\,dv - \int_\Omega f^-\,dv. \tag{33}$$

Thus one shows with the arguments of the standard Lebesgue theory adapted to the present case that f is (order) integrable for v iff $|f|$ is and that the class of v-integrable functions is a vector space. If $\mathscr{X} = L^p$, $1 \leq p < \infty$ (or $\mathscr{X} = L^\varphi$ and φ is moderated), then it has an absolutely continuous norm, i.e., for any $\{f_n \downarrow 0\} \subset \mathscr{X}$ we have $\|f_n\| \to 0$, and one notes that the order and norm continuities are equivalent. From this one concludes that the general D–S integral and the present order integral are the same and that the result of Theorem 14 holds in this case. However, for the general order integral as defined, we also have a monotone convergence (and hence the dominated convergence and Fatou's inequality) result as follows.

15. Proposition *Let $0 \leq f_n \uparrow f$ be a sequence of measurable functions on (Ω, Σ) and $v: \Sigma \to \mathscr{X}$ (a Banach lattice) a vector valued measure. Then we have*

$$\int_\Omega f\,dv = \lim_n \int_\Omega f_n\,dv = \sup_n \int_\Omega f_n\,dv. \tag{34}$$

Proof If $\int_\Omega f_n\,dv = \infty$ for some n, then the fact that $\int_\Omega f\,dv \geq \int_\Omega f_n\,dv$ implies the truth of (34) trivially. Suppose then $\int_\Omega f_n\,dv < \infty$ for all n. By the structure theorem, for each n, there exist step measurable f_{mn} such that $0 \leq f_{mn} \uparrow f_n$ as $m \uparrow \infty$. Define $g_{mn} = \max\{f_{kn} : 1 \leq k \leq m\}$. Then $0 \leq g_{nn} \uparrow f$, and each g_{nn} is bounded. Hence by definition of the order integral one has

$$\int_\Omega f\,dv = \sup_n \int_\Omega g_{nn}\,dv = \lim_n \int_\Omega g_{nn}\,dv. \tag{35}$$

On the other hand, $g_{nn} \leq f_n \leq f$ for each n. Hence

$$\int_\Omega g_{nn}\,dv \leq \int_\Omega f_n\,dv \leq \int_\Omega f\,dv. \tag{36}$$

Taking limits as $n \to \infty$ in (36) and using (35), we get (34).

Remark A general theory of order preserving mappings, subsuming the above account, can be found in McShane [2].

1.5 REMARKS ON MEASURABILITY AND LOCALIZABILITY

As seen in Section 2, the concept of localizability is an abstraction and a generalization of σ-finiteness of a measure. But the most interesting aspect of this notion is the ease in checking the measurability of collections of sets or functions. We illustrate this point in the following theorem, which will also be used in several applications. The treatment is a specialized version of the work of McShane [1].

1. Definition Let (Ω, Σ, μ) be a complete measure space with the finite subset property. Let $\mathscr{M} = \mathscr{M}(\Sigma)$ be the set of all extended real valued measurable functions on Ω, and $\mathscr{A} \subset \mathscr{M}$. Then $\mathscr{A}$ is said to have a *supremum* h in $\mathscr{M}$ if (i) $h \geq f$ a.e. (μ) for all $f \in \mathscr{A}$ and (ii) if g in $\mathscr{M}$ also has the same property as h in (i), then $g \geq h$ a.e. (μ).

The desired result is given by

2. Theorem *Let (Ω, Σ, μ) be a localizable space and $\mathscr{M}$ be the set of all extended real measurable functions on Ω. If $\mathscr{A} \subset \mathscr{M}$ is nonempty, then it has a supremum h. Moreover, if all elements of $\mathscr{A}$ are nonnegative and the set is directed upward, then*

$$\int_{\Omega} h\, d\mu = \sup\left\{\int_{\Omega} f\, d\mu : f \in \mathscr{A}\right\}. \tag{1}$$

Proof Consider the collection $\mathscr{D}_r \subset \Sigma$ consisting of sets $D_r^f = [f \geq r]$, $f \in \mathscr{A}$ and r rational. Since μ is localizable, $\mathscr{D}_r$ has a supremum, say, A_r. For $r' > r$, $D_{r'}^f \subset D_r^f$, and if $A_{r'}$ is the corresponding supremum of $\mathscr{D}_{r'}$, then $A_{r'} - A_r \in \mathscr{N}$, the set of all μ-null sets of Σ. If $B_r = \bigcup\{A_{r'} : r' \geq r, \text{rational}\}$, then $B_r \in \Sigma$ and $B_r - A_r \in \mathscr{N}$. But $B_r \supset B_{r'}$ for $r' > r$. Define $h_r = r$ on B_r, $h_r = -\infty$ on B_r^c. Thus $h_r \in \mathscr{M}$ and if $h = \sup\{h_r, r \text{ rational}\}$, then $h \in \mathscr{M}$, and we claim that this h satisfies the requirements of the statement.

To see that $h \geq f$ a.e. for all $f \in \mathscr{A}$, consider $C_r^f = \{\omega : h(\omega) < r < f(\omega)\}$ for any fixed but arbitrary f in $\mathscr{A}$. Then $C_r^f \subset D_r^f$, $C_r^f \in \Sigma$. But $C_r^f - A_r \subset D_r^f - A_r \in \mathscr{N}$ by definition of A_r. Since $C_r^f - B_r \subset C_r^f - A_r \in \mathscr{N}$ and $C_r^f \cap B_r = \varnothing$, it follows that $C_r^f \in \mathscr{N}$. Hence $f \leq h$ a.e. To see that h is the least such function, let $h' \in \mathscr{M}$ be another upper bound of $\mathscr{A}$, so that $f \leq h'$ a.e. for all $f \in \mathscr{A}$. Let $\bar{A}_r = [h' \geq r]$. Then $D_r^f \subset \bar{A}_r$ a.e., $f \in \mathscr{A}$, so that $\bar{A}_r$ is an upper bound of $\mathscr{D}_r$. Hence $A_r - \bar{A}_r \in \mathscr{N}$. Thus $B_r - \bar{A}_r \in \mathscr{N}$ and $\Omega_0 = \bigcup\{B_r - \bar{A}_r : r \text{ rational}\} \in \mathscr{N}$. We have

$$\begin{aligned} h(\omega) &= \sup\{r : \omega \in B_r, r \text{ rational}\} \leq \sup\{r : \omega \in \bar{A}_r, r \text{ rational}\} \\ &\leq h'(\omega). \end{aligned} \tag{2}$$

It follows that h is a supremum of $\mathscr{A}$ in the sense of Definition 1.

Suppose now $\mathscr{A} \subset \mathscr{M}^+$ and is an ascending net. To prove (1), let $\Sigma_0 = \{A \in \Sigma, \mu(A) < \infty\}$ and $a_f = \sup\{\|f\chi_A\|_1 : A \in \Sigma_0\}$. Clearly, $a_f \leq \|f\|_1$. We claim that there is equality here. If $a_f = \infty$, this is true; so let $a_f < \infty$. Then by definition there exists an increasing sequence $\{A_n, n \geq 1\} \subset \Sigma_0$ such that $a_f = \lim_n \|f\chi_{A_n}\|_1 = \|f\chi_A\|_1$, where $A = \bigcup_n A_n$. If the set $A^c \cap [f > 0]$ is not in $\mathscr{N}$, then by the finite subset property there exists a $B \in \Sigma_0$, $B \subset A^c \cap [f > 0]$ and $\mu(B) > 0$. Consequently,

$$\begin{aligned} a_f &= \|f\chi_A\|_1 < \|f\chi_A\|_1 + \|f\chi_B\|_1 \\ &= \|f\chi_{A \cup B}\|_1 \leq \sup\{\|f\chi_C\|_1 : C \in \Sigma_0\} = a_f. \end{aligned} \tag{3}$$

This contradiction shows that $[f > 0] \cap A^c \in \mathscr{N}$ and $a_f = \|f\|_1$. Suppose that we have proved (1) with $\mu(\Omega) < \infty$ so that $\|h\|_1 = \sup\{\|f\|_1 : f \in \mathscr{A}\}$. Then by (3), the result holds in general because

$$\begin{aligned} \|h\|_1 &= \sup\{\|h\chi_A\|_1 : A \in \Sigma_0\} \\ &= \sup\{\sup\{\|f\chi_A\|_1 : f \in \mathscr{A}\} : A \in \Sigma\} \\ &= \sup\{\sup\{\|f\chi_A\|_1 : A \in \Sigma_0\} : f \in \mathscr{A}\} \\ &= \sup\{\|f\|_1 : f \in \mathscr{A}\}. \end{aligned} \tag{4}$$

Let us therefore prove (1) when $\mu(\Omega) < \infty$. If $b = \sup\{\|f\|_1 : f \in \mathscr{A}\}$, then the fact that $h \geq f \geq 0$ a.e., implies $b \leq \|h\|_1$. The result is again true if $b = +\infty$. If $b < \infty$, then there exists a sequence $f_n \uparrow g$ a.e., and $b = \lim_n \|f_n\|_1 = \|g\|_1$. Using the argument leading to (3), we see that g is an upper bound of $\mathscr{A}$ so that $g \geq h$ a.e., $b = \|g\|_1 \geq \|h\|_1 \geq b$. Hence $\|h\| = b$, and the proof is complete.

Remark If the set $\mathscr{A} \subset \mathscr{M} \cap L^\infty$ is bounded, then h above is also essentially bounded. Further, if $\mathscr{A}$ is in a ball, then $\|h\|_1 = \sup\{\|f\|_1 : f \in \mathscr{A}\}$ is always finite in (1).

3. Definition A mapping $\rho: \mathscr{M} \to \mathscr{M}$ is said to be a *lifting* if we can continuously select a member from each equivalence class, i.e., the following conditions hold: (i) $\rho(f) = f$ a.e., (ii) $f = g$ a.e., implies $\rho(f) = \rho(g)$, (iii) $\rho(1) = 1$, (iv) $\rho(f) \geq 0$ a.e. if $f \geq 0$ a.e., (v) $\rho(af + bg) = a\rho(f) + b\rho(g)$, and (vi) $\rho(fg) = \rho(f)\rho(g)$.

Thus if N is the set of μ-null functions in $\mathscr{M}$ and $M = \mathscr{M}/N$ is the quotient space of equivalence classes, let $\pi: \mathscr{M} \to M$ be the quotient mapping. Then $\rho: M \to \mathscr{M}$ such that $\pi \circ \rho$ is the identity. In general such a ρ need not exist. The following important positive result on the existence is known, and it is due to Tulcea [1].

4. Theorem *Let (Ω, Σ, μ) be a strictly localizable measure space. Then there exists a lifting map on $L^\infty(\Omega, \Sigma, \mu)$ into bounded measurable real functions on Ω.*

The proof of this result involves many details. A relatively simple demonstration of it has recently been given by Traynor [1], to which we refer the reader. Another, somewhat simple proof is also included in the author's monograph [10].

Complements and Problems

1. If $F(\cdot, \ldots, \cdot)$ is an n-dimensional distribution function, its Fourier–Stieltjes transform φ defined by

$$\varphi(t_1, \ldots, t_n) = \int_{-\infty}^{\infty} \cdots \int_{-\infty}^{\infty} \exp[i(t_1 x_1 + \cdots + t_n x_n)]\, dF(x_1, \ldots, x_n),$$

is called the *characteristic function* of F. If τ is written for (it), then the integral, if it exists, defines the *moment generating function* of F. There is a one-to-one relation between the functions φ and F (the uniqueness theorem of Fourier–Stieltjes [or bilateral Laplace] transforms, or of characteristic functions). Use this uniqueness result and establish the following facts:

(a) If F is a Gaussian distribution, defined by Eq. (1.5), show that its characteristic function φ is given by

$$\varphi(t_1, \ldots, t_n) = \exp\left[i(t_1 \alpha_1 + \cdots + t_n \alpha_n) - \frac{1}{2} \sum_{p,q=1}^{n} k_{pq} t_p t_q \right].$$

(b) If T is a subset of $\mathbb{R}$, $\alpha\colon T \to \mathbb{R}$ is a function, and $r\colon T \times T \to \mathbb{R}$ is a mapping such that (i) $r(t, s) = r(s, t)$ and (ii) r is positive definite in the sense that for each finite set $\{t_1, \ldots, t_n\} \subset T$, the matrix $(r(t_i, t_j),\ 1 \leq i, j \leq n)$ is positive definite (or, for each $n \geq 1$, the determinant of the matrix is positive), show that there exists a probability space (Ω, Σ, P) and a real Gaussian stochastic process $\{X_t, t \in T\}$ on it with

$$P[X_{t_1} < x_1, \ldots, X_{t_n} < x_n] = F_{t_1, \ldots, t_n}(x_1, \ldots, x_n).$$

Here $F_{t_1, \ldots, t_n}$ is a Gaussian distribution (cf. Eq. (1.5)) with $\alpha = (\alpha(t_1), \ldots, \alpha(t_n))$ and $K = (r(t_i, t_j), 1 \leq i, j \leq n)$ *for each* $\{t_1, \ldots, t_n\} \subset T$, $n \geq 1$. The parameters α and r are called the *mean* (or *expected value*) *and covariance* of the Gaussian process. Verify also that

$$\alpha(t) = E(X_t) = \int_{-\infty}^{\infty} x\, dF_t(x) = \int_{\Omega} X_t\, dP$$

and

$$r(t_1, t_2) = \operatorname{cov}(X_{t_1}, X_{t_2}) = E((X_{t_1} - \alpha(t_1))(X_{t_2} - \alpha(t_2)))$$

$$= \int_{-\infty}^{\infty} \int_{-\infty}^{\infty} (x - \alpha(t_1))(x - \alpha(t_2))\, dF_{t_1, t_2}(x_1, x_2).$$

Conclude that a Gaussian process is uniquely determined by its mean and covariance functions. [*Hint*: Use Theorem 3.1 for existence.]

2. Let $T \subset \mathbb{R}$ and u, v be two positive functions on T such that $(u/v)(t_1) < (u/v)(t_2)$ for $t_1 < t_2$, $t_i \in T$, $i = 1, 2$. If $r(s, t) = u(\min(s, t))v(\max(s, t))$, show that $r\colon T \times T \to \mathbb{R}$ is a covariance function, i.e., properties (i), (ii) of Problem 1(b) hold. It is called a *triangular covariance*. Hence deduce that the Gaussian process with mean zero and this r as its covariance function has the following n-dimensional density (i.e., the integrand of Eq. (1.5)) $f_{t_1,\ldots,t_n}$ for $t_1 < t_2 < \cdots < t_n$, $t_j \in T$:

$$f_{t_1,\ldots,t_n}(x_1, \ldots, x_n) = \left[(2\pi)^n \prod_{j=1}^{n+1} (u_j v_{j-1} - u_{j-1} v_j)\right]^{-1/2}$$

$$\times \exp\left[-\frac{1}{2} \sum_{j=1}^{n} \frac{v_{j-1}(x_k - (v_j/v_{j-1})x_{j-1})^2}{v_j(u_j v_{j-1} - u_{j-1} v_j)}\right],$$

where $u_j = u(t_j)$, $v_j = v(t_j)$, $u_0 = 0 = v_{n+1} = x_0$, and $v_0 = 1 = u_{n+1}$, [In case $v(t) \equiv 1$, $u(t) = t$, and $T = [0, 1]$, the corresponding process is called the *Brownian motion* (and *Wiener if also* $X_0 = 0$) *process*. To prove the above, verify that $(r(t_i, t_j))^{-1}$ is a matrix with positive diagonal entries and has nonzero elements only in the line above (and below) the diagonal.]

3. This problem illustrates the need for Theorem 3.5. The following useful criterion for the continuity of sample paths has been given by A. Kolmogorov: If $\{X_t, t \in T\}$ is a stochastic process on (Ω, Σ, P), $T \subset \mathbb{R}$ is an interval, suppose there exist three positive constants C, α, and β such that for any s, t in T we have

$$E(|X_t - X_s|^\alpha) \le C|t - s|^{1+\beta}. \tag{$*$}$$

Then $X_{(\cdot)}(\omega)\colon T \to \mathbb{R}$ is continuous for almost all $\omega \in \Omega$. [Thus the condition is on all two-dimensional distributions. The proof is nontrivial.] Verify that the Brownian motion process (cf. Problem 2) satisfies the condition $(*)$ with $\alpha = 4$, $C = 3$, $\beta = 1$, so that it has almost all continuous sample paths. The difficulties noted in Theorem 3.5 are all present for this important process.

4. This problem illustrates, for the processes of the type considered in Problem 2, the kind of distribution questions that arise in applications. Thus let $\{X_t, t \in (0, 1]\}$ be a Gaussian process with mean zero and covariance r of the triangular type given there with $u(t) = 1$ and $v(t) = -\log t$. If we define $Y(\omega) = \int_0^1 X_t^2(\omega)\, dt$, which exists as a Riemann integral for a.a.

$\omega \in \Omega$, then show that Y is a random variable on Ω and its moment generating function φ is given by

$$\varphi(\tau) = E(\exp[-\tfrac{1}{2}\tau Y]) = [I_0(2\tau^{1/2})]^{-1/2}, \qquad \tau > 0, \qquad (+)$$

where

$$I_0(x) = 1 + \frac{x^2}{2^2} + \frac{x^4}{2^2 \cdot 4^2} + \frac{x^6}{2^2 \cdot 4^2 \cdot 6^2} + \cdots$$

is a modified Bessel function.

Sketch of Proof The method is typical for such problems, and we outline it. The idea is to express X_t as a series and then find $\varphi(\tau)$ for Y. For the given kernel r, consider the integral equation $(*)$ $u(t) = \lambda \int_0^1 r(s,t)u(s)\,ds$. If the value of $r(s,t)$ is substituted, $u(t)$ differentiated relative to t, one transforms $(*)$ into a differential equation ($u' = (du/dt)$, $u'' = (u')'$, etc.):

$$tu''(t) + u'(t) + \lambda u(t) = 0, \qquad u(1) = 0, \quad u'(0) = 0, \quad \lambda \geq 0. \qquad (++)$$

Setting $t = x^2/4\lambda$ and simplifying, this becomes

$$x\frac{d^2u}{dx^2} + \frac{du}{dx} + \lambda u = 0 \qquad \text{with} \quad u(1) = 0, \quad u(0) < \infty.$$

But this is a known differential equation for the Bessel functions. Its solution is $u(x) = C_0 J_0(x)$, where

$$J_0(x) = 1 - \frac{x^2}{2^2} + \frac{x^4}{2^2 \cdot 4^2} - \cdots.$$

Hence the solution for $(++)$ is $u(t) = C_0 J_0(2\sqrt{\lambda t})$. Since $u(1) = 0$, $J_0(2\sqrt{\lambda}) = 0$. Let its nonzero solutions be $\lambda = \lambda_n$. Then $k_n = 2\sqrt{\lambda_n}$ are the eigenvalues of $(*)$ $(n - \frac{1}{2})\pi < k_n < n\pi$, $n = 1, 2, \ldots$, and the corresponding eigenfunctions ψ_n are given by $\tilde{\psi}_n(t) = C_0 J_0(k_n\sqrt{t})$. It is verified that $\{\tilde{\psi}_n, n \geq 1\}$ are orthogonal and $\int_0^1 J_0^2(k_n\sqrt{t})\,dt = (J_0'(k_n))^2$. Hence the normalized eigenfunctions ψ_n (i.e., $\|\psi_n\|_2 = 1$ on $L^2(0,1)$) are $\psi_n(t) = J_0(k_n\sqrt{t})/J_0'(k_n)$, $n \geq 1$. We expand X_t in a Fourier series, with $a_k = \lambda_k \int_0^1 X_t \psi_k(t)\,dt$. $X_t = \sum_{k=1}^{\infty} a_k[\psi_k(t)/\sqrt{\lambda_k}]$ converges in $L^2(0,1)$ for almost all ω. One also verifies that each a_k is a Gaussian random variable with mean zero and variance $= E(a_k^2) = 1$. Next show that $E(a_k a_{k'})$ is zero, for $k \neq k'$, so that their characteristic functions factor, i.e.,

$$E(\exp[i(t_1 a_{k_1} + \cdots + t_j a_{k_j})]) = \prod_{l=1}^{j} E(e^{it_l a_{k_l}}) \qquad \text{for all} \quad j \geq 1.$$

This property is called *mutual independence* of a_k's. Hence the series for X_t is also convergent in $L^2(P)$. Thus the following computation is legitimate:

$$\varphi(\tau) = E\left[\exp\left(-\frac{\tau}{2}\int_0^1 X_t^2\,dt\right)\right] = E\left[\lim_{n\to\infty}\exp\left(-\frac{\tau}{2}\sum_{k=1}^{n}\frac{a_k^2}{\lambda_k}\right)\right]$$

$$= \lim_{n\to\infty}\prod_{k=1}^{n} E\left(\exp\left[-\frac{\tau}{2}\frac{a_k^2}{\lambda_k}\right]\right)$$

$$= \prod_{k=1}^{\infty}\left(1+\frac{\tau}{\lambda_k}\right)^{-1/2} = [D(-\tau)]^{-1/2} \quad \text{(say).}$$

This $D(\cdot)$ is known as the *Fredholm determinant*. Also $\tau = -\lambda_k$ are zeros of this function. Since $J_0(2\sqrt{\lambda_k})=0$, one has $D(-\tau)=MJ_0(2i\sqrt{\tau})=MI_0(2\sqrt{\tau})$, where I_0 is defined in the statement and M is a constant. Since $D(0) = 1$, one has $M = 1$, and (+) follows. Inverse Laplace transform of this φ gives the distribution F of Y, which cannot, however, be obtained in a closed form here. The result admits extensions to all continuous covariances r on $[0, 1] \times [0, 1]$; but then a more involved analysis is needed.

5. For the validity of Theorems 3.1 and 3.2, the topology of the spaces Ω_ts and (hence) the regularity of P_ts were crucial. If this hypothesis is dropped, the results need not hold as the following example shows. Let $(\Omega, \mathscr{A}, Q)$ be the Lebesgue unit interval, $I_n = (\frac{1}{2} - (1/2^{n+1}), \frac{1}{2})$, $n \geq 1$, and $X_n = 2^{n+1}\chi_{I_n}$. Let $\mathscr{A}_n = \mathscr{A}(I_n) = \{A \cap I_n : A \in \mathscr{A}\}$, the trace σ-algebra, and $P_n(A) = \int_A X_n\,dQ$. Then for $n > m \geq 1$, $P_m = P_n | \mathscr{A}_m$ and $P_n(I_n) = 1$. Consider for $\alpha = (1, \ldots, n)$, $\Omega_\alpha = \times_{k=1}^{n} I_k$, $\mathscr{A}_\alpha = \bigotimes_{k=1}^{n} \mathscr{A}_k$, and $P_\alpha = P_n \circ \Delta_n^{-1}$, where Δ_n: $I_n \to \Omega_\alpha$ is the diagonal mapping, i.e., $\Delta(\omega) = (\omega, \omega, \ldots, \omega) \in \Omega_\alpha$. Then show that the preceding construction implies that $\{(\Omega_\alpha, \mathscr{A}_\alpha, P_\alpha, \pi_{\alpha\beta})_{\alpha<\beta}\}$ is a compatible family of probability spaces where $\alpha < \beta$ means $\alpha \subset \beta$ with $\beta = (1, 2, \ldots, n + k)$, $k \geq 1$. If $\Omega_{\mathbb{N}} = \times_{k=1}^{\infty} I_k$ and $\mathscr{A}_{\mathbb{N}}$ is similarly defined, then using the necessity proof of Theorem 3.2 (cf. Eq. (4)), verify that there exists an additive set function P on $\mathscr{A}_{\mathbb{N}}$ such that $P | \mathscr{A}_\alpha = P_\alpha$, but P is *not* σ-additive. [*Hint*: If not, the cylinder sets $C_n = \{\omega = \{\omega_n, n \geq 1\} : \omega_i = \omega_0,$ $1 \leq i \leq n$, $\omega_0 \in \Omega_n\}$ have the property that $C_n \in \mathscr{A}_{\mathbb{N}}$, $P(C_n) = 1$ for all n, and $C_n \downarrow \varnothing$, giving a contradiction. Note that P_αs here are *not* regular on Ω_α. This example is essentially due to Doob [1, p. 681]. It shows that one needs to impose some further conditions on the measures P_α or algebras $\mathscr{A}_\alpha$ or both in exchange to regularity. We shall analyze this point in Chapter III.]

6. The proof of Theorem 3.2 admits the following simplification if $\Omega_t = S$, $t \in T$, where S is a compact Hausdorff space. Let $\Omega = \Omega_T = S^T$, where T is an index set, and Ω is given the product (Tychonov) topology. Let $C(\Omega)$ be the Banach space of all real continuous functions on Ω under the uniform norm. Using the notation of Theorem 3.2 so $\Omega_\alpha = S^\alpha$, $\alpha \in D$, the (directed by inclusion) family of all finite subsets of T, identify $C(\Omega_\alpha)$ as a subalgebra of $C(\Omega)$ such that $f \in C(\Omega_\alpha)$ iff there is an $\tilde{f} \in C(\Omega)$ such that $\tilde{f}|_{\Omega_\alpha} = f$. If $C_0 = \bigcup_{\alpha \in D} C(\Omega_\alpha)$, then the algebra C_0 separates the points of Ω, has a unit, and is dense in $C(\Omega)$ so that the uniform closure of C_0 is $C(\Omega)$ by the Stone–Weierstrass theorem. Let $\mathscr{B}_\alpha$ be the Borel σ-algebra of Ω_α and $P_\alpha: \mathscr{B}_\alpha \to \mathbb{R}^+$ be a regular probability. Set $\mathscr{B} = \bigotimes_{\alpha \in D} \mathscr{B}_\alpha$. Show that there exists a regular probability on $\mathscr{B}$ iff for each α, β in D with $\alpha < \beta$, $P_\alpha = P_\beta \circ \pi_{\alpha\beta}^{-1}$ where $\pi_{\alpha\beta}: \Omega_\beta \to \Omega_\alpha$ are coordinate projections. [*Hints*: Let $l(f) = \int_\Omega f\,dP_\alpha$, $f \in C(\Omega_\alpha)$. Then the compatibility condition implies that l is well defined on C_0 and is independent of α, $|l(f)| \leq \|f\|$ so that $l(\cdot)$ is continuous on C_0 and has a unique continuous extension to $C(\Omega)$ preserving its bound which is equal to $l(1) = 1$. Then by the classical Riesz–Markov theorem there is a regular probability P on $\mathscr{B}$, and $l(f) = \int_\Omega f\,dP, f \in C(\Omega)$. In this context, see Nelson [1].]

CHAPTER

II

Conditional Expectations and Probabilities

This chapter is devoted to a detailed study of conditional expectations in regard to both the structural and functional properties. The work includes characterizations of conditional expectations as a class of contractive projections, as well as averaging operators on the L^p-spaces. Then the conditional probabilities are studied as function space valued measures. The questions of regularity and a comparison of the present point of view with Rényi's new axiomatic approach are included. Among applications of the work, an integral representation of Reynolds operators and several complements may be noted.

2.1 INTRODUCTION OF THE CONCEPT

The notion of a conditional expectation is one of the most fundamental, not quite intuitive, and somewhat involved ideas in probability. To motivate the definition, consider the following situation. Given a probability space (Ω, Σ, P) describing a physical phenomenon, suppose that a particular event A $(\in \Sigma)$ has been observed by an experimenter. The problem is to assign probabilities to all other events B in Σ, having known everything about A. This means one should consider the class $\{A \cap B : B \in \Sigma\} = \Sigma(A)$ of events and assign P_A to members of $\Sigma(A)$ using P on Σ. A natural candidate is then $P_A \colon \Sigma(A) \to [0, 1]$ given by $P_A(C) = P(C)/P(A)$ for $C \in \Sigma(A)$ so that $P_A(B) = P(A \cap B)/P(A)$ for all $B \in \Sigma$, provided $P(A) \neq 0$. The expectation $E_A(\cdot)$, on (Ω, Σ, P_A) of a random variable X in $L^1(P)$, becomes

$$E_A(X) = \int_\Omega X \, dP_A = \frac{1}{P(A)} \int_A X \, dP. \tag{1}$$

If A is replaced by A^c, then $E_{A^c}(X)$ is generally a different number from that of (1) so that it is a function of the sets A, A^c. This computation is easily extended to a countable collection of sets $\{A_i, i \geq 1\} \subset \Sigma$, $P(A_i) > 0$. Thus if Π denotes a partition $\{A_n, n \geq 1\}$ of Ω, $A_n \in \Sigma$, $P(A_n) > 0$, then we define

$$E_\Pi(X) = \sum_{n=1}^{\infty} \frac{1}{P(A_n)} \left(\int_{A_n} X \, dP \right) \chi_{A_n}, \tag{2}$$

which is the sum of the individual expected values on each A_n. In the above $P_A(B)$ is termed the conditional probability of B given A. In the case of a partition $\Pi = \{A_n, n \geq 1\}$ this becomes

$$P_\Pi(B) = \sum_{n=1}^{\infty} P_{A_n}(B) \chi_{A_n}, \qquad B \in \Sigma, \tag{3}$$

where again we "add" the individual probabilities on each A_n. These definitions, from the absolute (or unconditional) concepts of Chapter I (cf. I.6.1) are natural extensions and are entirely sufficient as long as conditioning only with partitions is desired. However, more complicated situations are common. For instance, we may want to define these concepts if the known event is based on a random variable that takes more than countably many values, such as the problem of defining a conditional expectation of X given Y, where X, Y have a joint continuous (e.g., Gaussian) distribution. In such a case, the preceding procedure clearly fails and one has to proceed differently.

The alternative definition is based on the observation that if $A \in \Pi$, then (2) becomes

$$\int_A E_\Pi(X) \, dP = \sum_{n=1}^{\infty} \frac{1}{P(A_n)} \left(\int_{A_n} X \, dP \right) \cdot \int_A \chi_{A_n} \, dP = \int_A X \, dP. \tag{4}$$

Thus if $\nu_X(A) = \int_A X \, dP$ and $\mathscr{B}$ is written for the σ-algebra generated by Π, then it is clear that (4) can be expressed as

$$\nu_X(A) = \int_A E^{\mathscr{B}}(X) \, dP, \qquad A \in \mathscr{B}, \tag{5}$$

where $E^{\mathscr{B}}(X)$ is a constant on each A_n in Π, or each "generator" of $\mathscr{B}$. Taking $X = \chi_B$, we get $P_\Pi(B) = E_\Pi(\chi_B)$, and in the new notation this is written $P^{\mathscr{B}}(B) = E^{\mathscr{B}}(\chi_B)$. We abstract this and introduce the general concept following Kolmogorov [1]:

1. Definition Let (Ω, Σ, P) be a probability space, and let $X \in L^1(P)$. If $\mathscr{B} \subset \Sigma$ is any σ-algebra, then a $\mathscr{B}$-measurable function $E^{\mathscr{B}}(X)$ satisfying the functional equation

$$\int_B E^{\mathscr{B}}(X)\,dP_{\mathscr{B}} = \int_B X\,dP, \qquad B \in \mathscr{B}, \tag{6}$$

is called the *conditional expectation* relative to $\mathscr{B}$, and $P^{\mathscr{B}}(\cdot)$ on Σ defined by $P^{\mathscr{B}}(A) = E^{\mathscr{B}}(\chi_A)$ is termed the *conditional probability measure* relative to $\mathscr{B}$ (or given $\mathscr{B}$), where $P_{\mathscr{B}}$ is the restriction of P to $\mathscr{B}$.

Writing $\nu_X(B) = \int_B X\,dP$ and noting that ν_X is a signed measure on $\mathscr{B}$ which is $P_{\mathscr{B}}$ continuous, we see that $E^{\mathscr{B}}(X)$ exists P-uniquely as a $\mathscr{B}$-measurable function $(= d\nu_X/dP_{\mathscr{B}})$ by the Radon–Nikodým theorem (cf. I.2.7). Hence the definition is meaningful. Moreover, it clearly reduces to (2) and (3) if $\mathscr{B}$ is generated by a partition. For a conditional probability $P^{\mathscr{B}}$, (6) reduces to the *functional equation*

$$\int_B P^{\mathscr{B}}(A)\,dP_{\mathscr{B}} = \int_B \chi_A\,dP = P(A \cap B), \qquad A \in \Sigma, \quad B \in \mathscr{B}. \tag{7}$$

We first analyze the operator $E^{\mathscr{B}}$ on $L^1(P)$ and later consider $P^{\mathscr{B}}$, which is also (by definition) a function $A \mapsto P^{\mathscr{B}}(A)$ satisfying Eq. (7). Hereafter the underlying triple (Ω, Σ, P) is fixed without further mention, and all random variables and σ-algebras are based on it, $E(X) = \int_\Omega X\,dP$.

We can list some immediate properties of the operation $E^{\mathscr{B}}$ that follow from definition and the functional equation (6). The mapping $E^{\mathscr{B}}: X \mapsto E^{\mathscr{B}}(X)$ is a positive linear operator defined on $L^1(P)$. Moreover, $E(E^{\mathscr{B}}(X)) = E(X)$, and if $\mathscr{B} = \{\varnothing, \Omega\}$, then $E^{\mathscr{B}}(X) = E(X)$ a.e. Since $-|X| \le X \le |X|$ so that $-E^{\mathscr{B}}(|X|) \le E^{\mathscr{B}}(X) \le E^{\mathscr{B}}(|X|)$ for any $\mathscr{B} \subset \Sigma$, we note that $|E^{\mathscr{B}}(X)| \le E^{\mathscr{B}}(|X|)$ a.e., and

$$\int_A |E^{\mathscr{B}}(X)|\,dP_{\mathscr{B}} \le \int_A E^{\mathscr{B}}(|X|)\,dP_{\mathscr{B}} = \int_A |X|\,dP, \qquad A \in \mathscr{B}. \tag{8}$$

Taking $A = \Omega$, one deduces from (8) that $\|E^{\mathscr{B}}(X)\|_1 \le \|X\|_1$, i.e., $E^{\mathscr{B}}$ is a contraction on $L^1(P)$ into itself. A similar statement is also true on other L^p-spaces, $p \ge 1$, but it needs further analysis and proof. It is already clear from these remarks that $E^{\mathscr{B}}$ has a nontrivial structure, and we intend to study various aspects of this operator in some detail.

Let us record some other properties that are not entirely obvious.

2. Proposition *Let $X_i \in L^1(P)$, $i = 1, 2$, such that $X_1 X_2$ is also integrable. Then the following properties of $E^{\mathscr{B}}$ hold:*

(a) (Averaging) *If $\mathscr{B} \subset \Sigma$ is a σ-algebra, then*

$$E^{\mathscr{B}}(X_1 E^{\mathscr{B}}(X_2)) = E^{\mathscr{B}}(X_1) E^{\mathscr{B}}(X_2) \quad \text{a.e.}$$

(b) (Commutativity) *If $\mathscr{B}_1 \subset \mathscr{B}_2 \subset \Sigma$ are σ-algebras, then*

$$E^{\mathscr{B}_1}E^{\mathscr{B}_2}(X_1) = E^{\mathscr{B}_2}E^{\mathscr{B}_1}(X_1) = E^{\mathscr{B}_1}(X_1) \quad \text{a.e.}$$

In particular, if $\mathscr{B}_1 = \mathscr{B}_2$, then $E^{\mathscr{B}_1}E^{\mathscr{B}_1} = E^{\mathscr{B}_1}$ so that it is a projection operator.

Proof (a) Let $Y = E^{\mathscr{B}}(X_2)$, which is $\mathscr{B}$-measurable. Then $Y \in L^1(P)$. So it is to be shown that

$$\int_A E^{\mathscr{B}}(X_1 Y)\, dP_{\mathscr{B}} = \int_A X_1 Y\, dP = \int_A YE^{\mathscr{B}}(X_1)\, dP_{\mathscr{B}}, \qquad A \in \mathscr{B}. \tag{9}$$

This is true if $Y = \chi_B$, $B \in \mathscr{B}$, by (6) since $A \cap B \in \mathscr{B}$. Hence by linearity of the integral, (9) is again true if $Y = \sum_{i=1}^{n} a_i \chi_{B_i}$, $B_i \in \mathscr{B}$. By the basic structure theorem of measurable functions, there exists a sequence of simple $\mathscr{B}$-measurable Y_n such that $Y_n \to Y$ pointwise and $|Y_n| \leq |Y|$. Hence by the dominated convergence theorem ($|X_1 Y|$ and $|YE^{\mathscr{B}}(X_1)|$ are the dominating functions), (9) is true in general.

(b) Since $E^{\mathscr{B}_1}(X_1)$ is $\mathscr{B}_1$-measurable (hence $\mathscr{B}_2$-measurable) and $E^{\mathscr{B}_2}(1) = 1$ a.e., we get $E^{\mathscr{B}_2}(E^{\mathscr{B}_1}(X_1)) = E^{\mathscr{B}_1}(X_1)$ a.e. But by (6),

$$\int_A E^{\mathscr{B}_1}(X_1)\, dP_{\mathscr{B}_1} = \int_A X_1\, dP = \int_A E^{\mathscr{B}_2}(X_1)\, dP_{\mathscr{B}_1}$$

$$= \int_A E^{\mathscr{B}_1}(E^{\mathscr{B}_2}(X_1))\, dP_{\mathscr{B}_1}, \qquad A \in \mathscr{B}_1.$$

Since A is arbitrary and the integrands of the first and last integrals are $\mathscr{B}_1$-measurable, they can be identified a.e. The last statement is now obvious. This proves (b).

The following consequence is often used in applications.

3. Corollary *Let X, Y and XY be integrable random varibles. If $X_1, \ldots, X_n$ are any random variables and $\mathscr{B}_k = \sigma(X_i, k \leq i \leq n)$ is the σ-algebra generated by $X_k, X_{k+1}, \ldots, X_n$, then one has*

(i) $E^{\mathscr{B}}(XY) = E^{\mathscr{B}}(YE^{\tilde{\mathscr{B}}}(X))$ *a.e., for any $\mathscr{B} \subset \tilde{\mathscr{B}} \subset \Sigma$ and Y is $\tilde{\mathscr{B}}$-measurable;*

(ii) $E(X) = E(E^{\mathscr{B}_n}(E^{\mathscr{B}_{n-1}} \cdots (E^{\mathscr{B}_1}(X)) \cdots))$ *a.e.*

For the operator $E^{\mathscr{B}}$ the standard Lebesgue type convergence statements hold. (This is analogous to Proposition I.4.15, and later this connection is explored further.)

4. Proposition *Let $\{X_n, n \geq 1\} \subset L^1(P)$ and $\mathscr{B} \subset \Sigma$, a σ-algebra. Then*

(i) [monotone convergence] *if $X_n \uparrow X$ a.e., we have $E^{\mathscr{B}}(X_n) \uparrow E^{\mathscr{B}}(X)$ a.e.;*

(ii) [Vitali, dominated convergence] *if $X_n \to X$ a.e. and the set is uniformly integrable, or $|X_n| \leq Y$ with $Y \in L^1(P)$, then $E^{\mathscr{B}}(X_n) \to E^{\mathscr{B}}(X)$ in L^1-mean; also a.e. in the dominated case.*

(iii) [Fatou's inequality] *all $X_n \geq 0$ imply $\lim_n \inf E^{\mathscr{B}}(X_n) \geq E^{\mathscr{B}}(\lim_n \inf X_n)$ a.e.*

Proof All these statements are proved with simple modifications of the unconditional statements. Let us establish, for instance, (ii).

If $|X_n| \leq Y$ and $Y \in L^1(P)$, then the set $\{X_n, n \geq 1\}$ is uniformly integrable since it is bounded and $E(|X_n|\chi_A) \leq E(Y\chi_A) \to 0$ as $P(A) \to 0$ uniformly in n. Hence in both cases by (classical) Fatou's lemma $X \in L^1(P)$ and $\{|X_n - X|, n \geq 1\}$ is then uniformly integrable. Since $|X_n - X| \to 0$ a.e., we get $E(|X_n - X|) \to 0$. Also (cf. (8))

$$\begin{aligned} E(|E^{\mathscr{B}}(X_n) - E^{\mathscr{B}}(X)|) &= E(|E^{\mathscr{B}}(X_n - X)|) \\ &\leq E(E^{\mathscr{B}}(|X_n - X|)) \\ &= E(|X_n - X|) \qquad \text{by Corollary 3 (iii).} \end{aligned}$$

But the right-hand side tends to zero as $n \to \infty$. Hence $E^{\mathscr{B}}(X_n) \to E^{\mathscr{B}}(X)$ in $L^1(P)$. [If L^1 is replaced by L^p, $p \geq 1$, then a similar result is true, and this follows from the Jensen inequality established below.] The proofs of the a.e. convergence and other parts follow by similar modifications of classical cases.

If $\mathscr{B}_1, \mathscr{B}_2$ are two σ-algebras, they are said to be *independent* if for all $A \in \mathscr{B}_1$, $B \in \mathscr{B}_2$ we have $P(A \cap B) = P(A)P(B)$. A random variable X is independent of a σ-algebra $\mathscr{B}$ if $\mathscr{B}_X = \sigma(X)$, the generated σ-algebra of X, and $\mathscr{B}$ are independent. A family of σ-algebras $\{\mathscr{B}_\alpha, \alpha \in I\}$ is *mutually independent* if each finite subfamily $\{\mathscr{B}_{\alpha_i}, i = 1, \ldots, n\}$ is independent (for $n \geq 2$) in the sense that $A_i \in \mathscr{B}_{\alpha_i}$ implies $P(\bigcap_{i=1}^n A_i) = P(A_1)P(A_2) \cdots P(A_n)$. Consequently, if $X_1, \ldots, X_n$ are independent integrable random variables, then $E(X_1 X_2 \cdots X_n) = E(X_1)E(X_2) \cdots E(X_n)$, which follows for simple functions, and then generally. Alternately, the independence of X_i implies that of $|X_i|, i = 1, \ldots, n$. So by Tonelli's theorem $E(|X_i \cdots X_n|) = \prod_{i=1}^n E(|X_i|) < \infty$, *and then* by Fubini's theorem (cf. Theorem I.1.8) the result follows.

5. Proposition *If X is an integrable random variable, and $\mathscr{B} \subset \Sigma$ is a σ-algebra, then (i) X is independent of $\mathscr{B}$ implies $E^{\mathscr{B}}(X) = E(X)$ a.e., and (ii)*

$\mathscr{B} = \sigma(Y)$, where Y is some random variable, implies the existence of a Borel function $g_Y: \mathbb{R} \to \mathbb{R}$ such that $E^{\mathscr{B}}(X) = g_Y(X)$ a.e.

Proof (i) Let $B \in \mathscr{B}$. Then since $E(X)$ is a constant,

$$\begin{aligned}\int_B E(X)\, dP_{\mathscr{B}} &= E(X)P(B) = E(X)E(\chi_B) \\ &= E(X\chi_B) \qquad \text{by the independence of} \quad \mathscr{B} \quad \text{and} \quad \sigma(X) \\ &= \int_B X\, dP = \int_B E^{\mathscr{B}}(X)\, dP_{\mathscr{B}}. \end{aligned} \tag{10}$$

Since the extreme integrands of (10) are $\mathscr{B}$-measurable and B ($\in \mathscr{B}$) is arbitrary, we have $E^{\mathscr{B}}(X) = E(X)$ a.e.

(ii) By Definition 1, $E^{\mathscr{B}}(X)$ is $\mathscr{B}$-measurable where $\mathscr{B} = \sigma(Y)$. Then the Doob–Dynkin lemma (cf. I.2.3) implies the existence of a Borel $g = g_Y: \mathbb{R} \to \mathbb{R}$ such that $E^{\mathscr{B}}(X) = g \circ X$ a.e., since all conditions are fulfilled.

Sometimes, when $\mathscr{B} = \sigma(Y)$, $E^{\mathscr{B}}(X)$ is denoted by $E(X \mid Y)$ or $E^Y(X)$. Part (ii) is then stated verbally as "the conditional expectation of X given Y is a (Borel) function of the function Y."

To prove the contractivity property of $E^{\mathscr{B}}$ on L^p- spaces, we establish the following useful result, called the *conditional Jensen inequality*.

6. Theorem *Let $\varphi: \mathbb{R} \to \overline{\mathbb{R}}$ be a (measurable) convex function and X be a random variable on (Ω, Σ, P) such that X, $\varphi(X)$ are integrable. Then*

$$E^{\mathscr{B}}(\varphi(X)) \geq \varphi(E^{\mathscr{B}}(X)) \quad \text{a.e.} \tag{11}$$

Moreover, if $\mathscr{B}$ is completed for P and φ is strictly convex, then equality holds in (11) *when and only when X is $\mathscr{B}$-measurable.*

Proof First recall from Section I.4 that φ is convex iff it has the representation

$$\varphi(x) = \varphi(a) + \int_a^x \varphi'(t)\, dt, \qquad \{a, x\} \subset \mathbb{R}, \tag{12}$$

where the right derivative φ' of φ is nondecreasing on $\mathbb{R}$ and is strictly increasing if φ is strictly convex. Then (12) implies

$$\varphi(x) - \varphi(a) \geq (x - a)\varphi'(a), \qquad \{a, x\} \subset \mathbb{R}. \tag{13}$$

For any $\omega \in \Omega$, let $x = X(\omega)$ and $a = E^{\mathscr{B}}(X)(\omega)$. [Note that on a probability space the mere integrability of $\varphi(X)$ implies that $E(X^+)$ or $E(X^-)$ is finite (support line property of φ) and so $E^{\mathscr{B}}(X)$ always exists.] Substitution in (13) then yields

$$\varphi(X) - \varphi(E^{\mathscr{B}}(X)) \geq (X - E^{\mathscr{B}}(X))\varphi'(E^{\mathscr{B}}(X)) \quad \text{a.e.} \tag{14}$$

Applying the operator $E^{\mathscr{B}}$ to both sides of (14) and using Proposition 2 (and the order preserving nature of $E^{\mathscr{B}}$), one finds

$$E^{\mathscr{B}}(\varphi(X)) - \varphi(E^{\mathscr{B}}(X)) \geq 0 \quad \text{a.e.} \tag{15}$$

It is not hard to show that the integrability of $\varphi(X)$ implies that of $X\varphi'(X)$. This is a consequence of the Hölder inequality if φ is a Young function (i.e., the inequality (I.4.7)). Alternately, this is true if X is simple, and the general case follows by approximation. Now (15) is just (11). It remains to prove the equality condition of the last part. The following simple argument is due to N. Dinculeanu.

Let φ be strictly convex and $E^{\mathscr{B}}(\varphi(X)) = \varphi(E^{\mathscr{B}}(X))$ a.e. So integrating (14) one obtains

$$\begin{aligned} \int_\Omega \varphi(X)\,dP &\geq \int_\Omega \varphi(E^{\mathscr{B}}(X))\,dP + \int_\Omega \varphi'(E^{\mathscr{B}}(X))(X - E^{\mathscr{B}}(X))\,dP \\ &= \int_\Omega E^{\mathscr{B}}(\varphi(X))\,dP_{\mathscr{B}} + \int_\Omega E^{\mathscr{B}}[\varphi'(E^{\mathscr{B}}(X))(X - E^{\mathscr{B}}(X))]\,dP_{\mathscr{B}} \\ &= \int_\Omega \varphi(X)\,dP, \end{aligned} \tag{16}$$

using Proposition 2(a), the hypothesis of equality in (11), and (6). So there is equality throughout. Hence (16) and (13) yield

$$0 \leq \int_\Omega [\varphi(X) - \varphi(E^{\mathscr{B}}(X)) - \varphi'(E^{\mathscr{B}}(X))(X - E^{\mathscr{B}}(X))]\,dP = 0, \tag{17}$$

and so the nonnegative integrand must vanish a.e. But φ is strictly convex, so that there is strict inequality in (13) unless $x = a$. Hence in (17) we must have, if $A = \{\omega : X(\omega) \neq E^{\mathscr{B}}(X)(\omega)\}$, $P(A) = 0$. Thus $X = E^{\mathscr{B}}(X)$ a.e., so that X is $\mathscr{B}$-measurable by the completeness of $\mathscr{B}$. This proves the theorem.

As a consequence, we have the important property:

7. Theorem *For any σ-algebra $\mathscr{B} \subset \Sigma$, $E^{\mathscr{B}}: L^p \to L^p$, $1 \leq p \leq \infty$, is a contractive projection with range $L^p(\Omega, \mathscr{B}, P_{\mathscr{B}})$.*

Proof Only contraction property remains, because of Proposition 2(b). If $\varphi(x) = |x|^p$, $1 \leq p < \infty$, then by the preceding theorem for $X \in L^p$,

$$\|X\|_p^p = \int_\Omega \varphi(X)\,dP \geq \int_\Omega \varphi(E^{\mathscr{B}}(X))\,dP = \|E^{\mathscr{B}}(X)\|_p^p,$$

with strict inequality for $1 < p < \infty$ unless X is $P_{\mathscr{B}}$-measurable. If $p = +\infty$, then $|X| \le \|X\|_\infty$ a.e. So $|E^{\mathscr{B}}(X)| \le E^{\mathscr{B}}(|X|) \le E^{\mathscr{B}}(\|X\|_\infty) = \|X\|_\infty$ a.e. Hence $\|E^{\mathscr{B}}(X)\|_\infty \le \|X\|_\infty$. Since $E^{\mathscr{B}}|L^p(\Omega, \mathscr{B}, P_{\mathscr{B}})$ is the identity, the proof is complete.

2.2 SOME CHARACTERIZATIONS OF CONDITIONAL EXPECTATIONS

The preceding result shows that conditional expectations form a subclass of projection operators defined on L^p into itself. With a view to understanding the structure of these operators, we present some characterizations of this class as functional transformations. These results have several applications in diverse parts of analysis. The treatment here, as well as in the rest of this chapter, is adapted from the author's paper [2]. Related works will also be cited at appropriate places. Unless stated otherwise, all function spaces are based on a probability space (Ω, Σ, P), fixed throughout the discussion, and all spaces considered are real.

Let us start with the following concept:

1. Definition (i) A linear mapping $T: L^p \to L^p$ is said to be an *averaging operator* if the following conditions hold:

(a) $T(f\,Tg) = (Tf)(Tg), f, g \in L^\infty$,
(b) $T1 = 1$, and
(c) $\|Tf\|_p \le \|f\|_p$.

(ii) The mapping T is called a *Šidák operator* if for any f, g in L^p,

$$T((Tf) \vee (Tg)) = (Tf) \vee (Tg), \tag{$*$}$$

where $f \vee g$ is the $\max(f, g)$ and (b) and (c) hold.

In order to avoid the boundedness condition on f, g in the definition of an averaging operator, Šidák [1] has formulated the (ii) $(*)$ with the maximum condition in the context of L^2-spaces. One of the purposes of this section is to show that the averaging and Šidák operators on L^p spaces, $1 \le p < \infty$, are characterized as conditional expectations since it is evident that they satisfy the conditions of the above definition because of Proposition 1.2. These characterizations are nontrivial and necessarily use results from abstract analysis, as for example, outlined in Section I.4.

Let us start with the following observation:

2. Lemma *Every averaging or Šidák operator on L^p is idempotent and thus is a contractive projection.*

Proof Consider the Šidák identity. Taking $g = 0$ in its definition,

$$T((Tf)^+) = T((Tf \vee 0)) = (Tf) \vee 0 = (Tf)^+. \tag{1}$$

Since $f^- = -(-f)^+$, one obtains $(Tf)^- = -(-Tf)^+ = -(T(-f))^+$. Thus (1) yields

$$T((Tf)^-) = T(-(T(-f))^+) = -T((T(-f))^+) = -(T(-f))^+ = (Tf)^-. \tag{2}$$

Hence for $f \in L^p$ we have

$$T^2 f = T(Tf) = T((Tf)^+ - (Tf)^-) = (Tf)^+ - (Tf)^- = Tf, \tag{3}$$

so that $T^2 = T$. Here conditions (b) and (c) are not used.

The case with averaging is simpler than the above. In fact, if $g = Tf$, then using the averaging identity and (b), one has

$$T^2 f = Tg = T(1 \cdot g) = T1 \cdot Tf = Tf, \qquad f \in L^p, \tag{4}$$

so that $T^2 = T$. Condition (c) is needed in both cases only to conclude contractivity of T. This completes the proof.

The character of these operators is reflected in the structure of their range spaces as seen in the following.

3. Proposition *Let $T: L^p \to L^p$, $1 \le p \le \infty$ be an averaging or a Šidák operator. Then the range space $\mathcal{R}_T$ $(= T(L^p))$ contains densely an algebra of bounded functions in the first case and is a vector lattice in the second case.*

Proof By Lemma 2, T is a contractive projection in either case. Hence its range $\mathcal{R}_T$ is a closed linear manifold of L^p. In fact, the linearity being evident, let $\{f_n, n \ge 1\} \subset \mathcal{R}_T$ be a Cauchy sequence. Then $f_n \to f$ in L^p and (by completeness) $f \in L^p$. If $g = Tf \in \mathcal{R}_T$, then one has by the continuity of T (by its contractivity)

$$\begin{aligned} \|f - g\|_p &\le \|f - f_n\|_p + \|f_n - g\|_p = \|f - f_n\|_p + \|T(f_n - f)\|_p \\ &\le 2\|f - f_n\|_p \to 0, \qquad n \to \infty. \end{aligned}$$

Thus $f = g \in \mathcal{R}_T$.

If T is a Šidák operator, let f_1, f_2 be in $\mathcal{R}_T$. So $Tf_i = f_i$, $i = 1, 2$, and since $f_1 \vee f_2 = (Tf_1) \vee (Tf_2) \in L^p$, the latter being a lattice, one has

$$T((Tf_1) \vee (Tf_2)) = (Tf_1) \vee (Tf_2) = f_1 \vee f_2 \in \mathcal{R}_T. \tag{5}$$

Hence $\mathcal{R}_T$ is a Banach lattice, and we need to consider the averaging operator case.

If $\{f_1, f_2\} \subset \mathcal{R}_T$, f_1 (or f_2) bounded, then $T(f_1 f_2) = T(f_1(Tf_2)) = (Tf_1)(Tf_2) = f_1 f_2 \in \mathcal{R}_T$. Hence bounded elements in $\mathcal{R}_T$ form an algebra. To prove the density, we assert that Tf is bounded if f is. Indeed, since the assertion is clear for $p = \infty$, let $p < \infty$. For any $f \in L^\infty \subset L^p$, let

$g = Tf$ $(= T^2 f = Tg)$. By the averaging property, $T(fg) = T(f \cdot Tg) = (Tf)\cdot(Tg) = g^2$, since f is bounded. By induction, $T(fg^{n-1}) = (Tf)\cdot(Tg^{n-1}) = g^n$. Dividing by a suitable constant, it may be assumed that $\|f\|_\infty = 1$. Hence using the last identity,

$$\|g^n\|_p = \|T(fg^{n-1})\|_p \leq \|fg^{n-1}\|_p \qquad \text{since } T \text{ is contractive on } L^p$$

$$= \left[\int_\Omega |f|^p |g^{n-1}|^p \, dP\right]^{1/p} \leq \left[\int_\Omega |g^{n-1}|^p \, dP\right]^{1/p} \qquad \text{since } |f| \leq 1 \quad \text{a.e.}$$

$$= \|g^{n-1}\|_p \leq \|g^{n-2}\|_p \leq \cdots \leq \|f\|_p. \tag{6}$$

If $A = \{\omega : |g(\omega)| > 1\}$, then (6) implies

$$\|g^n \chi_A\|_p \leq \|g^n\|_p \leq \|f\|_p < \infty, \qquad n \geq 1. \tag{7}$$

Letting $n \to \infty$ and interchanging the limit and norm by the "Fatou property" of the norm or by the dominated convergence theorem, one concludes that (7) holds only if $P(A) = 0$. Hence $|g| \leq 1$ a.e.

Now bounded functions in L^p are norm dense. Hence for any $h \in \mathscr{R}_T \subset L^p$, and $\varepsilon > 0$, there exists a bounded h_ε in L^p with $\|h - h_\varepsilon\|_p < \varepsilon$. Let $g_\varepsilon = Th_\varepsilon$. Then $g_\varepsilon \in \mathscr{R}_T \cap L^\infty$, $\|h - g_\varepsilon\|_p = \|T(h - h_\varepsilon)\|_p \leq \|h - h_\varepsilon\|_p < \varepsilon$ since T is a contraction. So bounded elements of $\mathscr{R}_T$ are dense in it. This completes the proof.

Because of the special nature of the range spaces, it will be necessary to analyze their structure for our characterization problems.

4. Definition A closed linear manifold $\mathscr{M} \subset L^p(\Omega, \Sigma, P)$ is called a *measurable subspace* if $\mathscr{M} = L^p(\Omega, \mathscr{B}, P_{\mathscr{B}})$ for some σ-algebra $\mathscr{B} \subset \Sigma$.

A solution to our problem is assisted by the following:

5. Theorem *Let $\mathscr{M}$ be a linear manifold containing constants where $\mathscr{M} \subset L^p = L^p(\Omega, \Sigma, P)$, $1 \leq p \leq \infty$. Then $\mathscr{M}$ is a measurable subspace iff either (i) $\mathscr{M}$ is a Banach lattice and (ii) $0 \leq f_n \in \mathscr{M}$, $f_n \uparrow f$ a.e., $f \in L^p$ implies $f \in \mathscr{M}$, or (i′) $\mathscr{M}$ is complete and the bounded elements $\mathscr{S}$ of $\mathscr{M}$ form a dense algebra and (ii′) $f_n \in \mathscr{M}$, $f_n \to f$ a.e. $|f_n| \leq g \in L^p$ implies $f \in \mathscr{M}$.*

Let us first remark that condition (ii) or (ii′) is automatic in L^p, $1 \leq p < \infty$, by the dominated convergence theorem since in these cases $0 \leq (f - f_n)^p \leq f^p$ or $|f_n - f|^p \leq 2^p g^p$, so f^p or $(2g)^p$ serve as the dominating functions, and $\mathscr{M}$ is closed. Thus only in L^∞ are these conditions critical. Before proving this technical result, let us employ it in obtaining the desired characterizations that give a better appreciation for the type of results considered.

With practically no additional work, we can present the following theorem for operators including (by Lemma 2) those of Definition 1. It is a consequence of certain general results obtained independently by Douglas [1], and the author [3] (cf. also Andô [1]).

6. Theorem *If $T: L^1(\Sigma) \to L^1(\Sigma)$ is a contractive projection such that $T1 = 1$ a.e., then $T = E^{\mathscr{B}}$ for a (completed) σ-algebra $\mathscr{B} \subset \Sigma$.*

Proof The result is established in three steps.

Step I The operator T is positive and $\int_\Omega Tf\,dP = \int_\Omega f\,dP, f \in L^1$.

For consider $0 \le f \le 1$ a.e. Then by using $T1 = 1$ and contractivity,

$$\int_\Omega (1 - f)\,dP = \|1 - f\|_1 \ge \|T(1 - f)\|_1 = \int_\Omega |(1 - Tf)|\,dP$$

$$\ge \int_\Omega dP - \int_\Omega Tf\,dP.$$

Hence

$$0 \le \int_\Omega f\,dP \le \int_\Omega Tf\,dP \le \int_\Omega |Tf|\,dP \le \int_\Omega f\,dP \tag{8}$$

by contractivity of T. It follows that $Tf \ge 0$ a.e., and $\int_\Omega Tf\,dP = \int_\Omega f\,dP$ in this case. Then by linearity the same result holds for every positive bounded and then (by continuity of T) for all $0 \le f$ in L^1. The linearity of T further implies that $\int_\Omega Tf\,d\mu = \int_\Omega f\,d\mu$ for all $f \in L^1$.

Step II The range $\mathscr{R}_T$ of T is a vector lattice containing constants.

For $T1 = 1$ implies $\mathscr{R}_T$ has all constants. To show that it is a lattice, let $f \in \mathscr{R}_T$. Let us first check that $|f| \in \mathscr{R}_T$. Now $Tf = f$ since T is identity on $\mathscr{R}_T$. Hence using the positivity of T, proved above, one has

$$|f| = |Tf| \le T(|f|) \Rightarrow \|f\|_1 \le \|T(|f|)\|_1 \le \|f\|_1, \tag{9}$$

since T is a contraction. Thus $|f| = T|f| \in \mathscr{R}_T$. If $f, g \in \mathscr{R}_T$, then

$$f \vee g = \frac{(f + g) + |f - g|}{2} \in \mathscr{R}_T$$

because $f \pm g \in \mathscr{R}_T$ by linearity. Thus $\mathscr{R}_T$ is a lattice (and complete). Hence by Theorem 5, there exists a σ-algebra $\mathscr{B} \subset \Sigma$ such that $\mathscr{R}_T = L^1(\mathscr{B})$, and it is a measurable subspace.

Step III $T(L^1) = L^1(\mathscr{B})$ implies that T coincides with the conditional expectation $E^{\mathscr{B}}$.

For, let $f \in L^1$ and $A \in \mathscr{B}$. Then Tf is $\mathscr{B}$-measurable and

$$\int_A Tf\,dP_{\mathscr{B}} = \int_\Omega \chi_A(Tf)\,dP_{\mathscr{B}} = \int_\Omega T(\chi_A(Tf))\,dP_{\mathscr{B}} \qquad \text{by Step I.} \tag{10}$$

On the other hand,

$$\int_A E^{\mathscr{B}}(f)\,dP_{\mathscr{B}} = \int_A f\,dP = \int_\Omega T(\chi_A f)\,dP_{\mathscr{B}} \tag{11}$$

by Step I and Definition 1.1. The result will follow if the right-hand sides of (10) and (11) are shown to be equal for all $A \in \mathscr{B}$. Since T is positive, the desired equality should be shown for $f \geq 0$ only. We first establish that

$$h = T(\chi_A f) - T(\chi_A Tf) = 0 \text{ a.e.,} \qquad \text{for any } A \in \mathscr{B}.$$

If $0 \leq f \leq \chi_A$, then $0 \leq Tf \leq T\chi_A = \chi_A$ since T is positive and $\chi_A \in \mathscr{R}_T = L^1(\mathscr{B})$. Thus Tf vanishes outside A. Hence

$$\chi_A \pm h = T[\chi_A(\chi_A \pm f \mp Tf)] \leq T[\chi_A \pm f \mp Tf] = \chi_A. \tag{12}$$

This can hold iff $h = 0$ a.e. Next let $0 \leq f \leq 1$. Then $f\chi_A, f\chi_{A^c}$ satisfy the above special result. Hence

$$\int_A T(f\chi_A)\,dP = \int_A f\chi_A\,dP \qquad \text{and} \qquad \int_{A^c} T(f\chi_{A^c})\,dP = \int_{A^c} (f\chi_{A^c})\,dP. \tag{13}$$

Since we already noted that $T(f\chi_A)$ vanishes outside A, $\int_{A^c} T(f\chi_A)\,dP = 0$, so that

$$\begin{aligned}\int_A Tf\,dP &= \int_A T(f\chi_A)\,dP + \int_A T(f\chi_{A^c})\,dP = \int_A T(f\chi_A)\,dP \\ &= \int_\Omega T(f\chi_A)\,dP = \int_\Omega f\chi_A\,dP = \int_A f\,dP.\end{aligned} \tag{14}$$

This shows the equivalence of (10) and (11) for $0 \leq f \leq 1$ and by linearity for all bounded f in L^1. Thus $Tf = E^{\mathscr{B}}(f)$ a.e. for all bounded $f \in L^1$. Since both T and $E^{\mathscr{B}}$ are continuous linear operators on L^1 and bounded functions are dense in L^1, the same holds for all f in L^1. This completes the proof.

Since both the averaging and Šidák operators on L^1 are contractive projections leaving constants invariant, we have the following consequence of the above result.

7. Corollary *Every averaging or Šidák operator on L^1 coincides with a conditional expectation.*

Since $E^{\mathscr{B}}$ is also a contractive projection on all L^p- spaces $p \geq 1$ by Theorem 1.7, it is of interest to know whether the result of the corollary holds on all L^p-spaces. We shall show below that the answer is decisively negative on L^∞ and essentially affirmative on L^p for $p < \infty$. These are nontrivial, and the work illuminates the intricate structure of conditional expectation operators (and hence the far-reaching nature of the Definition 1.1 in this context). Since the contractive projections unify both the averaging and Šidák cases, we proceed with these projections. Some of the work here extends to Orlicz spaces also. The details of this case that are more involved are given by the author [4, IV]. Here only the L^p case is treated.

The following result, due to Andô [1], uses techniques that are special to the L^p-spaces. There is an analogous result for a class of Orlicz spaces which, however, needs somewhat different methods.

8. Theorem *Let $T: L^p \to L^p$, $1 \le p < \infty$, be a contractive projection such that $T1 = 1$. If $p \neq 2$, then T is equivalent to a conditional expectation $E^{\mathscr{B}}$, and the same result holds for $p = 2$ if further T is positive. In all these cases T can be extended to be a contractive projection on L^1 so that the result reduces to that of Theorem 6, i.e., $T = E^{\mathscr{B}}$.*

Proof Since $p = 1$ was treated in Theorem 6, let $1 < p < \infty$. Recall that the equation $(p^{-1} + q^{-1} = 1)$

$$\langle Tf, g\rangle = \int_\Omega (Tf)g\,dP = \int_\Omega f(T^*g)\,dP = \langle f, T^*g\rangle, \qquad f \in L^p, \quad g \in L^q,$$

defines an operator $T^*: L^q \to L^q$ which is again a projection, i.e., $(T^*)^2 = T^*$ since $T^2 = T$. Moreover,

$$\begin{aligned}\|T^*\| &= \sup\{\|T^*g\|_q : \|g\|_q \le 1\}\\ &= \sup\{\sup\{|\langle f, T^*g\rangle| : \|f\|_p \le 1\} : \|g\|_q \le 1\}\\ &= \sup\{\sup\{|\langle Tf, g\rangle| : \|g\|_q \le 1\} : \|f\|_p \le 1\}\\ &= \sup\{\|Tf\|_p : \|f\|_p \le 1\} = \|T\|. \end{aligned} \tag{15}$$

Hence T^* is also a contractive projection on $(L^p)^* = L^q$ (cf. Theorem I.4.1).

Let $\varphi(x) = |x|^p/p$ so that $\psi(x) = |x|^q/q$, $p^{-1} + q^{-1} = 1$. Let $\varphi'[\psi']$ be the right derivative of $\varphi[\psi]$. If $f \in \mathscr{R}_T = T(L^p)$, then we claim that $\varphi'(f) \in \mathscr{R}_{T^*} = T^*(L^q)$. In fact, since $Tf = f$ $(\varphi'(f) = |f|^{p-1}\operatorname{sgn}(f))$ and clearly $\varphi'(f) \in L^q$, we have on using the contractivity of T^* (since $\operatorname{sgn}(x) = -1$ if $x < 0$, $=0$ if $x = 0$, $= +1$ if $x > 0$),

$$\begin{aligned}\|f\|_p^p &= \int_\Omega f \cdot \varphi'(f)\,dP = \int_\Omega Tf \cdot \varphi'(f)\,dP = \int_\Omega f \cdot T^*(\varphi'(f))\,dP\\ &\le \|f\|_p \|T^*(\varphi'(f))\|_q \le \|f\|_p \|\varphi'(f)\|_q \qquad \text{by Hölder's inequality}\\ &= \|f\|_p \|f^{(p-1)}\|_q = \|f\|_p^p \qquad \text{since } (p-1)q = p. \end{aligned} \tag{16}$$

So there is equality in Hölder's inequality. Since $1 < p < \infty$, this happens for only one function, namely, $T^*\varphi'(f) = |f|^{p/q}\operatorname{sgn}(f) = \varphi'(f)$. Hence $\varphi'(f) \in \mathscr{R}_{T^*}$; and $\varphi'(1) = 1 = T^*1$. We first prove the results for $1 < p < 2$, and then the case $2 < p < \infty$ can be reduced to this by a duality argument.

Case 1 $1 < p < 2$. Then $L^p \supset L^q \supset \mathscr{R}_{T^*}$. We assert that if $f \in \mathscr{R}_T$, then $\varphi'(f) \in \mathscr{R}_T$ also (in addition to being in $\mathscr{R}_{T^*}$ as shown above).

For since $T1 = 1 \in \mathscr{R}_T$, so $\varphi'(1) = 1 \in \mathscr{R}_{T^*}$, we see that $1 + \varepsilon\varphi'(f) \in \mathscr{R}_{T^*}$ for all ε. Let $h_\varepsilon = [\psi'(1 + \varepsilon\varphi'(f)) - 1]/\varepsilon$ for $\varepsilon > 0$. Since $\psi'(1) = 1$ and ψ', φ' are differentiable except at the origin, $\psi''(1) = q - 1 > 0$, it follows that $\lim_{\varepsilon\downarrow 0} h_\varepsilon(\omega) = \psi''(1)\varphi'(f)(\omega)$. Next observe that this limit also exists in the L^p-norm. Indeed, $T^*(1 + \varepsilon\varphi'(f)) = 1 + \varepsilon\varphi'(f)$. So that by the above result and the fact that $(T^*)^* = T$ (by reflexivity of L^q), we deduce that $\psi'(1 + \varepsilon\varphi'(f)) \in \mathscr{R}_T$. Then $h_\varepsilon \in \mathscr{R}_T$ for each $\varepsilon > 0$. If the limit is in L^p-norm also, then it belongs to $\mathscr{R}_T$ so that $\varphi'(f)$ is in $\mathscr{R}_T$. But now $q > 2$, so ψ' is convex and ψ'' is increasing. It follows that $(\psi'(1 + \varepsilon x) - 1)/\varepsilon$ tends to a limit monotonely as $\varepsilon \downarrow 0$. Hence $|h_1|^p$ dominates $|h_\varepsilon|^p$ for $0 < \varepsilon < 1$. Consequently, by the dominated convergence the limit does exist in L^p, and hence $\varphi'(f) \in \mathscr{R}_T \cap \mathscr{R}_{T^*}$. We now iterate the procedure.

Thus $\varphi'(\varphi'(f)) \in \mathscr{R}_{T^*}$. However,

$$\varphi'(\varphi'(f)) = (|f|^{p-1})^{p-1}\operatorname{sgn}(f) = |f|^{(p-1)^2}\operatorname{sgn}(f).$$

By induction then $\varphi^n(f) \equiv \varphi^{n-1}(\varphi'(f)) \equiv |f|^{(p-1)^n}\operatorname{sgn}(f) \in \mathscr{R}_{T^*}$ for each $n \geq 1$. Since $1 < p < 2$, $0 < p - 1 < 1$ and hence $\varphi^n(f) \to \operatorname{sgn}(f)$ a.e., as $n \to \infty$. Moreover, $|\varphi^n(f)| \leq 1 + |\varphi'(f)| \in L^q$ so that the convergence is also in the norm of L^q. If $g = Th$, for $h \in L^p$ we have $g \in \mathscr{R}_T$ and

$$\begin{aligned}
\int_\Omega |Th|\,dP &= \int_\Omega Th \cdot \operatorname{sgn}(g)\,dP = \int_\Omega Th \cdot \lim_{n\to\infty} \varphi^n(g)\,dP \\
&= \lim_{n\to\infty} \int_\Omega Th \cdot \varphi^n(g)\,dP \qquad \text{by the dominated convergence} \\
&= \lim_{n\to\infty} \int_\Omega h \cdot T^*(\varphi^n(g))\,dP \\
&= \lim_{n\to\infty} \int_\Omega h \cdot \varphi^n(g)\,dP \qquad \text{since} \quad \varphi^n(g) \in \mathscr{R}_{T^*} \\
&= \int_\Omega h\operatorname{sgn}(g)\,dP \leq \int_\Omega |h|\,dP, \qquad (17)
\end{aligned}$$

where the interchange of limit and integral is again justified by the dominated convergence theorem. Since L^p is dense in L^1, (17) proves that T

is a contractive projection in the L^1-norm and has a unique extension to all of L^1 preserving the same properties (and hence T is also positive). The result now follows from Theorem 6.

Case 2 $2 < p < \infty$. By the first part $T^*: L^q \to L^q$ is a contractive projection such that $T^*1 = 1$. Since $1 < q < 2$ $[q = p/(p-1)]$, Case 1 implies that T^* is contractive in L^1 and hence $T^* = E^{\mathscr{B}}$ for some $\mathscr{B} \subset \Sigma$ by Theorem 6 as above. However, $(E^{\mathscr{B}})^* = (T^*)^* = T$. Also $(E^{\mathscr{B}})^* = E^{\mathscr{B}}$ on $L^\infty \cap L^p = L^\infty$. To see this, consider, for f, g in $L^\infty \subset L^p$, $1 \le p < \infty$,

$$\begin{aligned}\int_\Omega f \cdot E^{\mathscr{B}}(h)\, dP &= \int_\Omega E^{\mathscr{B}}(f \cdot E^{\mathscr{B}}(h))\, dP_{\mathscr{B}} \\ &= \int_\Omega E^{\mathscr{B}}(f) \cdot E^{\mathscr{B}}(h)\, dP_{\mathscr{B}} \qquad \text{by Proposition 1.2(a)} \\ &= \int_\Omega E^{\mathscr{B}}(h \cdot E^{\mathscr{B}}(f))\, dP_{\mathscr{B}} = \int_\Omega h \cdot E^{\mathscr{B}}(f)\, dP. \end{aligned} \tag{18}$$

Hence $\langle f, E^{\mathscr{B}}(h) \rangle = \langle E^{\mathscr{B}}(f), h \rangle$ using the duality pairing notation. It follows that $((E^{\mathscr{B}})^* =)E^{\mathscr{B}} = T$ on L^∞ and then they agree on L^p itself.

Case 3 Finally, let T be also positive. Then T^* is a positive contractive projection with $T1 = 1 = T^*1$, and $|T^*h| \le T^*(|h|)$ for all $h \in L^2$. Hence if $h = \operatorname{sgn}(Tf) \in L^2$, we have

$$\begin{aligned}\int_\Omega |Tf|\, dP &= \int_\Omega Tf \cdot h\, dP = \int_\Omega f \cdot T^*h\, dP \le \int_\Omega |f| \cdot T^*(|h|)\, dP \\ &= \int_\Omega |f|\, dP \qquad \text{since} \quad T^*1 = 1 \quad \text{and} \quad |h| = 1. \end{aligned} \tag{19}$$

This shows that T is again a contractive projection on L^1 and Theorem 6 applies. Thus the result is proved in all cases.

9. Remark In the last part of the above theorem, $\mathscr{R}_T = \mathscr{R}_{T^*}$ since a classical result says that in a Hilbert space contractivity of a projection is equivalent to its self-adjointness (hence $T = T^*$). This fact has nothing to do with conditional expectations. Let us include a proof for completeness.

Recall that a projection T on a Hilbert space $\mathscr{H} = (\mathscr{H}, (,))$ is orthogonal if the range and null spaces $\mathscr{R}_T$ and $\mathscr{N}_T$ satisfy $\mathscr{H} = \mathscr{R}_T \oplus \mathscr{N}_T$ and $y \in \mathscr{R}_T$, $z \in \mathscr{N}_T \Rightarrow (y, z) = 0$. So $\mathscr{R}_T = \mathscr{N}_T^\perp$. Hence $x \in \mathscr{H}$, $x = y + z \Rightarrow Tx = y$, and $\|x\|^2 = \|y\|^2 + \|z\|^2 = \|Tx\|^2 + \|z\|^2 \ge \|Tx\|^2$. Thus T is a contraction. Conversely, if T is a contractive projection with $\mathscr{R}_T$, $\mathscr{N}_T$ as its range and null spaces so that $\mathscr{H} = \mathscr{R}_T + \mathscr{N}_T$, let $\mathscr{N}_T^\perp$ be the orthogonal complement of $\mathscr{N}_T$. Then $\mathscr{H} = \mathscr{N}_T^\perp \oplus \mathscr{N}_T = \mathscr{R}_T + \mathscr{N}_T$. If $x_0 \in \mathscr{N}_T^\perp$ and $y_0 = x_0 - Tx_0$, then $y_0 \in \mathscr{N}_T$. By contractivity of T, we have

$$\|x_0\|^2 \geq \|Tx_0\|^2 = \|x_0 - y_0\|^2 = \|x_0\|^2 + \|y_0\|^2$$

since $y_0 \in \mathscr{N}_T$ and $x_0 \in \mathscr{N}_T^\perp$. Thus $y_0 = 0$ and so $x_0 = Tx_0 \in \mathscr{R}_T \Rightarrow \mathscr{N}_T^\perp \subset \mathscr{R}_T$. On the other hand, if $x_0 \in \mathscr{R}_T$ and $x_0 = y + z$ where $y \in \mathscr{N}_T^\perp$, $z \in \mathscr{N}_T$, then $x_0 = Tx_0 = Ty + Tz = Ty$. But we showed that $\mathscr{N}_T^\perp \subset \mathscr{R}_T$ so $Ty = y$. Hence $x_0 = y$ and $\mathscr{R}_T \subset \mathscr{N}_T^\perp$, i.e., $\mathscr{R}_T = \mathscr{N}_T^\perp$, and T is orthogonal. Now if x, x' are two elements of $\mathscr{H}$, $x = y + z$ and $x' = y' + z'$ are orthogonal decompositions relative to $\mathscr{R}_T$ and $\mathscr{N}_T$, then $(y, z') = 0$, $(z, y') = 0$, so that

$$\begin{aligned}(Tx, x') &= (y, x') = (y, x') - (y, z') = (y, x' - z') = (y, y') \\ &= (y, y') + (z, y') = (y + z, y') = (x, Tx') = (T^*x, x').\end{aligned}$$

Hence $T = T^*$. Thus T is contractive implies $T = T^*$.

If $T = T^*$, then for any $x \in \mathscr{H}$ the decomposition $x = (x - Tx) + Tx$ is orthogonal since

$$(x - Tx, Tx) = (Tx - T^2x, x) = (Tx - Tx, x) = 0.$$

So $\mathscr{R}_T = \mathscr{N}_T^\perp$, and T is an orthogonal projection. We already showed that this implies T is a contraction and hence $\|T\| \leq 1$ iff $T = T^*$, as asserted.

Using Lemma 2, Proposition 3, and the above remark, we have the following consequence of the preceding theorem.

10. Corollary *If T is either an averaging or a Šidák operator on L^p, $1 \leq p < \infty$, then T is equivalent to a conditional expectation $E^{\mathscr{B}}$, $\mathscr{B} \subset \Sigma$.*

Proof Only the case $p = 2$ has to be identified since the contractive projection T is also not given to be positive to invoke the preceding theorem. However, by the above remark, $T = T^*$ on L^2, which is a Hilbert space. Also $T1 = T^*1 = 1$. But by Theorem 5, in both cases, the range $\mathscr{R}_T$ is a measurable subspace and so has the form $L^2(\mathscr{B})$ for a σ-algebra $\mathscr{B} \subset \Sigma$. Hence for $f \in L^2$, since $Tf \in \mathscr{R}_T = L^2(\mathscr{B})$,

$$\begin{aligned}\int_A Tf\, dP_{\mathscr{B}} &= \int_\Omega (T^*\chi_A) \cdot f\, dP = \int_\Omega (T\chi_A) f\, dP = \int_\Omega \chi_A f\, dP \\ &= \int_A E^{\mathscr{B}}(f)\, dP_{\mathscr{B}}, \qquad A \in \mathscr{B}.\end{aligned}$$

Since the integrands are $\mathscr{B}$-measurable, they may be identified. Thus $Tf = E^{\mathscr{B}}(f)$ a.e., $f \in L^2$. This completes the proof.

An independent proof of this corollary can be given on using Proposition 3 and Theorem 5. It reduces to showing that each contractive projection T on $L^p(\Sigma)$ with range $L^p(\mathscr{B})$ for some σ-algebra $\mathscr{B} \subset \Sigma$ coincides with a conditional expectation for each $1 \leq p < \infty$. The proof is an extension of the L^2 case above, needing a further computation when $1 \leq p < \infty$, $p \neq 2$, since $T = T^*$ cannot be used, and the details are omitted. (The author's first characterization given in [3], in an Orlicz space context, was precisely of this form.)

The special form $\varphi(x) = |x|^p/p$ for L^φ $(= L^p)$ allows interesting side results. The following is such a possibility.

11. Theorem *Let $T: L^p \to L^p$ be a contractive linear operator of the averaging type in the sense that* (a) *$T(f \cdot Tg) = (Tf) \cdot (Tg)$ for $f, g \in L^\infty$ and* (b) *$Tf_0 = f_0$ for some $0 < f_0 \in L^\infty$, where f_0 is not necessarily a constant. Then there exists a σ-algebra $\mathscr{B} \subset \Sigma$ relative to which f_0 is measurable and $T = E^{\mathscr{B}}$.*

Proof The fact that a contractive operator T satisfies the averaging identity (a) implies $T(L^\infty) \subset L^\infty$ and is a contraction there was shown in the proof of Proposition 3. However, a trivial modification of the proof of Lemma 2 shows that $T^2 = T$ when (a) and (b) hold (even though $T1 = 1$ is not known a priori). Let us reduce the result to the previous case.

Define a (finite) measure μ on Σ by $d\mu = f_0^p\, dP$. Then $f \in L^p(\Omega, \Sigma, \mu)$ iff $ff_0 \in L^p(\Omega, \Sigma, P)$. Since f_0 is bounded, it is clear that $L^p(\Omega, \Sigma, \mu) \supset L^p(\Omega, \Sigma, P)$, and in fact $L^p(\Omega, \Sigma, \mu) = f_0 L^p(\Omega, \Sigma, P)$. Define a new operator A on $L^p(\Omega, \Sigma, \mu)$ by the equation

$$Af = \frac{T(ff_0)}{f_0}, \qquad f \in L^p(\Sigma, \mu). \tag{20}$$

By (a) and (b), $Af = Tf$. Moreover, $A1 = 1$ as well as $Af_0 = f_0$. We note that A also satisfies the averaging identity.

$$\begin{aligned} A(f \cdot Ag) &= A\left(f \cdot \frac{T(gf_0)}{f_0}\right) = \frac{T(f \cdot T(gf_0))}{f_0} \\ &= T(f) \cdot \frac{T(gf_0)}{f_0} = T(f) \cdot T(g) = A(f) \cdot A(g) \end{aligned} \tag{21}$$

for all f, g in L^∞. To see that A is a contraction, we have

$$\begin{aligned} \int_\Omega |Af|^p\, d\mu &= \int_\Omega |Af|^p f_0^p\, dP = \int_\Omega |T(ff_0)|^p\, dP \\ &\leq \int_\Omega |ff_0|^p\, dP = \int_\Omega |f|^p\, d\mu, \qquad f \in L^p(\Omega, \Sigma, \mu). \end{aligned}$$

Thus A is an averaging operator on $L^p(\Omega, \Sigma, \mu)$, where $\mu(\Omega) < \infty$. Hence by Corollary 10 for which μ may be any finite measure (same is true of Theorem 8), we conclude that $A = E^{\mathscr{B}}$ for some σ-algebra $\mathscr{B} \subset \Sigma$. Since $f_0 = Af_0 = E^{\mathscr{B}}(f_0), f_0$ is $\mathscr{B}$-measurable. Hence $T(f) = E^{\mathscr{B}}(f)$ on $L^p(\Omega, \Sigma, \mu)$.

To show that the same relation holds on $L^p(\Omega, \Sigma, P)$, let f be an element of this space. Then $h = f/f_0 \in L^p(\Omega, \Sigma, \mu)$ since $hf_0 \in L^p(\Omega, \Sigma, P)$. Consequently,

$$T(f) = T(hf_0) = f_0 A(h) = f_0 E^{\mathscr{B}}(h) = E^{\mathscr{B}}(f_0 h) = E^{\mathscr{B}}(f) \tag{22}$$

since f_0 is $\mathscr{B}$-measurable. This completes the proof.

In each of the proofs of the above results, Theorem 5 was used at an important point. We now present its proof.

Proof of Theorem 5 Let us first consider the case that $\mathscr{M} \subset L^p$ is a lattice. Then it contains constants and satisfies conditions (i) and (ii). Define $\mathscr{B} = \{A : \chi_A \in \mathscr{M}\}$. Thus $A \in \mathscr{B} \Rightarrow A^c \in \mathscr{B}$ since $1 \in \mathscr{M}$, and $\{\emptyset, \Omega\} \subset \mathscr{B}$. If $\{A, B\} \subset \mathscr{B}$, then $\chi_{A \cap B} = \min(\chi_A, \chi_B) \in \mathscr{M}$ by the lattice property. Hence $A \cap B \in \mathscr{B}$. Moreover,

$$\chi_{A-B} = (\chi_A - \chi_{A \cap B}) \in \mathscr{M}, \qquad \chi_{A \cup B} = (\chi_A + \chi_{B-A}) \in \mathscr{M}. \tag{23}$$

Hence $\{A - B, A \cup B\} \subset \mathscr{B}$ and $\mathscr{B}$ is an algebra. If $A_n \in \mathscr{B}$, $A_n \uparrow A$, then $\chi_{A_n} \in \mathscr{M}$ and $\chi_{A_n} \uparrow \chi_A \in L^p$, so by (ii) $\chi_A \in \mathscr{M}$ and $A \in \mathscr{B}$. Thus $\mathscr{B}$ is a σ-algebra. We assert that $\mathscr{M} = L^p(\mathscr{B})$ for this $\mathscr{B} \subset \Sigma$.

In fact, if $f = \sum_{i=1}^n a_i \chi_{A_i}$, $A_i \in \mathscr{B}$, $a_i \in \mathbb{R}$, then $f \in L^p(\mathscr{B})$ and since $\chi_{A_i} \in \mathscr{M}$, $f \in \mathscr{M}$. If $1 \le p < \infty$, then such functions are dense in $L^p(\mathscr{B})$, and since $\mathscr{M}$ is closed, it follows that $L^p(\mathscr{B}) \subset \mathscr{M}$. If $p = \infty$ and $0 \le f \in L^p(\mathscr{B})$, then there is a sequence of $\mathscr{B}$-measurable step functions $f_n \ge 0$ such that $f_n \uparrow f$ (pointwise), and since $f_n \in \mathscr{M}$ by (ii), one concludes that $f \in \mathscr{M}$. Since both the spaces are lattices, it follows that $L^p(\mathscr{B}) \subset \mathscr{M}$ for $1 \le p \le \infty$.

For the opposite containment, let $0 \le f \in \mathscr{M}$. If $\alpha > 0$, $A \in \mathscr{B}$, consider $f_n = \min[n(f - \alpha\chi_A)^+, \chi_A]$. By the lattice property and the fact that $\chi_A \in \mathscr{M}$, we have $f_n \in \mathscr{M}$. Since $f_n \le f_{n+1}$, let $\tilde{f}$ be the limit of $\{f_n\}_1^\infty$. But $f_n \le \chi_A$ so $\tilde{f} \le \chi_A$ and by (ii) $\tilde{f} \in \mathscr{M}$. If $A_\alpha = \{\omega \in A : f(\omega) - \alpha\chi_A(\omega) > 0\}$, then $\tilde{f} = \chi_{A_\alpha}$ so that $A_\alpha \in \mathscr{B}$. Thus $A \cap [f > \alpha] \in \mathscr{B}$ for each $\alpha > 0$. Taking $A = \Omega$, we conclude that f is $\mathscr{B}$-measurable so that $f \in L^p(\mathscr{B})$. Thus $\mathscr{M} \subset L^p(\mathscr{B})$ and here $1 \le p \le \infty$, so $\mathscr{M} = L^p(\mathscr{B})$. The result is true in the lattice case.

The algebra part can be reduced to the lattice case by means of the following lemma. Its elementary proof for the L^p-case ($1 \le p < \infty$) is due to Kopp *et al.* [1], which is an extension of the classical result for $p = \infty$ (cf., e.g., Loomis [1], p. 9).

12. Lemma *Let $\mathscr{M} \subset L^p$, $1 \le p \le \infty$, satisfy the algebra conditions of Theorem 5. Thus let $\mathscr{S}$ be the set of bounded functions of $\mathscr{M}$; then the L^p-*

closure of $\mathcal{S}$, which is $\mathcal{M}$, is a lattice which satisfies conditions (i) *and* (ii) *of that theorem.*

Proof of Lemma 12 First consider the case $p = \infty$. Let $f \in \mathcal{S}$. It suffices to show that $|f| \in \mathcal{M} \subset L^\infty$. Since $\mathcal{M}$ is linear, we may assume $|f| \leq 1$ a.e., for convenience. Then by the Taylor series expansion of $(1 - t)^{1/2}$ we have

$$|f|(\omega) = (1 - (1 - f^2(\omega)))^{1/2} = 1 + \sum_{n=1}^{\infty} \binom{\frac{1}{2}}{n} (f^2(\omega) - 1)^n, \tag{24}$$

where the series on the right-hand side converges uniformly for $\omega \in \Omega - \Omega_0$, $P(\Omega_0) = 0$. But the right-hand side series is then a uniformly convergent sequence of functions,

$$1 + \sum_{k=1}^{n} \binom{\frac{1}{2}}{k} (f^2 - 1)^k \in \mathcal{S} \subset \mathcal{M},$$

because $\mathcal{S}$ is an algebra. Since $\bar{\mathcal{S}} = \mathcal{M}$, $|f| \in \mathcal{M}$. Then (i) and (ii) are obvious in this case. Thus the result holds for $p = \infty$.

Next consider $1 \leq p < \infty$. Again let $f \in \mathcal{S}$, and for convenience, $|f| \leq 1$. It is to be shown that $|f| \in \mathcal{M}$. If $\varepsilon > 0$, then

$$(f^2 + \varepsilon^2)^{1/2} = (1 + \varepsilon^2)^{1/2} \left(1 + \frac{f^2 - 1}{1 + \varepsilon^2}\right)^{1/2} = (1 + \varepsilon^2)^{1/2} \sum_{n=0}^{\infty} \binom{\frac{1}{2}}{n} \left(\frac{f^2 - 1}{1 + \varepsilon^2}\right)^n. \tag{25}$$

Since $\mathcal{S}$ is an algebra, $f \in \mathcal{S}$ implies $[(f^2 - 1)/(1 + \varepsilon^2)]^n \in \mathcal{S}$. But the series in (25) converges in L^p to $(f^2 + \varepsilon^2)^{1/2}$ because (using $f^2 \leq 1$)

$$\sum_{n=0}^{\infty} \binom{\frac{1}{2}}{n} \left\| \left(\frac{f^2 - 1}{1 + \varepsilon^2}\right)^n \right\|_p \leq \sum_{n=0}^{\infty} \binom{\frac{1}{2}}{n} \frac{1}{(1 + \varepsilon^2)^n} = \left(\frac{2 + \varepsilon^2}{1 + \varepsilon^2}\right)^{1/2} < \infty, \tag{26}$$

and so $(f^2 + \varepsilon^2)^{1/2} \in \mathcal{M} = \bar{\mathcal{S}}$. Letting $\varepsilon \downarrow 0$ through a sequence, we deduce that $|f| \in \mathcal{M}$. Thus $\mathcal{M}$ is a lattice, and the lemma follows. Consequently the proof of Theorem 5 is also complete, since the converse part is trivial.

Remark The above proofs can be (slightly) modified to show that the result holds if the measure space is not finite and if $1 \in \mathcal{M}$ is replaced by a suitable sequence $\{\chi_{A_\alpha}, \alpha \in I\} \subset \mathcal{M}$, where $\bigcup A_\alpha = \Omega - \Omega_0$, and Ω_0 is μ-null; A_α is of finite μ-measure making the measure strictly localizable. But these extensions are not difficult and are left to the reader.

It is of interest to observe that the full force of linearity of $\mathcal{M}$ (and L^p) was not used in the above proof. Only that there are enough bounded functions; in fact, the following facts are used. If L^p is replaced by the collection $\mathcal{N}^+$ of all $\bar{\mathbb{R}}^+$-valued measurable functions on (Ω, Σ, μ), then $\mathcal{N}^+$ is closed under lattice operations, monotone limits, and generates Σ. Suppose $\mathcal{M}$ is replaced by a collection $\mathcal{C}$ ($\subset \mathcal{N}^+$) satisfying

(I) $\{f_1, f_2\} \subset \mathcal{C} \Rightarrow a_1 f_1 + a_2 f_2 \in \mathcal{C}$ for $a_i \geq 0$, $i = 1, 2$; if also $f_1 \leq f_2$ and f_2 is finite a.e., then $f_2 - f_1 \in \mathcal{C}$, and let $1 \in \mathcal{C}$,

(II) $\{f_1, f_2\} \subset \mathcal{C} \Rightarrow f_1 \cdot f_2 \in \mathcal{C}$,

(III) $f_n \in \mathcal{C}, f_n \to f$ a.e., $f_n \leq g$, g bounded $\Rightarrow f \in \mathcal{C}$,

(I′) same as (I),

(II′) $\{f_1, f_2\} \subset \mathcal{C} \Rightarrow f_1 \vee f_2 \in \mathcal{C}$,

(III′) $f_n \in \mathcal{C}, f_n \uparrow f$ a.e. $\Rightarrow f \in \mathcal{C}$.

If $\mathcal{S}_0$ ($\subset \mathcal{C}$) is the set of bounded functions satisfying (I)–(III) or (I′)–(III′), then the σ-algebra $\mathcal{B} \subset \Sigma$ generated by $\mathcal{S}_0$ is the same as that generated by the set of all bounded functions $\mathcal{S}$ of $\mathcal{C}$. This is a consequence of the fact that $\mathcal{S}$ is the uniform closure of $\mathcal{S}_0$, as the proofs of the above Theorem 5 and Lemma 12 show. Since there are no "density" conditions, we cannot say that $\mathcal{B}$ is also the same σ-algebra generated by $\mathcal{C}$. The latter clearly holds true iff every element of $\mathcal{C}$ is a pointwise limit of some sequence from $\mathcal{S}$. We can state these observations in the form of the following proposition for reference.

13. Proposition *If $\mathcal{S}_0 \subset \mathcal{C} \subset \mathcal{N}^+$ is a collection of $\bar{\mathbb{R}}^+$-valued measurable functions on (Ω, Σ, P) such that $\mathcal{S}_0$ and $\mathcal{C}$ satisfy conditions (I)–(III) or (I′)–(III′) above, then the σ-algebra $\mathcal{B}$ generated by $\mathcal{S}_0$ is given by $\mathcal{B} = \{A \in \Sigma : \chi_A \in \mathcal{C}\}$. If $\mathcal{M}$ is the class of all $\bar{\mathbb{R}}^+$-valued $\mathcal{B}$-measurable functions, then $\mathcal{M} \subset \mathcal{C}$ and there is equality iff every element of $\mathcal{C}$ is a pointwise limit of some bounded sequence from $\mathcal{S}_0$.*

It is also easy to check directly that $\mathcal{B}$ is a σ-algebra and has the stated properties. As an application, we can give a characterization of an averaging operator (under a continuity condition) on $\mathcal{N}^+$; this together with some of its consequences is due to Moy [1]. They extend some earlier work of Birkhoff [1]. Since this class of ideas has generated a great deal of interest and recent research, starting with Reynolds and principally continued by Kampé de Fériet in the 1930s, and since it was one of the connecting links of probability theory and functional operators, we give a key result here. The intimate relation between such operators and conditional expectations will become clear. (Cf. also Section 6.)

14. Theorem *Let $T: \mathcal{N}^+ \to \mathcal{N}^+$ be a generalized averaging operator in the sense that* (a) *$T(a_1 f + a_2 g) = a_1 Tf + a_2 Tg$ for all $a_1, a_2 \geq 0$ and $f, g \in \mathcal{N}^+$,*

(b) *Tf is bounded if f is bounded,* (c) *$T(f\cdot Tg)=(Tf)\cdot(Tg), f, g\in\mathcal{N}^+$, and* (d) *$T$ is continuous in the following sense: $f_n\in\mathcal{S}, f_n\uparrow f$ implies $Tf_n\uparrow Tf$. Here $\mathcal{N}^+$ is the previously introduced set of all $\overline{\mathbb{R}}^+$- valued measurable functions on (Ω,Σ,P). Suppose either (i) T is just as above, or (ii) since there are constants in $\mathcal{N}^+$, let T be invariant on them, i.e., $T1=1$. Then there exist a (not necessarily unique) σ-algebra $\mathcal{B}\subset\Sigma$ and a P-unique function $h\in\mathcal{N}$ such that if $E^{\mathcal{B}}$ is the conditional expectation operator, we have*

$$Tf=E^{\mathcal{B}}(fh),\qquad f\in\mathcal{N}^+, \tag{27}$$

where $E^{\mathcal{B}}(h)$ is essentially bounded. In case (ii), if $\mathcal{M}=T(\mathcal{N}^+)$, then $\mathcal{B}$ is generated by $\mathcal{M}$, T is a projection, i.e., $T^2=T$, and $E^{\mathcal{B}}(h)=1$ a.e.

Proof If $\mathcal{M}=T(\mathcal{N}^+)$, it should be shown that $\mathcal{M}$ satisfies conditions (I)–(III) of Proposition 13. However, there is no obvious way of checking this since there is little control over the growth rate of T. So we proceed indirectly. The key idea here is to find a set $\mathcal{C}\supset\mathcal{M}$ that satisfies conditions (I)–(III). The following set is such a candidate. Define $\mathcal{C}=\{f\in\mathcal{N}^+: T(f\cdot g)=f\cdot Tg$ for all $g\in\mathcal{N}^+\}$. Clearly, $1\in\mathcal{C}$. Moreover, by the averaging identity (c), $T(f\cdot Tg)=(Tg)(Tf)$ for all $f,g\in\mathcal{N}^+$, so that $Tf\in\mathcal{C}$. Thus one can conclude that $\mathcal{M}\subset\mathcal{C}$. We need to check conditions (I)–(III) of Proposition 13 for $\mathcal{C}$.

Since for any f_1,f_2 in $\mathcal{C}$, $T(f_1f_2g)=f_1T(f_2g)=f_1f_2\cdot T(g)$ for all $g\in\mathcal{N}^+$, $f_1\cdot f_2\in\mathcal{C}$, (II) is immediate. Similarly, $a_1f_1+a_2f_2\in\mathcal{C}$ for any $a_1,a_2\geq 0$. Let $f_1\leq f_2$ and f_2 be finite a.e., so that $f_2-f_1\in\mathcal{N}^+$ and is finite a.e. If $g\in\mathcal{N}^+$ is any bounded function, then

$$f_2T(g)=T(g(f_2-f_1+f_1))=T(g(f_2-f_1))+T(gf_1)$$

by (a). Hence $T(f_1g)=f_1T(g)$ being finite, we have

$$(f_2-f_1)T(g)=T(g(f_2-f_1)). \tag{28}$$

This is true for every bounded $g\in\mathcal{N}^+$, and since every element of $\mathcal{N}^+$ is a monotone limit of a sequence of bounded elements of $\mathcal{N}^+$, it follows that (28) holds for all g by (d). Thus $f_2-f_1\in\mathcal{C}$ and (I) holds. Finally, let $f_n\in\mathcal{C}$, $f_n\to f$ a.e., and $f_n\leq g$ bounded. Then for any bounded $\tilde{h}\in\mathcal{N}^+, f_n\tilde{h}\leq g\tilde{h}$ and $f_n\tilde{h}\to f\tilde{h}\leq g\tilde{h}$ a.e. We claim that this implies

$$T(f\tilde{h})=\lim_n T(f_n\tilde{h})=\lim_n[f_nT(\tilde{h})]=fT(\tilde{h})$$

so that we can conclude the same equation for all $\tilde{h}\in\mathcal{N}^+$, and hence $f\in\mathcal{C}$. Thus (III) will follow if it is shown that $T(f_n)\to T(f)$. This needs a further computation. Let $h_n=\inf\{f_i, i\geq n\}$. Then $h_n\uparrow f=\lim_n f_n$ and $h_n\leq f_n\leq g$. But $T(f_n)=T(f_n-h_n+h_n)=T(h_n)+T(f_n-h_n)\geq T(h_n)$ and by (d)

$$T(f)=\lim_n T(h_n)\leq\liminf_n T(f_n). \tag{29}$$

On the other hand, $g - f = \lim_n (g - f_n) \geq 0$ a.e., and thus by (29)

$$T(g - f) \leq \liminf_n T(g - f_n). \tag{30}$$

Since $f \leq g$ and $f_n \leq g$ and g is bounded,

$$T(g) = T(g - f) + T(f), \qquad \text{or} \qquad T(g) - T(f) = T(g - f),$$

and (30) simplifies to

$$\begin{aligned} T(g) - T(f) &\leq \liminf_n [T(g) - T(f_n)] \\ &= -\limsup_n [T(f_n) - T(g)] \leq T(g) - \limsup_n T(f_n). \end{aligned}$$

Since $T(g)$ is bounded, it can be canceled from the above inequality so that with (29) we have

$$\limsup_n T(f_n) \leq T(f) \leq \liminf_n T(f_n), \tag{31}$$

and there is equality throughout in (31), as claimed.

Thus by Proposition 13, the bounded functions of $\mathscr{C}$ generate a σ-algebra $\mathscr{B}$, and if $\mathscr{M}_1$ is the set of all $\bar{\mathbb{R}}^+$-valued $\mathscr{B}$-measurable functions, then $\mathscr{M}_1 \subset \mathscr{C}$. By conditions (b) and (d), if $h \in \mathscr{N}^+$ is bounded, Th is a bounded element of $\mathscr{C}$ so that it is $\mathscr{B}$-measurable, i.e., $Th \in \mathscr{M}_1$, and since every $h \in \mathscr{M}$ is a monotone limit of bounded elements, we deduce that Th is $\mathscr{B}$-measurable and $\mathscr{M} \subset \mathscr{M}_1$. Thus far the measure P has not been crucial in the arguments. This will be essential now.

We can complete the proof as follows. Let $\nu(A) = \int_\Omega T(\chi_A)\, dP$, $A \in \Sigma$. It is clear that $\nu: \Sigma \to \bar{\mathbb{R}}^+$ is additive. If $\{A_n\}_1^\infty \subset \Sigma$ is a disjoint sequence, then $T(\chi_{\cup_{i=1}^n A_i}) \uparrow T\chi_{\cup_{i=1}^\infty A_i}$, by (d), and by monotone convergence

$$\nu\left(\bigcup_{n=1}^\infty A_n\right) = \lim_n \int_\Omega T\chi_{\cup_{i=1}^n A_i}\, dP = \sum_{n=1}^\infty \nu(A_i).$$

Also ν is P-continuous. Since in both cases P is a finite (i.e., probability) measure, the classical (cf. Theorem I.2.7) Radon–Nikodým theorem is applicable, and there is a unique $h \in \mathscr{N}^+$, which may take both 0 and $+\infty$ values on sets of positive P-measure, such that

$$\int_\Omega T\chi_E\, dP = \nu(E) = \int_E h\, dP = \int_\Omega \chi_E h\, dP, \qquad E \in \Sigma. \tag{32}$$

By the linearity of T and of the integral, this can be extended first for simple functions and then by (d) and the monotone convergence theorem for all f of $\mathscr{N}^+$ so that

$$\int_\Omega Tf\, dP = \int_\Omega fh\, dP, \qquad f \in \mathscr{N}^+. \tag{33}$$

On the other hand, for all $B \in \mathscr{B}$, $\chi_B \in \mathscr{M}_1 \subset \mathscr{C}$ and hence $T(f\chi_B) = \chi_B T(f)$, all $f \in \mathscr{N}^+$, by definition of $\mathscr{C}$. Consequently with (33), on remembering the element $Tf \in \mathscr{M} \subset \mathscr{M}_1$ so that it is $\mathscr{B}$-measurable, we have

$$\begin{aligned}\int_B Tf\,dP_{\mathscr{B}} = \int_\Omega \chi_B(Tf)\,dP_{\mathscr{B}} &= \int_\Omega T(\chi_B f)\,dP_{\mathscr{B}} \\ &= \int_\Omega \chi_B f h\,dP \quad \text{by (33)} \\ &= \int_B fh\,dP = \int_B E^{\mathscr{B}}(fh)\,dP_{\mathscr{B}}. \qquad (34)\end{aligned}$$

Since the extreme integrands are $\mathscr{B}$-measurable, and $B \in \mathscr{B}$ is arbitrary, one has $Tf = E^{\mathscr{B}}(fh)$, $f \in \mathscr{N}^+$, and (27) follows. Taking $f = 1$ a.e., we conclude that $E^{\mathscr{B}}(h)$ is bounded.

Finally, in case (ii) since $T1 = 1$ by Lemma 2, it follows that $T^2 f = Tf$ for all $f \in \mathscr{N}$ so that $T^2 = T$ on $\mathscr{N}^+$ to $\mathscr{M} = T(\mathscr{N}^+)$. Taking $f = 1$ in

$$T(f) = E^{\mathscr{B}}(fh) \quad \text{a.e.,} \qquad (35)$$

one has that $E^{\mathscr{B}}(h) = 1$ a.e. Next, to show that $\mathscr{M} = \mathscr{M}_1$, let $f \in \mathscr{M}_1$. Then

$$Tf = E^{\mathscr{B}}(fh) = fE^{\mathscr{B}}(h) = f \quad \text{a.e.} \qquad (36)$$

Thus $f \in \mathscr{M}$ so that $\mathscr{M} \supset \mathscr{M}_1$. Since $\mathscr{M} \subset \mathscr{M}_1$ always holds, the last part follows and $\mathscr{B}$ is generated by $\mathscr{M}_1 = \mathscr{M}$ (and completed). Thus the theorem is completely proved.

Remark We note that if $A = \{\omega : h(\omega) = 0\}$, $B = \{\omega : h(\omega) = +\infty\}$, and if at least one of them has positive measure, then $\mathscr{B}$ need not be unique in the representation (27). For instance, suppose $P(A) > 0$ and $A \in \mathscr{B}$. Consider the trace σ-algebra $\mathscr{B}(A^c) \subset \mathscr{B}$. Let $\mathscr{B}_1 = \sigma(A \cup F : F \in \mathscr{B}(A^c))$ and $\mathscr{B}_0 = \mathscr{B}_1 \cup \mathscr{B}(A^c)$, which is a σ-algebra, and where A is an atom of $\mathscr{B}_0$. If A is not an atom of $\mathscr{B}$, then $\mathscr{B} \neq \mathscr{B}_0$. But $E^{\mathscr{B}}(fh) = E^{\mathscr{B}_0}(fh) = T(f)$, $f \in \mathscr{N}^+$. Usually $P(A) > 0$, $P(B) = 0$ is the case. This important result is in essence due to Moy [1].

By using (I′)–(III′), a similar result can be proved if T satisfies the Šidák identity instead of the averaging:

$$T(\max(Tf, Tg)) = \max(Tf, Tg). \qquad (c')$$

Since by Lemma 2 it is true that $T^2 = T$ (the conditions (I′)–(III′) may be verified for $\mathscr{M} = T(\mathscr{N}^+)$, though (I′) needs some computation), we can make the following precise statement in this case corresponding to the above theorem, which, however, is slightly weaker than the former result.

15. Proposition *Let $T: \mathcal{N}^+ \to \mathcal{N}^+$ be an operator satisfying (a), (b), (d) of Theorem 14, and (c') above. If, moreover, $T1 = 1$, then there exists a σ-algebra $\mathscr{B} \subset \Sigma$ and a unique $h \in \mathcal{N}^+$ such that*

$$\int_\Omega Tf\,dP = \int_\Omega E^{\mathscr{B}}(fh)\,dP, \qquad f \in \mathcal{N}^+ \tag{37}$$

and $E^{\mathscr{B}}(h) = 1$ a.e. Here again the σ-algebra $\mathscr{B}$ is generated by the bounded elements of $\mathscr{M} = T(\mathcal{N}^+)$.

The proof is similar to and simpler than the above result and will be omitted. In general $h = 1$ need not hold in (27) or (37), even if $\mathcal{N}^+ \subset L^p(\Sigma)$, when T is merely a bounded operator.

Continuing the preceding study, we present a characterization of the averaging operators on $L^\infty(\Sigma)$. Except for certain implicit results contained in Theorem 5, this question was not discussed since the proper place for it is after the preceding result. The general form of the result resembles that of the preceding and is essentially due to Rota [1].

16. Proposition *Let $T: L^\infty(\Sigma) \to L^\infty(\Sigma)$ be an averaging operator for which $T1 = 1$, i.e., T satisfies the averaging identity, is a contraction, and constants are invariant. Moreover, suppose T has the following property: For each sequence $\{f_n\}_1^\infty \subset L^\infty(\Sigma)$ such that $f_n \to f$, a.e. and boundedly, it is true that $Tf_n \to Tf$ a.e. Then there exists a unique σ-algebra $\mathscr{B} \subset \Sigma$ such that $T(L^\infty(\Sigma)) = L^\infty(\mathscr{B})$. Furthermore, we have $Tf = E^{\mathscr{B}}(fh), f \in L^\infty(\Sigma)$ for a unique nonnegative measureable h (relative to Σ) such that $E^{\mathscr{B}}(h) = 1$ a.e. Moreover, $h = 1$ a.e. iff $T^*P = P$, where T^* is the adjoint of T on $(L^\infty)^*$.*

Proof Since by hypothesis: (i) $T(f \cdot Tg) = (Tf) \cdot (Tg)$, (ii) $\|T\| \le 1$, and (iii) $T1 = 1$, we deduce that $\mathscr{M} = T(L^\infty(\Sigma))$ is a norm closed subalgebra since T is a bounded projection on $L^\infty(\Sigma)$. The present hypothesis implies that $\mathscr{M}$ satisfies the conditions of Theorem 5. Hence we deduce from that theorem that there exists a unique σ-algebra $\mathscr{B} \subset \Sigma$ such that $\mathscr{M} = L^\infty(\mathscr{B})$. Since $1 \in \mathscr{M}$, the last part of the proof of Theorem 14 again applies (cf. (32)), and we conclude that $E^{\mathscr{B}}$ exists, $Tf = E^{\mathscr{B}}(fh)$, $f \in L^\infty(\Sigma)$, for a P-unique nonnegative h. Since $T1 = 1$ a.e., we also have $E^{\mathscr{B}}(h) = 1$ a.e.

Finally, to prove that $h = 1$ a.e. iff $T^*P = P$, we note that $T^*: (L^\infty(\Sigma))^* \to (L^\infty(\Sigma))^*$ is a mapping on bounded additive set functions on Σ, vanishing on P-null sets. So by (33), since $P \in (L^\infty(P))^*$,

$$\int_\Omega hf\,dP = \int_\Omega Tf\,dP = \int_\Omega f d(T^*P), \qquad f \in L^\infty(\Sigma). \tag{38}$$

If $h = 1$ a.e., then taking $f = \chi_A$, $A \in \Sigma$, we get from (38) that $(T^*P)(A) = P(A)$ or $T^*P = P$. Conversely, if this last equation holds, then again (38) implies that $\int_A h\,dP = P(A)$, $A \in \Sigma$. Hence $h = 1$ a.e. This completes the proof.

In the above result the bounded sequential convergence condition on T, which is automatic on the $L^p(\Sigma)$-spaces for $1 \le p < \infty$, is clearly a restriction on $L^\infty(\Sigma)$ for the class of averaging operators for which the above proof holds. Is it the proof or the result itself at fault? To answer this, we shall analyze the condition and explain its significance in this case. Since T is a contractive averaging operator on $L^\infty(\Sigma)$, it is automatically positive. *For*, suppose the statement is false. Then there must exist an $\alpha > 0$ and an $f \in L^\infty(\Sigma)$ satisfying $0 < f < 1$ a.e. such that $Tf \le -\alpha < 0$ on a set B $(\in \Sigma)$ of positive measure. Then with $T1 = 1$ we have $T(1-f) \ge 1 + \alpha$ on B. But now $1 \ge \|1-f\|_\infty \ge \|T(1-f)\|_\infty \ge 1 + \alpha > 1$, which is a contraction. (Here we used the contractivity of T in the second inequality.) Thus T is positive and hence $\mathcal{M} = T(L^\infty(\Sigma))$ is a Banach lattice. If $f_n \in \mathcal{M}$, $0 \le f_n \uparrow f$ a.e., $f \in L^\infty(\Sigma)$, then $f_n = T(f_n) \le T(f)$ so that $f \le Tf$ a.e. Also $\|f\|_\infty \le \|Tf\|_\infty \le \|f\|_\infty$ so that there is equality. If $\|\cdot\|_\infty$ were a strictly monotone norm (as in the case of $\|\cdot\|_p$, $1 \le p < \infty$), we could infer $f = Tf$ a.e., and then the bounded sequential convergence would be automatic. Unfortunately, $\|\cdot\|_\infty$ is not a strictly monotone norm and this conclusion does not hold. On the other hand, if $\mathcal{B}_0 = \{A \in \Sigma : \chi_A \in \mathcal{M}\}$, then $\mathcal{B}_0$ is clearly an algebra. If $\mathcal{B} = \sigma(\mathcal{B}_0)$, then $(T^*P)_{\mathcal{B}} = P_{\mathcal{B}}$ as seen from

$$\int_A dP = \int_\Omega T\chi_A\,dP = \int_\Omega \chi_A\,d(T^*P) = \int_A d(T^*P), \qquad A \in \mathcal{B}_0, \tag{39}$$

so that P and T^*P agree on $\mathcal{B}_0$ and hence, by the Hahn extension theorem, also on $\mathcal{B}$. But this does not imply $T^*P = P$ on Σ (which is false in general). At least that T^*P is σ-additive on Σ is needed to conclude that $f = Tf$ a.e. by using the following extension of (39):

$$\begin{aligned}\int_\Omega f\,dP &= \lim_n \int_\Omega f_n\,dP \quad \text{by monotone convergence}\\ &= \lim_n \int_\Omega (Tf_n)\,dP = \lim_n \int_\Omega f_n\,d(T^*P)\end{aligned} \tag{40}$$

(here the limit and integral cannot be interchanged if T^*P is not σ-additive on Σ). Thus an additional condition is *necessary*. The last needed condition can be stated as: $T\colon L^\infty(\Sigma) \to L^\infty(\Sigma)$ is weak-star continuous, i.e., $T = \tau^*$ where $\tau\colon L^1(\Sigma) \to L^1(\Sigma)$ is a continuous mapping so that $T^* = \tau^{**}$; one needs $T^*(L^1(\Sigma)) \subset L^1(\Sigma)$. This can also be stated as follows: for each $g \in L^1(\Sigma)$, $h \in L^\infty(\Sigma)$ we have

$$\int_\Omega g\,Tf\,dP = \int_\Omega (T^*g)\cdot f\,dP = \int_\Omega (\tau g)\cdot f\,dP. \tag{41}$$

It is then clear that taking $g = 1$ a.e., (41) implies (in (40)) that the last limit equals $\int_\Omega f(T^*1)\,dP$. Thus $\int_\Omega f\,dP = \int_\Omega fT^*1\,dP = \int_\Omega Tf\,dP$ yields, from $f \leq Tf$, that $f = Tf$ a.e. Consequently, an additional condition such as that given in Proposition 16 on T cannot be omitted. It may be stated in any of the above-discussed alternative forms. The problem would not arise if $(L^\infty(\Sigma))^*$ were a space of functions, which it is not, and this sufficient condition is satisfied whenever the $\|\cdot\|$ on the space is absolutely continuous [i.e., $A_n \in \Sigma$, $A_n \downarrow \varnothing$ implies $\|f\chi_{A_n}\| \downarrow 0$ for all f for which $\|f\| < \infty$. This is automatic for $L^p(\Sigma)$, $p < \infty$, and false for $L^\infty(\Sigma)$].

Remark In the above proof, the fact that T is an averaging (and $T1 = 1$) was used in deducing (i) T is a positive projection and (ii) $\mathcal{M} = T(L^\infty(\Sigma))$ is an algebra. However, the analysis essentially shows that the result holds if T is merely a positive contractive projection on $L^\infty(\Sigma)$ with 1 as a fixed point.

This remark illustrates the fact that contractive projections and conditional expectations on $L^p(\Sigma)$ are closely related. Another aspect of the work of Section 1 shows that $E^{\mathcal{B}}(\cdot)$ behaves almost like an integral, and if $P^{\mathcal{B}}(A) = E^{\mathcal{B}}(\chi_A)$, then $P^{\mathcal{B}}(\cdot)(\omega)\colon \Sigma \to [0, 1]$ is what was called the conditional probability function. Does $P^{\mathcal{B}}(\cdot)(\omega)$ behave as a measure? We analyze this problem in Section 3 in detail.

2.3 CONDITIONAL PROBABILITIES

The concept of conditional probability has already been introduced in the preceding sections. Let us first state some elementary properties following from the definition.

1. Proposition *Let (Ω, Σ, P) be a probability space and $\mathcal{B} \subset \Sigma$ be a σ-algebra. The function $P^{\mathcal{B}}\colon A \mapsto P^{\mathcal{B}}(A)$, $A \in \Sigma$, defined by $P^{\mathcal{B}}(A) = E^{\mathcal{B}}(\chi_A)$, which satisfies the functional equation:*

$$\int_B P^{\mathcal{B}}(A)\,dP_{\mathcal{B}} = P(A \cap B), \qquad A \in \Sigma, \quad B \in \mathcal{B}, \tag{1}$$

has the properties holding on a suitable set $\Omega - \Omega_0$ with $P(\Omega_0) = 0$: (i) $0 \leq P^{\mathcal{B}}(A) \leq 1$, $A \in \Sigma$; (ii) $P(A) = 0$ or $= 1 \Rightarrow P^{\mathcal{B}}(A) = 0$ or $= 1$, respectively; (iii) $A_n \in \Sigma$, $A_n \uparrow A$ (or $A_n \downarrow A$) $\Rightarrow P^{\mathcal{B}}(A_n) \to P^{\mathcal{B}}(A)$; and (iv) $\{A_n, n \geq 1\} \subset \Sigma$, disjoint $\Rightarrow P^{\mathcal{B}}(\bigcup_{n=1}^\infty A_n) = \sum_{n=1}^\infty P^{\mathcal{B}}(A_n)$.

Proof All the properties are immediate consequences of the corresponding ones for the conditional expectations. As an illustration, let us prove (iii) and (iv).

If $A_n \uparrow A$, then $E^{\mathcal{B}}(\chi_{A_n}) \uparrow E^{\mathcal{B}}(\chi_A)$ a.e. by the conditional monotone convergence theorem. A similar result holds for the decreasing sequence. This is (iii). Regarding (iv), let us present an alternate method. Using (1), we have for any $B \in \mathscr{B}$:

$$\int_B \left(\sum_{n=1}^{\infty} P^{\mathscr{B}}(A_n) \right) dP_{\mathscr{B}} = \sum_{n=1}^{\infty} \int_B P^{\mathscr{B}}(A_n)\, dP_{\mathscr{B}}$$

by the dominated convergence

$$= \sum_{n=1}^{\infty} P(A_n \cap B) = P\left(B \cap \bigcup_{n=1}^{\infty} A_n \right)$$

since P is a measure

$$= \int_B P^{\mathscr{B}}\left(\bigcup_{n=1}^{\infty} A_n \right) dP_{\mathscr{B}} \qquad \text{by (1).} \tag{2}$$

The extreme integrands of (2) are $\mathscr{B}$-measurable and hence can be identified. This completes the proof.

Even though the statements of the above proposition are formally the same as those of P, there is an exceptional P-null set with each result. In fact, $P^{\mathscr{B}}(A)$ is a function with values in $L^{\infty}(\mathscr{B})$. Thus the mapping $A \mapsto P^{\mathscr{B}}(A)$ should properly be interpreted as a function space valued additive mapping on Σ, the convergencies in (iii), and (iv) of the above proposition being interpreted as order limits (cf. Eq. (31) of Section I.4). It is clearly tempting to consider $P^{\mathscr{B}}(\cdot)(\omega)\colon \Sigma \to [0, 1]$ as a probability measure for each $\omega \in \Omega - \Omega_0$, $(\mathscr{B} \subset \Sigma)$ with $P(\Omega_0) = 0$. However, as examples show (one is given below), this is not always the case. There are two possibilities: (a) find conditions on $\mathscr{B}$ (or Σ) and P such that $P^{\mathscr{B}}(\cdot)(\omega)$ may be regarded as a (scalar) measure and (b) treat $P^{\mathscr{B}}$ as a vector valued (i.e., $L^{\infty}(\mathscr{B})$-valued) set function and develop the theory of integration. We shall explore both problems here since this will clarify the structure of these functions besides being useful for applications.

Let us start with possibility (b). If $P^{\mathscr{B}}\colon \Sigma \to L^{\infty}(\mathscr{B})$ is considered as in Proposition 1, then the σ-additivity is in the sense of order convergence, as noted above. However, if $P^{\mathscr{B}}\colon \Sigma \to L^{p}(\mathscr{B})$, $1 \le p < \infty$, then it is strongly σ-additive and hence is a vector measure. Indeed, if $A = \bigcup_{n=1}^{\infty} A_n$, $A_n \in \Sigma$, disjoint, then

$$\left\| P^{\mathscr{B}}(A) - \sum_{k=1}^{n} P^{\mathscr{B}}(A_k) \right\|_p^p = \int_{\Omega} \left[P^{\mathscr{B}}\left(\bigcup_{k>n} A_k \right) \right]^p dP_{\mathscr{B}}$$

$$= \int_{\Omega} |E^{\mathscr{B}}(\chi_{\bigcup_{k>n} A_k})|^p \, dP_{\mathscr{B}}$$

$$\le \int_{\Omega} E^{\mathscr{B}}(\chi_{\bigcup_{k>n} A_k})\, dP_{\mathscr{B}} \qquad \text{by Theorem 1.6}$$

$$= P\left(\bigcup_{k>n} A_k \right) \to 0 \qquad \text{as} \quad n \to \infty. \tag{3}$$

But in L^p, $1 \le p < \infty$, the order and norm convergence are equivalent. Thus $P^{\mathscr{B}}$ is a vector measure in $L^p(\mathscr{B})$, $1 \le p < \infty$, and a vector valued (order convergent) σ-additive positive set function in $L^\infty(\mathscr{B})$. Thus the work of Section I.4 on integration of scalar functions with $P^{\mathscr{B}}$ will be relevant.

To aid the calculation in (3) further, let us evaluate the variation and semivariation of $P^{\mathscr{B}}$. By definition in Section I.4, these are given as

$$\begin{aligned} |P^{\mathscr{B}}|(A) &= \sup\left\{\sum_{i=1}^{n} \|P^{\mathscr{B}}(A_i)\|_1 : A_i \subset A, \text{ disjoint}, A_i \in \Sigma\right\} \\ &= \sup\left\{\sum_{i=1}^{n} \int_\Omega E^{\mathscr{B}}(\chi_{A_i})\, dP_{\mathscr{B}} : A_i \subset A, \text{ disjoint}, A_i \in \Sigma\right\} \\ &= P(A), \qquad A \in \Sigma \end{aligned} \tag{4}$$

If we use the L^p-norm, $1 < p < \infty$, then the computation is messy. However, for the semivariation it is obtained for $A \in \Sigma$ as follows:

$$\begin{aligned} \|P^{\mathscr{B}}\|(A) &= \sup\left\{\left\|\sum_{i=1}^{n} a_i P^{\mathscr{B}}(A_i)\right\|_p : A_i \subset A, \text{ disjoint}, A_i \in \Sigma, |a_i| \le 1\right\} \\ &\le \sup\left\{\left\|\sum_{i=1}^{n} P^{\mathscr{B}}(A_i)\right\|_p : A_i \subset A, \text{ as above}\right\} \\ &\le \sup\{\|E^{\mathscr{B}}(\chi_A)\|_p\} = [P(A)]^{1/p}. \end{aligned} \tag{5}$$

Thus $\|P^{\mathscr{B}}\|(\Omega) < \infty$, and $P(\cdot)$ serves as a "control measure" for $P^{\mathscr{B}}$ on L^p, and if $p = 1$, then $|P^{\mathscr{B}}|(A) = \|P^{\mathscr{B}}\|(A)$, which follows on taking $a_1 = a_2 = \cdots = a_n = 1$ for (5). Having P therefore one can use the D–S integration relative to $P^{\mathscr{B}}$ (or also using the order integral of Section I.4). With Theorem I.4.14 (or I.4.15) we set for a simple function $f = \sum_{i=1}^{n} a_i \chi_{A_i}$,

$$\begin{aligned} \int_A f\, dP^{\mathscr{B}} &= \sum_{i=1}^{n} a_i P^{\mathscr{B}}(A_i \cap A) = \sum_{i=1}^{n} a_i E^{\mathscr{B}}(\chi_{A_i \cap A}) \\ &= E^{\mathscr{B}}\left(\left(\sum_{i=1}^{n} a_i \chi_{A_i}\right)\chi_A\right) = E^{\mathscr{B}}(f\chi_A), \qquad A \in \Sigma \end{aligned} \tag{6}$$

If $0 \le f_n \uparrow f$, $f \in L^p$, then the results of Section I.4 imply $\{\int_A f_n\, dP^{\mathscr{B}}\}_1^\infty$ is Cauchy in L^p, $p < \infty$, or with the order definition if $p = \infty$, and one has

$$E^{\mathscr{B}}(f\chi_A) = \int_A f\, dP^{\mathscr{B}}, \qquad A \in \Sigma, \quad f \in L^p. \tag{7}$$

The value of this D–S (or order) integral is an element of $L^p(\mathscr{B})$. The definition may be extended to all $f \in L^p$, using $f = f^+ - f^-$ and $f^\pm \in L^p$. We state it as follows:

2. Proposition *Let (Ω, Σ, P) be a probability space, $1 \le p < \infty$, and $\mathscr{B} \subset \Sigma$ a σ-algebra. Then the mapping $P^{\mathscr{B}}: A \mapsto P^{\mathscr{B}}(A)$ is a vector measure from Σ*

into $L^p(\mathscr{B})$ *and is of finite semivariation with* P *as its "control measure." If* $p = 1$, *then* $P^{\mathscr{B}}$ *has* P *as its variation measure. Moreover, if* $f \in L^p$, *then*

$$E^{\mathscr{B}}(f\chi_A) = \int_A f\, dP^{\mathscr{B}} \qquad \in L^p(\mathscr{B}), \qquad A \in \Sigma, \quad 1 \le p \le \infty, \tag{8}$$

where the integral exists in the Dunford–Schwartz (or order convergence) sense. Thus all elements of L^1 *are* $P^{\mathscr{B}}$*-integrable and the monotone and dominated convergence theorems hold for the integral in* (8).

As a consequence of this representation, the result of Corollary 1.3 takes the following form:

3. Corollary *Let* $\mathscr{B}_1 \subset \mathscr{B}_2 \subset \Sigma$ *be* σ*-algebras and* f, g *be integrable random variables on* (Ω, Σ, P) *such that* fg *is also integrable. If* g *is* $\mathscr{B}_2$*-measurable, then*

$$\int_\Omega fg\, dP^{\mathscr{B}_1} = \int_\Omega g\, dP^{\mathscr{B}_1} \int_\Omega f\, dP^{\mathscr{B}_2} \qquad \in L^1(\mathscr{B}_1), \tag{9}$$

where all the integrals are taken in the D–S sense. Moreover, one has

$$\int_\Omega fg\, dP = \int_\Omega dP \int_\Omega g\, dP^{\mathscr{B}_1} \int_\Omega f\, dP^{\mathscr{B}_2}, \tag{10}$$

the integrals relative to P *being in the sense of Lebesgue.*

In (10), or (8), all integrals can be defined as the abstract Lebesgue integrals if it is possible to take $P^{\mathscr{B}}(\cdot)(\omega)$ as a measure for almost all ω. This means there is a function $P(\cdot,\cdot)\colon \Sigma \times \Omega \to \mathbb{R}^+$ such that $P(A,\cdot)$ $(=P^{\mathscr{B}}(A))$ is $\mathscr{B}$-measurable, $A \in \Sigma$, and $P(\cdot,\omega)$ is a probability for each $\omega \in \Omega - \Omega_0$ with $P(\Omega_0) = 0$. When such a choice of $P^{\mathscr{B}}$ is possible (i.e., the vector measure $P^{\mathscr{B}}$ should behave like a scalar measure for almost all evaluations $P^{\mathscr{B}}(\cdot)(\omega)$), then it is called a *regular* conditional probability. If $\mathscr{B}$ is generated by a partition, then, by Eqs. (2) and (3) of Section 1, such a choice is possible. An example below implies, on the other hand, that this is not the rule. We start with a simple result on regularity (cf. Loève [1, p. 356]).

4. Lemma *Let* (Ω, Σ, P) *be a probability space and* $\mathscr{B} \subset \Sigma$ *be a* σ*-algebra that is countably generated. If the conditional probability function* $P^{\mathscr{B}}$ *is regular and if* $P^{\mathscr{B}}(A)$ *is a constant for all* $A \in \Sigma(B)$ $(=\{B \cap D : D \in \Sigma\})$, *where* B *is any* $P_{\mathscr{B}}$*-atom, then* $\Omega = \bigcup_{\alpha \in I} B_\alpha \cup \Omega_0$, *with* $P_{\mathscr{B}}(\Omega_0) = 0$ *and each* B_α *is a* $P_{\mathscr{B}}$*-atom. Moreover,*

$$P^{\mathscr{B}}(A) = \sum_{\alpha \in I} P_{\mathscr{B}_\alpha}(A)\chi_{B_\alpha}, \qquad A \in \Sigma, \quad \omega \in \Omega - \Omega_0, \tag{11}$$

where $P_{B_\alpha}(A) = P(A)/P(B_\alpha)$ *for all* $A \in \Sigma(B_\alpha)$, $\alpha \in I$. *In other words,* $P^{\mathscr{B}}$ *is essentially like* $P_\Pi(\cdot)$, $\Pi = \{\Omega_0, B_\alpha, \alpha \in I\}$, *of Eq.* (3) *of Section* 1.

Proof Since by definition two $P_{\mathscr{B}}$-atoms are either a.e. disjoint or coincident, and $P_{\mathscr{B}}$ is finite, there exist $\{B_\alpha, \alpha \in I\}$, disjoint atoms, $P_{\mathscr{B}}(B_\alpha) > 0$, such that $\Omega = \Omega_0 \cup \bigcup_{\alpha \in I} B_\alpha$, where $P_{\mathscr{B}}(\Omega_0) = 0$. This follows from the strict localizability of the finite $P_{\mathscr{B}}$. (See Definition I.2.6.) Then $\mathscr{B}$ is generated by $\{B_\alpha, \alpha \in I\}$ and completed for $P_{\mathscr{B}}$. So $\Omega_0 \in \mathscr{B}$. By Eq. (7) of Section 1,

$$\int_{B_\alpha} P^{\mathscr{B}}(A)\, dP_{\mathscr{B}} = P(A \cap B_\alpha), \qquad A \in \Sigma. \tag{12}$$

If we denote by $f_{B_\alpha}(A)$ the constant value $P^{\mathscr{B}}(A)$, $A \in \Sigma(B_\alpha)$, then (12) implies $P(A \cap B_\alpha) = f_{B_\alpha}(A)P(B_\alpha)$ a.e. $[P_{\mathscr{B}}]$ for $A \in \Sigma$ so that $f_{B_\alpha}(\cdot)$ is a probability by the regularity of $P_{\mathscr{B}}$. Now $P_{\mathscr{B}}(B_\alpha) > 0$, $\alpha \in I$. If N_α is the $P_{\mathscr{B}}$-null set so that $N = \bigcup_{\alpha \in I} N_\alpha \subset \Omega_0$ is $P_{\mathscr{B}}$-null, then the above equation implies $f_{B_\alpha}(B_\alpha) = 1$ on $\Omega - \Omega_0$. Hence $P^{\mathscr{B}} = \sum_{\alpha \in I} f_{B_\alpha}\chi_{B_\alpha}$ a.e. $[P_{\mathscr{B}}]$ and then $f_{B_\alpha} = P_{B_\alpha}$ as described in the statement. The rest is immediate.

The question now is whether the regularity hypothesis can be suppressed from the lemma since $\mathscr{B}$ is countably generated. But every open subset of $\mathbb{R}$ is a countable union of disjoint open intervals, and hence its Borel algebra is countably generated; the ensuing example gives a negative answer to this as well as to the earlier queries.

5. Counterexample The following example is designed to show the nonexistence of regular conditional probabilities. Thus let (Ω, Σ, P) be the Lebesgue unit interval, $\mathscr{B} \subset \Sigma$ the Borel σ-algebra, and A a thick set in the sense of Theorem I.3.5 (i.e., it has outer measure one and inner measure zero). Let Σ_1 be the σ-algebra generated by $(\mathscr{B}, A)$. Then each set M of Σ_1 is of the form $(A \cup B_1) \cup (A^c \cap B_2)$ with $B_i \in \mathscr{B}$, $i = 1, 2$. Define $Q: \Sigma_1 \to [0, 1]$ by the rule

$$Q(M) = Q((A \cap B_1) \cup (A^c \cap B_2)) = P(B_1). \tag{13}$$

One may also define $\tilde{Q}$ on Σ_1 by $\tilde{Q}(M) = P(B_2)$. Then both Q and $\tilde{Q}$ are extensions of P on $\mathscr{B}$ since A and A^c are both thick sets and Theorem I.3.5 applies to either extension. We work with Q. Note that even though P is regular (cf. Definition I.2.9), neither Q nor $\tilde{Q}$ is regular. Let $Q^{\mathscr{B}}$ be the conditional probability function. $Q^{\mathscr{B}}: \Sigma_1 \to L^\infty(\Omega, \mathscr{B}, P)$ exists (since $Q_{\mathscr{B}} = P_{\mathscr{B}}$), and

$$\int_B Q^{\mathscr{B}}(C)\, dP_{\mathscr{B}} = Q(C \cap B), \qquad C \in \Sigma_1, \quad B \in \mathscr{B}. \tag{14}$$

We claim that $Q^{\mathscr{B}}_{\omega}(\cdot)$ $(=Q^{\mathscr{B}}(\cdot)(\omega))$ is not a probability measure on Σ_1 for $\omega \in \Omega - N$, $Q(N) = 0$ even though both $\mathscr{B}$ and Σ_1 are countably generated. It follows from (14) that

$$\int_{\Omega} Q^{\mathscr{B}}(A)\, dP_{\mathscr{B}} = Q(A \cap \Omega) = 1, \tag{15}$$

and hence $Q^{\mathscr{B}}(A) = 1$ a.e. $[P_{\mathscr{B}}]$. Also note that $Q^{\mathscr{B}}(A^c) > 0$. In fact, if $Q^{\mathscr{B}}(A^c) = 0$, then replacing A by A^c in (15) we get $Q(A^c) = 0$ and hence $A^c \in \Sigma$ and $P(A^c) = 0$ so that $A \in \Sigma$, and this contradicts the fact that A is a thick set. Now, if $Q^{\mathscr{B}}_{\omega}$ is a probability measure for $\omega \in \Omega - N$, then the upper and lower integrals of $f = \chi_A$ are found to be

$$\int_{\Omega}^{*} f\, dQ^{\mathscr{B}}_{\omega} = 1 > 0 = \int_{*\Omega} f\, dQ^{\mathscr{B}}_{\omega}, \qquad \omega \in \Omega - N. \tag{16}$$

So the measurable bounded function f is not integrable, contradicting a classical theorem of Lebesgue. Thus $Q^{\mathscr{B}}_{\omega}$ is not a measure. However, $Q^{\mathscr{B}}\colon \Sigma_1 \to L^p(\mathscr{B})$ is a vector valued σ-additive set function by Proposition 2, and the D–S integral $\int_{\Omega} f\, dQ^{\mathscr{B}} = 1$ a.e. holds.

In view of the above example, it is advantageous to treat conditional probability functions as vector measures, without further restrictions. Many results of the Lebesgue theory extend. For ease in handling with this new point of view, let us include a few more elementary results and remarks.

The definition of semivariation of a vector measure $v\colon \Sigma \to \mathscr{X}$, given in Eq. (I.4.29) may be recast in the following, more suggestive form, for $A \in \Sigma$:

$$\|v\|(A) = \sup\left\{\left\|\int_A f\, dv\right\|_{\mathscr{X}} : f = \sum_{i=1}^{n} a_i \chi_{A_i},\ A_i \subset A,\ A_i \text{ disjoint},\ \|f\|_{\infty} \le 1\right\}, \tag{17}$$

where the integral of a simple function is the usual sum, and $f \in L^{\infty}(\Omega, \Sigma, P)$. This may be generalized by replacing L^{∞} with L^p. Using the fact that $\|x\|_{\mathscr{X}} = \sup\{|\langle x^*, x\rangle| : x^* \in \mathscr{X}^*, \|x^*\| \le 1\}$, where $\langle x^*, x\rangle = x^*(x)$ is the duality pairing, we have an extension as follows.

6. Definition Let $v\colon \Sigma \to \mathscr{X}$ be a vector measure, and $1 \le p < \infty$. If $v(A) = 0$ for $P(A) = 0$, then v is said to be of *q-semivariation* finite $(q = p/(p-1))$ provided $\sup\{\|v\|_q(A) : A \in \Sigma\} < \infty$, where for $x^* \in \mathscr{X}^*$ $(\Sigma(A) = \{A \cap B : B \in \Sigma\})$, we have

$$\|v\|_q(A) = \sup\left\{\sup\left\{\left|\left\langle x^*, \int_A f\, dv\right\rangle\right| : \|f\|_p \le 1, f \in L^p(A, \Sigma(A), P)\right\} : \|x^*\| \le 1\right\}. \tag{18}$$

Even though (17) and (18) appear slightly different, it may be verified without difficulty that $\|v\|(A) = \|v\|_1(A)$ (so $p = \infty$ in (18)). The explicit form (18) is useful in computations. The following estimate is needed later.

7. Lemma *Let (Ω, Σ, P) be a probability space and $\mathscr{B} \subset \Sigma$ a σ-algebra. Then for the conditional probability function $P^{\mathscr{B}}: \Sigma \to L^p$, we have $\|P^{\mathscr{B}}\|_q (= \|P^{\mathscr{B}}\|_q(\Omega)) \leq 1$, $q \geq 1$, $p = q/(q-1)$.*

Proof Let $f = \sum_{i=1}^n a_i \chi_{A_i}$, $A_i \in \Sigma$, disjoint, $\|f\|_p \leq 1$, and $x^* = \sum_{j=1}^m b_j \chi_{B_j}$, $B_j \in \mathscr{B}$, disjoint, and $\|x^*\|_q \leq 1$, $p^{-1} + q^{-1} = 1$. Then we may simplify, since $\mathscr{X} = L^p(\mathscr{B})$ and $L^q(\mathscr{B}) \subset (L^p(\mathscr{B}))^*$, with equality if $q > 1$, the expressions in (18) as follows:

$$\left\langle x^*, \int_\Omega f\, dP^{\mathscr{B}} \right\rangle = \left\langle x^*, \sum_{i=1}^n a_i P^{\mathscr{B}}(A_i) \right\rangle = \sum_{i=1}^n a_i \sum_{j=1}^m b_j \int_\Omega \chi_{B_j} E^{\mathscr{B}}(\chi_{A_i})\, dP_{\mathscr{B}}$$

$$= \sum_{i=1}^n a_i \int_{A_i} x^*\, dP = \sum_{i=1}^n a_i G_{x^*}(A_i), \text{ say.} \tag{19}$$

Then $G_{x^*}: \Sigma_0 \to \mathbb{R}$ is a signed measure and $\|G_{x^*}\|_q = \|x^*\|_q \leq 1$ (by a simple calculation, since $\|G_{x^*}\|_q = \int_\Omega |x^*|^q\, dP$). Thus, (18) becomes

$$\sup\left\{ \left| \sum_{i=1}^n a_i G_{x^*}(A_i) \right| : f = \sum_{i=1}^n a_i \chi_{A_i}, \|f\|_p \leq 1 \right\} = \|G_{x^*}\|_q. \tag{20}$$

Hence (18) is simplified, on noting that the simple functions of $L^q(\mathscr{B})$ are norm determining for $L^p(\mathscr{B})$, as follows:

$$\|P^{\mathscr{B}}\|_q = \sup\left\{ \sup\left\{ \left| \left\langle x^*, \int_\Omega f\, dP^{\mathscr{B}} \right\rangle \right| : \|f\|_p \leq 1, f \text{ simple} \right\} : \|x^*\|_q \leq 1 \right\}$$

$$= \sup\{\|G_{x^*}\|_q : \|x^*\|_q \leq 1\} \leq 1.$$

This completes the proof.

The preceding estimate is used in the following type of computations employed, for instance, in the characterization problems of Section 4. If $v_f(\cdot) = \int_{(\cdot)} f\, dP^{\mathscr{B}}$, $f \in L^p$, then with $f_n = \sum_{i=1}^n a_i \chi_{A_i}$, A_i disjoint in Σ, we have

$$\left\| \sum_{i=1}^n a_i v_f(A_i) \right\|_p = \left\| \sum_{i=1}^n a_i \int_\Omega \chi_{A_i} f\, dP^{\mathscr{B}} \right\|_p \leq \|ff_n\|_p \|P^{\mathscr{B}}\|_q \leq \|ff_n\|_p \tag{21}$$

by Lemma 7, and then $\|v_f\|(\Omega) \leq \sup\{\|ff_n\|_p : \|f_n\|_\infty \leq 1\} \leq \|f\|_p < \infty$.

The following result is a different form of Proposition 2 and has independent interest.

8. Proposition *Let $\mathscr{B} \subset \Sigma$ be a σ-algebra of a probability space (Ω, Σ, P). Then the mapping $f \mapsto \int_A f\, dP^{\mathscr{B}}$ is order preserving and faithful (i.e.,*

$\int_\Omega f\, dP^{\mathscr{B}} = 0, f \geq 0 \Rightarrow f = 0$ *a.e.* $[P]$). *The integral* (8) *can also be defined for each* $0 \leq f \in L^p(\Sigma)$, $1 \leq p < \infty$, *as*

$$\int_\Omega f\, dP^{\mathscr{B}} = \sup\left\{ \sum_{i=1}^n a_i P^{\mathscr{B}}(A_i) : 0 \leq h \leq f \text{ a.e., } h = \sum_{i=1}^n a_i \chi_{A_i}, A_i \in \Sigma \right\}, \quad (22)$$

where the equality holds a.e.

Proof The first part is clear. To prove (22), note that if $L(h) = \sum_{i=1}^n a_i P^{\mathscr{B}}(A_i)$, then $L(L^p(\Sigma)) = L^p(\mathscr{B})$, $L = E^{\mathscr{B}}$. Thus L is a positive contractive projection. If $0 \leq f_n \uparrow f$ a.e., then $L(f_n) \uparrow L(f)$ a.e., and since $f \in L^p(\Sigma)$, we have $\|L(f_n) - L(f)\|_p \leq \|f_n - f\|_p \to 0$. Hence

$$\int_\Omega f\, dP^{\mathscr{B}} = \lim_{n\to\infty} L(f_n) = \sup\{L(h) : 0 \leq h \leq f, h \text{ simple}\}.$$

Since f_n can be taken to be simple, this proves (22).

This result is a restatement of Proposition I.4.15 in the context of conditional probability measures. The interest in this form will soon be evident.

Let us present another simple result on the image conditional probabilities under a random variable. In considering the regularity problem, this will be useful.

9. Proposition *If* $\mathscr{B} \subset \Sigma$ *is a complete* σ*-algebra of* (Ω, Σ, P) *and* $(\tilde{\Omega}, \tilde{\Sigma})$ *is a measurable space, let* $f: \Omega \to \tilde{\Omega}$ *be a measurable mapping and* $\tilde{P}^{\mathscr{B}} = P^{\mathscr{B}} \circ f^{-1}: \tilde{\Sigma} \to L^p(\mathscr{B})$ *be the image of the conditional probability* $P^{\mathscr{B}}: \Sigma \to L^p(\mathscr{B})$, *relative to* f. *If* $\psi(A, \omega) = P^{\mathscr{B}}(A)(\omega)$ *and* $\tilde{\psi}(\tilde{A}, \tilde{\omega}) = \psi(f^{-1}(\tilde{A}), \tilde{\omega})$, *then* $\tilde{P}^{\mathscr{B}}$ *is a vector valued measure and for any* $\tilde{P}^{\mathscr{B}}$*-integrable real* g *on* $\tilde{\Omega}$ *we have*

$$\int_{\tilde{\Omega}} g(\tilde{\omega}) \tilde{\psi}(d\tilde{\omega}, \cdot) = \int_\Omega g \circ f(\omega) \psi(d\omega, \cdot), \qquad g \in L^1(\Sigma), \quad (23)$$

where the integrals are taken either in the D–S or order sense.

The result is clear if $g = \chi_A$ and then for a simple function. In the general case, it again follows from dominated or monotone convergence theorems, which are true for the D–S (or order) integrals since the "control measures" of ψ and $\tilde{\psi}$ are seen to be P and $P \circ f^{-1}$. This yields (23).

The point of this proposition is that it allows us to transfer the regularity problem from the abstract spaces to the topological spaces since $(\tilde{\Omega}, \tilde{\Sigma})$ will be a space such as $(\mathbb{R}^n, \mathscr{B}_n)$—the usual n-dimensional Borelian space. We now analyze the regularity problem for conditional probabilities in some detail and present sufficient (as well as necessary) conditions for regularity. This will be handy in analytical probability, such as in Markov processes. A great deal of the abstract theory can however be developed if one uses the D–S theory of integration, without further restrictions.

Let us introduce a definition of regularity of a vector measure and then show in the following theorem that for conditional probabilities this is the same as our earlier concept under the same name. The present definition is more "operational." In what follows, for the definition and work, all σ-algebras are (again) assumed complete.

10. Definition (a) Let (Ω, Σ, P) be a complete space, $\mathscr{B} \subset \Sigma$ a completed σ-algebra, and $(\tilde{\Omega}, \tilde{\Sigma})$ a topological measurable space, where $\tilde{\Omega}$ is Hausdorff and $\tilde{\Sigma}$ is its Borel σ-algebra. If $f: \Omega \to \tilde{\Omega}$ is measurable and $\psi: \Sigma \to L^\infty(\mathscr{B})$ is the image under f of $P^{\mathscr{B}}: \Sigma \to L^\infty(\mathscr{B})$, then ψ, and also $P^{\mathscr{B}}$, will be called *inner regular* relative to f if the following condition holds. For each relatively compact open set $A \subset \tilde{\Omega}$, we have

$$\psi(A) = \sup\{\psi(C) : C \subset A, C \text{ compact}\}. \tag{24}$$

The same concept is also termed *quasi-regular* and *wide sense conditional distribution* of f. [Note that since P is finite, hence localizable, the supremum in (24) exists by Theorem I.5.2, and $0 \leq \psi(A) \leq 1$ a.e.]

(b) The function $\psi(\cdot)$ or $P^{\mathscr{B}}$ above is said to be *outer regular*, or *strict sense conditional distribution*, relative to f, if for each relatively compact Borel set $A \subset \tilde{\Omega}$,

$$\psi(A) = \inf\{\psi(B) : B \supset A, B \text{ open}\}, \tag{25}$$

where again the infimum exists for the same reason. If ψ is both inner and outer regular, then it is called *regular* relative to f.

Replacing Borel sets by Baire sets in the above definition, the corresponding Baire concepts are obtained. If (Ω, Σ, P) is topological, then, taking f as identity, we get the concepts for $P^{\mathscr{B}}$ itself. The following result contains a characterization of regularity and related information and shows its equivalence with the notion introduced prior to Lemma 4. So there is no conflict in terminology. Later simpler sufficient conditions on (Ω, Σ, P) will be given for the regularity of $P^{\mathscr{B}}$.

11. Theorem *Let (Ω, Σ, P) be a complete probability space, $\mathscr{B} \subset \Sigma$ a completed σ-algebra, $\tilde{\Omega}$ a metric space, and $\tilde{\Sigma}$ its Borel algebra. If $f: \Omega \to \tilde{\Omega}$ is measurable, then $P^{\mathscr{B}}$ or $\psi = P^{\mathscr{B}} \circ f^{-1}: \tilde{\Sigma} \to L^\infty(\mathscr{B})$ is always inner regular relative to f. The D–S integral relative to $P^{\mathscr{B}}$ on Ω can be taken in the Lebesgue sense iff ψ is regular relative to f (i.e., iff it is outer regular).*

Proof For the first part, on inner regularity, let $A \subset \tilde{\Omega}$ be a relatively compact open set. The openness of A implies χ_A is lower semicontinuous (l.s.c.). Recall that a mapping $g: \tilde{\Omega} \to \mathbb{R}$ is l.s.c. iff the set $\{\tilde{\omega} : g(\tilde{\omega}) > a\}$ is open for each $a \in \mathbb{R}$. We now use a classical result on the structure of these functions, namely, that each positive finite valued l.s.c. function on a metric space is the pointwise limit of an increasing sequence of *continuous* functions. [A proof of this is outlined in Remark 12(3) below.] Thus there are continuous f_n on $\tilde{\Omega}$ such that $0 \leq f_n \uparrow \chi_A$. Since real continuous functions on $\tilde{\Omega}$ are Borel measurable, we have (with the notation of the proof of Proposition 8):

$$\psi(A) = \tilde{L}(\chi_A) = \sup\left\{\int_{\Omega} f_n \, d\psi : 0 \leq f_n \uparrow \chi_A, f_n \text{ continuous}\right\}$$

$$= \sup\left\{\int_{\Omega} f \, d\psi : 0 \leq f \leq \chi_A, f \text{ continuous}\right\}, \qquad (26)$$

where in the first line we used the monotone convergence theorem, which was noted to be valid for ψ (cf. Proposition 2 above). Let f_0 be an arbitrary continuous function with $0 \leq f_0 \leq \chi_A$. If $C_n = \{\tilde{\omega} : 1/n \leq f_0(\tilde{\omega}) \leq 1\}$, then $C_n \subset A \subset \bar{A}$ for all n. Since C_n is closed and $\bar{A}$ is compact, it follows that C_n is compact, $C_n \uparrow$ and $\lim_n C_n = \bigcup_{n=1}^{\infty} C_n \subset A$. Hence

$$\psi(A) \geq \psi\left(\lim_n C_n\right) = \lim_n \psi(C_n) = \lim_n \int_{\tilde{\Omega}} \chi_{C_n} \, d\psi$$

$$\geq \lim_n \int_{\tilde{\Omega}} \chi_{C_n} f_0 \, d\psi = \int_{\tilde{\Omega}} f_0 \, d\psi, \qquad (27)$$

by the monotone convergence theorem again. From (26) and (27) we get

$$\psi(A) \geq \sup\{\psi(C) : C \subset A \text{ compact}\}$$

$$\geq \sup\left\{\int_{\tilde{\Omega}} f_0 \, d\psi : 0 \leq f_0 \leq \chi_A\right\} = \psi(A).$$

Hence $\psi(\cdot)$ is inner regular, proving the first part.

For the last part, suppose that the D–S integral of a bounded Borel u can also be defined in the Lebesgue sense. Then it may be approximated from above and below. Thus for any relatively compact Borel set A, with Proposition 8, we have

$$\int_{\tilde{\Omega}} \chi_A \, d\psi = \sup\left\{\int_{\tilde{\Omega}} f \, d\psi : 0 \leq f \leq \chi_A\right\} = \inf\left\{\int_{\tilde{\Omega}} h \, d\psi : h \geq \chi_A\right\},$$

where f and h are simple or continuous functions with compact support. Hence ψ is both inner and outer regular, so $P^{\mathscr{B}}(\cdot)$ is regular on $f^{-1}(\tilde{\Sigma})$, according to Definition 10.

If on the other hand ψ is regular in the sense of Definition 10, then we have to show that, given $\varepsilon > 0$ and any simple Borel $u \in L^1(\tilde{\Sigma}) = L^1(\tilde{\Omega}, \tilde{\Sigma}, P \circ f^{-1})$, there exist simple functions g and h such that $g \leq u \leq h$ a.e. and $\int_{\Omega}(h - g)\, d\psi \leq \varepsilon$ a.e. For this, by linearity of the integral, it is sufficient to show, for any relatively compact Borel set A and $u = \chi_A$, that there exist $g = \chi_G, h = \chi_H$, where G is compact and H is open and $G \subset A \subset H$, satisfying the above inequality. This implies the equivalence of the upper and lower approximations and so the D–S integral can be obtained by Lebesgue's method. Here we are using the fact that the order continuity is equivalent to norm convergence in the vector lattice $L^1(\tilde{\Sigma})$.

Let $\varepsilon > 0$ be given and select an open H such that, by the outer regularity of ψ (cf. (25)), $H \supset A$ and $\psi(H) < \psi(A) + \varepsilon/2$ a.e. But $H^c \subset A^c$, A^c is a Borel set, and H^c is closed. Then $A^c - H^c = A^c \cap H = H - A$ and

$$0 \leq \psi(A^c) - \psi(H^c) = \psi(A^c - H^c) = \psi(H - A) = \psi(H) - \psi(A) < \varepsilon/2 \quad \text{a.e.}$$

This shows that every Borel set A^c can be approximated from below by a closed set. In particular, if A is a relatively compact Borel set, then there is a $G \subset A \subset \bar{A}$, G is closed so that it is compact such that $\psi(A) - \psi(G) < \varepsilon/2$ a.e. Thus there exist $G \subset A \subset H$ such that

$$\psi(H) - \psi(G) \leq (\psi(H) - \psi(A)) + (\psi(A) - \psi(G)) < \varepsilon \quad \text{a.e.}$$

Since the last statement is a rewording of this inequality, the theorem is proved.

With regard to the statement and proof of the above theorem, several remarks are in order.

12. Remarks 1. In the definition of regularity, the test sets were required to be relatively compact. As the proof shows, this may be dropped if the approximating sets are closed for the inner regular case and open for the outer regular case. However, it can be shown that when $\tilde{\Omega}$ is locally compact or metric (as here), the above more general case of regularity is the same as that of Definition 10. Also the strong additivity of $P^{\mathscr{B}}$ has not been crucially used in the proof. Thus the result holds for any conditional measure $P^{\mathscr{B}}$ which is σ-additive in the order topology.

2. For a complete metric space, the Borel or Baire σ-algebra is the one generated by the class of all its closed sets which are G_δs. [If $\tilde{\Omega}$ is locally compact, then the word "closed" is replaced by "compact" in the above definitions by some authors; cf. Halmos [1]. A G_δ set is one that is a

countable intersection of open sets.] In fact, if $C \subset \tilde{\Omega}$ is any closed set, let $S(x, 1/n)$ be the open ball of radius $1/n$ for each $x \in C$ and let $S_n = \bigcup_{x \in C} S(x, 1/n)$ be the open cover of C. Then $C \subset S_n$ for each n, and so $C \subset \bigcap_{n=1}^{\infty} S_n$. Conversely, if $x \in \bigcap_{n=1}^{\infty} S_n$, then there is an $x_n \in C$ such that $d(x, x_n) < 1/n$, where $d(\cdot, \cdot)$ is the distance. Thus $\{x_n\}_1^{\infty} \subset C$ is a Cauchy sequence and since C, being a closed set, is complete, $x_n \to x$ and $x \in C$. Hence $C = \bigcap_{n=1}^{\infty} S_n$, and C is a G_δ set. Thus what we proved in Theorem 11 is that $\psi(\cdot)$ is always inner regular relative to the compact Baire (equal Borel) sets when $\tilde{\Omega}$ is a metric space, and that it is regular with respect to the compact Baire sets iff the two integration processes are equivalent. Since a regularity property for a given class does not imply the same property for an arbitrarily altered one, ψ need not be regular relative to other algebras than $\tilde{\Sigma}$ itself. In this context the following result of Dinculeanu and Kluvanek [1] is of interest.

A Regularity Theorem *Let Ω_0 be a locally compact space and Σ_0 be the σ-ring generated by the relatively compact G_δ-sets; or the Baire σ-ring in Halmos [1]. If $v_0\colon \Sigma_0 \to \mathscr{X}$ is a σ-additive set function (called a Baire measure) into a vector space $\mathscr{X}$ which has a locally convex topology, then v_0 is regular in the sense that for each $A \in \Sigma_0$, and a neighborhood of the origin $V \subset \mathscr{X}$, there exist a compact C and open G in Σ_0 with $C \subset A \subset G$, and $v_0(B) \in V$ for each B in Σ_0 with $B \subset G - C$. Moreover, every Baire measure $v_0\colon \Sigma_0 \to \mathscr{X}$ can be uniquely extended to a regular Borel measure $v\colon \tilde{\Sigma} \to \tilde{\mathscr{X}}$, where $\tilde{\Sigma}$ is the (Halmos [1]) Borel σ-ring of Ω_0 and $\tilde{\mathscr{X}}$ is the completion of $\mathscr{X}$.*

[Since $\tilde{\Omega}$ of Theorem 11 is not necessarily locally compact, that theorem is not directly implied by this one.]

3. We have used this fact: If f is a nonnegative lower semicontinuous function on a metric space X, then there exist continuous $f_n\colon X \to \mathbb{R}^+$ such that $f_n \uparrow f$. Here we add a proof of this classical result for completeness.

Let us define f_n on X as

$$f_n(x) = \inf\{f(y) + nd(x, y) : y \in X\}, \tag{28}$$

where $d(\cdot, \cdot)$ is again the distance in X. Then $0 \le f_n(x) \le f_{n+1}(x) \le f(x)$ since $d(x, x) = 0$. Let $g(x) = \lim_n f_n(x) \le f(x)$, $x \in X$. We claim that (i) $g = f$ and (ii) the f_n are continuous.

To prove (i), clearly $g(x) \le f(x)$. For the opposite inequality, let $x \in X$. For any $a < f(x)$, by the lower semicontinuity of f, the set $V = \{y : f(y) > a\}$ is open, containing x, and hence there is a neighborhood N of x such that $N \subset V$. Since $d(x, \cdot)$ is continuous for any $\varepsilon > 0$, the set $N_1 = N \cap [d(x, y) < \varepsilon] \ne \varnothing$. From (28) we have

$$f(y_0) + nd(x, y_0) > a + 0 = a, \qquad y_0 \in N_1, \quad n \ge 1, \tag{29}$$

and consider $N_2 = N \cap [d(x, y) \geq \varepsilon]$. Then

$$f(y_0) + nd(x, y_0) \geq n\varepsilon, \qquad y_0 \in N_2. \tag{30}$$

Hence for large enough n, $n\varepsilon \geq a$; taking the infimum over y_0 $(\in X)$ in (29) and (30), we see that $f_n(x) \geq a$. Letting first $n \to \infty$, and then $a \uparrow f(x)$, we deduce that $g(x) \geq f(x)$.

To prove (ii), let $\{x_1, x_2\} \subset X$. Then using the triangle inequality for $d(\cdot, \cdot)$, we get from (28), for any $n \geq 1$ and $y \in X$,

$$f_n(x_1) \leq f(y) + nd(x_1, y) \leq f(y) + nd(x_1, x_2) + nd(x_2, y).$$

Thus

$$f_n(x_1) - nd(x_1, x_2) \leq f(y) + nd(x_2, y),$$

and so $f_n(x_1) - nd(x_1, x_2) \leq f_n(x_2)$. Interchanging x_1 and x_2, we get a similar inequality which together imply

$$|f_n(x_1) - f_n(x_2)| \leq nd(x_1, x_2).$$

Thus $f_n(x_1) \to f_n(x_2)$ as $x_1 \to x_2$. This proves (ii).

The following result is a consequence of Theorem 11 [see also Remark 12(2)] above. When $\mathscr{X} = L^p(\mathscr{B})$, $p < \infty$, both regularities are the same.

13. Corollary *Let (Ω, Σ, P) be a probability space and $\mathscr{B} \subset \Sigma$ be a σ-algebra. If $X: \Omega \to \tilde{\Omega}$ is a measurable mapping relative to $(\Sigma, \tilde{\Sigma})$, where $\tilde{\Sigma}$ is the Baire σ-algebra of the locally compact metric space $\tilde{\Omega}$, then $P^{\mathscr{B}} \circ X^{-1}: \tilde{\Sigma} \to L^p(\mathscr{B})$, $1 \leq p < \infty$, is Baire regular, and hence a conditional distribution of X given $\mathscr{B}$ exists in the sense that there is a function $\psi(\cdot) = P^{\mathscr{B}} \circ X^{-1}$ so that it can be treated as a function of two variables and that $\psi = \psi(\cdot, \cdot): \tilde{\Sigma} \times \Omega \to \mathbb{R}^+$, where $\psi(M, \cdot)$ is equivalent to a $\mathscr{B}$-measurable function for each $M \in \tilde{\Sigma}$, and $\psi(\cdot, \omega)$ is a probability distribution on $\tilde{\Sigma}$ for all $\omega \in \Omega - N$, $P(N) = 0$. In particular, the result holds true if $\tilde{\Omega} = \mathbb{R}^n$ and $\tilde{\Sigma}$ is the σ-algebra of Borel sets.*

The last part of this corollary was proved earlier by Doob [1] when $X(\Omega) \in \tilde{\Sigma}$. In this case, $\psi(\cdot, \cdot)$ has additional (nice) properties. In the counterexample, X is the identity, $\mathscr{B}$ is the Borel σ-algebra of $\Omega = [0, 1]$, and $\Sigma = \Sigma_1$ is not Borel but is a large σ-algebra containing the Lebesgue σ-algebra. For the validity, it is important that $\tilde{\Sigma}$ be a Baire (equal Borel here) σ-algebra of the metric space $\tilde{\Omega}$, the range space of the random variables. Another result is given in the Complements section. (See Problem 3 or 10.)

As an application of the above corollary, an alternative proof of the *conditional Jensen inequality* (Theorem 1.6) can be obtained for regular conditional probabilities. This is simply a reduction to the classical scalar case. Thus consider for any continuous convex function $\varphi: \mathbb{R} \to \mathbb{R}$, and a random variable X such that X, $\varphi(X) \in L^1(\Omega, \Sigma, P)$ and $\tilde{\mathscr{B}} \subset \Sigma$; X being a real random variable, let $\tilde{\mathscr{B}}$ and Σ be complete for P. If $\tilde{\Sigma} = \mathscr{B}$, where $\mathscr{B}$ is the Borel σ-algebra of the line $\mathbb{R}$, then, writing $\tilde{P}(\cdot, \omega) = (P^{\mathscr{B}} \circ X^{-1}(\cdot))(\omega): \mathscr{B} \to [0, 1]$ for the image measure, we have

$$\varphi\left(\int_{-\infty}^{\infty} x\tilde{P}(dx, \omega)\right) \leq \int_{-\infty}^{\infty} \varphi(x)\tilde{P}(dx, \omega), \tag{31}$$

for almost all $\omega \in \Omega$. If φ is strictly convex, there is equality in (31) only when $X \in L^1(\tilde{\mathscr{B}})$. The first condition is a consequence of the fact that the integral is defined over the module $L^p(\tilde{\mathscr{B}})$ (and hence every element of $L^p(\tilde{\mathscr{B}})$ is a constant for the integral, and (31) becomes $\varphi(x \cdot 1) = \varphi(x) \cdot 1$); the second condition is the classical one since $\tilde{P}(\cdot, \omega)$ is a probability for almost all $\omega \in \Omega$, by the regularity hypothesis.

It should be noted that the inequality (31) can also be given directly on the original spaces. Then the result takes the form

$$\varphi\left(\int_{\Omega} X\, dP^{\mathscr{B}}\right) \leq \int_{\Omega} \varphi(X)\, dP^{\mathscr{B}} \tag{31'}$$

a.e., where the integrals *must* now be interpreted in the sense of Dunford–Schwartz or of order. To prove this result, we note that it is true for simple functions, as an easy consequence of the definition of convexity, and the general result is then obtained by considering $X_n \to X$ a.e. and in $L^1(\Sigma)$-norm so that the validity of the definition of order (or D–S) integral is verified. The details are straightforward. The equality conditions can again be translated from those of Theorem 1.6. *Thus these results are valid without qualifications when the integrals are interpreted in the Dunford–Schwartz sense* but may be false if the integrals are considered in the abstract Lebesgue sense.

Let us present some sufficient conditions for the existence of conditional distributions because of their interest in applications. We consider the regularity of $P^{\mathscr{B}}(\cdot)$ directly on (Ω, Σ), and not on its image in the topological space $(\tilde{\Omega}, \tilde{\Sigma})$. This will be useful since in probability theory, generally Ω is a point set without topology. Thus a direct treatment will complement the above work.

First we recall the topological definition (cf. I.2.9) of regularity to motivate the nontopological definition. If Ω is a Hausdorff space and Σ is a ring of subsets of Ω and $\mu: \Sigma \to \mathscr{X}$ (a Banach space or the real line) is σ-additive, then μ is Dunford–Schwartz (or D–S) *regular* on Σ if for each $\varepsilon > 0$, $A \in \Sigma$, there exist F, G in Σ satisfying the inclusions ($\bar{F}$ = closure (F)) $\bar{F} \subset A \subset \operatorname{int}(G)$, where F is relatively compact and, for any $C \subset G - F$, C in

Σ, we have $\|\mu(C)\| < \varepsilon$. Similarly μ is *inner regular* if for each $\varepsilon > 0$, $A \in \Sigma$ there is a relatively compact $F \in \Sigma$ such that $\bar{F} \subset A$ and for any $C \subset A - F$, $C \in \Sigma$, we have $\|\mu(C)\| < \varepsilon$; and μ is *outer regular* if there is a $G \in \Sigma$ such that $A \subset \text{int}(G)$ and for any $C \subset G - A$, $C \in \Sigma$ we have $\|\mu(C)\| < \varepsilon$. ($\|\cdot\|$ is the norm in $\mathscr{X}$.) Thus μ is D–S regular iff it is both inner and outer regular.

These concepts are connected as follows (cf. Dinculeanu [1]). All these definitions agree in L^p (or L^φ) spaces.

14. Proposition *Let Ω be a locally compact space and Σ be the σ-ring generated by the compact (or compact G_δ) subsets of Ω (or Baire σ-ring, Halmos [1]). If $\mu: \Sigma \to \mathscr{X}$, a Banach space, is a σ-additive set function, then μ is D–S regular iff it is inner regular, or equivalently, iff it is outer regular on Σ. A sufficient condition for the D–S regularity of μ is that its variation $|\mu|(\cdot)$ be finite on each compact set and it be regular on Σ. Moreover, $\mu: \Sigma \to \mathscr{X}$, with its variation $|\mu|$ finite on compact sets, is inner regular iff $|\mu|$ has the same property on Σ.*

Before proving this result, it will be useful to specialize it to the case of conditional probability functions.

15. Proposition *Let Ω be a locally compact metric space, Σ the Borel σ-algebra of Ω, and P an outer regular probability measure on Σ. Then, for any σ-algebra $\mathscr{B} \subset \Sigma$ with $\hat{\Sigma}$ and $\hat{\mathscr{B}}$ as completions of Σ and $\mathscr{B}$ for P, the conditional probability function $P^{\hat{\mathscr{B}}}: \hat{\Sigma} \to L^p(\hat{\mathscr{B}})$, $1 \le p < \infty$ is regular and hence a conditional distribution $\psi(\cdot, \cdot) = (P^{\hat{\mathscr{B}}}(\cdot))(\cdot)$ exists. Moreover, P itself is regular on Σ.*

Proof We already noted that $P^{\hat{\mathscr{B}}}: \hat{\Sigma} \to L^\infty(\hat{\mathscr{B}}) \subset L^1(\hat{\mathscr{B}})$ is well defined and, taking $\mathscr{X} = L^1(\hat{\mathscr{B}})$, the vector measure $P^{\hat{\mathscr{B}}}(\cdot)$ has finite variation with its variation as P itself (cf. (4)).

The regularity concepts, required in Proposition 14, are for σ-rings. They are given here to be σ-algebras, and by hypothesis P is outer regular. Then by that result $P^{\hat{\mathscr{B}}}(\cdot)$ is also outer regular. On the other hand, by Theorem 11 above [taking the map $X =$ identity there], $P^{\hat{\mathscr{B}}}(\cdot)$ is always inner regular. Thus $P^{\hat{\mathscr{B}}}(\cdot)$ is regular so that a conditional distribution exists since both regularity concepts are seen to be the same.

To prove the last part, again by the preceding proposition, $P^{\hat{\mathscr{B}}}(\cdot)$ is inner regular iff its variation measure $P(\cdot)$ is inner regular. Hence $P(\cdot)$ is also regular on $\hat{\Sigma}$. A direct proof of regularity, without using the above proposition, for the probability measure P is also easy. This completes the proof.

Proof of Proposition 14 Since by definition, the D–S regularity of $\mu: \Sigma \to \mathscr{X}$ implies both the inner and outer regularity, we need to prove the converse implication. Thus let μ be inner regular. Then for any $\varepsilon > 0$, $A \in \Sigma$, there exists a relatively compact $F \in \Sigma$ such that $\bar{F} \subset A$, and for any $C \subset A - F$, $C \in \Sigma$, we have $\|\mu(C)\| < \varepsilon$. Let us show that this implies μ is outer regular so that the result of D–S regularity follows from definition.

Now let $A \in \Sigma$, and let $\bar{A}$ be compact. Since Ω is locally compact, such sets generate Σ. Also there exists a relatively compact open set $U \supset \bar{A}$. [In fact, since relatively compact open sets $\{V_i\}_{i \in I}$ form a neighborhood basis of Ω by local compactness, there exist $\{V_i\}_{i=1}^n$ such that $\bar{A} \subset \bigcup_{i=1}^n V_i = U$ serves our purpose.] Thus $U \in \Sigma$, and so $U - A \in \Sigma$. By inner regularity there exists a compact set $C_0 \subset U - A$ such that for any $B \in \Sigma$ with $B \subset (U - A) - C_0$, $\|\mu(B)\| < \varepsilon$. Let $G = U - C_0 \in \Sigma$. Then G is open and $G = U - C_0 \supset U - (U - A) = A$. Also, $B \subset G - A = (U - C_0) - A = (U - A) - C_0$. Since $B \in \Sigma$ is arbitrary in this last inclusion, we deduce that μ is outer regular.

Next suppose μ is outer regular. To show that it is also inner regular let $A \in \Sigma$ be relatively compact. If $\varepsilon > 0$, then as in the preceding paragraph there is a relatively compact open set U with $\bar{A} \subset U$, having the following properties. If $C = \bar{U}$, then $C \in \Sigma$, $C - A \in \Sigma$. The latter set is relatively compact so that by the outer regularity of μ, there is a relatively compact open set $G \in \Sigma$, $G \supset C - A$ and, for any $B \in \Sigma$ with $B \subset G - (C - A)$, $\|\mu(B)\| < \varepsilon$. Let $C_0 = \bar{A} - G \in \Sigma$. Then C_0 is compact, and since $G \supset (C - A)$, $C_0 \supset A - G$, and $\bar{A} \supset C_0$, we have, for any $B \in \Sigma$ such that $B \subset A - C_0$,

$$B \subset A - C_0 \subset A - (A - G) = A \cap G \subset (G \cap A) \cup (G \cap C^c) = G - (C - A), \tag{32}$$

so that $\|\mu(B)\| < \varepsilon$. Thus μ is inner regular.

The sufficiency condition is immediate since $\|\mu(A)\| \leq |\mu|(A)$ always holds for $A \in \Sigma$.

To prove the last part, since the above inequality clearly shows that the inner regularity of $|\mu|$ implies the same property for μ, suppose conversely that $\mu(\cdot)$ is inner regular. Let $A \in \Sigma$ be relatively compact. By definition of $|\mu|$ (cf. I.4.29) for any $\varepsilon > 0$, since $|\mu|(A) < \infty$, there exists a disjoint collection $\{A_i\}_1^n \subset \Sigma$, $A_i \subset A$ such that

$$|\mu|(A) - \frac{\varepsilon}{2} < \sum_{i=1}^{n} \|\mu(A_i)\|. \tag{33}$$

By the inner regularity of μ there exist $C_i \in \Sigma$, compact, $C_i \subset A_i$, such that $\|\mu(A_i) - \mu(C_i)\| < \varepsilon/2n$. Then $C = \bigcup_{i=1}^n C_i \subset A$, is compact, $C \in \Sigma$, and C_i are disjoint. We have, by (33) and $|\mu|(C) = \sum_{i=1}^n |\mu|(C_i) \geq \sum_{i=1}^n \|\mu(C_i)\|$, that

$$\begin{aligned} 0 \leq |\mu|(A) - |\mu|(C) &\leq \sum_{i=1}^n \|\mu(A_i)\| + \frac{\varepsilon}{2} - \sum_{i=1}^n \|\mu(C_i)\| \\ &= \sum_{i=1}^n [\|\mu(A_i)\| - \|\mu(C_i)\|] + \frac{\varepsilon}{2} \\ &\leq \sum_{i=1}^n \|\mu(A_i) - \mu(C_i)\| + \frac{\varepsilon}{2} \quad \text{by the triangle inequality} \\ &\leq \varepsilon. \end{aligned}$$

It follows that $|\mu|(\cdot)$ is inner regular. This completes the proof of the proposition since the Baire ring case is similar.

We deduce a useful property of vector measures. In the above proof we only used the ring property of Σ and really did not use the algebra property. Thus the proof implies the following result.

16. Corollary *Let Ω be a locally compact space and Σ be a σ-ring containing the open sets of Ω. Then a vector measure $\mu: \Sigma \to \mathscr{X}$, a Banach space, is D–S regular iff it is inner regular. In particular, if the variation measure $|\mu|$ of μ is such that $|\mu|$ is finite on Σ, then μ is D–S regular iff $|\mu|$ is regular.*

The preceding two propositions and the proof of Theorem 11 show how to formulate the regularity concept for probabilities on abstract measure spaces. This can be stated precisely as follows.

17. Definition Let (Ω, Σ) be a measurable space and $\mathscr{C}, \mathscr{G}$ be classes of subsets of Ω (corresponding to the compact and open classes if Ω is a topological space). If $\mu: \Sigma \to \mathbb{R}^+$ is a measure and $\mathscr{A} \subset \Sigma$ is a σ-ring, then μ is *inner regular on $\mathscr{A}$ relative to $\mathscr{C}$* (or *approximable from below on $\mathscr{A}$ by $\mathscr{G}$*) if for any $\varepsilon > 0$, $A \in \mathscr{A}$, there exist $B \in \Sigma$, $C \in \mathscr{C}$ with $B \subset C \subset A$, and $\mu(A) - \mu(B) < \varepsilon$. It is *outer regular on $\mathscr{A}$ relative to $\mathscr{G}$* (or *approximable from above on $\mathscr{A}$ by $\mathscr{G}$*) if for any $\varepsilon > 0$, $A \in \mathscr{A}$, there exist $B \in \Sigma$, $G \in \mathscr{G}$ with $B \supset G \supset A$, and $\mu(B) - \mu(A) < \varepsilon$. If μ is both inner regular relative to $\mathscr{C}$ and outer regular relative to $\mathscr{G}$ on $\mathscr{A}$, then it is said to be *regular* on $\mathscr{A}$ relative to the pair $(\mathscr{C}, \mathscr{G})$. If $\mu: \Sigma \to \mathscr{X}$, a Banach space, is a σ-additive set function, then it is inner (outer) *regular* on $\mathscr{A}$ relative to $\mathscr{C}(\mathscr{G})$ if for any $\varepsilon > 0$, $A \in \mathscr{A}$, there is a $C \in \mathscr{C}$ $(G \in \mathscr{G})$ such that $C \subset A$ $(A \subset G)$, and for every $B \subset A - C$ $(B \subset G - A)$ with $B \in \Sigma$ we have $\|\mu(B)\| < \varepsilon$, $\|\cdot\|$ being the norm of $\mathscr{X}$. If μ is both inner and outer regular on $\mathscr{A}$ relative to $\mathscr{C}$ and $\mathscr{G}$, then it is *regular* on $\mathscr{A}$ relative to the pair $(\mathscr{C}, \mathscr{G})$.

It should be noted that $\mathscr{C}$ or $\mathscr{G}$ need not be contained in Σ. In case $\mathscr{A} = \mathscr{G} = \Sigma$ and $\mathscr{C} \subset \Sigma$, and is a "compact" class, i.e., for any $\{A_i\}_1^\infty \subset \mathscr{C}$ with $\bigcap_{i=1}^k A_i \neq \varnothing$, all $k < \infty$ imply that $\bigcap_{i=1}^\infty A_i \neq \varnothing$, then the above definition for $\mu: \Sigma \mapsto \mathbb{R}^+$ reduces to saying:

$$\mu(A) = \sup\{\mu(C) : C \in \mathscr{C}\}, \qquad A \in \Sigma. \tag{34}$$

In this case μ is termed a "compact measure" in the literature. Since the outer regularity is then trivial, μ is regular in the sense of the above definition. We can also define a vector measure to be *compact* in a similar manner. These definitions were abstracted, from the topological case, by Jiřina [1]. All these concepts agree in suitable spaces, e.g., $\mathscr{X} = L^p$ or $\mathscr{X} = \mathbb{R}^+$.

Since for any vector measure $\mu: \Sigma \to \mathscr{X}$, a Banach space, if $|\mu|(\cdot)$ is its variation measure, we have

$$\|\mu(A)\| \leq |\mu|(A), \qquad A \in \Sigma, \tag{35}$$

the following result is a consequence of the proof of Proposition 15, in the abstract case.

18. Proposition *Let (Ω, Σ, P) be a probability space, and $\mathscr{A} \subset \Sigma$ and $\mathscr{B} \subset \Sigma$ be σ-algebras. Suppose P on $\mathscr{A}$ is regular relative to a pair $(\mathscr{C}, \mathscr{G})$ of subsets of Ω in the sense of Definition* 17. *Then, for $(\mathscr{C}, \mathscr{G})$, the conditional probability function $P^{\mathscr{B}}(\cdot): \mathscr{A} \to L^\infty(\mathscr{B})$ is regular. If $\hat{\mathscr{B}}$, $\hat{\Sigma}$ are completions under P of $\mathscr{B}$ and Σ, then $\psi(\cdot, \cdot) = (P^{\hat{\mathscr{B}}}(\cdot))(\cdot): \mathscr{A} \to L^\infty(\hat{\mathscr{B}})$ is a conditional distribution of $P^{\hat{\mathscr{B}}}$. If, in particular, $\mathscr{A} = \Sigma = \mathscr{G}$ and $\mathscr{C} \subset \Sigma$ is a compact class and P is a compact measure on Σ relative to $\mathscr{C}$, then $P^{\hat{\mathscr{B}}}$ is always regular and its conditional probability distribution exists.*

Since by (4) the inequality (35) always holds for the conditional probability measures when $L^1(\mathscr{B})$ is used for $\mathscr{X}$, the result follows immediately as in Proposition 15. Note that in the absence of a topology in Ω the part on the inner regularity of $P^{\mathscr{B}}$ given in Theorem 11 does not have a natural counterpart here. For this reason it is preferable to give sufficient conditions on P itself and then (35) transfers the result to $P^{\mathscr{B}}(\cdot)$ because of (4).

The above proposition, together with various other versions, was directly proved by Jiřina [1] under the hypothesis that $\mathscr{A}$ be countably generated.

It is clear that when $P^{\mathscr{B}}(\cdot)$ is regular, one may extend the calculus of absolute probabilities to the conditional case. Some are given as problems. We now discuss an application.

Consider the mapping $h: \Omega \to \mathbf{E}$, where (Ω, Σ, P) is a probability space, $(\mathbf{E}, \mathscr{E})$ is a measurable space, and h is $(\Sigma, \mathscr{E})$-measurable. Then $Q = P \cdot h^{-1}: \mathscr{E} \to [0, 1]$ is a probability measure and if $\mathscr{B} = \mathscr{B}_h = h^{-1}(\mathscr{E}) \subset \Sigma$, then for any P-integrable X, the equation

$$\int_B g_X(e)\, dQ(e) = \nu_X(B) = \mu_X \circ h^{-1}(B) = \int_{h^{-1}(B)} X\, dP_{\mathscr{B}_h}, \qquad B \in \mathscr{E}, \tag{36}$$

defines the conditional expectation of X relative to $\mathscr{B}$, i.e.,

$$g_X(e) = E^{\mathscr{B}}(X)(h(\omega) = e). \tag{36a}$$

Thus if $X = \chi_A$, $A \in \Sigma$, let

$$\psi(A, e) = g_A(e) = E^{\mathscr{B}}(\chi_A)(h(\omega) = e). \tag{37}$$

Then $\psi(\cdot)(\cdot) = \psi(\cdot, \cdot): \Sigma \times \mathbf{E} \to [0, 1]$ has the following properties. $\psi(\cdot): \Sigma \to L^\infty(\mathscr{E}) \subset L^p(\mathscr{E})$, $1 \leq p < \infty$ is a vector measure, and for each $A \in \Sigma$, $\psi(A)$ is essentially bounded and measurable on $\mathbf{E}$. In this notation (36) can also be expressed as follows (compare with (7)):

$$\tilde{P}(A \times B) = P(A \cap h^{-1}(B)) = \int_B \psi(A, e)\, dQ(e), \qquad B \in \mathscr{E}. \tag{38}$$

Conversely, let a positive (order) measure $\psi(\cdot): \Sigma \to L^\infty(\mathscr{E})$ be given. Then (38) defines a measure $\tilde{P}$ on the product σ-algebra $\Sigma \otimes \mathscr{E}$ such that the "marginal" measure $\tilde{P}(\Omega, \cdot)$ is $\{Q(B), B \in \mathscr{E}\}$. This leads to an extension of the Fubini theorem when $\psi(\cdot)$ is regular on Σ relative to some pair $(\mathscr{C}, \mathscr{G})$. We have the following more precise statement.

19. Proposition *Let (Ω, Σ) and $(\mathbf{E}, \mathscr{E})$ be measurable spaces and let $Q: \mathscr{E} \to [0, 1]$ be a probability measure and $\psi(\cdot): \Sigma \to L^\infty(\mathscr{E}) \subset L^p(\mathscr{E})$, $1 \leq p < \infty$ be σ-additive such that $0 \leq \psi(A) \leq 1$, and $\psi(\Omega) = 1$ a.e. (Q) and has finite variation when ψ is considered as a measure into $L^1(\mathscr{E})$. Then there exists a unique probability measure P on the product measurable space $(\Omega \times \mathbf{E}, \Sigma \otimes \mathscr{E})$ whose marginal measure is Q and which satisfies the functional equation*:

$$P(A \times B) = \int_B \psi(A)(e)\, dQ(e), \qquad A \in \Sigma, \quad B \in \mathscr{E} \quad \text{(Lebesgue integral)}. \tag{39}$$

Further, if $f: \Omega \times \mathbf{E} \to \mathbb{R}$ *is* $(\Sigma \otimes \mathscr{E})$*-measurable and integrable either (i) relative to the product measure* $|\psi| \otimes Q$, *where* $|\psi|$ *is the variation measure of* ψ, *or (ii) relative to P, and* $\psi(\cdot)$ *is regular for some pair* $(\mathscr{C}, \mathscr{G})$, *then*

$$g(e) = \int_{\Omega} f(\omega, e)\psi(d\omega)(e) \tag{40}$$

is defined (a.e. Q) on $\mathbf{E}$ *and is integrable relative to Q. Moreover,*

$$\int_{\Omega \times \mathbf{E}} f\, dP = \int_{\mathbf{E}} \left[\int_{\Omega} f(\omega, e)\psi(d\omega)(e) \right] Q(de) \tag{41}$$

holds true where in case (i) the inner integral is defined in the D–S sense and the outer integral as well as the left integral are in the abstract Lebesgue sense, and in case (ii) all are Lebesgue integrals.

The proof of this result follows from a simple modification of the standard method in real analysis (cf., e.g., Royden [1, p. 265 ff]) and is left to the reader.

Before closing this section, we should note an intimate relation between the class of arbitrary conditional probability functions and the class of regular ones. There is an isometric order-preserving isomorphism between these two classes, as given by the following result which is a consequence of a fundamental isomorphism theorem (cf. Theorem I.4.8).

20. Theorem *Let* (Ω, Σ, P) *be a complete probability space and* $\mathscr{B} \subset \Sigma$ *be a* σ*-algebra. Consider the conditional probability function* $P^{\mathscr{B}}: \Sigma \to L^{\infty}(\mathscr{B})$. *Then there exists a compact Hausdorff space S, which is totally disconnected (i.e., the Stone space), and a regular measure* ν *on the Borel* σ*-algebra* $\mathscr{S}$ *of S to* $L^{\infty}(\mathscr{B})$ *with the following properties: If* $L^p(\mathscr{S})$ *is the Lebesgue space on* $(S, \mathscr{S}, |\nu|)$ *and* $L^p(\Sigma)$ *is the given one,* $p \geq 1$, *they are isometrically and lattice isomorphic. Moreover, if* $f \in L^p(\Sigma)$ *and* $\hat{f} \in L^p(\mathscr{S})$ *correspond to each other, then* $\int_{\Omega} f\, dP^{\mathscr{B}} = \int_S \hat{f}\, d\nu$, *where the integrals are understood in the D–S sense. Hence* ν *can be expressed as* $\nu = \hat{P}^{\mathscr{B}}: \mathscr{S} \to L^{\infty}(\mathscr{B})$. [*The* ν*-integral is also a Lebesgue integral.*]

This result is a consequence of the isomorphism theorem noted already. We shall sketch its proof for convenience.

Since Σ is a σ-algebra, (Ω, Σ) is algebraically isomorphic to $(S, \mathscr{S}_0)$, where S is a totally disconnected compact Hausdorff space and $\mathscr{S}_0$ is the algebra of its closed–open subsets of S. If $\tau: A \mapsto \hat{A} \in \mathscr{S}_0$ is the isomorphism, then define ν on $\mathscr{S}_0$ to $L^{\infty}(\mathscr{B})$ by the equation $P^{\mathscr{B}}(A) = \nu(\hat{A})$. Let $\mathscr{S} = \sigma(\mathscr{S}_0)$. Since the mapping τ preserves finite algebraic operations, it follows that $\nu(\cdot)$ is an additive function on $\mathscr{S}_0$, and since P is the variation of $P^{\mathscr{B}}$ and if $|\nu|(\cdot)$ is that of ν, it can be shown without much difficulty that $P(A) = |\nu|(\hat{A})$, i.e., $P = |\nu| \circ \tau^{-1}$. But $|\nu|(\cdot)$ is a Baire regular measure on $\mathscr{S}_0$ and so is σ-additive (Theorem I.2.10). Then $|\nu|$ and ν have unique σ-additive extensions to $\mathscr{S}$ (by Theorem I.2.2). With this we have that for $f = \sum_{i=1}^{n} a_i \chi_{A_i}$ and $\hat{f} = \sum_{i=1}^{n} a_i \chi_{\hat{A}_i}$ as the corresponding simple functions,

$$\int_{\Omega} f \, dP^{\mathscr{B}} = \sum_{i=1}^{n} a_i P^{\mathscr{B}}(A_i) = \sum_{i=1}^{n} a_i \nu(\hat{A}_i) = \int_{S} \hat{f} \, d\nu. \tag{42}$$

On the other hand, we already observed that the monotone convergence theorem holds for these integrals, so that the result (42) is true for positive f and then for any f that is integrable relative to $P^{\mathscr{B}}$ by Proposition 2. Since the same statement holds if $P^{\mathscr{B}}$ and ν are replaced by their variations above (then the integrals are the abstract Lebesgue integrals), the isometry of $L^p(\Sigma)$ and $L^p(\mathscr{S})$ is an easy deduction. Consequently, all the statements of the theorem hold as given. Since ν is regular, by Proposition 18 the integral relative to ν can also be defined pointwise in the Lebesgue sense, and the two functions thus obtained in (42) are equal a.e. (P), as elements of $L^1(\mathscr{B})$.

Application Jensen's inequality (31) or (31′) is true for any conditional probability function and equality holds there iff the conditions discussed under (31) are fulfilled. In fact, by the above theorem, this follows on $(S, \mathscr{S}, \hat{P}^{\mathscr{B}})$ by the scalar case, and then the map τ gives the result on $(\Omega, \Sigma, P^{\mathscr{B}})$. Further details are unnecessary.

Remark Because of this theorem, even though in general conditional probability distributions need not exist on the given measure space, it is always true that every such conditional probability function determines uniquely (up to isomorphism) a Baire and then Borel *regular* conditional probability and hence a conditional distribution on some "nice" measurable space $(S, \mathscr{S})$. Even when this space cannot be used for direct and concrete applications, it is quite useful for the structure theory, as in the case of the above application and elsewhere. Essentially it tells us that conditional distributions exist in abundance.

2.4 SOME CHARACTERIZATIONS OF CONDITIONAL PROBABILITIES

To gain further insight into conditional probability functions, complementing the work of the preceding section, we present a few characterizations of these from a class of vector measures. The first result is due to

Olson [1], and we give a different proof basing it on the work of Section 2—in particular on Theorems 2.6, 2.7, and 2.8. The other results are taken from the author's paper [5].

1. Theorem *Let* (Ω, Σ, P) *be a probability space and* $v: \Sigma \to L^\infty(\Omega, \Sigma, P)$ *be* σ*-additive pointwise a.e. Then* v *is a conditional probability function relative to a* σ*-algebra* $\mathscr{B} \subset \Sigma$ *iff the following three conditions hold:* (*a*) $0 \le v(A) \le 1$ *a.e.,* $A \in \Sigma$, (*b*) $v(\Omega) = 1$ *a.e., and* (*c*) *for all* A, B, C *in* Σ, v *satisfies the functional equation:*

$$\int_A v(B)v(C)\,dP = \int_B v(A)v(C)\,dP = \int_C v(A)v(B)\,dP. \tag{1}$$

The σ*-algebra* $\mathscr{B}$ *is given by* $\mathscr{B} = \{A \in \Sigma : v(A) = \chi_A \text{ a.e.}\}$.

Proof For the necessity, taking $f = \chi_A$, $g = \chi_B$ in the averaging property of conditional expectations: $E^{\mathscr{B}}(fE^{\mathscr{B}}(g)) = E^{\mathscr{B}}(f) \cdot E^{\mathscr{B}}(g)$ (cf. Proposition 1.2(a)), we get, since $E^{\mathscr{B}}(\chi_A) = P^{\mathscr{B}}(A)$, for $C \in \mathscr{B}$,

$$\begin{aligned}\int_C P^{\mathscr{B}}(A)P^{\mathscr{B}}(B)\,dP_{\mathscr{B}} &= \int_C E^{\mathscr{B}}(\chi_A P^{\mathscr{B}}(B))\,dP_{\mathscr{B}} \\ &= \int_C \chi_A P^{\mathscr{B}}(B)\,dP_{\mathscr{B}} \\ &= \int_A P^{\mathscr{B}}(C)P^{\mathscr{B}}(B)\,dP_{\mathscr{B}} \qquad \text{since} \quad P^{\mathscr{B}}(C) = \chi_C.\end{aligned} \tag{2}$$

Similarly,

$$\int_C P^{\mathscr{B}}(A)P^{\mathscr{B}}(B)\,dP_{\mathscr{B}} = \int_B P^{\mathscr{B}}(C)P^{\mathscr{B}}(A)\,dP_{\mathscr{B}}. \tag{3}$$

Combining (2) and (3), we get (1) for all $\{A, B\} \subset \Sigma$ and $C \in \mathscr{B}$. But observe that if $C \in \Sigma$ is arbitrary, then

$$\int_C P^{\mathscr{B}}(A)P^{\mathscr{B}}(B)\,dP = \int_\Omega P^{\mathscr{B}}(A)P^{\mathscr{B}}(B)P^{\mathscr{B}}(C)\,dP_{\mathscr{B}}. \tag{4}$$

By symmetry (3) or (4) holds for all A, B, C in Σ. We thus see that the identity (1) is a consequence of the averaging property of conditional expectations, and perhaps this is how (1) is discovered.

Conversely, let a pointwise σ-additive v, $v: \Sigma \to L^\infty(\Sigma)$, satisfy (a)–(c) of the theorem. First note that taking $B = C = \Omega$, we get

$$\|v(A)\|_1 = \int_\Omega v(A)\,dP = P(A), \qquad A \in \Sigma. \tag{5}$$

But this equation implies that the variation of v is P, just as for $P^{\mathscr{B}}$.

Now define $T: L^1(\Sigma) \to L^1(\Sigma)$, using ν, as $Tf = \int_\Omega f\, d\nu$, where the integral is in the Dunford–Schwartz sense, first for simple functions and then for all $f \in L^1(\Sigma)$ $(= L^1(\Omega, \Sigma, P))$. In fact if $f = \sum_{i=1}^n a_i \chi_{A_i}$, $A_i \in \Sigma$, then

$$\begin{aligned} \|Tf\|_1 &= \left\| \int_\Omega f\, d\nu \right\|_1 \le \sum_{i=1}^n |a_i| \, \|\nu(A_i)\|_1 \le \sum_{i=1}^n |a_i| \, |\nu|(A_i) \\ &= \sum_{i=1}^n |a_i| P(A_i) \qquad \text{by} \quad (5) \\ &= \int_\Omega |f| \, dP = \|f\|_1. \end{aligned} \tag{6}$$

Thus, from the density of simple functions in $L^1(\Sigma)$, we deduce that the above inequality holds for all $f \in L^1(\Sigma)$ and that T is a contraction. Since $\nu(A) \ge 0$ a.e., T is also a positive operator. Moreover, $\nu(\Omega) = 1$ implies $T1 = \int_\Omega d\nu = 1$. Since $0 \le \nu(A) \le 1$ a.e., we deduce from the definition that $\|Tf\|_\infty \le \max_i |a_i| = \|f\|_\infty$ and hence that $T: L^1 \cap L^\infty \to L^1 \cap L^\infty = L^\infty$. Thus $T(L^\infty) \subset L^\infty(\Sigma)$.

To prove the averaging identity for T, so that with the preceding paragraph and Corollary 2.7, $T = E^{\mathscr{B}}$, we proceed as follows. Equations (1) can be written

$$\int_\Omega \chi_A (T\chi_B)(T\chi_C)\, dP = \int_\Omega \chi_B (T\chi_A)(T\chi_C)\, dP = \int_\Omega \chi_C (T\chi_A)(T\chi_B)\, dP. \tag{7}$$

If we fix B and C in (7), the same equation is true by linearity, when χ_A is replaced by any simple f in $L^1(\Sigma)$. Thus

$$\int_\Omega f (T\chi_B)(T\chi_C)\, dP = \int_\Omega (Tf)\chi_B (T\chi_C)\, dP = \int_\Omega (Tf)(T\chi_B)\chi_C\, dP. \tag{8}$$

Similarly, with f and C fixed, we see that the above equation is true if χ_B is replaced by a simple function g, and χ_C is then replaced by a simple function h, from $L^1(\Sigma)$. Hence (8) becomes

$$\int_\Omega f(Tg)(Th)\, dP = \int_\Omega (Tf)g(Th)\, dP = \int_\Omega (Tf)(Tg)h\, dP. \tag{9}$$

That this equation is well defined for all $f, g, h \in L^\infty(\Sigma)$ is a consequence of the fact that Tf is bounded for bounded f, which we noted after (6). Thus (9) yields a pair of equations:

$$\int_\Omega T^*(f(Tg))h\, dP = \int_\Omega (Tf)(Tg)h\, dP, \tag{10}$$

where T^* is the adjoint of T, and similarly

$$\int_\Omega T^*(f(Tg))h\,dP = \int_\Omega T^*((Tf)g)h\,dP. \tag{11}$$

Since h is arbitrary in $L^\infty(\Sigma)$ and this is a dense subspace of $L^1(\Sigma)$, we can identify the integrands:

(i) $T^*(f(Tg)) = (Tf)(Tg)$ a.e.
(ii) $T^*(f(Tg)) = T^*(g(Tf))$ a.e.

Taking $g = 1$ and noting that $T1 = 1$ a.e., we get from (i) that $T^*f = Tf$ for all bounded f. Since $f(Tg)$ is bounded for $f, g \in L^\infty(\Sigma)$, we get from (i) again, with the just established fact, that $T(f(Tg)) = (Tf)(Tg)$ for all $f, g \in L^\infty(\Sigma)$. But T is continuous in the L^1-norm. So this holds on L^1 for f or g bounded, and hence T satisfies the hypothesis of Corollary 2.7, and hence $T = E^{\mathscr{B}}$, for a unique σ-algebra $\mathscr{B} \subset \Sigma$. It is then trivial that $\mathscr{B}$ is as given in the theorem. Thus $\nu(A) = T\chi_A = E^{\mathscr{B}}(\chi_A)$, $A \in \Sigma$, completing the proof.

The preceding proof suggests other characterizations. Thus we have the following result.

2. Theorem *Let (Ω, Σ, P) be a probability space. Then a pointwise a.e. σ-additive set function $\nu: \Sigma \to L^\infty(\Sigma)$ is a conditional probability function $P^{\mathscr{B}}(\cdot)$ relative to a unique σ-algebra $\mathscr{B} \subset \Sigma$ iff (i) $\nu(\Omega) = 1$ a.e., (ii) $|\nu|(A) \leq P(A)$, $A \in \Sigma$, where $|\nu|(\cdot)$ is the variation measure of ν, and (iii) for any $A \in \Sigma$ we have the functional equation:*

$$\nu(A)(\cdot) = \int_\Omega \nu(A)(\omega)\nu(d\omega, \cdot). \tag{12}$$

The integral is to be understood in the Dunford–Schwartz sense (to integrate the scalar function $\nu(A)(\cdot)$ relative to the vector measure $\nu(\cdot)$). The σ-algebra $\mathscr{B}$ is given by $\mathscr{B} = \{A \in \Sigma : \nu(A) = \chi_A \text{ a.e.}\}$.

Proof If $\nu = P^{\mathscr{B}}$, the conditional probability function, then (i), (ii) are true by Propositions 3.1 and 3.2, and (iii) follows from the projection property of the conditional expectation. To see the latter, let $A \in \Sigma$. Then

$$P^{\mathscr{B}}(A) = E^{\mathscr{B}}(\chi_A) = E^{\mathscr{B}}(E^{\mathscr{B}}(\chi_A)) = \int_\Omega E^{\mathscr{B}}(\chi_A)\,dP^{\mathscr{B}} \qquad \text{by Proposition 3.2}$$

$$= \int_\Omega P^{\mathscr{B}}(A)(\omega)\,dP^{\mathscr{B}}(\omega) \qquad \text{by definition of } P^{\mathscr{B}}.$$

Thus (12) holds, and only the converse part is nontrivial.

So for this part, let $\nu: \Sigma \to L^\infty(\Sigma)$ be a vector valued function as given in the theorem. Then we define a linear mapping $T: L^1(\Sigma) \to L^1(\Sigma)$, as in (5), by the equation

$$Tf = \int_\Omega f \, d\nu, \tag{13}$$

where the integral is taken in the sense of Dunford–Schwartz as before. By (i) and (ii), the integral is well defined for all $f \in L^1(\Sigma)$. We assert that T is a contraction since T is clearly a linear operator and by (i) $T1 = 1$ a.e.

Thus let f be a simple function, $f = \sum_{i=1}^n a_i \chi_{A_i}$, $A_i \in \Sigma$, and disjoint. Then as in (6),

$$\begin{aligned} \|Tf\|_1 &= \left\| \int_\Omega f \, d\nu \right\|_1 \le \sum_{i=1}^n |a_i| |\nu|(A_i) \le \sum_{i=1}^n |a_i| P(A_i) \\ &= \int_\Omega |f| \, dP = \|f\|_1 \qquad \text{by (ii).} \end{aligned} \tag{14}$$

Thus (14) implies that T is a contraction on $L^1(\Sigma)$. As yet, (iii) was not involved. Using this, we show that $T^2 = T$, i.e., T of (13) is a contractive projection operator on $L^1(\Sigma)$ satisfying $T1 = 1$.

Again from the density of simple functions in $L^1(\Sigma)$, it suffices to show that $T^2 f = Tf$, for all simple f. So let $f = \sum_{i=1}^n a_i \chi_{A_i}$, as in the preceding paragraph. Then

$$\begin{aligned} (T^2 f)(\omega') &= \int_\Omega (Tf)(\omega) \nu(d\omega, \omega') \\ &= \int_\Omega \left[\int_\Omega f(\omega'') \nu(d\omega'', \omega) \right] \nu(d\omega, \omega') \\ &= \int_\Omega \left[\sum_{i=1}^n a_i \nu(A_i, \omega) \right] \nu(d\omega, \omega') \\ &= \sum_{i=1}^n a_i \int_\Omega \nu(A_i, \omega) \nu(d\omega, \omega') \qquad \text{the integral being linear} \\ &= \sum_{i=1}^n a_i \nu(A_i, \omega') \qquad \text{by (iii)} \\ &= \int_\Omega f(\omega) \nu(d\omega, \omega') = (Tf)(\omega'). \end{aligned}$$

Hence $T^2 f = Tf$ for all simple functions of $L^1(\Sigma)$. So $T^2 = T$.

We have shown that $T: L^1(\Sigma) \to L^1(\Sigma)$ given by (13) satisfies the hypothesis of Theorem 2.6, and hence there exists uniquely a σ-algebra $\mathscr{B} \subset \Sigma$, such that $Tf = E^{\mathscr{B}}(f), f \in L^1(\Sigma)$. Now taking $f = \chi_A$, $A \in \Sigma$, we get from (13)

$$\nu(A) = T\chi_A = E^{\mathscr{B}}(\chi_A) = P^{\mathscr{B}}(A). \tag{15}$$

Since A in Σ is arbitrary, it follows that $\nu = P^{\mathscr{B}}$, and then $\mathscr{B} = \{A \in \Sigma : \nu(A) = P^{\mathscr{B}}(A) = \chi_A \text{ a.e.}\}$. This completes the proof of the theorem.

It is now clear that various other characterizations of the conditional expectations of Section 2 can be used in giving conditions on the vector measures, in order that they be conditional probability functions. Such a result, corresponding to the Šidák identity, will be stated without proof.

3. Proposition *Let* (Ω, Σ, P) *be a probability space and* $\nu: \Sigma \to L^\infty(\Sigma)$ *be pointwise a.e.* σ*-additive. Then* ν *is a conditional probability function* $P^{\mathscr{B}}$ *relative to some* σ*-algebra* $\mathscr{B} \subset \Sigma$ *iff* (*i*) $\nu(\Omega) = 1$, *and* $|\nu|(A) \le P(A)$, $A \in \Sigma$, *where* $|\nu|(\cdot)$ *is the variation measure of* ν, *and* (*ii*) *for any* A, B, C *in* Σ *we have the functional equation*:

$$\int_\Omega \chi_A(\nu(B) \vee \nu(C))\, dP = \int_\Omega \nu(A)(\nu(B) \vee \nu(C))\, dP. \tag{16}$$

The integrals are taken in the sense of Lebesgue. The σ*-algebra* $\mathscr{B}$ *is given by* $\mathscr{B} = \{A \in \Sigma \colon \nu(A) = \chi_A \text{ a.e.}\}$.

The difficulties encountered with the conditional probability measures elaborated and analyzed in this and the preceding sections are clearly due to the nonregularity of these measures. After this study, it is legitimate to ask whether it is not possible to define, axiomatically or otherwise, conditional probabilities so as to build in the regularity and "simplify" the treatment. This point has been elaborated by Rényi [1]. We now discuss his approach and compare it with our treatment, noting that everything above holds also for localizable measures, as shown by the author [2, 5].

2.5 RELATIONS WITH RÉNYI'S NEW AXIOMATIC APPROACH

The basis of the whole work in the preceding sections is a simple formula for the conditional probability of B given A, $P(B|A) = P(A \cap B)/P(A)$, whenever $P(A) > 0$. We noted that this result was not adequate for many problems, and this led to the general theory of the preceding sections. However, the above formula is based on the ratio of two measures. This suggests a theory of conditional probabilities, perhaps formulated axiomatically, so as to coincide with the simple model. This point of view has been pursued and developed by Rényi [1]. In this section we include a discussion of his approach for comparison and illustration.

Let (Ω, Σ) be a measurable space, and let $\mathscr{B} \subset \Sigma$ be a nonempty class. Suppose that a function $P(\cdot|\cdot)\colon \Sigma \times \mathscr{B} \to \mathbb{R}^+$ is given, satisfying the following three axioms.

Axiom I For any $A \in \Sigma$, $B \in \mathscr{B}$, $0 \le P(A|B) \le 1$ and $P(B|B) = 1$.

Axiom II For each $B \in \mathscr{B}$, $P(\cdot|B)$ is σ-additive, i.e., it is a measure.

Axiom III (a) For each $A \in \Sigma$, $B \in \mathscr{B}$, $P(A|B) = P(A \cap B|B)$, and (b) for each $A \in \Sigma$, $B \in \mathscr{B}$, $C \in \mathscr{B}$, with $A \subset B \subset C$,

$$P(A|B)P(B|C) = P(A|C). \tag{1}$$

Then $P(\cdot|\cdot)$ is called a *conditional probability function, in the sense of Rényi.* The last qualification will be omitted since only this function is discussed in this section.

The first observation to be made is that if $\mathscr{B} = \{\Omega\}$, in Axioms I and II, we have the ordinary (absolute) probability function on Σ. Axiom III, which distinguishes this approach, can be given the following two equivalent forms which are useful in proofs and other discussions.

1. Proposition *For* $P(\cdot|\cdot)\colon \Sigma \times \mathscr{B} \to [0, 1]$, *satisfying Axioms I and II, Axiom III is equivalent to either one of the following*:

Axiom III′ $A \in \Sigma$, $B \in \mathscr{B}$, $C \in \mathscr{B}$, $B \cap C \in \mathscr{B}$ *implies*

$$P(A|B \cap C)P(B|C) = P(A \cap B|C). \tag{2}$$

Axiom III″ $A \in \Sigma$, $B \in \mathscr{B}$, $C \in \mathscr{B}$, $B \subset C$ *implies*

$$P(A|B)P(B|C) = P(A \cap B|C). \tag{3}$$

Proof We prove that III$\Leftrightarrow$III′$\Leftrightarrow$III″. So let III be given. Then for $A \in \Sigma$, $\{B, C\} \subset \mathscr{B}$ such that $B \cap C \in \mathscr{B}$, we have $A \cap B \cap C \subset B \cap C \subset C$ and hence

$$\begin{aligned} P(A|B \cap C)P(B|C) &= P(A \cap B \cap C|B \cap C)P(B \cap C|C) \quad \text{by III(a)} \\ &= P(A \cap B \cap C|C) \quad \text{by (1)} \\ &= P(A \cap B|C) \quad \text{by III(a)}, \end{aligned}$$

so that (2) is true, and III′ holds.

Conversely, let III′ hold. Taking $B = C \in \mathscr{B}$ in (2), we get for any $A \in \Sigma$,

$$P(A|B)P(B|B) = P(A \cap B|B).$$

Since $P(B|B) = 1$, this yields III(a). Now let $A \subset B \subset C$ and $A \in \Sigma$, $B \in \mathscr{B}$, $C \in \mathscr{B}$. Then (2) reduces to III(b). Thus III′ implies III, and so these two are equivalent.

If $B \subset C$ is set in (2), then clearly it reduces to (3) so that III′ implies III″ trivially. Conversely, to get III′ from III″, let $A \in \Sigma$, $\{B, C\} \subset \mathscr{B}$, and $B \cap C \in \mathscr{B}$.

If $P(B|C) = 0$, then from the facts that $P(\cdot|C)\colon \Sigma \to [0, 1]$ is a measure so that it is monotone and $A \cap B \in \Sigma$, we have $0 \le P(A \cap B|C) \le P(B|C) = 0$. Hence $0 = P(A \cap B|C) = P(A|B \cap C)P(B|C)$, i.e., (2) is true. So let $P(B|C) > 0$. Taking $B = C$, we get from (3) and Axiom I, $P(A|B) = P(A \cap B|B)$ which is III(a). Hence, from the fact that $B \cap C \subset C$, the preceding equation applied to (3) yields

$$P(A|B \cap C)P(B \cap C|C) = P(A \cap B \cap C|C). \tag{4}$$

This implies (1) and hence, by the first part, III′. But we can also deduce (2) independently. In fact, $P(B \cap C|C) = P(B|C)$ and $P(A \cap B \cap C|C) = P(A \cap B|C)$ by the already established equation. This gives, from (4), Eq. (2) at once. Hence III″ implies III′. This completes the proof.

The main point of these axioms is, in the light of the work of Sections 3 and 4, to introduce axiomatically the "regular" conditional probability functions so that $P(\cdot|B)$ for each $B \in \mathscr{B}$ is a probability measure and the Lebesgue–Stieltjes integration applies. However, $P(\cdot|B)$ being a finite measure, we must have $P(\varnothing|B) = 0$ for each $B \in \mathscr{B}$. Since by Axiom I, $P(B|B) = 1$, it follows that $\varnothing \notin \mathscr{B}$. But one may think of adjoining $\varnothing$ to $\mathscr{B}$ and modifying Axiom I by saying that $P(B|B) = 1$ for all $B \in \mathscr{B}$ if $B \neq \varnothing$, and $P(\varnothing|\varnothing)$ is assigned a particular value to agree with the special case $P(A|B) = P(A \cap B)/P(B)$. This will immediately introduce the exceptional sets into the theory, however. The latter sets are, in fact, the source of most trouble in the work of the preceding sections. So we take the point of view that $\varnothing \notin \mathscr{B}$ and discuss the theory, calling $\{\Omega, \Sigma, \mathscr{B}, P(\cdot|\cdot)\}$ a *conditional probability space*.

Some immediate properties of $P(\cdot|\cdot)$ are listed below.

2. Proposition (a) *$P(A_n|B)\uparrow$ or $\downarrow$ according as $A_n\uparrow$ or $\downarrow$ for each $B \in \mathscr{B}$, and $P(A|B_n)\uparrow$ if $B_n\downarrow$ and $B_n \supset A$.*

(b) *If $\{A_n\}_1^\infty \subset \Sigma$ and $\{B_n\}_1^\infty \subset \mathscr{B}$ are monotone sequences, and $B = \lim_n B_n$ is in $\mathscr{B}$, then for each $A_0 \in \Sigma$ and $B_0 \in \mathscr{B}$ we have*

(i) $\lim_n P(A_n|B_0) = P(\lim_n A_n|B_0)$,

(ii) $\lim_n P(A_0|B_n) = P(A_0|\lim_n B_n)$ *if $B_n\uparrow$, and the same holds true for decreasing $\{B_n\}_1^\infty$ if there is an $\Omega_0 \in \mathscr{B}$ such that $P(B|\Omega_0) > 0$ and $B_n \subset \Omega_0$ for some n in addition.*

(c) *If* $\{B_n\}_1^\infty \subset \mathscr{B}$ *is a disjoint sequence and* $B = \bigcup_n B_n$, *then for any* $C \in \mathscr{B}$ *with* $C \subset B$ *and* $B_k \cap C \in \mathscr{B}$ *we have*

$$P(A \mid C) = \sum_{k=1}^{\infty} P(A \mid B_k \cap C) P(B_k \mid C). \tag{5}$$

Proof (a) Since $P(\cdot \mid B)$ is a measure, it is clearly monotone. If $B_n \downarrow$ and $B_n \supset A$, then

$$\begin{aligned} P(A \mid B_{n+1}) &= P(A \cap B_n \mid B_{n+1}) \qquad \text{since} \quad B_n \supset A \\ &= P(A \mid B_n \cap B_{n+1}) P(B_n \mid B_{n+1}) \qquad \text{by Axiom III}' \\ &\le P(A \mid B_n) P(B_{n+1} \mid B_{n+1}) \qquad \text{since} \quad B_{n+1} \supset B_n \\ &= P(A \mid B_n) \qquad \text{by Axiom I.} \end{aligned}$$

This proves (a).

(b) (i) is immediate since $P(\cdot \mid B_0)$ is a bounded measure. (ii) Let $B_n \uparrow B \in \mathscr{B}$. Then $P(B_n \mid B) \to 1$ as $n \to \infty$. So $P(B_n \mid B) > 0$ for all large enough n. Then by Axiom III′, taking $C = B$ there,

$$\begin{aligned} P(A \mid B_n) = P(A \mid B_n \cap B) &= \frac{P(A \cap B_n \mid \Omega_0)}{P(B_n \mid \Omega_0)} \to \frac{P(A \cap B \mid \Omega_0)}{P(B \mid \Omega_0)} \qquad \text{as} \quad n \to \infty \\ &= P(A \mid B \cap \Omega_0) = P(A \mid B). \end{aligned}$$

Thus (ii) is true, and so (b) is proved.

(c) This is also a consequence of Axiom III. In fact,

$$\begin{aligned} P(A \mid B_k \cap C) P(B_k \mid C) &= P(A \mid B_k \cap C) P(B_k \cap C \mid C) \qquad \text{by III(a)} \\ &= P(A \cap B_k \cap C \mid C) \qquad \text{by (2).} \end{aligned}$$

Hence

$$\begin{aligned} \sum_{k=1}^{\infty} P(A \mid B_k \cap C) P(B_k \mid C) &= P(A \cap B \cap C \mid C) \qquad \text{by (i) above} \\ &= P(A \cap C \mid C) = P(A \mid C) \qquad \text{since} \quad C \subset B. \end{aligned}$$

This completes the proof.

Remark Part (b) is the continuity property of $P(\cdot \mid \cdot)$, and part (c) may be considered as a "disintegration" property.

Since the basic motivation for $P(\cdot \mid \cdot)$ is that it is a ratio of two (probability) measures, we can ask for conditions on the (axiomatically defined) conditional probability function $P(\cdot \mid \cdot)$ in order that it may admit of a representation as a ratio of two finite measures. Such conditions are given in the next result.

3. Proposition *Let* $\{\Omega, \Sigma, \mathscr{B}, P(\cdot|\cdot)\}$ *be a conditional probability space. Suppose there exists a "covering" sequence* $\{B_n\}_1^\infty \subset \mathscr{B}$ *such that: (i)* $B_n \subset B_{n+1}$, $n \geq 0$, *(ii)* $P(B_0|B_n) > 0$, $n \geq 1$, *(iii) for any* $B \in \mathscr{B}$, *there is a* $B_n \in \{B_n\}_1^\infty$ *such that* $B \subset B_n$ *and* $P(B|B_n) > 0$. *If* $\Sigma_0 = \{A \in \Sigma$: *there is a* $B_n \in \{B_n\}_1^\infty$, $A \subset B_n\}$ *so that* $\mathscr{B} \subset \Sigma_0$, *suppose that* Σ_0 *is a ring. Then there exists a measure* $Q: \Sigma_0 \to \mathbb{R}^+$ *such that (a)* $Q: \mathscr{B} \to (0, \infty)$, *(b)* $P(A|B) = Q(A \cap B)/Q(B)$, $A \in \Sigma$, $B \in \mathscr{B}$. *If moreover (ii) is strengthened to (ii')* $P(B_0|B_n) \geq \varepsilon > 0$ *for some* $\varepsilon > 0$ *and all* $n \geq 1$, *then we can take* $\Sigma_0 = \Sigma$ *in the above. In this case, writing* $P = P(\cdot|\Omega)$, *we may consider the given conditional probability space as one obtained from* (Ω, Σ, P) *with* $P(A|B) = P(A \cap B)/P(B)$.

We omit the proof of this theorem, which is obtained in a straightforward way (lengthy computations) from the axioms and their consequences noted above. It is also possible to give a characterization of the above probability (ratio) representation in terms of a family of measures $\{Q_\alpha\}_{\alpha \in I}$ on Σ. This was proved by Császár [1]. The interest here is that the Q_α can be allowed to be infinite measures. This is similar to the Radon–Nikodým theorem for σ-finite measures. Leaving this point aside, we consider the (infinite) *product conditional probabilities* to understand another aspect of this theory.

Let $\{\Omega_i, \Sigma_i, \mathscr{B}_i, P_i(\cdot|\cdot)\}$, $i = 1, 2$, be a pair of conditional probability spaces. Then their product is defined as follows. Let $\Omega = \Omega_1 \times \Omega_2$, $\Sigma = \Sigma_1 \otimes \Sigma_2$ be the product σ-algebra, and $\mathscr{B} = \{B_1 \times B_2 : B_1 \in \mathscr{B}_1, B_2 \in \mathscr{B}_2\} \subset \Sigma$. Define, for any $A = A_1 \times A_2 \in \Sigma$, $B = B_1 \times B_2 \in \mathscr{B}$, the function $P(\cdot|\cdot)$ by the equation

$$P(A|B) = P_1(A_1|B_1)P_2(A_2|B_2). \tag{6}$$

Then $P(\cdot|B)$ has a unique extension to be a measure on Σ, satisfying (6) for each $B \in \mathscr{B}$, and is called the *product conditional probability measure*. If we have a finite number of spaces instead of two, then the result clearly extends. To see that $P(\cdot|\cdot)$ is a conditional probability, we have to check the validity of Axioms I–III. But this is not difficult, and we leave it to the reader. However, we shall add details for the infinite product case, from which the above also follows.

Let $\{(\Omega_i, \Sigma_i, \mathscr{B}_i, P_i(\cdot|\cdot)), i \geq 1\}$ be an infinite family of conditional probability spaces, and let $(\Omega, \Sigma) = \times_{i=1}^\infty(\Omega_i, \Sigma_i)$ be their usual cartesian product. If $\mathscr{B} = \{B = \times_{i=1}^\infty B_i : B_i \in \mathscr{B}_i\}$, then $\mathscr{B} \subset \Sigma$, and for any cylinder set

$$A = A_1 \times \cdots \times A_n \times \Omega_{n+1} \times \Omega_{n+2} \times \cdots \qquad (A_i \in \Sigma_i),$$

we define for each $B \in \mathscr{B}$

$$P(A|B) = \prod_{i=1}^n P_i(A_i|B_i) \tag{7}$$

and note that by the Jessen theorem (cf. Dunford–Schwartz [1], p. 203) $P(\cdot|B)$ has a unique extension to all of (Ω, Σ), satisfying (7). It is a probability there. We now assert that $\{\Omega, \Sigma, \mathscr{B}, P(\cdot|\cdot)\}$ is a conditional probability space. Since Axioms I and II are consequences of the definition (7) and the Jessen theorem, we need only check Axiom III. For this it suffices to show

$$P(A|B \cap C)P(B \cap C|C) = P(A \cap B \cap C|C) \tag{8}$$

for $\{A, B\} \subset \Sigma$, $C \in \mathscr{B}$, and $B \cap C \in \mathscr{B}$. In fact, taking $B = \Omega$, using $P(C|C) = 1$ and $P(A|C) = P(A \cap C|C)$, we get III(a). Then with $A \subset B \subset C$, (8) becomes $P(A|B)P(B|C) = P(A|C)$, which is III(b). We thus need only prove (8).

Since for fixed B, C the functions are measures in A, we may restrict to the case that A is a cylinder set in Σ. So let $A = A_1 \times \cdots \times A_n \times \Omega_{n+1} \times \Omega_{n+2} \times \cdots$. Then writing $(B \cap C)_i$ as the ith component in $\mathscr{B}_i$ of $B \cap C$, we have

$$P(A|B \cap C) = \prod_{i=1}^{n} P_i(A_i|(B \cap C)_i), \tag{9}$$

each of the other factors being unity. Let us set

$$p_k = \prod_{i=1}^{k} P_i((B \cap C)_i|C_i). \tag{10}$$

Then $p_k \downarrow p \geq 0$. But by definition $p = P(B \cap C|C)$. Thus if $p = 0$, then $0 \leq P(A \cap B \cap C|C) \leq P(B \cap C|C) = 0$ and (8) is true. Suppose $p > 0$. Then by definition,

$$\begin{aligned} P(A \cap B \cap C|C) &= \left(\prod_{i=1}^{n} P_i(A_i \cap (B \cap C)_i|C_i)\right) \prod_{i=n+1}^{\infty} P_i((B \cap C)_i|C_i) \\ &= \left(\prod_{i=1}^{n} P_i(A_i|(B \cap C)_i)P_i((B \cap C)_i|C_i)\right) \prod_{i=n+1}^{\infty} P_i((B \cap C)_i|C_i) \\ &\qquad \text{since each } P_i \text{ satisfies Axiom III}' \\ &= P(A|B \cap C)P(B \cap C|C) \quad \text{by (9), (10), and the value of } p. \end{aligned}$$

This is (8), and the result is proved.

Since $P(\cdot|B)$ is a (probability) measure, we can define conditional expectations and moments, etc., in the usual way. Thus, if $X: \Omega \to \mathbb{R}$ is a $P(\cdot|B)$-integrable random variable, let us define $E(X|B)$ for $B \in \mathscr{B}$ by

$$E(X|B) = \int_{\Omega} X(\omega)P(d\omega|B) \tag{11}$$

and call it the *conditional expectation* of X given B. This is a function of B and is not necessarily a constant. The conditional distribution of X is then given by

$$F_X(x \mid B) = P(\{\omega : X(\omega) < x\} \mid B),$$

and the theory proceeds quickly as in the case of scalar measures (or as in the case of *regular* conditional probability measures), and some of the classical results such as the law of large numbers have been discussed by Rényi [1, 2] in this framework.

It should be noted that the conditional probabilities here can also be considered as vector valued mappings. Thus $P: \mathscr{B} \to \mathscr{M}(\Sigma)$ is a mapping onto the space of all bounded σ-additive functions on Σ. These set functions in general are not σ-additive, though by Proposition 2(b) they have certain monotonicity properties. However, it is possible to generalize the theory of integration of point functions to include the vector valued functions $P(\cdot)$ on $\mathscr{B}$ relative to a σ-additive set function $P(\cdot \mid C)$, thus generalizing the formula (5). The latter is a sort of a reproducing property. The integration mentioned here proceeds along the lines known as the Burkill–Kolomogorov theory of integration, for cell functions. The use of these ideas and their implications for $P(\cdot \mid \cdot)$ are not clear at this time since the corresponding theory has not yet been developed. It will be interesting to explore these possibilities, which will add to Rényi's work. Such a point of view is seen to lead this approach again to the general theory of Section 3. *Thus the latter appears to have (with Dunford–Schwartz type vector integration) a certain global character.*

4. An Example In the preceding considerations, as well as in many other places, the treatment of the quantities such as $P(A \cap B)/P(B)$ was not discussed when $P(B) = 0$. In several interesting problems conditional expectations are calculated using (11), when $P(A \mid B) = [P(A \cap B)/P(B)]$. Thus

$$E(X \mid B) = \frac{1}{P(B)} \int_B X \, dP.$$

If $B_\lambda = \{\omega : a \le X(\omega) < a + \lambda\} \in \Sigma$ and $B_0 = \bigcap_{\lambda > 0} B_\lambda$, then it can happen that $P(B_\lambda) > 0$, $B_0 \in \Sigma$, but $P(B_0) = 0$. In this case a natural definition of $E(X \mid B_0)$ is to take it as $\lim_{\lambda \to 0} E(X \mid B_\lambda)$. However, the result may depend on the limit procedure and $E(X \mid X(\cdot) = a)$ may have different values for different sequences. We illustrate this pathology by the following example, due to Kaç and Slepian [1], for the case $X = \chi_A$.

Let $\{X(t), Y(t), 0 \le t \le 1\}$ be two Gaussian processes defined on (Ω, Σ, P). Denote the joint distribution by $F_t(\cdot, \cdot)$ with density $f_t(\cdot, \cdot)$. Thus

$$P\{\omega : X(t, \omega) < x, \ Y(t, \omega) < y\} = F_t(x, y) = \int_{-\infty}^{x} \int_{-\infty}^{y} f_t(u, v) \, du \, dv, \quad (12)$$

where an explicit form of the (Gaussian) density $f_t(\cdot, \cdot)$ can be given. Let $E(X(t)) = 0$, $E(Y(0)) = 0$, $E(Y^2(0)) = \alpha > 0$. In some problems of interest $Y(t)$ is the (pointwise) derivative of $X(t)$, and $Y(0)$ and $X(0)$ are independent. This is the case in the following example. Suppose now that it is desired to find the conditional density $p(\cdot, a)$ of $Y(0)$, given that $X(0) = a$ and the X-process is "ergodic", e.g., X has independent increments and X_{t+h}, $h > 0$ fixed, have the same distribution for all t; cf. Doob [1, 512]. Its discussion is unnecessary here. Thus

$$p(y, a) = \frac{d}{dy}(P[\{\omega : Y(0, \omega) < y\} \mid X(0, \cdot) = a]). \tag{13}$$

Let $g_0(\cdot)$ be the density of $X_0(\cdot)$. Since $P[X(0) = a] = 0$, the quantity in (13) is not defined. We can replace the conditioning set $B = \{\omega : X(0, \omega) = a\}$ by several approximating sequences. For instance, let $\delta > 0$ and consider (i) $B_{1\delta} = \{\omega : a \le X(0, \omega) \le a + \delta\}$, (ii) $B_{2\delta} = \{\omega : X(t, \omega) = a$ for some $t \in [0, \delta]\}$, (iii) $B_{3\delta} = \{\omega : (X(t, \omega) - a)^2 + t^2 \le \delta^2\}$. Note that as $\delta \to 0$, all these sets differ from B by a set of measure zero. Since the probability measure can be taken completed, no real measurability problems arise here. We now show that each of these approximations gives a different result for (13). By ergodicity, any $a \in \mathbb{R}$ gives the same $p(\cdot, a)$.

For (i), we have, with $h_0(\cdot)$ as the density of $Y(0)$

$$p(y, a) = \lim_{\delta \to 0}\left(\int_a^{a+\delta} f_0(y, x)\, dx \Big/ \int_a^{a+\delta} g_0(x)\, dx\right) = e^{-y^2/2\alpha}/\sqrt{2\pi\alpha}, \tag{14}$$

where we used the independence of $X(0)$ and $Y(0)$, which implies that $f_0(y, x) = h_0(y) g_0(x)$, where $h_0(y) = (\sqrt{2\pi\alpha})^{-1} e^{-y^2/2\alpha}$.

For (ii), if $y \ge 0$, we have

$$\begin{aligned} p(y, a) &= \lim_{\delta \to 0} \frac{\int_{a-y\delta}^{a} f_0(y, x)\, dx}{\int_0^\infty dz \int_{a-z\delta}^{a} f_0(z, x)\, dx + \int_{-\infty}^{0} dz \int_a^{a-z\delta} f_0(z, x)\, dx} \\ &= \frac{y}{2\alpha} e^{-y^2/2\alpha} \end{aligned}$$

because $B_{2\delta}$ is satisfied for a given $y > 0$ only if $a - y\delta \le X(0, \omega) \le a$, and similarly for $y < 0$. Thus we get

$$p(y, a) = \frac{|y|}{2\alpha} e^{-y^2/2\alpha}. \tag{15}$$

For (iii), on the other hand, we can similarly calculate the result to obtain

$$p(y, a) = \sqrt{1 + y^2}\, e^{-y^2/2\alpha} \Big/ \int_{-\infty}^{\infty} \sqrt{1 + x^2}\, e^{-y^2/2\alpha}\, dx. \tag{16}$$

It is clear that there is no inherent reason that any one value of (14), (15), or (16) should be preferred. Kaç and Slepian point out that (ii) has an interpretation, due to Smoluchowski, in statistical physics since the path through which $X(t)$ approaches $X(0)$ has a physical meaning. In general, however, for the Radon–Nikodým derivative (if it exists) all such evaluations must agree outside of a null set. The present divergence shows that the latter derivative does not exist here. For such situations, several weaker derivatives have been introduced in the literature. For instance, the Besicovitch derivative uses the sets of the form given in (iii) above, the others in many statistical applications use the sets of the form (i). When any of these many definitions are used, one has to point out the type of differentiation because of the divergence of the results. It is evident that Rényi's approach is of little help in such situations. Under further conditions, in the regular case, Tjur [1] has discussed an interesting application of differential geometry in determining conditional distributions. This does not apply to the above example either. For this and other reasons, we resume the work with the Kolmogorov model as discussed prior to this section.

2.6 APPLICATIONS TO REYNOLDS OPERATORS

The results and techniques of the preceding sections find a nice application in the structure theory of Reynolds operators, which are a natural extension of the averaging operators of Section 2. This generalized class will be found useful for a unified formulation of ergodic and martingale theories to be discussed in Chapter V. The unification has been a hope from at least the middle 1940s. However, the operator identity itself was discovered by an engineer, Osborne Reynolds, in a study of the dynamical theory of incompressible viscous fluids in 1895, and the averagings are a key component and are the specialized Reynolds operators. A great deal of effort has been expended to understand the structure of these transformations, particularly by Kampé de Fériet [1, 2], and later by Rota [2] and others. Only recently it was possible to present a reasonably satisfactory account of this structure since the theory of conditional expectations formed an integral part of these operators, and the latter were completely characterized only *after* 1950, as evidenced in the work of Section 2. It is thus appropriate that we present the structure theory here as an application of the above work and use it in Chapter V for a unified study of martingale and ergodic theories. We shall employ certain standard results from linear analysis and give their precise statements (with exact references) in using them in our proofs since it is impossible to include a complete account of the latter though an attempt is made to be reasonably detailed in most respects.

In Section 2 we defined an averaging operator $T: L^p(\Sigma) \to L^p(\Sigma)$ to be a contractive linear mapping such that $T(fTg) = (Tf)(Tg)$ is true whenever f or g is bounded, and that $Tf_0 = f_0$ a.e. for some $0 < f_0 \in L^p \cap L^\infty$. (Usually $f_0 = 1$.) A natural extension of this is the following:

1. Definition A continuous linear mapping $R: L^p(\Omega, \Sigma, P) \to L^p(\Omega, \Sigma, P)$, $p \geq 1$, is called a *Reynolds operator* if it satisfies (i) $R(L^p \cap L^\infty(P)) \subset L^p \cap L^\infty(P)$ and (ii) the algebraic equation

$$R(fg) = (Rf)(Rg) + R[(f - Rf)(g - Rg)], \qquad f, g \quad \text{in} \quad L^p \cap L^\infty(\Omega, \Sigma, P). \tag{1}$$

Sometimes (1) is called the *Reynolds identity*.

It was noted in Lemma 2.2 that an averaging operator is always a projection. Such an operator satisfies the identity (1). In fact (1) can be written, on expansion, in another form:

$$R(fRg) + R(gRf) = RfRg + R[Rf \cdot Rg]. \tag{2}$$

But for any averaging, $Rf \cdot Rg = R(fRg)$. Since $R^2 = R$ is also true by definition, it is clear that (2) is satisfied by an averaging projection R, and hence Reynolds operators form a larger class. That this class is strictly larger follows from the fact that a Reynolds operator need not be idempotent. The generalization contained in the Reynolds class will be shown in Chapter V to be essential for a unified formulation of the key martingale convergence and of some ergodic theorems. Because of this need, we consider the structure of these operators on a general σ-finite measure space.

We remark that positive Reynolds operators on L^p, if $p \in [1, 2)$, have an immediate extension to the Hilbert space L^2. In fact, taking $f = g$ ($\in L^p \cap L^\infty$) in (1), we get by the positivity of R (using $\vee$ for "max")

$$Rf^2 = (Rf)^2 + R[(f - Rf)^2] \geq \{(Rf)^2 \vee R[(f - Rf)^2]\}. \tag{3}$$

But $f \in L^p \cap L^\infty$ implies $f^2 \in L^1 \cap L^p$, and then (3) shows $(Rf) \in L^2$. Since $L^p \cap L^\infty$ is dense in L^2, we may conclude that R is a positive linear mapping on a dense subspace of L^2. In particular, if $p = 1$ (or $\mu(\Omega) < \infty$ and R^*1 is bounded), then R is also continuous on all of L^2. This follows from the fact that $R^*1 \in L^{p'} \cap L^\infty$ ($p' = p/(p-1)$ if $p > 1$, $p' = +\infty$ if $p = 1$) in these two cases, where R^* is the adjoint of R and $L^p \cap L^\infty$ is norm dense in L^2. Thus from (3) we have

$$\|Rf\|_2^2 \leq \int_\Omega R^*1 \cdot f^2 \, d\mu \leq k\|f\|_2^2, \qquad f \in L^p \cap L^\infty, \tag{4}$$

where $\infty > k \geq R^*1 \geq 0$ a.e. However, for most applications we cannot assume the positivity of R and the fact that it is defined on all of L^1 (hence continuous by positivity). Thus it is natural to make assumptions on the norm of R and then work out its structure theory. The ultimate result shows that R is also continuous on L^2 (in fact on all L^p, $1 \leq p \leq \infty$) and positive, but this cannot be deduced before the structure is completely determined. We shall show later that the positivity on L^1 also implies essentially the hypothesis we impose on R. Thus these are found to be two different ways of looking at the same problem.

We now proceed to establish the basic structure theorem of R.

2. Theorem *Let $L^p(\Sigma)$, $1 \leq p < \infty$, be a Lebesgue space on a σ-finite measure space (Ω, Σ, μ), and let f_0 be a weak unit, i.e., $f_0 = \{\chi_{A_n}, n \geq 1\}$, $0 < \mu(A_n) < \infty$ and $\bigcup_{n=1}^{\infty} A_n = \Omega$ (A_n may and will be assumed disjoint outside of μ-null sets). Let R: $L^p(\Sigma) \to L^p(\Sigma)$ be a Reynolds operator (satisfying Definition 1). Suppose that (i) $\|Rf\|_p \leq \|f\|_p$, i.e., R is contractive; (ii) $Rf_0 = f_0$, i.e., $R\chi_{A_n} = \chi_{A_n}$, $n \geq 1$; and (iii) $R^*f_0 = f_0$, where R^* is the adjoint of R and f_0 is the weak unit (which is also in $(L^p(\Sigma))^*$). Then there exist uniquely (a) a σ-algebra $\mathscr{B} \subset \Sigma$, $\mu_{\mathscr{B}}$ σ-finite, and (b) a strongly continuous semigroup $\{V(t), t \geq 0\}$ of linear operators on $L^p(\mathscr{B})$, induced by a measure preserving endomorphism τ on $(\Omega, \mathscr{B}, \mu)$, i.e., $V(t): f \mapsto f \circ \tau^t$ for $f \in L^p(\mathscr{B})$ and $\mu \circ \tau^{-1} = \mu$, in terms of which the following (strong) integral representation holds (here τ^t is defined, $t \geq 0$, through the so-called operational calculus):*

$$Rf = \int_0^{\infty} e^{-t} V(t) E^{\mathscr{B}}(f)\, dt, \qquad f \in L^p(\Sigma), \tag{5}$$

where the range $R(L^p(\Sigma))$ is dense in $L^p(\mathscr{B})$ and on which R and $E^{\mathscr{B}}$, the conditional expectation, commute; in fact, $Rf = RE^{\mathscr{B}}(f) = E^{\mathscr{B}}R(f), f \in L^p(\Sigma)$.

3. Remarks 1. The integral in (5) can be taken simply in the Riemann sense for vector valued functions, but it is easier to regard it as the strong or Bochner integral, i.e., the norm limit of the sums for simple functions as defined in Section I.4 (cf. for elementary properties Theorems I.4.10 and I.4.11). Note that the representation (5) implies, since both $E^{\mathscr{B}}$ and $V(t)$ are positive operators, that R is also positive as remarked before. But this follows only after (5) is established, and this is definitely nontrivial.

2. It may seem unnecessary to bring the adjoint R^* of R into the statement of the theorem. However, if $1 < p < \infty$, it can be shown that since $f_0 \in L^p \cap L^{p'} \cap L^{\infty}(\Sigma)$, the hypothesis $Rf_0 = f_0$ implies $R^*f_0 = f_0$ also. (See Problem 8.) But when $p = 1$, some weak compactness hypothesis on the operator R is needed for this conclusion. In case $p = 2$ there is an elegant and simple method due to Riesz and Sz.-Nagy [1], which in fact

shows more generally the identity of fixed points of a contractive operator and of its adjoint in any Hilbert space. An argument of this is contained in the proof below.

Proof of Theorem 2 We present the details in steps for convenience and clarity. As before, we only consider the real L^p-space though the result extends to the complex case. Theorem 2.5 holds for σ-finite μ also, as one finds easily.

I. Let $\mathscr{M}_0 = R(L^p \cap L^\infty(\Sigma))$, $\mathscr{M}_1 = R(L^p(\Sigma))$, and $\mathscr{M}$ be the closure of $\mathscr{M}_1$ in L^p. Then $\mathscr{M}_0$ is an algebra dense in $\mathscr{M}$. *For*, since $\mathscr{M}_0 \subset L^p \cap L^\infty(\Sigma)$ is true by hypothesis and the set $\mathscr{M}_0$ is clearly linear, let $\{f_1, f_2\} \subset \mathscr{M}_0$. Then there exist $g_i \in L^p \cap L^\infty(\Sigma)$ such that $f_i = Rg_i$, $i = 1, 2$. By the identity (1), $f_1 f_2 = R(g_1 g_2) - R[(g_1 - f_1)(g_2 - f_2)] \in \mathscr{M}_0$ since g_i, f_i are bounded. Thus $\mathscr{M}_0$ is an algebra. Regarding density, let $g \in \mathscr{M}_1$ so that $g = Rh$ for some $h \in L^p(\Sigma)$. But, from the fact that $1 \leq p < \infty$, given $\varepsilon > 0$, there exists $h_\varepsilon \in L^p(\Sigma) \cap L^\infty$ such that $\|h - h_\varepsilon\|_p < \varepsilon$. If $g_\varepsilon = Rh_\varepsilon \in \mathscr{M}_0$, then $\|g - g_\varepsilon\|_p \leq C\|h - h_\varepsilon\|_p < C\varepsilon$, where C is the bound of R ($C \leq 1$ by the contractivity hypothesis). Hence $\mathscr{M}_0$ is dense in $\mathscr{M}_1$, which implies the same in $\mathscr{M}$. Note also that the weak unit $f_0 \in \mathscr{M}_0$.

By the preceding analysis, if $\mathscr{B} = \{A \in \Sigma : \chi_A \in \mathscr{M}\}$, then Theorem 2.5 applies and we can conclude that $\mathscr{B}$ is a σ-algebra and $\mathscr{M} = L^p(\mathscr{B})$. Since $f_0 \in \mathscr{M}$, it follows that for $\mathscr{B}$, $\mu_{\mathscr{B}}$ is σ-finite and $E^{\mathscr{B}}: L^p(\Sigma) \to L^p(\mathscr{B})$ is the conditional expectation with range $L^p(\mathscr{B})$. Thus $E^{\mathscr{B}}R = R$ trivially. We shall now show after much work that $RE^{\mathscr{B}} = R$ also holds. For this we first prove, with all the hypothesis, that R is defined and is a contraction on $L^2(\Sigma)$.

II. R is defined on $L^2(\Sigma)$ and is a contraction there. *For*, let $\{f, g\} \subset L^p(\Sigma) \cap L^\infty$. Since we may express the Reynolds identity as (2) for any $h \in L^{p'}(\Sigma)$, the following equation holds:

$$\int_\Omega hR(fRg + gRf)\, d\mu = \int_\Omega h[Rf \cdot Rg + R(Rf \cdot Rg)]\, d\mu. \tag{6}$$

But clearly the weak unit $f_0 \in L^p(\Sigma) \cap L^{p'}(\Sigma) \cap L^\infty$. Taking $h = f_0$ in (6) and using the hypothesis that $R^*f_0 = f_0$ (and $Rf_0 = f_0$), we may rewrite (6), after taking adjoints ($\int_\Omega Rf \cdot g\, d\mu = \int_\Omega f \cdot R^*g\, d\mu$), as

$$\int_\Omega R^*f_0[fRg + gRf]\, d\mu = \int_\Omega [f_0 Rf \cdot Rg + R^*f_0(Rf \cdot Rg)]\, d\mu. \tag{7}$$

Noting that $f_0 R(h) = R(f_0 h)$ for $h \in L^p(\Sigma)$ by (1), and $R^*f_0 = f_0$, we get

$$\int_\Omega [fR(f_0 g) + f_0 gRf]\, d\mu = 2\int_\Omega Rf \cdot R(f_0 g)\, d\mu. \tag{8}$$

Let $\mathcal{S} \subset L^p(\Sigma)$ be the class of all simple (or step) functions. Then $\mathcal{S}$ is dense in $L^p(\Sigma)$ for $1 \le p < \infty$. $\mathcal{S} \subset L^p \cap L^{p'} \cap L^\infty(\Sigma)$, $R(\mathcal{S}) \subset L^\infty(\Sigma)$, and $R^*(\mathcal{S}) \subset L^{p'}(\Sigma)$. We may assume $f_0 \in \mathcal{S}$ also (by adding it if necessary). Then by taking adjoints of R in (8) appropriately, we get

$$\int_\Omega (R^*f + Rf) f_0 g \, d\mu = 2 \int_\Omega R^*Rf \cdot f_0 g \, d\mu, \qquad g \in L^p \cap L^\infty. \tag{9}$$

Since g is arbitrary and $f_0 > 0$ a.e., (9) implies the useful equation:

$$(R^* + R)f = 2R^*Rf \quad \text{a.e.}, \qquad f \in L^p \cap L^\infty(\Sigma). \tag{10}$$

Let $U = I - 2R$. Then for any $h \in \mathcal{S}$ it follows that $Uh \in L^p(\Sigma) \cap L^\infty$ and, by (10), U^*Uh is well defined. Substitution and use of (10) show that $U^*Uh = h$ a.e. Now $h \in \mathcal{S} \subset L^p(\Sigma) \cap L^{p'}(\Sigma) \cap L^2(\Sigma)$. Thus

$$\int_\Omega h^2 \, d\mu = \int_\Omega h \cdot U^*Uh \, d\mu = \int_\Omega Uh \cdot Uh \, d\mu = \int_\Omega (Uh)^2 \, d\mu. \tag{11}$$

But (11) implies that U is an isometry on the dense set $\mathcal{S} \subset L^2(\Sigma)$, and hence has a unique extension, to remain an isometry, to all of $L^2(\Sigma)$. Since $R = \frac{1}{2}(I - U)$, we conclude that R is defined on $L^2(\Sigma)$, and with (11) it is a contraction there.

III. The operator U on $L^2(\Sigma)$ can be used to obtain various properties of R. We first prove that $(1/n)\sum_{i=0}^{n-1} U^i \to Q_1$, a contractive projection, strongly and that the range $\mathcal{M}_2$ of Q_1 is the set of fixed points of U which is thus the null space of R in $L^2(\Sigma)$. (This is a special case of J. von Neumann's (1932) mean ergodic theorem, and we use an elegant argument of Riesz and Sz.-Nagy [1] for the following proof.)

Indeed, if $f \in \mathcal{M}_2$ so that $Uf = f$, it is clear that $(1/n)\sum_{i=0}^{n-1} U^i f = f$. Since $\mathcal{M}_2$ is a closed subspace of $L^2(\Sigma)$, there exists a contractive projection Q_1 such that $Q_1(L^2(\Sigma)) = \mathcal{M}_2$ from the classical Hilbert space theory. Thus for $f \in \mathcal{M}_2$, $(1/n)\sum_{i=0}^{n-1} U^i f \to f = Q_1 f$ in norm as $n \to \infty$. The key idea now is to observe that in the representation $g = Ug + (g - Ug)$, $(1/n)\sum_{i=0}^{n-1} U^i(g - Ug) = (1/n)(g - U^n g)$ so that $\|(1/n)\sum_{i=0}^{n-1} U^i(g - Ug)\| \le 2(\|g\|/n) \to 0$ as $n \to \infty$ ($\|\cdot\| = \|\cdot\|_2$) since U is a contraction. Hence the linear subspace $\mathcal{N}$ generated by the elements $\{(g - Ug) : g \in L^2(\Sigma)\}$ is a *convergence manifold* for the operator sequence $\{(1/n)\sum_{i=0}^{n-1} U^i\}_1^\infty$. If $h \in \bar{\mathcal{N}}$, the closure, then for any $\varepsilon > 0$ there is an $h_\varepsilon \in \mathcal{N}$ such that $\|h - h_\varepsilon\| < \varepsilon/2$, and hence we deduce that

$$\left\| \frac{1}{n} \sum_{i=0}^{n-1} U^i h \right\| \le \left\| \frac{1}{n} \sum_{i=0}^{n-1} U^i h_\varepsilon \right\| + \frac{1}{n} \sum_{i=0}^{n-1} \|U^i(h - h_\varepsilon)\|$$

$$\le \|h - h_\varepsilon\| + \left\| \frac{1}{n} \sum_{i=0}^{n-1} U^i h_\varepsilon \right\|.$$

Since this goes to zero as $n \to \infty$ (and ε being arbitrary), $\bar{\mathcal{N}}$ is a convergence manifold. We claim $\mathcal{M}_2 \oplus \bar{\mathcal{N}} = L^2(\Sigma)$ (i.e., $\bar{\mathcal{N}} = \mathcal{M}_2^{\perp}$); this proves that $\bar{\mathcal{N}}$ is the range $(I - Q_1)$, and since every $f \in L^2(\Sigma)$ can be uniquely written as $f = f_1 + f_2$ for $f_1 \in \mathcal{M}_2, f_2 \in \bar{\mathcal{N}}$, and since $\mathcal{M}_2, \bar{\mathcal{N}}$ are convergence manifolds, it follows that $(1/n)\sum_{i=0}^{n-1} U^i f \to Q_1 f$ strongly. It is obvious that $U((1/n)\sum_{i=0}^{n-1} U^i f) = (1/n)\sum_{i=0}^{n-1} U^i g \to Q_1 g = Q_1 U f = U Q_1 f$ $(g = Uf)$, i.e., U and Q_1 commute. We thus have to prove: $\mathcal{N}^{\perp} = \bar{\mathcal{N}}^{\perp} = \mathcal{M}_2^{\perp\perp} = \mathcal{M}_2$.

Let $g \in \mathcal{N}$. Then $h \in \mathcal{N}^{\perp}$ iff $(h, g) = 0$ where $(\cdot, \cdot)$ is the inner product of $L^2(\Sigma)$. But $g = v - Uv$ is a generating element of $\mathcal{N}$ for some $v \in L^2(\Sigma)$. Hence $(h, v - Uv) = 0$ for all $v \in L^2(\Sigma)$. We may express this as

$$0 = (h, v) - (h, Uv) = (h, v) - (U^*h, v) = (h - U^*h, v), \qquad v \in L^2(\Sigma).$$

This is true iff $U^*h = h$. We now observe that this is true iff $Uh = h$ or $h \in \mathcal{M}_2$. In fact, consider $Uh - h \in L^2(\Sigma)$, where we identify $L^2(\Sigma)$ and $(L^2(\Sigma))^*$,

$$\begin{aligned} 0 \le \|Uh - h\|^2 &= (Uh - h, Uh - h) \\ &= \|Uh\|^2 - (Uh, h) - (h, Uh) + \|h\|^2 \\ &\le 2\|h\|^2 - (h, U^*h) - (U^*h, h) = 2\|h\|^2 - 2\|h\|^2 = 0 \end{aligned} \tag{12}$$

since $U^*h = h$. Thus $Uh = h$ (and conversely if $Uh = h$, we may interchange U and U^* here to conclude $U^*h = h$). So $\mathcal{M}_2 = \mathcal{N}^{\perp}$, as asserted.

IV. Consider $\mathcal{M}_2$ the subspace of fixed points of $U = I - 2R$. Then $\mathcal{M}_2$ is the annihilator of R. Consequently, $\mathcal{M}_2^{\perp} = \bar{\mathcal{N}}$ is the closure of the range of R. In fact, $\mathcal{N}$ is the linear span of elements $g - Ug = (I - U)g = 2Rg$ so that $\bar{\mathcal{N}} =$ closure of range R, and this is the range of $I - Q_1$. But by Step I, $\mathcal{M}_2^{\perp} = \bar{\mathcal{N}} = L^2(\mathcal{B})$, where $\mathcal{B} \subset \Sigma$ is a σ-algebra such that the weak unit $f_0 \in L^2(\mathcal{B})$. Since Q_1 and U commute, it follows that Q_1 and R, or $I - Q_1$ and R, commute also. On the other hand, $I - Q_1 \colon L^2(\Sigma) \to L^2(\mathcal{B})$ is a projection such that $(I - Q_1)f_0 = f_0$, and being an orthogonal projection is contractive. Hence by Corollary 2.10 (see the last half of its proof), we conclude that $I - Q_1 = E^{\mathcal{B}}$. Thus $E^{\mathcal{B}}$ and R commute and the range of $E^{\mathcal{B}}$ is the closure of the range of R. This proves the last assertion of the theorem since both $E^{\mathcal{B}}$ and R are defined on $L^p(\Sigma)$, $E^{\mathcal{B}}R = RE^{\mathcal{B}} = R$, and the properties established above clearly hold on $L^p(\mathcal{B})$.

We also note that $R \mid L^2(\mathcal{B})$ is one-to-one. If not, there exists $0 \neq f \in L^2(\mathcal{B})$ such that $Rf = 0$ a.e. But then $Uf = f \in \mathcal{M}_2 \cap \mathcal{M}_2^{\perp} = \{0\}$ giving a contradiction. Thus R^{-1} exists on $\mathcal{N} \subset L^2(\mathcal{B})$ and is densely defined.

V. Since R is continuous on $L^2(\mathcal{B})$, it is a closed operator (i.e., $f_n \in L^2(\mathcal{B})$, $g_n = Rf_n$, and $f_n \to f$, $g_n \to g$ in norm implies $f \in L^2(\mathcal{B})$ and

$Rf = g \in \mathcal{N}$). So R^{-1} is also closed but not bounded unless its domain $\mathcal{N}$ is closed (in which case by the above analysis $R = E^{\mathcal{B}}$ which in general is clearly false), by the classical closed graph theorem. Thus the operator $D = I - R^{-1}$ is a closed densely defined (on $L^2(\mathcal{B})$) operator. If now we introduce R_λ by setting $R_\lambda = (\lambda I - D)^{-1}$, then R_λ is well defined for $\lambda > 0$ since the spectrum of the operator D lies in the left half plane (by the spectral mapping theorem) and R_λ is the resolvent of D. This follows from general results in Dunford–Schwartz [1, Section VII.3]. But here is a simple proof of our special case. Since $\lambda I - D$ is densely defined in $L^2(\mathcal{B})$, we now claim that $(\lambda I - D)$ is one-to-one on $\mathcal{N}$. Indeed, if $0 \neq f \in \mathcal{N}$ and $(\lambda I - D)f = 0$ a.e., then $\lambda f = Df = f - R^{-1}f$ so that $(1 - \lambda)f = R^{-1}f$. Since R is injective so that R^{-1} is also one-to-one on $\mathcal{N}$, $Rf = (1 - \lambda)^{-1}f$. If $\lambda = 1$, then this is impossible since $Rf \in L^2(\mathcal{B})$ and hence is finite a.e. If $\lambda \neq 1$, $\lambda > 0$, then (1) implies $Rf^2 = (Rf)^2 + R[(f - Rf)^2]$. Substitution of $Rf = f/(1 - \lambda)$ yields, on simplification, $Rf^2 = f^2/(1 - 2\lambda)$. Since $R^*f_0 = f_0$, this gives

$$\int_\Omega f_0 f^2 \, d\mu = \int_\Omega f_0 R f^2 \, d\mu = \frac{1}{1 - 2\lambda} \int_\Omega f_0 f^2 \, d\mu. \tag{13}$$

But this is impossible since $\lambda > 0$ and $f \neq 0$ a.e. Thus $\lambda I - D$ is one-to-one for $\lambda > 0$, and R_λ is called the *resolvent operator* of D on $L^2(\mathcal{B})$.

VI. We claim that $\|\lambda R_\lambda\| \leq 1$ for all $\lambda > 0$. This means $\|\lambda R_\lambda f\|_2 \leq \|f\|_2$ for all $f \in \text{range}(\lambda I - D)$ in $L^2(\mathcal{B})$, which is dense, i.e., $f \in (\lambda I - D)(\mathcal{M}_1)$. On this space the inverse exists. For by definition there exists $h \in L^2(\Sigma)$ such that $f = (\lambda I - D)Rh$. Then $f = ((\lambda - 1)I + R^{-1})Rh = (\lambda - 1)Rh + h$. Moreover, for such an f we have $R_\lambda f = (\lambda I - D)^{-1}(\lambda I - D)Rh = Rh$. Also by (8) (taking f, g to be h there) we get

$$\int_\Omega (hRh) f_0 \, d\mu = \int_\Omega (Rh)^2 f_0 \, d\mu. \tag{14}$$

But a weak unit by definition is composed of $\{\chi_{A_n}, n \geq 1\}$, where $0 < \mu(A_n) < \infty$, $\bigcup_{n=1}^\infty A_n = \Omega$, and A_n are disjoint a.e. Also $Rf_0 = f_0$ is equivalent to $R\chi_{A_n} = \chi_{A_n}$ a.e., $n \geq 1$. This implies that we could replace f_0 by χ_{A_n} for each n in Eqs. (7)–(10). In particular, (14) holds if f_0 is replaced by χ_{A_n} for each n, and adding on n. Thus we get, on using $\Omega = \bigcup_{n=1}^\infty A_n$ a.e.,

$$(h, Rh) = \int_\Omega hRh \, d\mu = \int_\Omega (Rh)^2 \, d\mu = \|Rh\|_2^2, \qquad h \in L^2(\Sigma). \tag{15}$$

Consider the equation connecting f and h. We have clearly

$$\begin{aligned}\|f\|_2\|Rh\|_2 \geq (f, Rh) &= (\lambda - 1)(Rh, Rh) + (h, Rh) \\ &= (\lambda - 1)\|Rh\|_2^2 + \|Rh\|_2^2 \qquad \text{by} \quad (15) \\ &= \lambda\|Rh\|_2^2. \end{aligned} \tag{16}$$

Since $Rf_0 = f_0$ implies $R = 0$ is impossible, we get $\|f\|_2 \geq \lambda\|Rh\|_2 = \lambda\|R_\lambda f\|_2$, where $R_\lambda f = Rh$. Thus $\|\lambda R_\lambda\| \leq 1$. If $\lambda = 1$, this is immediate from (14).

VII. In the preceding two steps we showed that $D = I - R^{-1}$ is a closed densely defined linear operator on $L^2(\mathscr{B})$ and for each $\lambda > 0$ its resolvent R_λ satisfies the norm condition $\|\lambda R_\lambda\| \leq 1$. We can now apply the classical Hille–Yosida theorem according to which R_λ admits the following integral representation:

$$R_\lambda f = \int_0^\infty e^{-\lambda t} V(t) f \, dt, \qquad f \in L^2(\mathscr{B}), \quad \lambda > 0, \tag{17}$$

where $\{V(t), t \geq 0\}$ is a strongly continuous contractive semigroup of operators on $L^2(\mathscr{B})$ whose infinitesimal generator is D, i.e., $\|V(t)f - f\|_2 \to 0$ as $t \to 0$, $V(t + s) = V(t)V(s)$ and $\|([V(t)f - f]/t) - Df\|_2 \to 0$ as $t \to 0$, t, $s \geq 0$. We can write this as $dV(t)/dt|_{t=0} = D$, in the strong topology. Moreover, for $0 \leq t < \infty$, we have $(d/dt)V(t)f = DV(t)f = V(t)Df$ for all $f \in \text{domain}(D)$. (See Dunford–Schwartz [1, VIII.1.16]; we are going to use these and related properties of the operational calculus below.)

VIII. In order to extend the representation to the L^p-case where $p \neq 2$ is possible, we analyze the family $\{V(t), t \geq 0\}$ and show that for any $f, g \in \mathscr{M} \cap L^\infty(\Sigma)$, $V(t)(fg) = (V(t)f) \cdot (V(t)g)$. When this is established, all the other assertions of the theorem follow easily.

Let h_1, h_2 be in $L^2(\Sigma) \cap L^\infty$. Since $V(t)R$ is defined on $L^2(\Sigma)$, and the above-noted differential calculus implies

$$(d/dt)V(t)R = V(t)DR = V(t)(R - I),$$

consider the vector valued function $F(t, s) = V(s)((V(t)Rh_1)(V(t)Rh_2))$, $s \geq 0$, $t \geq 0$. We first show that $F(t, s) = \tilde{F}(t + s)$. This would imply $F(t, 0) = F(0, t)$ and since $V(0) = I$, it proves the multiplicative property of $V(t)$, for each $t \geq 0$, on $\mathscr{M} \cap L^\infty(\Sigma)$.

Recalling that $V(t)$ and D (hence $V(t)$ and R) commute, we can make the following calculation of $\partial F/\partial s$, $\partial F/\partial t$:

$$\frac{\partial F}{\partial s} = V(s)D((V(t)Rh_1)(V(t)Rh_2)) = V(s)D((RV(t)h_1)(RV(t)h_2)) \tag{18}$$

and

$$\begin{aligned}\frac{\partial F}{\partial t} &= V(s)[(V(t)(R-I)h_1)(V(t)Rh_2) + (V(t)(Rh_1))(V(t)(R-I)h_2)]\\ &= V(s)[(RV(t)h_1)(RV(t)h_2) + (D-I)R\{(RV(t)h_2)V(t)h_1\\ &\quad + (R(V(t)h_1))(V(t)h_2) - (R(V(t)h_1))(R(V(t)h_2))\}]\\ &= V(s)[(RV(t)h_1)(RV(t)h_2) + (D-I)\{(R(V(t)h_1))(RV(t)h_2)\}]\\ &\qquad \text{by} \quad (2)\\ &= V(s)D((RV(t)h_1)(RV(t)h_2)). \end{aligned} \tag{19}$$

From (18) and (19) we deduce that $\partial F/\partial s = \partial F/\partial t$. This implies (from elementary theory) that F is of the form $\tilde{F}(t+s)$, as asserted. Thus $V(t)(fg) = (V(t)f)(V(t)g)$, $t \geq 0$ on $\mathscr{M} \cap L^\infty(\Sigma)$.

From this, if $f = \chi_A \in \mathscr{M}_1$, then $(V(t)\chi_A) = (V(t)\chi_A)^2$ so that it is also a characteristic function $= \chi_B$ a.e. (say). Hence the mapping $T: A \mapsto B$ is measurable and induces $V(t)$, i.e., $V(t)f = f \circ T$. The smigroup property of $V(t)$s implies (if $V(1)f = f \circ \tau$) that $V(t)f = f \circ \tau^t$ for each integer $t \geq 0$ and then by continuity for all $t \geq 0$ (by the operational calculus). Now $\|V(t)\| \leq 1$ implies that $\mu(\tau^{-1}(A)) \leq \mu(A)$, $A \in \mathscr{B}$, $\chi_A \in L^2(\mathscr{B})$. However, choosing $A \in \{A_n, n \geq 1\}$ determining the weak unit so that $V(t)\chi_{A_n} = \chi_{A_n}$ a.e., we get $\mu(\tau^{-1}(A_n)) = \mu(A_n)$, $n \geq 1$. If we consider the trace σ-algebra $\mathscr{B}(A_n)$ for $A_n \in \{A_k, k \geq 1\}$ and if there exists $A \in \mathscr{B}(A_n)$ such that $\mu(\tau^{-1}(A)) < \mu(A)$, then we must have

$$\begin{aligned}\mu(A_n) = \mu(A) + \mu(A_n - A) &> \mu(\tau^{-1}(A)) + \mu(\tau^{-1}(A_n) - \tau^{-1}(A))\\ &= \mu(\tau^{-1}(A_n)) = \mu(A_n).\end{aligned} \tag{20}$$

This contradiction shows that τ is measure preserving on each $\mathscr{B}(A_n)$ and hence on $\mathscr{B}$ since $\{\mathscr{B}(A_n), n \geq 1\}$ determines $\mathscr{B}$. Thus the semigroup $\{V(t), t \geq 0\}$ is in fact generated by the family $\{\tau^t, t \geq 0\}$ where τ is a measure-preserving transformation. But such an operator has a norm-preserving extension to all $L^p(\mathscr{B})$, $p \geq 1$, since the underlying measure space (and τ) is fixed. Thus we deduce that (17) holds for all $f \in L^p(\Sigma)$, $p \geq 1$.

IX. We have already shown that R and $E^{\mathscr{B}}$ commute on $L^2(\mathscr{B})$. However, R is given on $L^p(\Sigma)$ and $E^{\mathscr{B}}$ has a norm-preserving extension to all $L^p(\Sigma)$, $p \geq 1$, $E^{\mathscr{B}}(L^p(\Sigma)) = L^p(\mathscr{B})$ for a $\mathscr{B} \subset \Sigma$, $\mu_{\mathscr{B}}$ σ-finite. Since $\mathscr{M}_1 = R(L^p(\Sigma)) = RE^{\mathscr{B}}(L^p(\Sigma)) = R(L^p(\mathscr{B}))$, we have the extended representation (17) as

$$R_\lambda f = \int_0^\infty e^{-\lambda t} V(t) E^{\mathscr{B}}(f)\, dt, \qquad f \in L^p(\Sigma), \quad \lambda > 0, \tag{21}$$

where the integral is the Bochner or strong integral. This is (5) with $\lambda = 1$. Note that after the representation of (21) we may now conclude that R, with the hypothesis on some $L^p(\Sigma)$, satisfies (21) for all $L^p(\Sigma)$, $1 \leq p < \infty$. This completes the proof of the theorem.

4. Discussion It is useful to make some observations on this theorem. The result is adapted from the author [2] and Rota [2].

(a) Taking $\lambda = 1$ so that $R_1 = R$ is a Reynolds operator, then its range $\mathcal{M}_1$ is norm closed iff R is a projection. In fact, if $\mathcal{M}_1$ is closed, then $\mathcal{M}_1 = L^p(\mathcal{B})$ by Steps I and II. To show that $R^2 = R$, since simple functions are dense in L^p, it suffices (by linearity) to show that $R\chi_A = \chi_A$, for $\chi_A \in L^p(\mathcal{B})$. But $\chi_A \in \mathcal{M}_1 \Rightarrow \chi_A = Rf$ for some $f \in L^p(\Sigma)$. Then by (2) and $(Rf)^2 = R(f)$, we have

$$\chi_A = Rf = (Rf)^2 = 2R(fRf) - R((Rf)^2). \tag{22}$$

But by Step IV, on $\mathcal{M}_1$, R^{-1} exists, so that (22) implies

$$f = 2f \cdot Rf - (Rf)^2 = 2f \cdot \chi_A - \chi_A,$$

or equivalently

$$\chi_A = (2\chi_A - 1)\,\mathrm{f} = (\chi_A - \chi_{A^c})\,\mathrm{f}. \tag{23}$$

Hence $f\chi_{A^c} = 0$ a.e., and $\chi_A = \chi_A f$ or $f = \chi_A$ and $R\chi_A = \chi_A$, or $R^2 = R$. Since the range of a continuous projection in a Banach space is closed (cf. Proposition 2.3), the converse follows.

(b) Since by Corollary 2.10 every contractive projection T on $L^p(\Sigma)$ with range $L^p(\mathcal{B})$, $\mathcal{B}$ a complete σ-algebra in Σ, is the conditional expectation, it follows that (by (a)) every contractive Reynolds operator such that $Rf_0 = f_0$ for a weak unit f_0 is a conditional expectation iff it has a closed range. In this case the semigroup $\{V(t), t \geq 0\}$ is trivial, i.e., $V(t) = I$ for all $t \geq 0$, and (5) thus yields this result again. [We may see this by noting that $R \,|\, L^2(\mathcal{B}) = I$ so $D = I - R^{-1} = 0$ in Step V.] However, if the contractivity hypothesis is dropped, this result is false. To see this, consider an averaging operator $T\colon L^p(\Sigma) \to L^p(\Sigma)$ on a probability space with $T1 = 1$ a.e. Then $T^2 = T$. If T is merely continuous, then by Moy's theorem (cf. Theorem 2.14) we have $Tf = E^{\mathcal{B}}(fg)$ for a unique $g \in L^{p'}(\Sigma)$ such that $E^{\mathcal{B}}(g) = 1$ a.e. But $g = 1$ a.e. is not necessarily true. If T is assumed positive, then we get $g \geq 0$ a.e. but not necessarily $g = 1$ a.e. Since every averaging projection satisfies the Reynolds identity, the assertion is true.

(c) Under the hypothesis of the theorem, since $V(t)f = f \circ \tau^t$ and so $V(t)$ is a positive operator, it follows that R is a positive operator by the representation (5). Hence R^* is also positive. Instead of the hypothesis that

$R^*f_0 = f_0$ for a weak unit, we may assume that R is positive on $L^p(\Sigma)$ and R^*f_0 is bounded for a weak unit $f_0 \in L^p \cap L^{p'}(\Sigma)$. Under these conditions (3) yields, on setting $d\nu = \tilde{f}_0 \, d\mu$, where $\tilde{f}_0 = R^*f_0$, on the space $L^p(\Omega, \Sigma, \nu)$:

$$\begin{aligned} \|(I - R)f\|_{2,\nu}^2 &= \int_\Omega \tilde{f}_0((I - R)f)^2 \, d\mu = \int_\Omega f_0 R[(I - R)f]^2 \, d\mu \\ &\leq \int_\Omega f_0 R f^2 \, d\mu \qquad \text{by} \quad (3) \\ &= \int_\Omega \tilde{f}_0 f^2 \, d\mu = \|f\|_{2,\nu}^2, \qquad f \in L^p \cap L^\infty. \end{aligned} \tag{24}$$

Thus on this new space $L^2(\Omega, \Sigma, \nu)$, $I - R$ is contractive and hence R is a continuous operator here. The proof can be extended to this case with various modifications. We shall sketch the special case that $p = 1$ in the Complements and Problems section. The general case will not be pursued here. If we drop $Rf_0 = f_0$ for a weak unit, then, in the corresponding representation (5), $\{V(t), t \geq 0\}$ *will not necessarily arise from a measure-preserving transformation*. (Equation (20) breaks down.) We shall touch on the latter point in Chapter V on the unification of the ergodic and martingale theories where this work will find one of its chief applications.

To illuminate the above remarks further, we prove two simple extensions of Theorem 2 above by weakening the hypothesis on the weak unit. The argument is adapted from the paper of Dinculeanu and the author [1].

5. Theorem *Let $R: L^p(\Sigma) \to L^p(\Sigma)$, $1 \leq p < \infty$, be a contractive Reynolds operator. If $\mathcal{M}_1$ is the range of R and $\mathcal{M}$ its closure, suppose there is a bounded support function $0 \leq f_0 \in \mathcal{M}$ which is a fixed point of R and of its adjoint R^*. This means if $f \in \mathcal{M}$, the support of f is contained in the support B_0 of f_0 except for a null set, and $Rf_0 = f_0$, $R^*f_0 = f_0$. Then there exists a σ-ring $\mathcal{B} \subset \Sigma$ with B_0 as its maximal element (i.e., $\mathcal{B} = \mathcal{B}(B_0) \subset \Sigma(B_0)$ in the notation of trace σ-algebras), a "generalized" conditional expectation $E^{\mathcal{B}}: L^p(\Sigma) \to L^p(\mathcal{B})$, and a strongly continuous contractive semigroup $\{V(t), t \geq 0\}$ of operators on $L^p(\mathcal{B})$ which are induced by a $(\mathcal{B}, \mathcal{B})$-measurable point mapping $\tau: B_0 \to B_0$ such that f_0 is invariant ($f_0 = f_0 \circ \tau$ a.e.) in terms of which we have*

$$Rf = \int_0^\infty e^{-t} V(t) E^{\mathcal{B}}(f) \, dt, \qquad f \in L^p(\Sigma). \tag{25}$$

Moreover, $\mathcal{M} = L^p(\mathcal{B})$ and, on $\mathcal{M}_1$, R and $E^{\mathcal{B}}$ commute, i.e., $Rf = RE^{\mathcal{B}}f = E^{\mathcal{B}}Rf$, for $f \in L^p(\Sigma)$.

Remark We must first explain the concept of a "generalized conditional expectation" in the statement of the theorem. This is a particular case of a general result of Dinculeanu [2]. Since $\mathscr{B}$ has a maximal element B_0 and since $f_0 \in L^p(\Sigma)$ implies B_0 is σ-finite, we conclude that $\mu | \mathscr{B}(B_0)$ is σ-finite. Hence by the Radon–Nikodým theorem (exactly as in Definition 1.1) the equation $\int_A E^{\mathscr{B}}(f)\, d\mu_{\mathscr{B}} = \int_A f\, d\mu$, $A \in \mathscr{B}$, μ-uniquely defines the contractive positive mapping $f \mapsto E^{\mathscr{B}}(f) \in L^p(\mathscr{B})$, the latter space being identified as the subspace of $L^p(\Sigma)$ of $\mathscr{B}$-measurable functions vanishing off B_0. It is also easy to deduce the averaging property: $E^{\mathscr{B}}(f\chi_A) = \chi_A E^{\mathscr{B}}(f)$ a.e., $A \in \mathscr{B}$, as well as most others. The prefix "generalized" is used to denote that (in contrast to Theorem 2.6 with the usual conditional expectations $E^{\mathfrak{F}}$ where $\mathfrak{F}$ is a σ-*algebra*) a contractive projection on $L^1(\Sigma)$ onto $L^1(\mathscr{B})$ need not be the conditional expectation as the following simple example indicates.

Let $\Omega = [0, 1]$, $\Sigma = \{\varnothing, A, A^c, \Omega\}$, and $\mu(A) = \mu(A^c) = \frac{1}{2}$. If $\mathscr{B} = \{\varnothing, A\}$, then $\mathscr{B}$ is a subring of Σ and $L^1(\Sigma) = \{a\chi_A + b\chi_{A^c} : a, b \text{ in } \mathbb{R}\}$, $L^1(\mathscr{B}) = \{a\chi_A : a \in \mathbb{R}\}$, and $E^{\mathscr{B}}(f) = a\chi_A$ for $f = a\chi_A + b\chi_{A^c}$. However, $Tf = (a + b)\chi_A$ also defines a contractive projection T on $L^1(\Sigma)$ onto $L^1(\mathscr{B})$, and $T \neq E^{\mathscr{B}}$. Thus the distinction is necessary. We again point out that the comment of Remark 3 about R^* applies here and will be used below.

Proof By a change of measures, we reduce this result to that of Theorem 2 and then translate it to the original space which implies (25). Such a technique has already been used in Theorem 2.11.

Since by hypothesis $B_0 = \{\omega : f_0(\omega) \neq 0\} \in \Sigma$ so that $Rf = \chi_{B_0} Rf$, we define a new measure ν on Σ as $d\nu = f_0^p\, d\mu$ and note that $f \in L^p(\Sigma, \nu)$ iff $|ff_0| \in L^p(\Sigma, \mu)$. Let $\tilde{f} = f_0$ on B_0 and $\tilde{f} = 1$ on $\Omega - B_0$. Define S: $L^p(\Sigma, \nu) \to L^p(\Sigma, \nu)$ by

$$Sf = R(ff_0)/\tilde{f} = \chi_{B_0} Rf = Rf \quad \text{a.e.}, \qquad f \in L^p(\Sigma, \mu), \tag{26}$$

where, by the Reynolds identity, $f_0 Rf = R(f_0 f)$ a.e. is used. Hence S is linear and $S | L^p(\Sigma, \mu) = R$. Since $\nu(\Omega) < \infty$, $1 \in L^p(\Sigma, \nu)$ and $S1 = 1$ a.e. Using (26) and evaluating the right and left sides of (2) separately for S and simplifying, it is immediate that S satisfies the Reynolds identity, and $S(L^p \cap L^\infty) \subset L^p \cap L^\infty$. Also $\|R\|_p \leq 1$. This implies S is a contraction in the $L^p(\Sigma, \nu)$-norm:

$$\int_\Omega |Sf|^p\, d\nu = \int_\Omega |R(f_0 f)|^p\, d\mu \leq \int_\Omega |f_0 f|^p\, d\mu = \int_\Omega |f|^p\, d\nu. \tag{27}$$

Note that by definition of $L^p(\Sigma, \nu)$, S is defined on all of this space, and 1 is a weak unit such that $S1 = 1$ a.e. In order to apply Theorem 2, we need only verfiy that $S^*1 = 1$ on using $R^*(f_0) = f_0$. This follows from the argument of Riesz–Nagy, for $p = 2$, as shown in the last part of Step III of Theorem 2.

If $p = 1$, the result is immediate since, for $f \in L^p(\Sigma, \nu) \cap L^\infty(\Sigma, \nu)$,

$$\int_\Omega (S^*1) f \, d\nu = \int_\Omega 1 \cdot S(f) \, d\nu = \int_\Omega \frac{R(f_0 f)}{\tilde{f}} f_0 \, d\mu = \int_\Omega R(f) f_0 \, d\mu$$

$$= \int_\Omega (R^* f_0) f \, d\mu = \int_\Omega f_0 f \, d\mu = \int_\Omega f \, d\nu, \tag{28}$$

so that $S^*1 = 1$ a.e., (ν). In the general case $1 < p < \infty$ one can show by a slightly involved argument (see Problem 8) that for the Reynolds operator S on $L^p(\Sigma, \nu)$ satisfying $S1 = 1$ a.e., we also have $S^*1 = 1$ a.e. Thus S satisfies the hypothesis of Theorem 2, so that $S(L^p(\Sigma, \nu))$ is dense in $L^p(\mathscr{B}, \nu)$ where $\mathscr{B} \subset \Sigma$ is a σ-ring (since the support of ν is B_0, $L^p(\Sigma, \nu) = L^p(\Sigma(B_0), \nu)$, the space of (Σ)-measurable functions vanishing off B_0 and pth-power integrable for ν). More explicitly, $\mathscr{B}$ is a σ-algebra in $\Sigma(B_0)$. It follows that (25) is a restatement of (5) from this reduction. We note that the semigroup $\{V(t), t \geq 0\}$ is generated by a ν-measure-preserving transformation τ on $B_0 \to B_0$ (i.e., $\nu \circ \tau^{-1} = \nu$), but this need not imply that it is also measure preserving for μ. Clearly $\nu(\tau^{-1}(A)) = \nu(A)$ for $A \in \mathscr{B}$ is the same statement as $f_0 \circ \tau = f_0$ a.e. This completes the proof of the theorem.

6. Remarks (a) The preceding result allows us to improve the Discussion 4(a) above. Namely, a contractive Reynolds operator $R: L^p(\Sigma, \mu) \to L^p(\Sigma, \mu)$, μ σ-finite, has a closed range iff $R = E^{\mathscr{B}}$, where $E^{\mathscr{B}}$ is a "generalized" conditional expectation with $\mathscr{B} \subset \Sigma(B_0)$, a σ-algebra, and B_0 is the maximal element of $\mathscr{B}$. In fact, the "if" being evident, let $\mathscr{M}_1 = R(L^p(\Sigma))$ be closed. Then (22) and (23) hold in the present case also, since $\mathscr{M}_1$ has bounded elements dense with the same proof as that of Step I of the proof of Theorem 2. It is then not difficult (with Theorem 2.5) to conclude that $\mathscr{M}_1 = L^p(\mathscr{B})$. We leave the by-now familiar argument to the reader. Hence $Rf = \chi_A \in \mathscr{M}_1$ implies $f = \chi_A$ a.e., so that R is a projection. It has been noted that $\mathscr{M}_1$ has a support f_0 which may be taken bounded since $\mu_{\mathscr{B}} = \mu | \mathscr{B}$ is σ-finite. Thus (25) holds for R, and as before, $R^2 = R$ implies $V(t) = I$ so that $R = E^{\mathscr{B}}$, where $E^{\mathscr{B}}$ is as described above.

(b) It will be useful to note that if a positive Reynolds operator on an $L^{p_0}(\Sigma)$ has a bounded fixed point f, then $|f|$ is also a fixed point (i.e., its range is a lattice), and if it is a contraction, then it is a contraction on all $L^p(\Sigma)$, $p \geq p_0$.

For by positivity, $Rf = f$ implies $|f| = |Rf| \leq R(|f|)$ and by the Reynolds identity $Rf^2 = (Rf)^2 + R[(f - Rf)^2] = f^2$ so that f^2 is a fixed point of R. But $f^2 = R(|f|)^2 \geq (R|f|)^2$. Hence $R(|f|) = |f|$ a.e. If R is a contraction on L^{p_0}, then by the classical Riesz interpolation theorem it is a contraction on all $L^p(\Sigma)$, $p_0 \le p \le \infty$ if it is so on $L^\infty(\Sigma)$ (cf. Dunford–Schwartz [1, Section VI.7]). However, this last statement about L^∞-contraction is immediate. Indeed, if $0 \le f \le 1$, then $f^2 \le f$ a.e., and so $Rf \ge Rf^2 \ge (Rf)^2$ (the last by (1)). Hence $0 \le Rf \le 1$ a.e. If $f \in L^\infty(\Sigma)$ is arbitrary and $\alpha = \|f\|_\infty > 0$, then $\tilde{f} = |f|/\alpha$ satisfies the above, and hence $R\tilde{f} \le 1$ or $|Rf| \le R|f| \le \alpha$ a.e. So $\|Rf\|_\infty \le \alpha = \|f\|_\infty$. We also note that if $\|R\| \le c$ on $L^{p_0}(\Sigma)$, then the interpolation theorem implies that for $p_0 \le p < \infty$, $\|R\| \le c^{p_0/p}$ on $L^p(\Sigma)$.

Before concluding this chapter, we give a simple extension of the above theory if the function spaces are vector valued. A general study involves new problems, but the following indicates the possibilities.

7. Corollary *Let $L^p(\Omega, \Sigma, \mu; \mathscr{X}) = L^p(\Sigma, \mathscr{X})$ be the Lebesgue space, with μ σ-finite, of functions $f\colon \Omega \to \mathscr{X}$, strongly measurable* (*cf.* Theorem I.4.11), *and $\|f\| \in L^p(\Omega, \Sigma, \mu)$ where $\|\cdot\|$ is the norm of the Banach space $\mathscr{X}$. Let $R\colon L^p(\Sigma, \mathscr{X}) \cap L^\infty(\Sigma, \mathscr{X}) \to L^p(\Sigma, \mathscr{X}) \cap L^\infty(\Sigma, \mathscr{X})$, $1 < p < \infty$ be a contractive Reynolds operator in the sense that for $f \in L^p(\Sigma, \mathscr{X})$ and $g \in L^p(\Sigma) \cap L^\infty(\Sigma)$ we have the identity* (1) *for R. Suppose R satisfies the following additional condition*: (∗) *for each $g \in L^p(\Sigma)$ there exists a scalar function $Sg \in L^p(\Sigma)$ such that*

(i) $R(gx) = (Sg)x$, $x \in \mathscr{X}$, *and*
(ii) $R(fSg) = (Rf)(Sg) + R[(f - Rf)(g - Sg)]$,
$f \in L^p(\Sigma, \mathscr{X}) \cap L^\infty(\Sigma, \mathscr{X})$.

If there exists $0 \le f_0 \in L^p(\Sigma) \cap L^{p'}(\Sigma) \cap L^\infty$ such that $R(f_0 x) = f_0 x$, $x \in \mathscr{X}$ where f_0 supports $R(L^p(\Sigma, \mathscr{X}))$, and $p^{-1} + p'^{-1} = 1$, then

$$R(fx) = \int_0^\infty e^{-t} V(t) E^{\mathscr{B}}(f)x\,dt, \qquad f \in L^p(\Sigma), \quad x \in \mathscr{X}, \tag{29}$$

where $E^{\mathscr{B}}$ is a generalized conditional expectation and $\{V(t), t \ge 0\}$ is a strongly continuous semigroup on $L^p(\mathscr{B}, \mathscr{X})$ on which $R = RE^{\mathscr{B}} = E^{\mathscr{B}}R$, the range of R being dense in $L^p(\mathscr{B}, \mathscr{X})$.

Proof Given R on $L^p(\Sigma, \mathscr{X})$, we can always define S^x for each $x \in \mathscr{X}$ on $L^p(\Sigma)$ by the equation $R(gx) = (S^x g)x$. Then $S^x\colon g \mapsto (S^x g)$ is contractive, but usually depends on x. Thus the hypothesis (∗) is a uniformity condition so that $S^x = S$. We check that S satisfies the hypothesis of Theorem 5 so that (29) is a consequence of (25).

If $x \in \mathscr{X}$ and $f \in L^p(\Sigma)$, then on noting that $\|f(\omega)x\| = |f(\omega)|\,\|x\|$, we have

$$\|Sf\|_p \|x\| = \|(Sf)x\|_p = \|R(fx)\|_p \le \|fx\|_p = \|f\|_p \|x\|. \tag{30}$$

Hence S is contractive on $L^p(\Sigma)$. If $f \in L^p(\Sigma) \cap L^\infty(\Sigma)$, $g \in L^p(\Sigma)$, then to show that S satisfies (1), we have

$$\begin{aligned} S(fg)x &= R(fgx) = R(fx)Sg + R[(fx - R(fx))(g - Sg)] \qquad \text{by} \quad (*) \\ &= (Sf)(Sg)x + R[(f - Sf)(g - Sg)x] \\ &= ((Sf)(Sg) + S[(f - Sf)(g - Sg)])x, \qquad x \in \mathscr{X}. \end{aligned} \tag{31}$$

Thus S is a Reynolds operator since $S(L^p \cap L^\infty(\Sigma)) \subset L^p \cap L^\infty(\Sigma)$ is evident. Also $R(f_0 x) = f_0 x = (Sf_0)x$. So f_0 is a fixed point of S. It is easy to see that f_0 supports the range of S in $L^p(\Sigma)$. We now use the result of Problem 8, with $1 < p < \infty$, to conclude that $S^* f_0 = f_0$ a.e., where S^* is the adjoint of S. Thus S satisfies the hypothesis of Theorem 5 for $1 < p < \infty$. Hence the representation (29) is an immediate consequence of (25). ($E^{\mathscr{B}}$ exists also on $L^p(\Sigma, \mathscr{X})$; cf. Chapter V.)

We present some applications of the general theory of this chapter and complements to it in the following section as problems.

Complements and Problems

1. The definition of conditional expectation can be generalized, and here is one such formulation. Let (Ω, Σ, P) be a probability space and for a σ-algebra $\mathscr{B} \subset \Sigma$, $\Omega_0 \in \Sigma$, define $\mathscr{B}_0 = \mathscr{B}(\Omega_0)$, the trace algebra. If $f\colon \Omega \to \mathbb{R}$ is P-measurable and $f\chi_{\Omega_0}$ is P-integrable, let $\nu\colon A \mapsto \int_A f\,dP$, $A \in \mathscr{B}_0$ be the signed measure. Then there exists a P-unique $\mathscr{B}_0$-measurable $\tilde{f}$ on Ω_0 such that $\bar{E}^{\mathscr{B}_0}\colon f \mapsto \tilde{f}$ defines a "modified" conditional expectation, having many of the properties of $E^{\mathscr{B}}$ given by Definition 1.1. If $\Omega_1 \subset \Omega_2 \subset \Omega$, $\Omega_i \in \Sigma$, $\mathscr{B}_i = \mathscr{B}(\Omega_i)$, $i = 1, 2$, are the traces, $\mathscr{B}_2(\Omega_1) \subset \mathscr{B}_1$, and fg, f are in $L^p(\Omega_1, \Sigma(\Omega_1), P)$ with $f = 0$ on $\Omega_2 - \Omega_1$, $f|\Omega_1$ is $\mathscr{B}_1$-measurable, show (with a simple modification of the textual proof) that

$$\bar{E}^{\mathscr{B}_2}(fg) = \bar{E}^{\mathscr{B}_2}(f\bar{E}^{\mathscr{B}_1}(g)) \qquad \text{a.e.} \quad \text{on} \quad \Omega_2.$$

With similar hypotheses on the varying domains and ranges of functions, extend the other results of Proposition 1.2. (This $\bar{E}^{\mathscr{B}_0}$ has applications in Markov processes and was discussed by Dynkin [1].)

2. Extensions of the classical Radon–Nikodým theorem (cf. Theorem I.2.7) to vector valued measures (cf. e.g., I.4.20) involve difficulties. Under certain conditions on these measures and on the Banach spaces, it is possible to obtain their derivatives, but now they are typically nonunique. For conditional probability measures, however, the result with a unique density is obtainable if the D–S integral is used as indicated below. Let (Ω, Σ, μ_i),

$i = 1, 2$, be finite measure spaces and $\mu_2 \ll \mu_1$ so that $\mu_2(A) = \int_A g\, d\mu_1$, $A \in \Sigma$, for a μ_1-unique (Σ)-measurable $g: \Omega \to \mathbb{R}^+$. If $\mathscr{B} \subset \Sigma$ is a σ-algebra, let $P_i^{\mathscr{B}}$ be the conditional probability function relative to μ_i, $i = 1, 2$. Show that

$$\int_B P_2^{\mathscr{B}}(A)\, d\mu_2 = \int_B E_1^{\mathscr{B}}(g) P^{\mathscr{B}}(A)\, d\mu_1, \qquad B \in \mathscr{B},$$

where $E_1^{\mathscr{B}}$ is the conditional expectation relative to μ_1. With this, prove that there is a positive μ_1-measurable, μ_1-unique g_1 such that $P_2^{\mathscr{B}}(A) = \int_A g_1\, dP_1^{\mathscr{B}}$, $A \in \Sigma$, where the integral is taken in the D–S sense. Verify that, in fact, $g_1 = g/E_1^{\mathscr{B}}(g)$ a.e. (μ_1). (Regarding nonuniqueness, and applications of this and other representations, one may consult the author's papers [6] and [7]).

3. A measurable space (Ω, Σ) is called a *standard Borel space*, if Σ is countably generated, $\{\omega\} \in \Sigma$ for each $\omega \in \Omega$, and there is a complete separable metric space $\tilde{\Omega}$ with $\tilde{\Sigma}$ as its Borel algebra such that Σ and $\tilde{\Sigma}$ are σ-isomorphic. If (Ω, Σ, P) is a probability space, (Ω, Σ) standard Borel, and $\mathscr{B} \subset \Sigma$ a σ-algebra, then show that a regular conditional probability $P^{\mathscr{B}}$ on Σ exists. [This can be obtained from Theorem 3.11 and the discussion there. On standard Borel spaces, see Parthasarathy [1] where another proof on the existence of regular $P^{\mathscr{B}}$ is found.]

4. Let T_1, T_2 be two operators on $L^p \to L^p$ that are bounded, linear, and satisfy the averaging identity (cf. Definition 2.1(i), a), so that T_i are continuous (not contractions). Suppose also that $T_i(L^\infty) \subset L^\infty$ and $T_i 1 = 1$ a.e., $i = 1, 2$. Show that there exist σ-algebras $\mathscr{B}_i \subset \Sigma$ such that $T_i f = E^{\mathscr{B}_i}(f h_i)$, for some h_i. [*Hint*: $\mathscr{B}_i = \{A \in \Sigma : \chi_A \in \mathscr{C}_i\}$, where $\mathscr{C}_i = \{f \in L^p : T_i(f Tg) = f(T_i g)$ for all $g \in L^\infty\}$, $i = 1, 2$.] Assume that T_i is also positive and that $T_1 f = 0$ a.e., $\Rightarrow T_2 f = 0$ a.e., and $\mathscr{B}_1 \supset \mathscr{B}_2$. Show that there is a ($\Sigma$)-measurable function $g: \Omega \to \mathbb{R}^+$ such that $T_2 f = E^{\mathscr{B}_2}(T_1(fg)) = T_1(E^{\mathscr{B}_2}(fg))$, $f \in L^p(\Sigma)$. [*Hint*: Use the chain rule for the Radon–Nikodým derivatives for the measures $\nu_i: A \mapsto \int_\Omega T_i \chi_A\, dP$, $A \in \Sigma$, $i = 1, 2$, and apply the first part. For related results, in the context of Stone algebra valued measures, see Wright [2].]

5. This problem discusses the "dependence" of $\mathscr{B}$ on the operator $E^{\mathscr{B}}$. Let $\{\mathscr{B}_\alpha, \alpha \in I\}$ be a net of sub σ-algebras of (Ω, Σ, P), and suppose that $\mathscr{B} = \sigma(\bigcup_{\alpha \in I} \mathscr{B}_\alpha)$. If $f \in L^p(\Sigma)$, $g_\alpha = E^{\mathscr{B}_\alpha}(f)$ and $g = E^{\mathscr{B}}(f)$, $p < \infty$. Show that $g_\alpha \to g$ in L^p-norm iff $g_\alpha \to g$ in P-measure, as $\alpha \uparrow$. [*Hints*: First consider the case that I is the set of integers. Verify with Theorem I.4.4 that $\{g_\alpha, g, \alpha \in I\} \subset L^p$ is uniformly integrable, on using the conditional Jensen inequality. Deduce the result in this case and show, by contradiction, that this must also hold for nets when it holds for all countable sequences.]

6. Using the above problem it is possible to show that the σ-algebras $\mathscr{B}_\alpha$ approach $\mathscr{B}$ in an "essential" sense as follows: For any net $\{\mathscr{B}_\alpha\}_{\alpha\in I} \subset \Sigma$ of σ-algebras of the probability space (Ω, Σ, P), we define the "lim sup" and "lim inf" as (the fact that lim inf $\subset$ lim sup below is clear):

$$\limsup_\alpha \mathscr{B}_\alpha = \bigcap_{\beta\in I} \sigma\Big(\bigcup_{\alpha\geq\beta} \mathscr{B}_\alpha\Big) \supset \liminf_\alpha \mathscr{B}_\alpha = \sigma\Big(\bigcup_{\beta\in I}\bigcap_{\alpha\geq\beta} \mathscr{B}_\alpha\Big).$$

If there is equality, we say that the $\{\mathscr{B}_\alpha\}_{\alpha\in I}$ converges essentially or in order sense. Assume that all σ-algebras are completed. The net $\{\mathscr{B}_\alpha\}_{\alpha\in I}$ in (Ω, Σ, P) is said to p-converge to $\mathscr{B}$ iff (i) $\mathscr{B} = \lim_\alpha \sup \mathscr{B}_\alpha$ and (ii) each $A \in \mathscr{B}$ can be p-approximated in the sense that for every collection $\{K_\alpha\}_{\alpha\in I}$ with $K_\alpha \in \mathscr{B}_\alpha$, $A \subset \lim_\alpha \sup K_\alpha$ and for any $\varepsilon > 0$, there exists a finite family $B_i \in \mathscr{B}_{\alpha_i}$, $B_i \subset K_{\alpha_i}$, $i = 1, \ldots, n$, $B = \bigcup_{i=1}^n B_i$, satisfying $\|\chi_B - \sum_{i=1}^n \chi_{B_i}\|_p < \varepsilon$, one has $P(A\Delta B) < \varepsilon$. Thus if each $A \in \mathscr{B}$ is covered by a $\{K_\alpha\}_{\alpha\in I}$ as above, then a finite subcollection can be extracted which "almost" covers A and is "almost" disjoint. Note that by finiteness of the measure P, lim sup K_α exists as a measurable set. This definition is due to Krickeberg [1]. Show that if $\mathscr{B}_\alpha$ p-converges to $\mathscr{B}$, then for any $f \in L^p(\Sigma)$, $1 \leq p < \infty$, $E^{\mathscr{B}_\alpha}(f) \to E^{\mathscr{B}}(f)$ in measure. Conversely, if this convergence in measure holds for every $f \in L^p(\Sigma)$, then $\mathscr{B}_\alpha$ p-converges to $\mathscr{B}$.

7. Consider a measure space (Ω, Σ, μ), and let $T: L^p(\Sigma) \to L^p(\Sigma)$ be a contractive projection, $1 \leq p < \infty$. If $\mathscr{B} = \{A \in \Sigma : T\chi_A = \chi_A \text{ a.e.}\}$ and T is also positive, show that $\mathscr{B}$ is a σ-ring and $T = E^{\mathscr{B}}$, the generalized conditional expectation in case we have: $\sup\{A : A \in \mathscr{B}\} = \Omega$ outside of a null set. Show, by an example, that this can be false if the additional hypothesis on $\mathscr{B}$ is dropped. Prove, however, that the result remains valid if the restriction on $\mathscr{B}$ is dropped but that T is also a contraction on $L^\infty(\Sigma)$ or only that $\|Tf\|_\infty \leq \|f\|_\infty$ for all $f \in L^p(\Sigma) \cap L^\infty(\Sigma)$. [*Hints*: First show that $\mathscr{B}$ is a σ-ring (easy part). Hence $L^p(\mathscr{B}) \subset \mathscr{M} = \text{range}\,(T)$. For the opposite inclusion show that $f \in \mathscr{M}$, $B \in \mathscr{B}$ imply $f\chi_B \in L^p(\mathscr{B})$ and then, with the additional condition, that $f \in L^p(\mathscr{B})$. So $\mathscr{M} = L^p(\mathscr{B})$. That $E^{\mathscr{B}} : L^p(\Sigma) \to L^p(\mathscr{B})$ is a positive contractive projection where $E^{\mathscr{B}}$ is a generalized conditional expectation is easily verified. Then $T = E^{\mathscr{B}}$ is a consequence, for $1 < p < \infty$, of the fact that L^p is a uniformly convex and "smooth" space since on such spaces there can be at most one contractive projection. This follows from a standard computation with ideas similar to those for the proof of Theorem 2.8. If $p = 1$, one has to show first that $T(f\chi_A) = \chi_A T(f)$ for $A \in \mathscr{B}, f \in L^1(\Sigma)$. When this is established, $Tf = E^{\mathscr{B}}(f)$, $f \in L^p(\Sigma)$ can be deduced from our earlier theorems since the above is essentially the averaging property. For the last part, after observing that $\mathscr{B}$ is a σ-ring, it should be shown that $\mathscr{M} = L^p(\mathscr{B})$ and then again that $T(f\chi_A) = \chi_A T(f)$, $A \in \mathscr{B}$, where $p = 1$.

These follow when we show that for $0 \leq f \in \mathcal{M}$, $\min(f, 1) \in \mathcal{M}$ (a condition analogous to the Stone–Daniell property in integration). This uses the L^∞-hypothesis crucially through $\min(f, 1) \in L^\infty(\Sigma)$. Next show that $T(f\chi_A) = 0$ for all $A \in \mathcal{B}$ implies (with the L^∞-hypothesis) that $Tf = 0$. Then deduce the averaging property so that the result follows as in the first part. Many of the details need careful computation. In this context, see the papers by Dinculeanu and the author [1] and by Wulbert [1].]

8. We present conditions on the fixed points of a Reynolds operator R and its adjoint R^* stated in Remark 6.3(2). Let $R: L^p(\Sigma) \to L^p(\Sigma)$, $1 \leq p < \infty$ be a contractive Reynolds operator which is weakly compact, i.e., R maps bounded sets into relatively weakly compact sets. [This is automatic when $p > 1$.] Suppose there exists a weak unit $f_0 \in L^p(\Sigma)$ such that $Rf_0 = f_0$ a.e. Then $f_0 \in (L^p(\Sigma))^*$ and $R^*f_0 = f_0$ a.e., where R^* is the adjoint of R. The same is true if f_0 is any bounded fixed point of R such that $f_0 \in (L^p(\Sigma))^*$, and (Ω, Σ, μ) is σ-finite (and complete). [*Hints*: We give an outline of proof. Observe that if $\mathcal{M} \subset L^p(\Sigma)$, $1 \leq p < \infty$, is a closed subspace and μ is σ-finite, then there is a σ-finite set $A_0 \in \Sigma$ which supports $\mathcal{M}$, i.e., for every $f \in \mathcal{M}$, $S_f = \operatorname{supp}(f) \subset A_0$ a.e. This follows from Theorem I.5.2. Hence there are $f_n \in \mathcal{M}$ and $S_{f_n} = B_n$, with $\bar{f} = \sum_{n=1}^{\infty} f_n(2^n \|f_n\|_p)^{-1} \in \mathcal{M}$ and $S_{\bar{f}} = A_0$, the support of $\mathcal{M}$. Thus there exists an element $f \in \mathcal{M}$ whose support is the same as that of $\mathcal{M}$. This fact will be useful both here and elsewhere. Note that if $\mathcal{M}$ is also a lattice, then, with f_n, $|f_n| \in \mathcal{M}$, and hence we may take $\bar{f} > 0$ on $S_{\bar{f}}$, the support of $\mathcal{M}$.

Since in the problem μ is σ-finite because of the existence of a weak unit in the first part and it is the hypothesis in the second (more general) part, we prove the latter. As above, a support set A_0 exists for $\mathcal{M} = \operatorname{cl}(\operatorname{range}(R))$ in $L^p(\Sigma)$. On the other hand by Step I of the proof of Theorem 6.2, $\mathcal{M}$ has bounded functions dense in it. (This implies that the $\bar{f} \in \mathcal{M}$ with $S_{\bar{f}} = A_0$ can be taken bounded also.) Thus by the same step, if $\mathcal{B} = \{A \in \Sigma : \chi_A \in \mathcal{M}\}$, then $\mathcal{B}$ is a σ-ring (since $p < \infty$) and A_0 is its maximal element, $\mathcal{M} = L^p(\mathcal{B})$. We remark that the same can be proved if we define $\mathcal{B}$ as $\mathcal{S} = \{S_f : f \in \mathcal{M}\}$, so that $\mathcal{S}$ is a σ-ring with A_0 as its maximal element and then using an argument of Theorem 2.5, we can deduce that $\mathcal{M} = L^p(\mathcal{B}) = L^p(\mathcal{S})$. Note that for $\mathcal{B}$ (or $\mathcal{S}$) $\mu_{\mathcal{B}}$ is σ-finite. Since R is a contractive weakly compact operator, the averages $(1/n)\sum_{i=0}^{n-1} R^i$ converge in the strong operator topology to a contractive projection Q on L^p onto $\mathcal{M}_0 = \{f : Rf = f\} \subset \mathcal{M}$, and $f_0 \in \mathcal{M}_0$ (cf. Dunford–Schwartz [1, p. 598]). Also by the same reference (cf. VI.4.8) the weak compactness of R implies a similar property of R^*. Hence by the above-quoted result $(1/n)\sum_{i=0}^{n-1} R^{*i} \to \bar{Q}$, in the strong operator topology, $\bar{Q}$ being a projection onto the fixed points of R^*. Next, considering weak operator topology for this result, we can conclude that $\bar{Q}$ is the adjoint Q^* of Q in $(L^p(\Sigma))^*$. By the Reynolds identity ($\mathcal{M}_0$ being complete) and the first part, bounded elements are dense in $\mathcal{M}_0$ and $\mathcal{M}_0 = L^p(\mathcal{B}_1)$ for a σ-ring $\mathcal{B}_1 \subset \mathcal{B}$ (with a maximal element), $\mu_{\mathcal{B}_1}$ is σ-finite, and f_0 is $\mathcal{B}_1$-measurable. Now the range and null space of Q^* are $\mathcal{N}_0^\perp$ and $\mathcal{M}_0^\perp$ where $\mathcal{N}_0^\perp$ ($\mathcal{M}_0^\perp$) is the annihilator of $\mathcal{N}_0$ ($\mathcal{M}_0$), the space $\mathcal{N}_0$ being the null space of Q. But $L^p(\Sigma) = \mathcal{M}_0 \oplus \mathcal{N}_0$, and $(L^p(\Sigma))^* = \mathcal{M}_0^\perp \oplus \mathcal{N}_0^\perp$. By classical results $\mathcal{N}_0^\perp = (\mathcal{M}_0)^* = (L^p(\mathcal{B}_1))^*$. Now $f_0 \in L^\infty(\Sigma) \cap L^p(\mathcal{B}_1) \cap (L^p(\Sigma))^*$ so that $f_0 \in \mathcal{N}_0^\perp$. Hence $R^* f_0 = f_0$ a.e. In this context, with more general function spaces, see the author's paper [2].]

9. In this problem we sketch a special extension of Theorem 6.2 for *positive* Reynolds operators on $L^1(\Sigma)$ with (Ω, Σ, μ) a σ-finite space. Thus R is continuous on $L^1(\Sigma)$ and by Remark 6.6(b) it is defined and continuous on all $L^p(\Sigma)$, $1 \le p \le \infty$, and is even a contraction on $L^\infty(\Sigma)$; but R may not be a contraction on other spaces. Then the adjoint $R^*: L^\infty(\Sigma) \to L^\infty(\Sigma)$ is defined and hence also acts on all $L^p(\Sigma)$, $1 < p \le \infty$. Let $\varphi = R^* 1$ have support $G \in \Sigma$, and let $Sf = \chi_G Rf$, $f \in L^1(\Sigma)$. Note that $R(\chi_G f) = \chi_G(Rf)$ a.e. Then S is a positive Reynolds operator on $L^p(\Sigma(G))$, $1 \le p \le \infty$, and has the same null space as R on $L^2(\Sigma)$. We thus may reduce the analysis to $L^2(\Sigma)$.

(a) Define $v(A) = \int_{G \cap A} \varphi \, d\mu$, $A \in \Sigma$. Then $v: \Sigma(G) \to \mathbb{R}^+$ is a measure and is equivalent to μ on it. If $L^2(G, \Sigma(G), v) = L^2(G, v)$ is the Hilbert space shown, prove that $L^2(G, \Sigma(G), \mu) = L^2(G, \mu)$ is norm dense in $L^2(G, v)$ and on this space the "new" Reynolds operator S is uniquely extendable to $L^2(G, v)$ and that $(I - S)$ is a contraction on the new space. [*Hint*: $R(f^2) \ge R([(I - R)f]^2)$.]

(b) Let $\tilde{S} = I - S$ be the contractive operator on $L^2(G, v)$ of (a). Then $(1/n) \sum_{k=0}^{n-1} \tilde{S}^k$ converges strongly to a contractive projection Q onto the fixed points of $\tilde{S}$ or equivalently onto the null space $\mathcal{N}$ of S in $L^2(G, v)$. Deduce that the projection $I - Q$ has for its range the closure of $(I - \tilde{S})(L^2(G, v))$, or equivalently the closure of $S(L^2(G, v))$, say, $\mathcal{M}$. Thus $L^2(G, v) = \mathcal{M} \oplus \mathcal{N}$, the direct sum. [*Hints*: The first statement is a consequence of the ergodic theorem used in Problem 8 above. The second statement (and hence the decomposition) is a consequence of another result in Dunford–Schwartz [1, VIII.5.2].]

(c) Show that $\mathcal{M} = L^2(G, \mathcal{B}, v)$ for some σ-ring $\mathcal{B} \subset \Sigma(G)$, and $S(L^2(G, v)) = \mathcal{M}$ (i.e., is closed) iff S is a (necessarily orthogonal) projection. [*Hints*: Recall that S is a Reynolds operator and hence $\mathcal{M}$ has an algebra of bounded functions dense in it. So the result follows as in Theorem 2.5 or in Step I of the proof of Theorem 6.2. From this the last part is deduced as in 6.4(a).]

(d) If R is a positive Reynolds operator on $L^1(\Sigma)$ and S is the unique extension of $\chi_G R$ to $L^2(G, \Sigma(G), v)$, then there exists a strongly continuous semigroup $\{V(t), t \geq 0\}$ on $L^2(G, \mathscr{B}, \nu)$ such that $V(t)(fg) = (V(t)f)(V(t)g)$ so that $V(t)f = f \circ \tau_t$ for a measurable but not necessarily measure-preserving point mapping τ_t: $G \to G$ such that

$$Sf = \int_0^\infty e^{-t} V(t) E^{\mathscr{B}}(f)\, dt, \qquad f \in L^2(G, \Sigma(G), \nu), \tag{$*$}$$

where $E^{\mathscr{B}}$ is the generalized conditional expectation on $L^2(G, \nu) \to L^2(G, \mathscr{B}, \nu)$. Moreover, if $\omega_0 = \frac{1}{2}(1 - \|R\|^{-1})$, where

$$\|R\| = \sup\{\|Rf\|_1 : \|f\|_1 \leq 1\},$$

then $\|V(t)\| \leq e^{\omega_0 t}$, $t \geq 0$, with $\|V(t)\| = \sup\{\|V(t)h\|_2 : \|h\|_2 \leq 1\}$, and, on the range of S, $E^{\mathscr{B}}$ and S commute. The representation $(*)$ then extends to all $L^p(G, \Sigma(G), \nu)$, $1 \leq p \leq \infty$. [*Hints*: First note that S is one-to-one on its range, as in the proof of Theorem 6.2, which is followed here. Then $D = I - S^{-1}$ is a closed densely defined operator on $L^2(G, \mathscr{B}, \nu)$. Since $\|Rf\|_{2,\mu}^2 \leq \|Rf^2\|_{1,\mu} \leq \|R\| \|f^2\|_{1,\mu}$, deduce by use of (2) of Section 6 that on $L^2(G, \nu)$ we have

$$2\int_G f(Sf)\, d\nu = \int_\Omega (Rf)^2\, d\mu + \int_\Omega R(Rf)^2\, d\mu \geq \int_\Omega (Rf)^2\, d\mu + \int_\Omega (Sf)^2\, d\nu$$

$$\geq \|Sf\|_{2,\varphi}^2 + \frac{1}{\|R\|}\|Rf\|_{2,\varphi}^2 \geq (1 + \|R\|^{-1})\|Sf\|_{2,\varphi}^2,$$

so that $\omega_0 \geq 1 - \|S\|_{2,\varphi}^{-1}$. Next consider the resolvent $S(\lambda, D)$ for $\lambda > \omega_0$ on the dense subspace $(\lambda I - D)(S(\mathscr{M}))$, and show, by an almost identical computation as in Step VI of the proof of Theorem 6.2, that $\|S(\lambda, D)\| \leq (\lambda - \omega_0)^{-1}$. This is sufficient to apply the Hille–Yosida theorem, and the rest of the proof is as in Theorem 6.2. This result was essentially given by Kopp *et al.* [1], with a somewhat different proof.]

10. This problem presents a simple sufficient condition for the regularity of the conditional probability measure, in the context of the discussion following Corollary 2.13. Let (Ω, Σ, P) be a probability space and $(S, \mathscr{S})$ a measurable space, where S is a metric space with $\mathscr{S}$ as its Borel σ-algebra. If $X: \Omega \to S$ is $(\Sigma, \mathscr{S})$-measurable, suppose $S_0 = X(\Omega) \in \mathscr{S}$, and let $\sigma(X) = X^{-1}(\mathscr{S}) \subset \Sigma$. Let $\mathscr{S}(S_0)$ be the trace of $\mathscr{S}$ on S_0. If $\mathscr{B} \subset \Sigma$ is a σ-algebra and $\mu_X(\cdot) = P^{\mathscr{B}} \circ X^{-1}: \mathscr{S} \to L^\infty(\mathscr{B})$ is the conditional probability function, then it is inner regular relative to X by Theorem 2.11. Show that there is a regular conditional probability $P(\cdot, \cdot): \sigma(X) \times \Omega \to [0, 1]$ such that $P(A, \omega) = \mu_X(B)(\omega)$ a.a. (ω), where $A = X^{-1}(B)$ for $B \in \mathscr{S}$. The point is that we are transferring the inner regularity of $\mu_X(\cdot)$ on $\mathscr{S}$ to the regularity of $P^{\mathscr{B}}$ on the sub-σ-algebra $\sigma(X)$ of Σ when the range S_0 of X is in $\mathscr{S}$, so that $\mathscr{S}$ may be replaced by $\mathscr{S}(S_0)$.

[*Hints*: Since $X(A) \subset S_0$ need not be in $\mathcal{S}(S_0)$, let B_1, B_2 in $\mathcal{S}(S_0)$ be such that $A = X^{-1}(B_1) = X^{-1}(B_2)$. Observe that $B_1 \cap S_0$ and $B_2 \cap S_0$ have the same $\mu_X(\cdot)(\omega)$-measure so that the $P(\cdot, \cdot)$ above is unambiguously defined on $\sigma(X)$ since $S_0 \in \mathcal{S}$ and $\mu_X(S_0) = 1$ a.e. Verify that this $P(\cdot, \cdot)$ satisfies the regularity conditions. The result is due to Doob [1], and it clarifies the problem at hand. This also shows that if $S = \mathbb{R}$ and $X(\cdot)$ is a coordinate function so that $X(\Omega) = \mathbb{R}$, a regular conditional probability $P(\cdot, \cdot)$ relative to X always exists on $\sigma(X)$ for any $\mathcal{B} \subset \Sigma$.]

CHAPTER

III

Projective and Direct Limits

In this chapter some abstract and topological extensions of Kolmogorov's existence theorem as projective limits of probability spaces are presented. This treatment also includes Prokhorov's theorem. In the abstract case, direct limits intervene, and a relevant part of the latter theory and a related projective limit characterization are presented. A brief account of these limits with conditional probabilities and vector measures is also discussed. Applications to entropy and information theory with extended sketches are included as complements and problems.

3.1 DEFINITION AND IMMEDIATE CONSEQUENCES

To motivate and introduce the basic ideas of this chapter, let us recall and recast the fundamental Kolmogorov–Bochner theorem (cf. Theorem I.3.2). Let $\{(\Omega_t, \mathcal{B}_t)_{t\in T}\}$ be a family of measurable spaces and let D be the directed set (by inclusion) of all finite subsets of the index set T. If $\Omega_\alpha = \times_{t\in\alpha}\Omega_t$, $\mathcal{B}_\alpha = \bigotimes_{t\in\alpha}\mathcal{B}_t$, let $P_\alpha: \mathcal{B}_\alpha \to [0, 1]$ be a probability for each $\alpha \in D$. For each pair $\alpha < \beta$ of D, let $\pi_{\alpha\beta}: \Omega_\beta \to \Omega_\alpha$ be the coordinate projection. The family $\{P_\alpha, \alpha \in D\}$ of probabilities on $\{\mathcal{B}_\alpha, \alpha \in D\}$ is said to be *compatible*, or *consistent*, iff $P_\alpha = P_\beta \circ \pi_{\alpha\beta}^{-1}$ for $\alpha < \beta$. Then the theorem considers a consistent family $\{(\Omega_\alpha, \mathcal{B}_\alpha, P_\alpha, \pi_{\alpha\beta}) : \alpha < \beta \text{ in } D\}$ and seeks to find a probability P on $(\Omega, \mathcal{B}) = \times_{t\in T}(\Omega_t, \mathcal{B}_t)$ such that its α-marginal is P_α, $\alpha \in D$, under some (topological) conditions. When $\Omega_t = \mathbb{R}$, $t \in T \subset \mathbb{R}$, and each P_α is determined by a distribution function $F_{t_1, \ldots, t_n}$ where $\alpha = (t_1, \ldots, t_n)$, then the classical Kolmogorov theorem (Theorem I.3.1) guarantees the existence of such a P on $(\Omega, \mathcal{B})$, $\Omega = \mathbb{R}^T$ and $\mathcal{B} = \bigotimes_{t\in T}\mathcal{B}_t$. The preceding observation on the compatible family $\{P_\alpha, \alpha \in D\}$, leading to a general measure problem with *projective* (or inverse) *systems* is due to Bochner [1]. This abstraction clarifies the structure of the existence problem and, as we indicate below, illuminates the close relation between Kolmogorov's ideas and (a set) martingale theory. Let us introduce the general concept of such a system.

1. Definition If $(D, <)$ is a directed set so that "$<$" defines a partial ordering on D and if α, β are any two elements of D, there is a $\gamma \in D$ such that $\gamma > \alpha$ and $\gamma > \beta$, let $\{(\Omega_\alpha, \Sigma_\alpha, P_\alpha): \alpha \in D\}$ be a family of real (or vector) measure spaces. Suppose there is given a set of mappings $\{g_{\alpha\beta}, \alpha < \beta, \alpha, \beta$ in $D\}$ such that (i) $g_{\alpha\beta}: \Omega_\beta \to \Omega_\alpha$ for each $\alpha < \beta$ and $g_{\alpha\beta}^{-1}(\Sigma_\alpha) \subset \Sigma_\beta$ (so $g_{\alpha\beta}$ is measurable for $(\Sigma_\beta, \Sigma_\alpha)$), and (ii) $g_{\alpha\beta}$s are compatible in the sense that for $\alpha < \beta < \gamma$, $g_{\alpha\beta} \circ g_{\beta\gamma} = g_{\alpha\gamma}$ and $g_{\alpha\alpha}$ is the identity. Then the abstract collection $\{(\Omega_\alpha, \Sigma_\alpha, P_\alpha, g_{\alpha\beta})_{\alpha<\beta}: \alpha, \beta$ in $D\}$ is called a *projective system of* (real or vector) *measure spaces* whenever $P_\alpha = P_\beta \circ g_{\alpha\beta}^{-1}$ for each $\alpha < \beta$. If all Ω_αs are topological spaces, then the $g_{\alpha\beta}$s are also required to be *continuous* mappings so that they are automatically measurable when each Σ_α is a Baire σ-algebra. And $\{(\Omega_\alpha, g_{\alpha\beta})_{\alpha<\beta}: \alpha, \beta$ in $D\}$ itself is called a *projective system of spaces.*

It is clear that in Theorem I.3.2, $g_{\alpha\beta} = \pi_{\alpha\beta}$, the coordinate projection, and in Theorem I.3.1, the P_αs are Lebesgue–Stieltjes probabilities. To proceed further, we need to define the concepts of a projective limit of a system of spaces and of probabilities. If $\{(\Omega_\alpha, g_{\alpha\beta})_{\alpha<\beta}: \alpha, \beta$ in $D\}$ is a projective system of spaces and $\Omega_D = \times_{\alpha\in D} \Omega_\alpha$ is their cartesian product, let $\Omega \subset \Omega_D$ be the set of elements, called *threads*, $\omega = \{\omega_\alpha, \alpha \in D\}$ such that for any $\alpha < \beta$ it is true that $\omega_\alpha = g_{\alpha\beta}(\omega_\beta)$. Then Ω is called the *projective limit* of the spaces $(\Omega_\alpha, g_{\alpha\beta})$ and denoted $\Omega = \lim_{\leftarrow}(\Omega_\alpha, g_{\alpha\beta})$. For this space we define $g_\alpha: \omega \mapsto \omega_\alpha$, $\alpha \in D$, $g_\alpha = g_{\alpha\beta} \circ g_\beta$ for $\alpha < \beta$. To compare this with familiar objects, note that in the case of Theorem I.3.1 or I.3.2, Ω can be identified with $\Omega_T = \times_{t\in T} \Omega_t$ in a natural way, i.e., there exists a bijective mapping $u: \Omega_T \to \Omega$ so that one does not need to distinguish between these two spaces. To see this, note that $g_{\alpha\beta} = \pi_{\alpha\beta}$, the coordinate projection for $\alpha < \beta$ (α, β in D—the class of all finite subsets of T). If $\Omega_\alpha = \times_{t\in\alpha} \Omega_t$, $\alpha \in D$, let $g_\alpha: \Omega \to \Omega_\alpha$ be as in the above description, and $\pi_\alpha: \Omega_T \to \Omega_\alpha$ be the coordinate projection. Then for $\alpha < \beta$, we have $\pi_\alpha = g_{\alpha\beta} \circ \pi_\beta$ since $g_{\alpha\beta} = \pi_{\alpha\beta}$. Now to find the desired u such that $g_\alpha \circ u = \pi_\alpha$, i.e.,

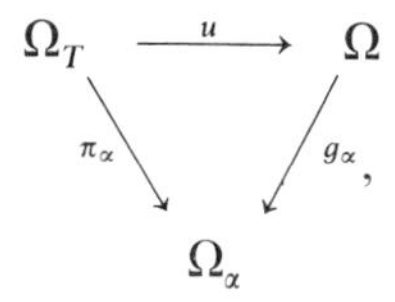

for each $\omega' = \{\omega_t, t \in T\} \in \Omega_T$ and each $\alpha \in D$, we need to show (since $\pi_\alpha(\omega) = \omega_\alpha = \{\omega_t, t \in \alpha\}$) that $g_\alpha(u(\omega')) = \omega_\alpha$. Considering $\{\omega_\alpha, \alpha \in D\} = \tilde{\omega}$ for each $\alpha < \beta$, one must have $g_{\alpha\beta}(g_\beta(u(\omega'))) = g_{\alpha\beta}(\omega_\beta) = \omega_\alpha$ due to the fact that $\pi_\alpha = g_{\alpha\beta} \circ \pi_\beta$. Hence for the element $\{\omega_\alpha, \alpha \in D\} = \tilde{\omega} \in \Omega$, a thread, one has $u: \omega' \mapsto \tilde{\omega}$ to be a mapping from Ω_T into Ω such that $g_\alpha \circ u = \pi_\alpha$. Also it is one-to-one since $u(\omega') = \tilde{\omega} = u(\omega'')$ implies the threads obtained from ω' and ω'' are equivalent in that each finite subset of ω' is the same as that of ω'' (i.e., all the coordinates of ω' are equal to all those of ω''). Hence $\omega' = \omega''$. The mapping is also onto. In fact, if $\tilde{\omega} = \{\omega_\alpha, \alpha \in D\} \in \Omega$ is an element, then for $t \in T$ we can find $\alpha \in D$ such that $\{t\} < \alpha$. Let $\omega_t = g_{\{t\}\alpha}(\omega_\alpha)$. Hereafter we write $g_{t\alpha}$ for $g_{\{t\}\alpha}$. Then ω_t is uniquely defined and does not depend on α since if $\beta \in D$ is another element satisfying $\{t\} < \beta$, there is a $\gamma \in D$ with $\gamma > \alpha$, $\gamma > \beta$, and

$$\begin{aligned} g_{t\alpha}(\omega_\alpha) &= g_{t\alpha} \circ g_{\alpha\gamma}(\omega_\gamma) = g_{t\gamma}(\omega_\gamma) = g_{t\beta} \circ g_{\beta\gamma}(\omega_\gamma) \\ &= g_{t\beta}(\omega_\beta) \qquad \text{since} \quad \tilde{\omega} \quad \text{is a thread.} \end{aligned} \tag{1}$$

Thus $g_{t\alpha}(\omega_\alpha) = g_{t\beta}(\omega_\beta) = \omega_t$ is independent of α. Also $g_{tt}(\omega_t) = \omega_t$. Let $\omega' = \{\omega_t, t \in T\}$. So $\omega' \in \Omega_T$ and $u(\omega') = \tilde{\omega}$ by the first part. This proves that Definition 1 is a generalization of Kolmogorov's notion of consistency.

If $\Omega = \lim_{\leftarrow}(\Omega_\alpha, g_{\alpha\beta})$ has sufficiently many points, i.e., if $\omega = \{\omega_\alpha, \alpha \in D\} \in \Omega$, $g_\alpha: \omega \mapsto \omega_\alpha$, $\alpha \in D$ is the corresponding mapping, then $g_\alpha(\Omega) = \Omega_\alpha$, let $\Sigma_\alpha^* = g_\alpha^{-1}(\Sigma_\alpha)$ be the σ-algebra on Ω. If $\alpha < \beta$, one has

$$\Sigma_\alpha^* = g_\alpha^{-1}(\Sigma_\alpha) = (g_{\alpha\beta} \circ g_\beta)^{-1}(\Sigma_\alpha) = g_\beta^{-1}(g_{\alpha\beta}^{-1}(\Sigma_\alpha)) \subseteq g_\beta^{-1}(\Sigma_\beta) = \Sigma_\beta^*, \tag{2}$$

so that $\Sigma_0 = \bigcup_{\alpha \in D} \Sigma_\alpha^*$ is an algebra even for the very general projective systems. Then the argument of the first part of the proof of Theorem I.3.2 (cf. Eq. (5) there) shows that the set function $P: \Sigma_0 \to \mathbb{R}^+$ (or $\mathscr{X}$ if all P_αs are $\mathscr{X}$-valued, $\mathscr{X}$ being a Banach space) given by $P(A) = P_\alpha(A_\alpha)$ for $A = g_\alpha^{-1}(A_\alpha)$, $A_\alpha \in \Sigma_\alpha$, is unambiguously defined and is additive. Let us state this as

2. Proposition *If $\{(\Omega_\alpha, \Sigma_\alpha, P_\alpha, g_{\alpha\beta})_{\alpha<\beta} : \alpha, \beta$ in $D\}$ is a general projective system of real, or vector (Banach) space $\mathscr{X}$, valued measures, if $\Omega = \lim_{\leftarrow}(\Omega_\alpha, g_{\alpha\beta})$ and if Ω is sufficiently rich (i.e., $g_\alpha(\Omega) = \Omega_\alpha$, $\alpha \in D$, where $g_\alpha: \omega \mapsto \omega_\alpha$ with $\omega = \{\omega_\alpha, \alpha \in D\}$), let $\Sigma_0 = \bigcup_{\alpha \in D} g_\alpha^{-1}(\Sigma_\alpha)$. Then there exists a $P_0: \Sigma_0 \to \mathbb{R}$ or $\mathscr{X}$, which is additive and which satisfies $P_\alpha = P \circ g_\alpha^{-1}$ for each $\alpha \in D$.*

In general P_0 is not σ-additive. In Kolmogorov's case each P_α is a Lebesgue–Stieltjes probability $\Omega_\alpha = \mathbb{R}^\alpha$, $\Omega = \mathbb{R}^T$, and Σ_0 is the algebra of cylindrical sets $\pi_\alpha^{-1}(A_\alpha)$ where A_α is a Borel set (its *base*). This made P_0 σ-additive, but Problem I.6.5 shows that such a conclusion may be false in the general case. If the function P_0 on Σ_0 is σ-additive, so that it has a unique σ-

additive extension P to $\Sigma = \sigma(\Sigma_0)$ by Theorem I.2.2, then P is called the *projective limit* of the consistent measures $\{P_\alpha, \alpha \in D\}$ and denoted by $P = \lim_{\leftarrow}(P_\alpha, g_{\alpha\beta})$. We then say that (Ω, Σ, P) is the *projective limit* of the projective system $\{(\Omega_\alpha, \Sigma_\alpha, P_\alpha, g_{\alpha\beta})_{\alpha<\beta} : \alpha, \beta \text{ in } D\}$, denoted $(\Omega, \Sigma, P) = \lim_{\leftarrow}(\Omega_\alpha, \Sigma_\alpha, P_\alpha, g_{\alpha\beta})$. Thus Kolmogorov's system admits a projective limit. Note that if this limit exists, it is unique.

It is to be observed that the generalization involved in Definition 1 is considerable. To see this, we first assert that $\Omega = \lim_{\leftarrow}(\Omega_\alpha, g_{\alpha\beta})$ can be small, or empty, even if each $g_{\alpha\beta}$ is onto and Ω_α has sufficiently many points. This is shown in the following two examples.

Example A Let $\Omega_n = (0, 1/n)$, $n \geq 1$, and let $g_{nr}: \Omega_r \to \Omega_n$ for $r > n$ be the inclusion map. Then it can easily be seen that $\{(\Omega_n, g_{nr}) : r > n \geq 1\}$ is a projective system and if $\Omega = \lim_{\leftarrow}(\Omega_n, g_{nr})$, then $\Omega = \varnothing$. In fact, if we set $K_{nr} = \{x = \{x_n, n \geq 1\} \in \times_{i=1}^{\infty} \Omega_i : g_{nr}(x_r) = x_n, n < r\}$, then by definition $\Omega = \bigcap\{K_{nr} : 1 \leq n < r < \infty\}$. If $y \in \Omega$, then $y = \{y_n, n \geq 1\}$ and $g_{nr}(y_r) = y_n$. But it is seen that $g_{nr}(x_r) = x_n \Rightarrow x_r = x_n$, $1 \leq n < r$. Hence $y_1 = y_2 = y_3 = \cdots = t$, and such a y cannot be in Ω since there is no $0 < t < 1/n$ for all n. Thus Ω is empty.

In this example the g_{nr}s are not onto. The next example, involving a little more construction, shows that the situation is unaltered even if they are onto. The example is based on a general result by Bourbaki [1].

Example B Let $T = [0, 1]$ and $(D, \prec)$ be the collection of all finite subsets of T, directed by inclusion. Then D has no countable cofinal subset. Indeed, if there existed such a set $\{\alpha_n, n \geq 1\}$, let $J = \bigcup_{n=1}^{\infty} \alpha_n \subset D$ so that J is countable, and if $\alpha \subset T - J$, then there can be no $\beta \in J$ such that $\alpha \prec \beta$ since $\alpha \cap \beta = \varnothing$, so we get a contradiction. Consider a set $X = \{x : x = (\alpha_1, \alpha_2, \ldots, \alpha_{2n}),\ \alpha_i \in D,\ \alpha_{2i-1} \prec \alpha_{2i} \text{ for } 1 \leq i \leq n, \text{ and } \alpha_{2i-1} \not\prec \alpha_{2j-1} \text{ for } 1 \leq j < i \leq n\}$. Evidently, X is nonempty. Define $r(x) = \alpha_{2n-1}$, $s(x) = \alpha_{2n}$ and call n the *length* of $x \in X$. For each $\alpha \in D$ define $\Omega_\alpha = \{x \in X : r(x) = \alpha\}$. Since one can take $x = \{\alpha, \alpha\}$, we note that $\Omega_\alpha \neq \varnothing$. If $x_\beta = (\alpha_1, \alpha_2, \ldots, \alpha_{2n}) \in \Omega_\beta$, $\alpha \prec \beta$, let $j = \inf\{i \geq 1 : \alpha \subset \alpha_{2i-1}\}$ and $j = n$ if the $\{\ \}$ is empty. Define $x_\alpha = (\alpha_1, \ldots, \alpha_{2j-2}, \alpha, \alpha_{2j})$. Since $r(x_\alpha) = \alpha$, $x_\alpha \in \Omega_\alpha$. Define $g_{\alpha\beta}: x_\beta \mapsto x_\alpha$. It is clear that $g_{\alpha\beta}(\Omega_\beta) \subset \Omega_\alpha$. To see it is onto, consider an element $x \in \Omega_\alpha$, so that $r(x) = \alpha$, $x = (\alpha_1, \ldots, \alpha_{2n-2}, \alpha, \alpha_{2n})$. If the lengths of α and β are the same, since $\alpha \prec \beta$, $x \in \Omega_\beta$ also, and if they are different, then they differ by two elements so that $x_\beta = (\alpha_1, \ldots, \alpha_{2n}, \beta, \alpha_{2n+2}) \in \Omega_\beta$ where $\alpha_{2n+2} \in D$, $\beta \subset \alpha_{2n+2}$. Since $\alpha \prec \beta$ ($j = 2n$ now), $g_{\alpha\beta}(x_\beta) = x$ and $g_{\alpha\beta}$ is onto. It is checked by a similar reasoning that $g_{\alpha\beta} \circ g_{\beta\gamma} = g_{\alpha\gamma}$ for any $\alpha \prec \beta \prec \gamma$ and $g_{\alpha\alpha}$ is the identity, so $\{(\Omega_\alpha, g_{\alpha\beta})_{\alpha<\beta} : \alpha, \beta \text{ in } D\}$ is a projective system. If $\Omega = \lim_{\leftarrow}(\Omega_\alpha, g_{\alpha\beta})$, then $\Omega = \varnothing$. This follows from the following two facts whose verification (as above) is left to the reader.

(a) If x_α, x_β is part of a thread, i.e., $x_\alpha \in \Omega_\alpha$, $x_\beta \in \Omega_\beta$ and there is $x_\gamma \in \Omega_\gamma$, $\gamma \succ \alpha$, $\gamma \succ \beta$ such that $x_\alpha = g_{\alpha\gamma}(x_\gamma)$, $x_\beta = g_{\beta\gamma}(x_\gamma)$, and if both x_α, x_β have the same length, then the preceding two relations, with x_γ and the definition of X, imply that $s(x_\alpha) = s(x_\beta)$, i.e., they must have the same last elements.

(b) If $x = \{x_\alpha, \alpha \in D\} \in \Omega$, then by the above result we have with $\beta = s(x_\alpha)$, where $x_\alpha = g_{\alpha\gamma}(x_\gamma)$ for some $\gamma \succ \alpha$, the set of βs to be countable and cofinal in D. Since we already observed that D has no such cofinal sequence, we must conclude that there can exist no such thread, i.e., $\Omega = \varnothing$.

These two examples show that an additional condition on the projective system is needed to proceed further and shows the extent of the generalization involved from the cartesian product systems of Section I.3. The following useful condition was introduced by Bochner [1].

3. Definition A projective system $\{(\Omega_\alpha, g_{\alpha\beta})_{\alpha<\beta} : \alpha, \beta \text{ in } D\}$ of spaces is *sequentially maximal* (s.m.) if for any sequence $\alpha_1 < \alpha_2 < \cdots$ in D and any point $\omega = \{\omega_\alpha, \alpha \in D\} \in \times_{\alpha \in D} \Omega_\alpha = \Omega_D$, satisfying $\omega_{\alpha_i} = g_{\alpha_i\alpha_{i+1}}(\omega_{\alpha_{i+1}})$ for $i \geq 1$, we have $\omega \in \Omega = \lim_{\leftarrow}(\Omega_\alpha, g_{\alpha\beta})$.

The s.m. condition is clearly satisfied if $g_{\alpha\beta}$s are coordinate projections. Here is an important known system with the s.m. property.

4. Proposition *Let $\{(\Omega_\alpha, g_{\alpha\beta})_{\alpha<\beta} : \alpha, \beta \text{ in } D\}$ be a projective system of Hausdorff topological spaces with Ω as their limit. Then Ω is a closed subset of Ω_D, and if each Ω_α is a nonvoid compact set, so also is Ω. In this case the s.m. condition holds automatically.*

Proof Let $C_{\alpha\beta} = \{\omega \in \Omega_D : \omega_\alpha = g_{\alpha\beta}(\omega_\beta), \alpha < \beta\}$. If $G_{\alpha\beta}$ is the graph of $g_{\alpha\beta}$ so that $G_{\alpha\beta} = \{(\omega_\alpha, \omega_\beta) : \omega_\alpha = g_{\alpha\beta}(\omega_\beta)\} \subset \Omega_\alpha \times \Omega_\beta$, then $C_{\alpha\beta} = G_{\alpha\beta} \times \Omega_{D-\{\alpha,\beta\}}$. Since a graph of a continuous mapping is closed and since clearly the limit space $\Omega = \bigcap\{C_{\alpha\beta} : \alpha < \beta, \alpha, \beta \text{ in } D\}$, we conclude that Ω is closed in Ω_D.

If each Ω_α is compact, then Ω_D is compact by Tychonov's theorem so that its closed subset Ω is also compact (but may be empty). Suppose now that each $\Omega_\alpha \neq \varnothing$ and compact. We show that the nonempty compact family $\{C_{\alpha\beta} : \alpha < \beta \text{ in } D\}$ has the finite intersection property, which then implies that Ω is nonempty by a classical result since $\Omega = \bigcap C_{\alpha\beta}$. Let $\alpha_i < \beta_i$, $i = 1, \ldots, n$ be n-pairs from D. By directedness there is $\gamma (\in D)$ such that γ dominates all α_i, β_i, $i = 1, \ldots, n$. Now define with any $\omega_\gamma \in \Omega_\gamma$ the elements $\omega_{\alpha_i} = g_{\alpha_i\gamma}(\omega_\gamma)$ and $\omega_{\beta_i} = g_{\beta_i\gamma}(\omega_\gamma)$. Let $\omega \in \Omega_D$ be any element with $2n$ of its coordinates as $\omega_{\alpha_i}, \omega_{\beta_i}$, $i = 1, \ldots, n$. Since $\alpha_i < \beta_i < \gamma$ and

$g_{\alpha_i\gamma} = g_{\alpha_i\beta_i} \circ g_{\beta_i\gamma}$ by compatibility, we get $\omega_{\alpha_i} = g_{\alpha_i\beta_i}(\omega_{\beta_i})$ and hence that $\omega \in C_{\alpha_i\beta_i}$, $i = 1, \ldots, n$ or $\omega \in \bigcap_{i=1}^{n} C_{\alpha_i\beta_i}$. So $\Omega = \bigcap_{\alpha,\beta} C_{\alpha\beta} \neq \varnothing$.

To see that the s.m. condition holds, let $E_\alpha = g_\alpha^{-1}(\{\omega_\alpha\})$, where $g_\alpha\colon \omega \mapsto \omega_\alpha$, $\omega = \{\omega_\alpha, \alpha \in D\} \in \Omega$ $(\neq \varnothing)$. By the continuity of g_α, E_α is a closed (hence compact) nonempty subset of Ω. Let $\alpha_1 < \alpha_2 < \cdots$ be a sequence of D. Observe that $E_{\alpha_n} \supset E_{\alpha_{n+1}}$. In fact, since $g_{\alpha_n} = g_{\alpha_n\alpha_{n+1}} \circ g_{\alpha_{n+1}}$, we have $\omega_{\alpha_n} = g_{\alpha_n\alpha_{n+1}}(\omega_{\alpha_{n+1}})$ and then

$$E_{\alpha_n} = g_{\alpha_n}^{-1}(\{\omega_{\alpha_n}\}) = g_{\alpha_{n+1}}^{-1}(g_{\alpha_n\alpha_{n+1}}^{-1}(\{\omega_{\alpha_n}\})) \supset g_{\alpha_{n+1}}^{-1}(\{\omega_{\alpha_{n+1}}\}) = E_{\alpha_{n+1}}. \tag{3}$$

Since $E_{\alpha_n} \neq \varnothing$, $n \geq 1$, and compact, $\bigcap_{n=1}^{\infty} E_{\alpha_n} \neq \varnothing$. If ω is any point of this intersection, then $g_{\alpha_n}(\omega) = \omega_{\alpha_n}$, $n \geq 1$ so that the s.m. condition holds. This completes the proof.

Remark Since every element of Ω_α can be considered as part of some such sequence with $\alpha = \alpha_1 < \alpha_2 < \cdots$, the preceding argument shows that the s.m. condition implies $g_\alpha(\Omega) = \Omega_\alpha$, $\alpha \in D$ (and a fortiori $g_{\alpha\beta}(\Omega_\beta) = \Omega_\alpha$).

It should be noted, however, that the s.m. condition is not necessary for $\Omega \neq \varnothing$ and $g_\alpha(\Omega) = \Omega_\alpha$, $\alpha \in D$. Recently Millington and Sion [1] gave examples of (topological) projective systems having the above property but not satisfying the s.m. condition. In the context of (Radon) measure systems the spaces there satisfy an "almost s.m." condition. Because of its simplicity, we shall use the s.m. hypothesis in this work.

The following important result on the existence of projective limit measures has been established by Bochner [1] immediately after formulating the abstract generalization of Kolmogorov's basic idea. The behavior of the limit measure is an additional problem, and it has been solved by Choksi [1] in the important compact case.

5. Theorem *Let $\{(\Omega_\alpha, \Sigma_\alpha, P_\alpha, g_{\alpha\beta})_{\alpha<\beta} : \alpha, \beta$ in $D\}$ be a projective system of Hausdorff topological probability measure spaces satisfying the s.m. condition where each Σ_α is the Borel σ-algebra of Ω_α (i.e., Σ_α is generated by the closed sets of Ω_α) and P_α is inner regular for compact sets. Then $(\Omega, \Sigma, P) = \lim_{\leftarrow}(\Omega_\alpha, \Sigma_\alpha, P_\alpha, g_{\alpha\beta})$ exists, and P is inner regular on the cylindrical algebra $\Sigma_0 = \bigcup_\alpha g_\alpha^{-1}(\Sigma_\alpha)$. If each Ω_α is compact, in the above, then P is a Baire regular probability on Σ and hence has a unique extension to be a Radon probability on Ω.*

A proof of the existence assertion can be given with a few modifications of that of Theorem I.3.2. The latter is in fact a specialization of the above theorem. That P is a Baire regular probability on Ω is nontrivial and needs an independent proof for which we refer the reader to Choksi [1]. However, it is a special case of Prokhorov's theorem which is proved in Section 2. So we omit a further discussion of the above result here. The point of this theorem, and of the general formulation of projective limits by Bochner, is that a whole new set of important results of considerable interest for stochastic analysis and elsewhere emerged. We wish to present some of these in this chapter. The following concept is pertinent.

6. Definition Let $\{\Sigma_\alpha, \alpha \in D\}$ be an (increasing) net of σ-algebras from a measurable space (Ω, Σ). If $P_\alpha: \Sigma_\alpha \to \mathbb{R}$ (*or* $\mathscr{X}$, a Banach space) is a σ-additive function, then the family $\{P_\alpha, \alpha \in D\}$ is called a *set martingale* if for each $\alpha < \beta$ in D, $P_\beta | \Sigma_\alpha = P_\alpha$ and $\{\Sigma_\alpha, \alpha \in D\}$ is its *base*. If each P_α is only finitely additive and each Σ_α is an algebra ($P_\beta | \Sigma_\alpha = P_\alpha$ again), then $\{P_\alpha, \alpha \in D\}$ is an *additive set martingale* with base $\{\Sigma_\alpha, \alpha \in D\}$.

We record the following simple but important observation.

7. Proposition *With every projective system* $\{(\Omega_\alpha, \Sigma_\alpha, P_\alpha, g_{\alpha\beta})_{\alpha<\beta} : \alpha, \beta \text{ in } D\}$ *having the s.m. property, there is associated a set martingale as follows. Let* $\Omega = \lim_{\leftarrow}(\Omega_\alpha, g_{\alpha\beta})$ *and* $g_\alpha: \Omega \to \Omega_\alpha$ *be the corresponding mappings. (Here s.m. may be replaced by the assumption that* $g_\alpha(\Omega) = \Omega_\alpha$*.) If* $\Sigma_\alpha^* = g_\alpha^{-1}(\Sigma_\alpha)$ *and* $P_\alpha^*: \Sigma_\alpha^* \to \mathbb{R}$ *(or* $\mathscr{X}$*) is given by* $P_\alpha^* \circ g_\alpha^{-1} = P_\alpha$*, so that* P_α^* *is uniquely defined, and if* $\Sigma = \sigma(\bigcup_\alpha \Sigma_\alpha^*)$*, then* $\{(\Omega, \Sigma_\alpha^*, P_\alpha^*) : \alpha \in D\}$ *is a set martingale where* $\{\Sigma_\alpha^*, \alpha \in D\}$ *is its base from* (Ω, Σ)*. On the other hand, if* $\{(\Omega, \mathscr{B}_\alpha, P_\alpha) : \alpha \in D\}$ *is a set martingale with* $\{\mathscr{B}_\alpha, \alpha \in D\}$ *as its base, then it is always a projective system with* $\Omega_\alpha \equiv \Omega$, g_α, $g_{\alpha\beta}$ *being identity mappings for all* α, β *in* D $(\alpha < \beta)$.

Since by (2) $\Sigma_\alpha^* \subset \Sigma_\beta^*$ if $\alpha < \beta$ and hence $P_\alpha^* = P_\beta^* | \Sigma_\alpha^*$, the proof is immediate. Suppose that there is a finite measure μ on Σ such that each P_α^* is μ-continuous. In the vector case, we need to assume that $\mathscr{X}$ is either a separable reflexive or adjoint space. Then by the Radon–Nikodým theorem (the vector case will be proved in Chapter V), there is a μ-unique Σ_α^*-measurable function $f_\alpha: \Omega \to \mathbb{R}$ (or $\mathscr{X}$) such that

$$P_\alpha^*(A) = \int_A f_\alpha \, d\mu_{\Sigma_\alpha^*}, \qquad A \in \Sigma_\alpha^*. \tag{4}$$

The family $\{f_\alpha, \Sigma_\alpha^*, \alpha \in D\}$ is a *martingale of point functions* on (Ω, Σ, μ). These processes are important (the next two chapters are therefore devoted to them), and the above proposition thus connects the projective systems of measures and general martingales. In view of this and other reasons, we now undertake a deeper analysis of projective systems and their limits than occurred thus far.

3.2 SOME CHARACTERIZATIONS OF PROJECTIVE LIMITS

Let us start with a projective system of abstract probability spaces and present sufficient conditions on the system in order that it admits a limit. In some ways it is based on (the proof of) Theorem I.3.2. To state the desired result, we introduce precisely the idea of a compact class, which already appeared in Section II.3, and it is due to Marczewski [1].

1. Definition (a) A family $\mathscr{C}$ of subsets of a point set Ω is called a *compact class* if for any $\{\mathscr{C}_n, n \geq 1\} \subset \mathscr{C}$ with $\bigcap_{n=1}^{\infty} C_n = \varnothing$ there exists an integer $n_0 \geq 1$ such that $\bigcap_{n=1}^{n_0} C_n = \varnothing$.

(b) An additive function $P: \Sigma_0 \to \mathbb{R}^+$, where Σ_0 is an algebra on Ω, is said to have the *approximation property* relative to a compact class $\mathscr{C} \subset \Sigma_0$ if for each $A \in \Sigma$, $P(A) = \sup\{P(C) : C \subset A, C \in \mathscr{C}\}$. Such a P is also called a *compact measure* relative to $\mathscr{C}$.

By the argument used in the proof of Theorem I.3.2, we see that P is σ-additive on Σ_0 if it is compact relative to some compact class $\mathscr{C} \subset \Sigma_0$. Then it has a unique σ-additive extension to $\Sigma = \sigma(\Sigma_0)$.

Using these concepts, we may present the following result on the existence of projective limits, which is sufficient for many applications. It is due to Choksi [1]. The fact that all the conditions are needed and, e.g., (iii) may not be dropped was observed by Métivier [1].

2. Theorem *Let $\{(\Omega_\alpha, \Sigma_\alpha, P_\alpha, g_{\alpha\beta})_{\alpha<\beta} : \alpha, \beta$ in $D\}$ be a projective system of probability spaces satisfying the s.m. condition. For each $\alpha \in D$, let $\mathscr{C}_\alpha \subset \Sigma_\alpha$ be a compact class relative to which P_α is compact. Suppose for each α, β in D $(\alpha < \beta)$ (i) $g_{\alpha\beta}(\mathscr{C}_\beta) \subset \mathscr{C}_\alpha$, (ii) $g_{\alpha\beta}^{-1}(\{\omega_\alpha\}) \cap \mathscr{C}_\beta$ is a compact class for each $\omega_\alpha \in \Omega_\alpha$, and (iii) for each α, β in D and $A_\alpha \in \mathscr{C}_\alpha$, $B_\beta \in \mathscr{C}_\beta$ there is a $\gamma \in D$ and $C_\gamma \in \mathscr{C}_\gamma$ such that $\gamma > \alpha$, $\gamma > \beta$, and $C_\gamma = g_{\alpha\gamma}^{-1}(C_\alpha) \cap g_{\beta\gamma}^{-1}(C_\beta)$. Then the projective limit (Ω, Σ, P) of the system exists, and the limit is a "compact measure space" in the following sense: if $\mathscr{C}^* = \bigcup_{\alpha\in D} g_\alpha^{-1}(\mathscr{C}_\alpha) \subset \Sigma_0$, then $\mathscr{C}^*$ is a compact class and for each $A \in \Sigma$ and $\varepsilon > 0$ there is a $C \in (\mathscr{C}^*)_\delta \subset \Sigma$, with $P(A - C) < \varepsilon$, where $\Sigma_0 = \bigcup_\alpha g_\alpha^{-1}(\Sigma_\alpha)$ and Σ is the σ-algebra generated by Σ_0.*

Proof Since g_α^{-1} is injective and preserves set operations, it follows that the compactness of $\mathscr{C}_\alpha$ implies that of $g_\alpha^{-1}(\mathscr{C}_\alpha) = \mathscr{C}_\alpha^*$ (say). Also $\mathscr{C}^* = \bigcup_\alpha \mathscr{C}_\alpha^* \subset \Sigma_0 = \bigcup_\alpha g_\alpha^{-1}(\Sigma_\alpha) = \bigcup_\alpha \Sigma_\alpha^*$. By Proposition 1.2, there is a (unique) additive set function P on Σ_0 such that $P_\alpha = P \circ g_\alpha^{-1}$ for all $\alpha \in D$.

So for any $A \in \Sigma_0$ (thus $A \in \Sigma_\alpha^*$ for some α and $A = g_\alpha^{-1}(A_\alpha)$, $A_\alpha \in \Sigma_\alpha$),

$$\begin{aligned} P(A) = P_\alpha(A_\alpha) &= \sup\{P_\alpha(C): C \subset A_\alpha, C \in \mathscr{C}_\alpha\} \\ &= \sup\{P(g_\alpha^{-1}(C)): g_\alpha^{-1}(C) \subset A, g_\alpha^{-1}(C) \in \mathscr{C}^*\}. \end{aligned} \tag{1}$$

Hence $\mathscr{C}^*$ is an approximating class for P on Σ_0. Consequently, by the remarks preceding the statement of the theorem, it suffices to show that $\mathscr{C}^*$ is a compact class in Σ_0 so that P is σ-additive on Σ_0, and then the compactness of P on Σ_0 is an immediate consequence of (1). From this it follows directly, as in Corollary I.3.4, that P has the approximation property relative to $(\mathscr{C}^*)_\delta$.

Let $(\mathscr{C}^*)_d = \{A = \bigcap_{i=1}^n A_i : A_i \in \mathscr{C}^*, i = 1, \ldots, n, \text{ some } n\}$. We note that $(\mathscr{C}^*)_d = \mathscr{C}^*$. In fact, since clearly $\mathscr{C}^* \subset (\mathscr{C}^*)_d$, let $A \in (\mathscr{C}^*)_d$ for the opposite inclusion. Then $A = \bigcap_{i=1}^n A_i$ for some n, $A_i \in \mathscr{C}^*$. Next consider $A_i \in \mathscr{C}_{\alpha_i}^*$, $i = 1, \ldots, n$, so that $A_i = g_{\alpha_i}^{-1}(A_{\alpha_i})$, $A_{\alpha_i} \in \mathscr{C}_{\alpha_i}$. Hence by (iii) there is a $\gamma > \alpha_i$, $i = 1, \ldots, n$, and an $A_\gamma \in \mathscr{C}_\gamma$ such that $A_\gamma = \bigcap_{i=1}^n g_{\alpha_i\gamma}^{-1}(A_{\alpha_i})$. But then

$$g_\gamma^{-1}(A_\gamma) = \bigcap_{i=1}^n g_\gamma^{-1} \circ g_{\alpha_i\gamma}^{-1}(A_{\alpha_i}) = \bigcap_{i=1}^n g_{\alpha_i}^{-1}(A_{\alpha_i}) = A$$

since $g_\alpha = g_{\alpha\gamma} \circ g_\gamma$. So $A \in \mathscr{C}_\gamma^*$ and $(\mathscr{C}^*)_d \subset \mathscr{C}^*$. Hence $\mathscr{C}^*$ is closed under finite intersections.

Thus it suffices to show that if $A_{\alpha_n} \in \mathscr{C}^*$, $A_{\alpha_n} \supset A_{\alpha_{n+1}}$, for $\alpha_n < \alpha_{n+1}$ is any sequence with the finite intersection property (i.e., $\bigcap_{i=1}^n A_{\alpha_i} = A_{\alpha_n} \neq \varnothing$, any n), then one has $\bigcap_{n=1}^\infty A_{\alpha_n} \neq \varnothing$ since any $\{A_{\alpha_n}\}_1^\infty \subset \mathscr{C}^*$ with $\bigcap_{i=1}^n A_{\alpha_i} \neq \varnothing$ can be replaced by a monotone sequence in $\mathscr{C}^*$ because $\mathscr{C}^* = (\mathscr{C}^*)_d$.

Let $g_{\alpha_n}(A_{\alpha_n}) = C_{\alpha_n} \in \mathscr{C}_{\alpha_n}$. By the s.m. condition g_{α_n}s are onto so that $C_{\alpha_n} \neq \varnothing$. Also by (i) of the hypothesis $g_{\alpha_n\alpha_{n+1}}(C_{\alpha_{n+1}})$ is an element of $\mathscr{C}_{\alpha_n}$. Moreover,

$$g_{\alpha_n\alpha_{n+1}}(C_{\alpha_{n+1}}) = g_{\alpha_n\alpha_{n+1}} \circ g_{\alpha_{n+1}}(A_{\alpha_{n+1}}) = g_{\alpha_n}(A_{\alpha_{n+1}}) \subset g_{\alpha_n}(A_{\alpha_n}). \tag{2}$$

Hence the set $\{g_{\alpha_n\alpha_j}(C_{\alpha_j})\}_{j \ge n}$ is a monotone decreasing family of nonempty sets, and by the compactness of $\mathscr{C}_n$, as well as the compatibility of $\{g_{\alpha\beta}\}_{\alpha,\beta}$, one has

$$\varnothing \neq \tilde{D}_n = \bigcap_{j=n}^\infty g_{\alpha_n\alpha_j}(C_{\alpha_j}) = \bigcap_{j=n}^\infty g_{\alpha_n}(A_{\alpha_j}). \tag{3}$$

By (ii) for each $\omega_n \in \tilde{D}_n$, $g_{\alpha_n\alpha_{n+1}}^{-1}(\{\omega_n\}) \cap \mathscr{C}_{\alpha_{n+1}}$ is a compact class. The idea of the proof from now on is to show that one can find $\omega_n \in \tilde{D}_n$ such that $g_{\alpha_n\alpha_{n+1}}(\{\omega_{n+1}\}) = \omega_n$ for each n so that by the s.m. condition there exists a point $\omega \in \Omega$ with $g_{\alpha_n}(\omega) = \omega_{\alpha_n} \in \tilde{D}_n$, and therefore also $\omega \in g_{\alpha_n}^{-1}(\{\omega_n\}) \subset A_{\alpha_n}$ for all n, i.e., $\omega \in \bigcap_{n=1}^\infty A_{\alpha_n}$, and hence $\mathscr{C}^*$ is compact as asserted.

Let $\omega_n \in \tilde{D}_n$ and consider $g_{\alpha_n\alpha_{n+1}}^{-1}(\{\omega_n\}) \cap g_{\alpha_{n+1}\alpha_j}(C_{\alpha_j})$ for $j \ge n+1$. This set is nonempty for every $j \ge n+1$, because $\omega_n \in g_{\alpha_n}(A_{\alpha_j}) =$

$g_{\alpha_n\alpha_{n+1}} \circ g_{\alpha_{n+1}}(A_{\alpha_j})$ for all $j \geq n+1$ so that there is an $\omega_0 \in g_{\alpha_{n+1}}(A_{\alpha_j}) = g_{\alpha_n\alpha_{n+1}}(C_{\alpha_j})$ (cf. (2)), with $\omega_n = g_{\alpha_n\alpha_{n+1}}(\omega_0)$, or $\omega_0 \in g_{\alpha_n\alpha_{n+1}}^{-1}(\{\omega_n\})$. This implies the finite intersection property and hence, by compactness in (ii), $g_{\alpha_n\alpha_{n+1}}^{-1}(\{\omega_n\}) \cap \tilde{D}_{n+1} \neq \emptyset$.

Since $\omega_n \in \tilde{D}_n$ is arbitrary, $g_{\alpha_n\alpha_{n+1}}^{-1}(\tilde{D}_n) \cap \tilde{D}_{n+1} \neq \emptyset$ for all $n \geq 1$. Consequently, choose any $\omega_1 \in \tilde{D}_1$ and then by this equation there is an $\omega_2 \in \tilde{D}_2$ such that $\omega_1 = g_{\alpha_1\alpha_2}(\omega_2)$. Next choose $\omega_3 \in \tilde{D}_3$ such that $\omega_2 = g_{\alpha_2\alpha_3}(\omega_3)$, etc., by induction. Thus $\omega_n \in \tilde{D}_n \subset \Omega_{\alpha_n}$ and $g_{\alpha_n\alpha_{n+1}}(\omega_{n+1}) = \omega_n$. This satisfies the requirements for the s.m. hypothesis, and hence the proof is complete.

The next result presents a characterization of projective limits in case the projective system is topological. It is a natural generalization of the last part of Theorem 1.5. Let us motivate the result sought for. When each Ω_α $(\neq \emptyset)$ is compact and P_α is regular on the Baire σ-algebra of Ω_α, then $\Omega = \lim_{\leftarrow}(\Omega_\alpha, g_{\alpha\beta})$ is compact and P on (Ω, Σ) is also a (Baire) regular probability. The reason for the regularity of P on the Baire algebra Σ of Ω is that it is only necessary to know its values on each compact Baire set $A \subset \Omega$, and this fact is implied by the values of P on $\{g_\alpha(A), \alpha \in D\}$ or by the values P_α on $g_\alpha(A)$, $\alpha \in D$. When each Ω_α is not necessarily compact, then one takes the above property as the defining condition for the inner regularity of P_α, i.e., one *assumes* that there is a single compact set $K_\varepsilon \subset \Omega$ such that $P_\alpha(\Omega_\alpha - g_\alpha(K_\varepsilon)) < \varepsilon$ for all $\alpha \in D$. If each Ω_α is compact, then we take $K_\varepsilon = \Omega$, so that $g_\alpha(\Omega) = \Omega_\alpha$ and the above condition is true automatically whatever $\varepsilon > 0$ is. The interesting point of the proposed result is that this condition characterizes P to be an inner regular measure on Ω. Let us turn to details.

The following fundamental result is due to Prokhorov [1], and we present it in a form due to Bourbaki [2].

3. Theorem *Let $\{(\Omega_\alpha, \Sigma_\alpha, P_\alpha, g_{\alpha\beta})_{\alpha<\beta} : \alpha, \beta \text{ in } D\}$ be a projective system of topological probability spaces where each Ω_α is an arbitrary Hausdorff space, Σ_α its Borel algebra (i.e., σ-algebra generated by its closed sets), P_α is inner regular (i.e., $P_\alpha(A) = \sup\{P_\alpha(C) : C \subset A,\ C \text{ compact}\}$, $A \in \Sigma_\alpha$), and the $g_{\alpha\beta}$ are continuous. Let $\Omega = \lim_{\leftarrow}(\Omega_\alpha, g_{\alpha\beta})$. Suppose that the space Ω is rich in the sense that the mappings $g_\alpha, g_\alpha\colon \Omega \to \Omega_\alpha$, $\alpha \in D$ (are continuous, $g_\alpha = g_{\alpha\beta} \circ g_\beta$, and) separate points of Ω (this is true in particular when $g_\alpha(\Omega) = \Omega_\alpha$, $\alpha \in D$). Then there exists a unique Radon probability P on $\Sigma_0 = \bigcup_{\alpha\in D} g_\alpha^{-1}(\Sigma_\alpha)$ (so the system admits a projective limit (Ω, Σ, P) where $\Sigma = \sigma(\Sigma_0)$) iff the following condition holds*:

For each $\varepsilon > 0$ there is a compact set $C_\varepsilon = C \subset \Omega$ such that $P_\alpha(\Omega_\alpha - g_\alpha(C)) < \varepsilon$, $\alpha \in D$. $(*)$

When $(*)$ *obtains, the* P *satisfies the equation* $\tilde{P}(K) = \inf_{\alpha \in D} P_\alpha(g_\alpha(K))$ *for each compact set* $K \subset \Omega$ *and* $\tilde{P}|\Sigma = P$, *where* $\tilde{P}$ *is a unique inner regular (or Radon) probability with respect to which each compact subset of* Ω *is measurable.*

Discussion We first note that the conditions on $(\Omega_\alpha, \Sigma_\alpha, P_\alpha)$ are satisfied if each Ω_α is locally compact and P_α is a Radon measure there. If each Ω_α is compact, and nonempty, then so is Ω (cf. Proposition 1.4). In this case $C_\varepsilon = \Omega$ for any $\varepsilon > 0$, and $(*)$ is true. Hence the last part of Theorem 1.5 is a consequence of this result since a Baire regular measure on Ω has a unique Borel regular (or Radon) extension on the same space. (Cf., e.g., Theorem I.2.10 on this relation.) If $\Omega_\alpha = \mathbb{R}^\alpha = \mathbb{R}^n$ ($n = \text{card}(\alpha) < \infty$), then each P_α is just a Lebesgue–Stieltjes measure. The theorem is thus another extension of the Kolmogorov–Bochner theorem considered earlier (Theorems I.3.1 or I.3.2). The proof is accomplished along the following lines. Since the $g_\alpha: \Omega \to \Omega_\alpha$, $\alpha \in D$, are continuous, for every compact set $C \subset \Omega$, $g_\alpha(C)$ is compact in Ω_α and hence $g_\alpha(C) \in \Sigma_\alpha$ by hypothesis. If $\mathscr{C}$ is the lattice of compact parts of Ω, then we define $\tilde{P}: \mathscr{C} \to [0, 1]$ by the equation

$$\tilde{P}(C) = \inf_{\alpha \in D} P_\alpha(g_\alpha(C)) \tag{4}$$

and prove the key fact that the set function $\tilde{P}$ is additive and continuous from above at $\varnothing$ so that it extends uniquely to an inner regular probability onto $\sigma(\mathscr{C})$, denoted by the same symbol. We then show that $\tilde{P} = P$ on Σ_0, and hence on Σ, iff condition $(*)$ holds. Thus the method of proof differs considerably from that of Theorem I.3.2, and the auxiliary fact about $\tilde{P}$ (cf. (4)) is nontrivial. We shall present the details in a series of steps and a proposition for ease. First we give a technical result (essentially due to Bourbaki [2]) on the extension of $\tilde{P}$ to $\sigma(\mathscr{C})$. Though this method of proof is standard, some of the techniques needed are tricky, and we include the complete details.

4. Proposition *Let* $\{(\Omega_\alpha, \Sigma_\alpha, P_\alpha, g_{\alpha\beta})_{\alpha<\beta} : \alpha, \beta \text{ in } D\}$ *be a projective system of Radon probability spaces, i.e.,* P_α *is inner regular for the compact sets of* Ω_α *(the latter are in* Σ_α*). Suppose that* $g_\alpha: \Omega\,(=\lim_{\leftarrow} \Omega_\alpha) \to \Omega_\alpha$ *are the associated (separating) canonical mappings. If* $\mathscr{C}$ *is the lattice of compact sets of* Ω *and* $\tilde{P}$ *is defined by* (4) *on* $\mathscr{C}$, *then there is a unique* σ*-additive* μ *on* $\sigma(\mathscr{C})$ *such that* $\mu|\mathscr{C} = \tilde{P}$ *and that* $\mu \circ g_\alpha^{-1} \leq P_\alpha$, $\alpha \in D$.

Proof We present the proof in a series of steps

I. The set function $\tilde{P}$ of (4) is increasing, subadditive, and additive on $\mathscr{C}$. Since the increasing nature of $\tilde{P}$ is obvious, we need to prove the remaining two properties. To this end, observe that the "inf" in (4) can be replaced by the limit since for each $C \in \mathscr{C}$, $P_\alpha(g_\alpha(C)) \geq P_\beta(g_\beta(C))$ for $\alpha < \beta$. In fact, since $P_\alpha = P_\beta \circ g_{\alpha\beta}^{-1}$ and $g_\alpha = g_{\alpha\beta} \circ g_\beta$, we have

$$g_\beta(C) \subset g_{\alpha\beta}^{-1}(g_{\alpha\beta}(g_\beta(C))) = g_{\alpha\beta}^{-1}(g_\alpha(C)), \qquad C \in \mathscr{C}. \tag{5}$$

Also $g_\beta(C) \in \Sigma_\beta$, $g_{\alpha\beta}^{-1}(g_\alpha(C)) \in \Sigma_\beta$. Hence

$$P_\beta(g_\beta(C)) \leq P_\beta(g_{\alpha\beta}^{-1}(g_\alpha(C))) = P_\alpha(g_\alpha(C)), \qquad \alpha \in D, \quad \alpha \leq \beta. \tag{6}$$

This shows $P_\alpha \circ g_\alpha$ is decreasing and $\inf_\alpha P_\alpha \circ g_\alpha(C) = \lim_\alpha P_\alpha \circ g_\alpha(C)$. If now C_1, C_2 are two sets from $\mathscr{C}$ (since $g_\alpha(C_1 \cup C_2) = g_\alpha(C_1) \cup g_\alpha(C_2)$ and $C_1 \cup C_2 \in \mathscr{C}$), we have from the subadditivity of P_αs (and (6))

$$P_\alpha(g_\alpha(C_1 \cup C_2)) \leq P_\alpha(g_\alpha(C_1)) + P_\alpha(g_\alpha(C_2)).$$

Taking limits on α in D, we get the subadditivity of $\tilde{P}$.

The additivity of $\tilde{P}$ is obtained in two stages. If C_1, C_2 are disjoint sets from $\mathscr{C}$, we assert that there is an $\alpha_0 \in D$ such that for $\alpha > \alpha_0$, $g_\alpha(C_1)$ and $g_\alpha(C_2)$ are disjoint (compact) subsets of Ω_α. When this is proved, the additivity of P_α and the fact that (4) is true imply the additivity of $\tilde{P}$, by taking limits for $\alpha > \alpha_0$ in D. To prove the above assertion, first note the trivial inclusion for $C_0 \in \mathscr{C}$, $g_\alpha^{-1}(g_\alpha(C_0)) \supset C_0$ and hence $C_0 \subset \bigcap_{\alpha \in D} g_\alpha^{-1}(g_\alpha(C_0))$. We claim that there is equality here. For let ω be a point in this intersection. So $\omega \in g_\alpha^{-1}(g_\alpha(C_0))$, the set being closed since g_αs are continuous. Let $C_\alpha = C_0 \cap g_\alpha^{-1}(\{g_\alpha(\omega)\})$. Then $C_\alpha \subset C_0$ is closed and hence compact. Also $\omega \in C_\alpha$, $\alpha \in D$, so it is nonvoid. But $g_\alpha = g_{\alpha\beta} \circ g_\beta$ for $\alpha < \beta$, and $g_\alpha^{-1}(g_\alpha(C)) \supset C$. Thus we have

$$\begin{aligned} C_\beta &= C_0 \cap g_\beta^{-1}(\{g_\beta(\omega)\}) \subset C_0 \cap [g_\beta^{-1} \circ g_{\alpha\beta}^{-1}(\{g_{\alpha\beta} \circ g_\beta(\omega)\})] \\ &= C_0 \cap [(g_\beta^{-1} \circ g_{\alpha\beta}^{-1})(\{g_\alpha(\omega)\})] = C_0 \cap [g_\alpha^{-1}(\{g_\alpha(\omega)\})] = C_\alpha. \end{aligned} \tag{7}$$

Hence $\bigcap_{\alpha \in D} C_\alpha \neq \varnothing$. Let ω' be a point of this intersection. Then $\omega' \in C_\alpha \subset C_0$, so that $\omega' \in g_\alpha^{-1}(\{g_\alpha(\omega)\})$, for all $\alpha \in D$. Thus $g_\alpha(\omega) = g_\alpha(\omega')$, and since the g_αs separate points of Ω, we deduce that $\omega = \omega'$. Hence $C_0 = \bigcap_{\alpha \in D} g_\alpha^{-1} \circ g_\alpha(C_0)$. (By a slight change of the argument, it can be shown that this equation holds also if C_0 is merely a closed set.)

We now use this to prove additivity. Let $B_\alpha = g_\alpha^{-1} \circ g_\alpha(C_1) \supset C_1$. Since B_α is closed (and may not be compact), define $A_\alpha = B_\alpha \cap C_2$, which is compact. By (5) or (7), for $\alpha < \beta$, we conclude that $A_\alpha \supset A_\beta$. Hence $\bigcap_{\alpha \in D} A_\alpha = C_2 \cap \bigcap_{\alpha \in D} B_\alpha = C_2 \cap C_1 = \varnothing$. But then the compact decreasing family $\{A_\alpha, \alpha \in D\}$ must vanish from some $\alpha_0 \in D$. So $A_\alpha = B_\alpha \cap C_2 = \varnothing$ for $\alpha > \alpha_0$. Thus if $\tilde{\omega} \in B_\alpha = g_\alpha^{-1} \circ g_\alpha(C_1)$, then $\tilde{\omega} \notin C_2$ and $g_\alpha(\tilde{\omega}) \notin g_\alpha(C_2)$. So $g_\alpha(C_1) \cap g_\alpha(C_2) = \varnothing$ for $\alpha > \alpha_0$. By the initial observation this proves the additivity of $\tilde{P}$ on $\mathscr{C}$.

II. The set function $\tilde{P}$ is continuous from above, i.e., $C_i \in \mathscr{C}$, $C_i \downarrow C \Rightarrow \tilde{P}(C) = \inf_{i \in I} \tilde{P}(C_i)$, for any ordered index I. For we observe that the monotonicity of C_i implies $g_\alpha(C) = \bigcap_{i \in I} g_\alpha(C_i)$ for each $\alpha \in D$. In fact, $g_\alpha(C) \subset g_\alpha(C_i)$ so that $g_\alpha(C) \subset \bigcap_{i \in I} g_\alpha(C_i)$. To prove the opposite inclusion, let ω be a point in the intersection. Then for each i, $E_i = g_\alpha^{-1}(\{\omega\}) \cap C_i \neq \varnothing$, and closed, hence compact. Since α is fixed and C_i decreasing, $E_j \subset E_i$ if $j > i$. So $\bigcap_{i \in I} E_i \neq \varnothing$, and if ω' is a point in this intersection, then $g_\alpha(\omega) = \omega' \in g_\alpha(C_i)$, $i \in I$. But $\bigcap_{i \in I} g_\alpha^{-1}(\{\omega\}) \cap C_i = g_\alpha^{-1}(\{\omega\}) \cap C \neq \varnothing$ and hence $g_\alpha(\omega) \in g_\alpha(C)$, so that $\bigcap_{i \in I} g_\alpha(C_i) \subset g_\alpha(C)$. Thus there is equality here.

Let us observe that for the above compact sequence $g_\alpha(C_i) \downarrow g_\alpha(C)$, $P_\alpha(g_\alpha(C)) = \inf_i P_\alpha(G_\alpha(C_i))$. Indeed, if a is the infimum of the right side, then $a \geq 0$, and so there exists a countable set $I' \subset I$ such that $\inf_{j \in I'} P_\alpha(g_\alpha(C_j)) = a$. Let $A = \bigcap_{j \in I'} g_\alpha(C_j)$. Then A is compact, and so $A \in \Sigma_\alpha$. Also since $g_\alpha(C_j)$ is decreasing, we get $a = P_\alpha(A)$. If $P_\alpha(A - g_\alpha(C_j)) = P_\alpha(A) - P_\alpha(A \cap g_\alpha(C_i)) > 0$ for some $i = i_0$ (say) in I, $a = P_\alpha(A) > P_\alpha(A \cap g_\alpha(C_{i_0})) \geq \inf_i P_\alpha(g_\alpha(C_i)) = a$, and this gives a contradiction. Hence $P_\alpha(A - g_\alpha(C_i)) = 0$ for all $i \in I$, so that A is the essential infimum of $\{g_\alpha(C_i), i \in I\}$. But $g_\alpha(C)$ is the infimum of this class. We need to show that $P_\alpha(A - g_\alpha(C)) = 0$. Since P_α is inner regular by hypothesis, this follows if $B \subset A - g_\alpha(C)$, compact, implies $P_\alpha(B) = 0$. But $\varnothing = B \cap g_\alpha(C) = B \cap \bigcap_{i \in I} g_\alpha(C_i)$, and since $g_\alpha(C_i) \downarrow$, and compact, there exists $i' \in I$ such that for all $i > i'$, $B \cap g_\alpha(C_i) = \varnothing$ by the last part of Step I. Hence $P_\alpha(B) \leq P_\alpha(A - g_\alpha(C_i)) = 0$, $i > i'$. Thus $P_\alpha(A) = P_\alpha(g_\alpha(C)) = \inf_i P_\alpha(g_\alpha(C_i))$. Now using (6), we deduce that (since the indexing sets I, D are independent, and sequences $\downarrow$)

$$\begin{aligned}\tilde{P}(C) &= \inf_\alpha P_\alpha(g_\alpha(C)) = \inf_\alpha \inf_i P_\alpha(g_\alpha(C_i)) \\ &= \inf_i \inf_\alpha P_\alpha(g_\alpha(C_i)) = \inf_i P(C_i).\end{aligned} \tag{8}$$

This proves the continuity of $\tilde{P}$ from above (or right continuity).

III. If we define P^* on $\mathscr{P}(\Omega)$, the power set of Ω, by

$$P^*(A) = \sup\{\tilde{P}(C) : C \subset A, C \in \mathscr{C}\}, \qquad A \in \mathscr{P}(\Omega), \tag{9}$$

then we claim that P^* is σ-additive on $\sigma(\mathscr{C})$, $P^*(C) = \tilde{P}(C)$, $C \in \mathscr{C}$, and is inner regular. This is the *key* step, which will be split into substeps.

We first prove this result for the case that $\Omega \cap \mathscr{C}$, i.e., Ω is *compact*. Let $\mathscr{D}$ be the class of all $A \subset \Omega$ such that $P^*(A) + P^*(A^c) = 1$. Clearly $\varnothing, \Omega$ are in $\mathscr{D}$, and it is closed under complementation.

III(a). We assert that $\mathscr{D}$ is an algebra containing $\mathscr{C}$. The procedure is similar to the Carathéodory extension, but the latter work is not directly applicable and we need to use critically the compactness of C in $\mathscr{C}$ and the regularity of $\tilde{P}$ for $\mathscr{C}$. By (9), for any $A \subset \Omega$, for $\varepsilon > 0$ there are $C_\varepsilon^i \in \mathscr{C}$, $A \supset C_\varepsilon^1$, $A^c \supset C_\varepsilon^2$ such that

$$\begin{aligned} P^*(A) + P^*(A^c) &= \sup\{\tilde{P}(C): C \subset A, C \in \mathscr{C}\} + \sup\{\tilde{P}(C): C \subset A^c, C \in \mathscr{C}\} \\ &< \tilde{P}(C_\varepsilon^1) + (\varepsilon/2) + \tilde{P}(C_\varepsilon^2) + (\varepsilon/2) = \tilde{P}(C_\varepsilon^1 \cup C_\varepsilon^2) + \varepsilon \le 1 + \varepsilon \end{aligned} \tag{10}$$

Thus the left side is ≤ 1, and for the opposite inequality let $C \in \mathscr{C}$. Let us show that $C \in \mathscr{D}$, using various topological properties of Ω and $\tilde{P}$. Since Ω is normal and C is compact (hence normal), there exists a decreasing family of compact neighborhoods $\{C_\alpha, \alpha \in I\}$ of C such that $\bigcap_{\alpha \in I} C_\alpha = C$. To see this for each $p \in \Omega - C$, there is a compact neighborhood C_p of C with $p \notin C_p$, i.e., $C \subset \text{interior}(C_p)$, C_p compact, and $p \notin C_p$. [This follows easily from the compactness of C.] Let I be the directed set formed of all finite subsets of $\Omega - C$ ordered by inclusion. Let $C_\alpha = \bigcap_{p \in \alpha} C_p$ for each $\alpha \in I$. Then $\{C_\alpha, \alpha \in I\}$ satisfies the requirements. Next by (8), given $\varepsilon > 0$, there is an α_0 such that $\tilde{P}(C) + \varepsilon > \tilde{P}(C_{\alpha_0})$, $C_{\alpha_0} \supset C$, and the closure of $(\Omega - C_{\alpha_0}) = E$ (say) is disjoint with C. But E is compact and $E \cup C_{\alpha_0} = \Omega$. Hence by step I

$$1 = \tilde{P}(E \cup C_{\alpha_0}) \le \tilde{P}(E) + \tilde{P}(C_{\alpha_0}) < \tilde{P}(C) + \varepsilon + \tilde{P}(E). \tag{11}$$

Since $E \subset C^c$, it follows now that $1 - \varepsilon \le \tilde{P}(C) + \tilde{P}(E) \le P^*(C) + P^*(C^c)$, and with (10) one has $C \in \mathscr{D}$. Thus $\mathscr{C} \subset \mathscr{D}$. However, we could have used any compact subset $F \subset \Omega$ for Ω in the above and get the following equation for all $C \in \mathscr{C}$ by considering $C \cap F$ and $C^c \cap F$ there. In other words, (10) and (11) take the following form:

$$P^*(C \cap F) + P^*(C^c \cap F) = P^*(F) \qquad (= \tilde{P}(F)). \tag{12}$$

We next show that, using (12), $\mathscr{D}$ is an algebra, i.e., with A_1, A_2 it has also $A_1 \cup A_2$ (since it is closed under complementation). If $\varepsilon > 0$ is given, again by (9) there exist compact sets $C_i \subset A_i$, $E_i \subset A_i^c$, $i = 1, 2$ such that

$$\tilde{P}(C_i) + (\varepsilon/2) > P^*(A_i), \qquad \tilde{P}(E_i) + (\varepsilon/2) > P^*(A_i^c), \qquad i = 1, 2. \tag{13}$$

If $F_1 = C_1 \cup E_1 \in \mathscr{C}$ is the disjoint union (so $F_1 \in \mathscr{D}$), we have

$$1 = P^*(F_1) + P^*(F_1^c), \qquad P^*(F_1) = \tilde{P}(C_1 \cup E_1) = \tilde{P}(C_1) + \tilde{P}(E_1). \tag{14}$$

From (13) and (14) it follows that $1 \ge P^*(A_1) + P^*(A_1^c) - \varepsilon + P^*(F_1^c) = 1 - \varepsilon + P^*(F_1^c)$ since $A_1 \in \mathscr{D}$. Hence $P^*(F_1^c) \le \varepsilon$. On the other hand, (12) yields, with $C = F_1$ there,

$$P^*(F) = P^*(F \cap F_1) + P^*(F_1^c \cap F) \le P^*(F \cap F_1) + \varepsilon.$$

But $F_1 \cap F = (C_1 \cap F) \cup (E_1 \cap F)$ is a compact disjoint union, and using the fact that $\tilde{P} = P^*$ on $\mathscr{C}$, we get

$$P^*(F) \le P^*(C_1 \cap F) + P^*(E_1 \cap F) + \varepsilon. \tag{15}$$

Letting $F = C_2$ first, then $F = E_2$ in (15), and using the result that $1 = P^*(A_2) + P^*(A_2^c) < \tilde{P}(C_2) + \tilde{P}(E_2) + \varepsilon$, we have

$$\begin{aligned} 1 - \varepsilon < \tilde{P}(C_2) + \tilde{P}(E_2) &\le P^*(C_1 \cap C_2) + P^*(E_1 \cap C_2) + P^*(C_1 \cap E_2) \\ &\quad + P^*(E_1 \cap E_2) + 2\varepsilon \\ &= P^*((C_1 \cap C_2) \cup (E_1 \cap C_2) \cup (C_1 \cap E_2)) + P^*(E_1 \cap E_2) + 2\varepsilon. \end{aligned} \tag{16}$$

This is because $C_1 \cap C_2 \subset A_1 \cap A_2$, $E_1 \cap C_2 \subset A_1^c \cap A_2$, and $C_1 \cap E_2 \subset A_1 \cap A_2^c$, they are disjoint (and in $\mathscr{C}$); so the additivity of P^* on $\mathscr{C}$ applies. If B is the union of these three disjoint compact sets, then $B \subset A_1 \cup A_2$. Hence (16) becomes

$$1 - \varepsilon < P^*(B) + P^*(E_1 \cap E_2) + 2\varepsilon < P^*(A_1 \cup A_2) + P^*(A_1^c \cap A_2^c) + 2\varepsilon. \tag{17}$$

Since $\varepsilon > 0$ is arbitrary, this proves $1 \le P^*(A_1 \cup A_2) + P^*((A_1 \cup A_2)^c)$. But by definition of P^* the opposite inequality is immediate so that $A_1 \cup A_2 \in \mathscr{D}$ and $\mathscr{D}$ is an algebra.

III(b). To see that P^* is additive on $\mathscr{D}$, note that from the first part of (17) and the fact that $A_1 \cup A_2 \in \mathscr{D}$, we can deduce at once the following inequalities (since $E_i \subset A_i^c$):

$$1 - \varepsilon = P^*(A_1 \cup A_2) + P^*(A_1^c \cap A_2^c) - \varepsilon < P^*(B) + P^*(E_1 \cap E_2) + 2\varepsilon.$$

Hence $P^*(B) > P^*(A_1 \cup A_2) - 3\varepsilon$. If $A_1 \cap A_2 = \varnothing$, then $C_1 \cap C_2 = \varnothing$ in the above and $(E_1 \cap C_2) \subset A_2$, $E_2 \cap C_1 \subset A_1$, so

$$P^*(B) = P(B) = P(E_1 \cap C_2) + P(E_2 \cap C_1) > P^*(A_1 \cup A_2) - 3\varepsilon.$$

Taking appropriate suprema, we deduce that $P^*(A_1 \cup A_2) \le P^*(A_1) + P^*(A_2)$. From the definition we get the opposite inequality. Hence P^* is additive on $\mathscr{D}$. We thus have shown that P^* defined by (9) is additive, increasing on the algebra $\mathscr{D} \supset \mathscr{C}$, and $P^* = \tilde{P}$ on $\mathscr{C}$. Let us now show that P^* on $\mathscr{D}$ has a σ-additive extension and that $\mathscr{D}$ is in fact a σ-algebra.

Since P^* is a finitely additive positive set function on the class $\mathscr{D}$, the σ-additivity follows if we show that for each sequence $\{A_n\}_1^\infty \subset \mathscr{D}$, $A_n \downarrow \varnothing$, $P^*(A_n) \downarrow 0$. Given $\varepsilon > 0$, for each n there exists a $C_n \in \mathscr{C}$, $C_n \subset A_n$, and $P^*(A_n) < \tilde{P}(C_n) + \varepsilon/2^n$ by (9). Since C_n need not be monotone, let $B_n = \bigcap_{i=1}^n C_i \in \mathscr{C}$. Then $B_n \subset A_n$, $B_n \downarrow \varnothing$. We assert that $P^*(A_n) < P^*(B_n) + \varepsilon(1 - 1/2^n)$. Clearly, this is true for $n = 1$ by choice. To use induction, suppose it is true for n. Then for $n + 1$ one has

$$\begin{aligned} P^*(B_{n+1}) &= P^*(B_n \cap C_{n+1}) = P^*(B_n) + P^*(C_{n+1}) - P^*(B_n \cup C_{n+1}) \\ &\qquad \text{by the additivity of } P^* \text{ on } \mathscr{D} \text{ proved above} \\ &> [P^*(A_n) - \varepsilon(1 - 1/2^n)] + [P^*(A_{n+1}) - (\varepsilon/2^{n+1})] - P^*(A_n) \\ &= P^*(A_{n+1}) - \varepsilon(1 - 1/2^{n+1}). \end{aligned} \tag{18}$$

Here we used the induction hypothesis and the fact that $B_n \cup C_{n+1} \subset A_n$. Hence $\lim_{n\to\infty} P^*(A_n) \le \lim_{n\to\infty}[P^*(B_n) + \varepsilon(1 - 1/2^n)] = \varepsilon$ since $\lim_n P^*(B_n) = \lim_n \tilde{P}(B_n) = \tilde{P}(\lim_n B_n)$ (cf. (8)). The arbitrariness of ε shows that $P^*(A_n) \downarrow 0$, and P^* is σ-additive, and so it is a measure. To see that the collection $\mathscr{D}$ is a σ-algebra, it suffices to show (by the monotone class theorem) that it is closed under increasing limits. Let $E_n \in \mathscr{D}$, $E_n \uparrow E$. Then $\lim_n P^*(E_n) = P^*(\lim_n E_n) = P^*(E)$, where $E \in \sigma(\mathscr{D})$. Similarly, $E_n^c \in \mathscr{D}$, $E_n^c \downarrow E$, and $\lim_n P^*(E_n^c) = P^*(\lim_n E_n^c) = P^*(E^c)$. But $P^*(E_n) + P^*(E_n^c) = 1$, and hence $P^*(E) + P^*(E^c) = 1$. Consequently $E \in \mathscr{D}$. Thus $\mathscr{D} = \sigma(\mathscr{D})$, as asserted.

Finally, $\mathscr{C} \subset \mathscr{D}$ implies $\sigma(\mathscr{C}) \subset \mathscr{D}$, and let $\bar{P} = P^* | \sigma(\mathscr{C})$. Then $\bar{P}$ is an extension of $\tilde{P}$ and has the property (9), i.e., $\bar{P}$ is the inner regular extension of $\tilde{P}$ to $\sigma(\mathscr{C})$. The uniqueness of $\bar{P}$ is immediate since any extension Q of $\tilde{P}$ that is inner regular must satisfy (9), and hence

$$\bar{P}(A) = Q(A) = \sup\{Q(C) : C \subset A, C \in \mathscr{C}\} = \sup\{\bar{P}(C) : C \subset A, C \in \mathscr{C}\}$$

since $Q(C) = P(C)$, $C \in \mathscr{C}$. [This also follows from the Hahn extension theorem, cf. I.2.2.]

III(c). It remains to consider the case that Ω is not compact. If $K \subset \Omega$ is a compact set, consider $\mathscr{C}_K$, the class of all compact subsets of K. Let $\tilde{P}_K = \tilde{P} | \mathscr{C}_K$. Then by the above work there is a unique $\bar{P}_K$ on $\sigma(\mathscr{C}_K)$ such that $\bar{P}_K = \tilde{P}_K$ on $\mathscr{C}_K$. By the uniqueness, we deduce that for any $K_1 \subset K$, $\bar{P}_{K_1} = (\bar{P}_K)_{K_1}$, i.e., $\bar{P}_K$ and $\bar{P}_{K_1}$ agree on the compact subsets of K_1 and hence on all $\sigma(\mathscr{C}_{K_1}) \subset \sigma(\mathscr{C}_K)$. So we may define unambiguously a measure μ: $K \mapsto \bar{P}_K(K)$, and this gives $\mu(K) = \bar{P}_K(K) = \tilde{P}(K)$, $K \in \mathscr{C}$, and the μ is an inner regular probability on $\sigma(\mathscr{C})$.

IV. To complete the proof, we only need to show that $\mu \circ g_\alpha^{-1} \le P_\alpha$.

Indeed, let $\mu_\alpha = \mu \circ g_\alpha^{-1}$. Then for any compact set $K \subset \Omega_\alpha$, let $g_\alpha^{-1}(K) \subset \Omega$ be the corresponding closed set. If $E = g_\alpha^{-1}(K)$, let us denote by $\mathscr{C}_E$ the class of all compact subsets of E. Then $\mathscr{C}_E \subset \mathscr{C}$, and by definition of μ $\mu(K) = \tilde{P}(K)$, so $\mu(E) = \sup\{\mu(C) : C \in \mathscr{C}_E\} = \sup\{\tilde{P}(C) : C \in \mathscr{C}_E\}$. But for each compact $C \subset \Omega$, $g_\alpha(C) \subset \Omega_\alpha$ is compact, and $g_\alpha(C) \subset g_\alpha(g_\alpha^{-1}(K)) \subset K$. Hence with (6) one has

$$\tilde{P}(C) = \inf_\alpha P_\alpha(g_\alpha(C)) \le P_\alpha(g_\alpha(C)) \le P_\alpha(K). \tag{19}$$

Taking supremum for $C \in \mathscr{C}_E$ in (19), we get $\mu_\alpha(K) = \mu(E) \le P_\alpha(K)$, $K \in g_\alpha(\mathscr{C})$. This implies $\mu_\alpha \le P_\alpha$ and completes the proof of the proposition.

Note It must be emphasized that the additivity of a positive set function on a lattice does *not* generally imply the same property on the ring generated by the lattice. The topological properties of $\mathscr{C}$ and $\tilde{P}$ are crucial for the truth in the present case. (See Pettis [1] for counterexamples.)

Proof of Theorem 3 We can now complete the proof quickly. From the above proposition $\{(\Omega_\alpha, \Sigma_\alpha, \mu_\alpha, g_{\alpha\beta})_{\alpha<\beta} : \alpha, \beta \text{ in } D\}$ is a projective system, $\mu_\alpha(\Omega_\alpha) \leq 1$, admitting the limit $\mu = \lim_{\leftarrow} \mu_\alpha$, with $\mu(\Omega) \leq 1$. As shown in (6), $\mu(C) = \inf_\alpha \mu_\alpha(g_\alpha(C))$, $C \subset \Omega$ is compact, and $\mu_\alpha \leq P_\alpha$, $\alpha \in D$. Since projective limits are unique, the set function P is σ-additive iff $P_\alpha = \mu_\alpha$, i.e., $P \circ g_\alpha^{-1} = \mu \circ g_\alpha^{-1}$ because by the preceding equation $\mu(C) = P(C)$, $C \in \mathscr{C}$, compact class, and hence by the proposition, μ being σ-additive on $\sigma(\mathscr{C})$, the assertion about P follows.

Now by the inner regularity of μ, for each $\varepsilon > 0$ there is a $C_\varepsilon \in \mathscr{C}$ such that $\mu(\Omega - C_\varepsilon) < \varepsilon$. Since $g_\alpha^{-1} \circ g_\alpha(C) \supset C$, this can be written

$$\mu_\alpha(\Omega_\alpha - g_\alpha(C_\varepsilon)) = \mu(\Omega - g_\alpha^{-1} \circ g_\alpha(C_\varepsilon)) \leq \mu(\Omega - C_\varepsilon) < \varepsilon. \tag{20}$$

Thus if $P = \lim_{\leftarrow} P_\alpha$, i.e., P is σ-additive (and inner regular), then $P(C) = \inf_\alpha P_\alpha(g_\alpha(C)) = \tilde{P}(C) = \mu(C)$, $C \in \mathscr{C}$, so that $P = \mu$ and $P \circ g_\alpha^{-1} = P_\alpha = \mu \circ g_\alpha^{-1} = \mu_\alpha$, $\alpha \in D$. Since (20) is simply (∗), the necessity follows.

Conversely, if $P_\alpha = \mu_\alpha$, $\alpha \in D$, then

$$\mu(C) = \inf_\alpha P_\alpha(g_\alpha(C)) = 1 - \sup_\alpha P_\alpha(\Omega - g_\alpha(C))$$

and so $\mu(C_\varepsilon) > 1 - \varepsilon$. Thus $\mu(\Omega) = 1$, $\mu = \lim_{\leftarrow}(\mu_\alpha, g_{\alpha\beta}) = \lim_{\leftarrow}(P_\alpha, g_{\alpha\beta}) = P$, and P is an inner regular probability on $\sigma(\mathscr{C})$. This shows that (∗) is sufficient, and the proof of the theorem is complete.

The following alternative version (and a slight extension) of the above theorem, also due to Prokhorov, is useful in some applications.

5. Theorem *Let* $\{(\Omega_\alpha, \Sigma_\alpha, P_\alpha, g_{\alpha\beta})_{\alpha<\beta} : \alpha, \beta \text{ in } D\}$ *be a projective system as in Theorem 3. Let T be a Hausdorff space and suppose that* $\{h_\alpha, \alpha \in D\}$ *is a family of continuous mappings on T such that (a)* $h_\alpha: T \to \Omega_\alpha$, $\alpha \in D$ *(and (b) for each* $\alpha < \beta$, $h_\alpha = g_{\alpha\beta} \circ h_\beta$. *Then there is an inner regular (or Radon) probability ν on the Borel σ-algebra of T such that* $\nu \circ h_\alpha^{-1} = P_\alpha$, $\alpha \in D$, *iff the following condition holds*:

$$\text{For each } \varepsilon > 0 \text{ there is a compact set } K_\varepsilon \subset T \text{ such that } P_\alpha(\Omega_\alpha - h_\alpha(K_\varepsilon)) < \varepsilon \text{ for all } \alpha \in D. \tag{∗∗}$$

If, moreover, the h_α separate points of T, then ν is unique.

Proof It is clear that if there is such a ν, (∗∗) necessarily holds. Conversely, let $\Omega = \lim_{\leftarrow}(\Omega_\alpha, g_{\alpha\beta})$. Then one can define a mapping $u: T \to \Omega$ such that $h_\alpha = g_\alpha \circ u$ just as in Example B of Section 1. This means that for each $t \in T$ we need to find an element $u(t) = \{u_\alpha(t), \alpha \in D\} \in \times_{\alpha \in D} \Omega_\alpha$ such that for any $\alpha < \beta$, $u_\alpha(t) = g_{\alpha\beta}(u_\beta(t))$ so that $u(t) \in \Omega$. Now if we set $u_\alpha = h_\alpha$, then the hypothesis implies $u: t \mapsto \{h_\alpha(t), \alpha \in D\}$ is such a mapping and u is also seen to be continuous from T into Ω. If the h_α separate points of T, then u is one-to-one. Let us now show that there is a Radon probability ν on T if (**) holds.

For any $\varepsilon > 0$, let $K_\varepsilon \subset T$ be as in (**) and set $\tilde{K}_\varepsilon = u(K_\varepsilon) \subset \Omega$. Then the continuity of u implies $\tilde{K}_\varepsilon$ is compact, and this satisfies (*) of Theorem 3. Hence by that theorem there is a Radon probability P on Ω, $P = \lim_{\leftarrow} P_\alpha$. We claim that ν on T defined by $P = \nu \circ u^{-1}$ satisfies the requirements. For it is evident that $\nu(T) = P(\Omega) = 1$. Let $l(f) = \int_T f\, dP \circ u^{-1}$, $f \in C(K_\varepsilon)$. Then $l(\cdot)$ is a continuous (positive) linear functional and $l(1) = l(\chi_{K_\varepsilon}) \geq P(\tilde{K}_\varepsilon) \geq 1 - \varepsilon$. By the Riesz representation theorem there is a unique Radon measure ν_ε supported by K_ε such that $l(f) = \int_T f\, d\nu_\varepsilon$ so that $\nu_\varepsilon(K_\varepsilon) = P(\tilde{K}_\varepsilon)$ (cf. Royden [1, p. 310]). If one lets $\varepsilon \downarrow 0$ through a sequence, then by the uniqueness of this representation it follows that there is a Radon measure ν on T such that $\nu(K_\varepsilon) = \nu_\varepsilon(K_\varepsilon)$ and ν is supported by a σ-compact subset of T. Moreover, $\nu = P \circ u^{-1}$. Such a ν will be unique if the h_α separate points of T since then u is one-to-one. This completes the proof.

This formulation is primarily of interest when the Ω_αs are vector spaces. The following simple consequence illustrates the point.

6. Proposition *Let $\mathscr{X}$ be a Banach space and $\mathfrak{F}$ be the directed set (under inclusion) of all finite dimensional subspaces of $\mathscr{X}$ and set $\mathscr{X}_\alpha = \mathscr{X}/\alpha^\perp$, where $\alpha^\perp$ is the annihilator of α so that $\mathscr{X}_\alpha$ is finite dimensional (isomorphic to α). If $\varphi_\alpha: \mathscr{X} \to \mathscr{X}_\alpha$ and $\varphi_{\alpha\beta}: \mathscr{X}_\beta \to \mathscr{X}_\alpha$ $(\alpha < \beta)$ are canonical projections on these factor spaces so that $\varphi_\alpha = \varphi_{\alpha\beta} \circ \varphi_\beta$ and $\varphi_{\alpha\beta} \circ \varphi_{\beta\gamma} = \varphi_{\alpha\gamma}$ for $\alpha < \beta < \gamma$, let Σ_α be the Borel algebra of $\mathscr{X}_\alpha$ and $P_\alpha: \Sigma_\alpha \to \mathbb{R}^+$ be a Radon probability such that $P_\alpha = P_\beta \circ \varphi_{\alpha\beta}^{-1}$ for $\alpha < \beta$ in D. Then the projective system $\{(\mathscr{X}_\alpha, \Sigma_\alpha, P_\alpha, \varphi_{\alpha\beta})_{\alpha<\beta} : \alpha, \beta$ in $\mathfrak{F}\}$ admits a Radon projective limit P $(=\lim_{\leftarrow} P_\alpha)$ whose support lies in $\mathscr{X}$ iff for each $\varepsilon > 0$ there exists a compact set $K_\varepsilon \subset \mathscr{X}$ such that*

$$P_\alpha(\mathscr{X}_\alpha - \varphi_\alpha(K_\varepsilon)) < \varepsilon \qquad \text{for all} \quad \alpha \in D. \tag{$*'$}$$

This is an immediate consequence of Theorem 5 since φ_α and $\varphi_{\alpha\beta}$ are continuous (even open) mappings and all its conditions are satisfied.

The essential point here is that $\Omega = \lim_{\leftarrow}(\mathscr{X}_\alpha, \varphi_{\alpha\beta})$ is not $\mathscr{X}$ but is a much larger space (without mention of the s.m. condition). In fact, one can show that $\mathscr{X} \subset \mathscr{X}^{**} \subset \Omega$ when Ω is endowed with the projective limit topology. For further discussion and applications of measures on topological spaces, the reader is referred to Schwartz [1, Part II] as well as Gel'fand and Vilenkin [1, Chapter IV].

The interest in Prokhorov's condition (*), or (**), stems from the fact that it characterizes both the existence and topological (i.e., Radon) properties of the projective limit measure simultaneously. It is the uniformity (in

$\alpha \in D$) that allows one to check this condition quickly in some applications. The inner regularity and the distinguished class $\mathscr{C}$ (cf. Proposition 4) are crucial in the proof. The result of Theorem 3 can be reinterpreted as follows. From the system $\{(\Omega_\alpha, \Sigma_\alpha, P_\alpha, g_{\alpha\beta})_{\alpha<\beta} : \alpha, \beta \text{ in } D\}$, consider $P_\alpha^* = P \circ g_\alpha^{-1}: \Sigma_\alpha^* \to [0,1]$, the restriction of P to $\Sigma_\alpha^* = g_\alpha^{-1}(\Sigma_\alpha) \subset \Sigma$. Then P_α^* is always σ-additive since P_α is (cf. Proposition 1.2). Suppose each P_α^* has a (perhaps nonunique) σ-additive extension $\overline{P_\alpha^*}$ to $\Sigma_0 = \bigcup_{\alpha \in D} \Sigma_\alpha^*$. Then the system $\{\overline{P_\alpha^*}, \alpha \in D\}$ $(=\mathscr{K}$, say) admits a limit iff for each $\varepsilon > 0$ there is a compact set $C_\varepsilon \subset \Omega$ such that $\sup_{\nu \in \mathscr{K}} \nu(\Omega - C_\varepsilon) < \varepsilon$. This can be regarded as a "uniform σ-additivity" of $\mathscr{K}$, and it has a close relation with some classical results in abstract analysis on the weakly compact subsets of $ca(\Omega, \Sigma_0)$, the Banach space (under the variation norm) of bounded σ-additive scalar functions on Σ. This observation allows us to consider the projective limit problem for abstract bounded measure systems and obtain different sets of conditions characterizing these limits. The theorem following below contains a result of this kind, and certain related results are included in the Complements and Problems section.

7. Definition Let M be a set of signed measures on a measurable space (Ω, Σ). Then M is *uniformly σ-additive* if for any sequence $\{A_n\}_1^\infty \subset \Sigma$, $A_n \downarrow \varnothing$, it is true that $|\mu(A_n)| \to 0$ uniformly in $\mu \in M$ as $n \to \infty$.

The following result has interest in relation to the work of Krickeberg and Pauc [1].

8. Theorem *Let $\{(\Omega_\alpha, \Sigma_\alpha, P_\alpha, g_{\alpha\beta})_{\alpha<\beta} : \alpha, \beta \text{ in } D\}$ be a projective system of probability spaces with $g_\alpha: \Omega \to \Omega_\alpha$ satisfying $g_\alpha(\Omega) = \Omega_\alpha$ where $g_{\alpha\beta} \circ g_\beta = g_\alpha$ is the canonical projection on $\Omega = \lim_{\leftarrow} \Omega_\alpha$. If P_α^* is the image measure of P_α, i.e., $P_\alpha^* \circ g_\alpha^{-1} = P_\alpha$, on $\Sigma_\alpha^* = g_\alpha^{-1}(\Sigma_\alpha)$ and P is the additive set function on $\Sigma_0 = \bigcup_{\alpha \in D} \Sigma_\alpha^*$ determined as in Proposition 1.2, then the system admits a projective limit (Ω, Σ, P) (i.e., P is σ-additive on Σ_0 and $\Sigma = \sigma(\Sigma_0)$) iff for each set of probability measures $\{\tilde{P}_\alpha^*, \alpha \in D\}$ on (Ω, Σ_0) such that (i) $\tilde{P}_\alpha^* | \Sigma_\alpha^* = P_\alpha^*$ and (ii) for each countable set $\alpha_1 < \alpha_2 < \cdots$ of indexes of D the family $\{\tilde{P}_{\alpha_n}^*, n \geq 1\}$ is uniformly σ-additive on Σ_0. When the limit exists, there are σ-extensions $\tilde{P}_\alpha^*$ of the P_α^* such that $\tilde{P}_\alpha^* \to P$ strongly, i.e., $\|\tilde{P}_\alpha^* - P\| \to 0$, where $\|\tilde{P}^* - P\| = \operatorname{var}(\tilde{P}^* - P)$, the total variation of $(\tilde{P}^* - P)$ on Ω. (If $\tilde{P}_\alpha^* \to P$ in this sense, then the given family is trivially uniformly σ-additive.)*

Proof First note that an extension $\tilde{P}_\alpha^*$ of P_α^* always exists, but it may be only finitely additive. In fact, if $\mathscr{M}(\Sigma_\alpha^*)$ and $\mathscr{M}(\Sigma_0)$ are the uniformly closed vector spaces of real Σ_α^*- and Σ_0-step functions, then $\mathscr{M}(\Sigma_\alpha^*) \subset \mathscr{M}(\Sigma_0)$, and if $l_\alpha(f) = \int_\Omega f \, dP_\alpha^*, f \in \mathscr{M}(\Sigma_\alpha^*)$, then $l_\alpha(1) = 1$ and is a positive linear functional.

It is continuous on $\mathscr{M}(\Sigma_\alpha^*)$. Let $\bar{l}_\alpha$ be its (nonunique) Hahn–Banach extension to $\mathscr{M}(\Sigma_0)$ and let $\tilde{P}_\alpha^*$ be the corresponding additive set function given by the Riesz representation. So $\bar{l}_\alpha(f) = \int_\Omega f\, d\tilde{P}_\alpha^*$ and $\tilde{P}_\alpha^* | \Sigma_\alpha^* = P_\alpha^*$. Here the integral relative to the additive set function $\tilde{P}_\alpha^*$ is defined in a standard manner. Note that $\tilde{P}_\alpha^*(\Omega) = P_\alpha^*(\Omega) = 1$. Thus the additional hypothesis of σ-additivity of $\tilde{P}_\alpha^*$ demanded in the theorem is not redundant.

Let μ^* be the outer measure generated by the pair (P, Σ_0) and $\mathscr{M}_{\mu^*}$ be the class of all μ^*-measurable subsets of Ω. Then the Carathéodory theorem (cf. Theorem I.2.1) implies that $\Sigma_0 \subset \mathscr{M}_{\mu^*}$ (since P is additive on Σ_0) and the largest σ-algebra on which μ^* is σ-additive is $\mathscr{M}_{\mu^*}$. Moreover, $\mu^* | \Sigma_0 = P$ iff P is σ-additive. We now suppose that the set $\{\tilde{P}_\alpha^*, \alpha \in D\}$ in the statement satisfies (i) and (ii) and prove that $\mu^* | \Sigma_0 = P$.

Since every set $A \in \Sigma_\alpha^*$ is trivially covered by itself (from Σ_0), and $P | \Sigma_\alpha^* = P_\alpha^*$, we always have $\mu^*(A) \leq P_\alpha^*(A)$. For the opposite inequality, let $\varepsilon > 0$ be given. Then there exists a sequence $\{A_n\}_1^\infty \subset \Sigma_0$ such that $A \subset \bigcup_{n=1}^\infty A_n$, and for $n \geq 1$,

$$\mu^*(A) + \varepsilon > \sum_{i=1}^{\infty} P(A_i) \geq P\left(\bigcup_{i=1}^{n} A_i\right) + \sum_{i=n+1}^{\infty} P(A_i). \tag{21}$$

Since $A_i \in \Sigma_0$ implies $A_i \in \Sigma_{\alpha_i}$ for some α_i, and $\alpha_{i+1} > \alpha_i$ implies $\Sigma_{\alpha_i}^* \subset \Sigma_{\alpha_{i+1}}^*$ as a consequence of the compatibility relations $g_{\alpha_i} = g_{\alpha_i \alpha_{i+1}} \circ g_{\alpha_{i+1}}$, we see that $A_i \in \Sigma_{\alpha_m}^*$ for $\alpha_m > \alpha_i$, $i = 1, \ldots, n$ (note that the existence of such an α_m $[= \alpha_{mn}]$ is a consequence of the fact that D is directed). Hence (21) can be expressed as (since $\tilde{P}_{\alpha_n}^*$ is $P_{\alpha_n}^*$ on $\Sigma_{\alpha_n}^*$)

$$\mu^*(A) + \varepsilon > \tilde{P}_{\alpha_m}^*\left(\bigcup_{i=1}^{n} A_i\right). \tag{22}$$

But the set $\{\tilde{P}_{\alpha_n}^*, n \geq 1\}$ is uniformly σ-additive on Σ_0 by hypothesis. Hence (22) becomes, for $\alpha_m > \alpha$ (which may be taken),

$$\mu^*(A) + 2\varepsilon > \tilde{P}_{\alpha_m}\left(\bigcup_{i=1}^{\infty} A_i\right) \geq \tilde{P}_{\alpha_m}^*(A) = P_\alpha^*(A). \tag{23}$$

Since $\varepsilon > 0$ is arbitrary, this proves, together with the earlier inequality, that $\mu^* | \Sigma_\alpha^* = P_\alpha^*$, $\alpha \in D$. However, for each $E \in \Sigma_0$ we have

$$P(E) = \lim_\alpha P_\alpha^*(E) = \lim_\alpha (\mu^* | \Sigma_\alpha^*)(E) = \lim_\alpha \mu^*(E) = \mu^*(E), \tag{24}$$

so that $\mu^* | \Sigma_0 = P$ and P is σ-additive. Consequently, the projective limit of the system exists.

The converse is immediate, and this is also a simple consequence of the (somewhat harder) last part of the theorem. We first present a proof of the special assertion for convenience. Let the projective limit exist so that P is

σ-additive on Σ_0 and has a unique extension to Σ. Then clearly P is a σ-additive extension of each P_α^* from Σ_α^* to Σ_0. Suppose $\tilde{P}_\alpha^*$ is another such extension of P_α^* to Σ_0, possibly different from P. Let $\{A_n\}_1^\infty \subset \Sigma_0$ and $A_n \downarrow \varnothing$. Then for any $\varepsilon > 0$ there is an $n_0(=n_0(\varepsilon))$ such that for $n \geq n_0$, $P(A_n) < \varepsilon$. But $A_{n_0} \in \Sigma_{\alpha_0}^*$ for some α_0 and $P|\Sigma_{\alpha_0}^* = \tilde{P}_{\alpha_0}^*|\Sigma_{\alpha_0}^*$. So we have, from the hypothesis that $A_n \subset A_{n_0}$ for $n \geq n_0$ and $\Sigma_\alpha^* \supset \Sigma_{\alpha_0}^*$ for $\alpha > \alpha_0$,

$$\tilde{P}_\alpha^*(A_n) \leq \tilde{P}_\alpha^*(A_{n_0}) = P_{\alpha_0}^*(A_{n_0}) = P(A_{n_0}) < \varepsilon. \tag{25}$$

Since $\varepsilon > 0$ is arbitrary and the last term does not involve $\alpha > \alpha_0$, it follows from (25) that $\{\tilde{P}_\alpha^*, \alpha > \alpha_0\}$ is uniformly σ-additive. In this argument, if $D = \mathbb{N}$, the positive integers, then $\alpha_0 \in \mathbb{N}$ is finite, and since every finite set of finite measures is trivially uniformly σ-additive, we can take $\alpha_0 = 1$ here, i.e., the whole set $\{\tilde{P}_\alpha^*, \alpha \in D\}$ is uniformly σ-additive.

The last part of the theorem is somewhat more involved in that we have to use the following two results: (i) Kakutani's theorem on the concrete representation of an abstract (L-)space and (ii) a mean martingale convergence theorem of Chapter V (Section V.2). The latter will be proved independently of this result, and the present statements will also serve as one of the motivations for the work of Section V.2. For the results used from linear analysis, the reader may consult Dunford–Schwartz [1, Section IV.9].

Thus suppose the projective system admits the limit (Ω, Σ, P). Consider the set $K = \{\tilde{P}_\alpha^*, P, \alpha > \alpha_0\}$, which is a subset of the (Banach) space of real bounded σ-additive set functions $ca(\Omega, \Sigma)$ where $\alpha_0 \in D$ is a fixed index. Here the norm $\|\cdot\|$ in $ca(\Omega, \Sigma)$ is the total variation on Ω. It is a known and easily verifiable fact that $ca(\Omega, \Sigma)$ is an abstract L- (or AL-)space (i.e., the norm is additive on positive elements). Let K_1 be the closed sub (AL-)space generated by K in $ca(\Omega, \Sigma)$. Then the fixed α_0 may be taken to be any one of D such that $\tilde{P}_{\alpha_0}^*(\in K_1)$ is a unit in that for any $Q \in K_1$, $Q > 0$ implies $Q \wedge \tilde{P}_{\alpha_0}^* = \min(Q, \tilde{P}_{\alpha_0}^*) > 0$. This is true since we may assume that Ω is the support of $\tilde{P}_{\alpha_0}^*$ (by replacing Ω with the support of $\tilde{P}_{\alpha_0}^*$, which exists as an element of $\Sigma_{\alpha_0}^*$ because of the finiteness of the measure, cf. Section I.5). Then by the classical Kakutani representation theorem, K_1 is isometrically lattice isomorphic to an $L^1(S, \mathscr{A}, \lambda)$ $(=L^1(\lambda))$ on some regular measure space $(S, \mathscr{A}, \lambda)$, $\lambda(S) < \infty$ with S as a compact totally disconnected Hausdorff space and $\mathscr{A}$ as its Baire σ-algebra. If for each $f \in L^1(\lambda)$ we associate $v_f \in K_1$ by $v_f(A) = \int_A f\,d\lambda$, $A \in \mathscr{A}$, and $\tau: f \mapsto v_f$, then τ is an isometric and lattice isomorphism between $L^1(\lambda)$ and K_1 (cf. also Theorem I.4.8).

If f is an indicator function in $L^1(\lambda)$ ($f = \chi_A$ for $A \in \mathscr{A}$), and $\lambda_0(A) = \tau f$ so that $\lambda_0(\cdot) = \lambda(T^{-1}(\cdot))$ where T is the induced (by τ) mapping between Σ and $\mathscr{A}$, then λ_0 is a finite measure on Σ_0, the "image" of λ, and every element of K_1 is λ_0-continuous. In particular, $\tilde{P}_\alpha^*$ and P are λ_0-continuous for $\alpha > \alpha_0$.

Hence by the Radon–Nikodým theorem (by considering the restrictions P_α^* of $\tilde{P}_\alpha^*$) there exist λ_0-unique Σ_α^* (and Σ)-measurable functions f_α and f such that $f_\alpha = d\tilde{P}_\alpha^*/d\lambda_0, f = dP/d\lambda_0$. Since $P|\Sigma_\alpha^* = P_\alpha^*$, we also have

$$P_\alpha^*(A) = \int_A f_\alpha \, d\lambda_0 = \int_A f \, d\lambda_0 = P(A), \qquad A \in \Sigma_\alpha^*, \quad \alpha > \alpha_0. \tag{26}$$

This is the most important relation in the proof. It is the defining equation of the martingale $\{f_\alpha, \Sigma_\alpha^*, \alpha > \alpha_0\}$ where one may express this more particularly as $f_\alpha = E_\alpha(f) = E^{\Sigma_\alpha^*}(f)$ in the notation and terminology of Chapter II and Proposition 1.7. [The operator $E_\alpha: f \mapsto f_\alpha$ is the conditional expectation.] Now according to the mean martingale convergence theorem of Section V.2 (cf. Corollary V.2.3), using (26) and the fact that $\Sigma = \sigma(\Sigma_0)$, we deduce that $f_\alpha \to f$ in $L^1(\Omega, \Sigma, \lambda_0)$. But

$$\|\tilde{P}_\alpha^* - P\| = \mathrm{var}(\tilde{P}_\alpha^* - P)(\Omega) = \int_\Omega |f_\alpha - f| \, d\lambda_0 \to 0 \tag{27}$$

by definition of the total variation and the convergence deduced above. Thus $\tilde{P}_\alpha^* \to P$, and the proof of the theorem is complete.

Note that the uniform σ-additivity of the set $\{\tilde{P}_\alpha^*, \alpha > \alpha_0\}$ for any $\alpha_0 \in D$ is a trivial consequence of (27). The intimate relation between the set martingales and projective systems is also brought out in the last part of the above theorem. This is made more transparent by the following general result. Its proof will be omitted since the theorem will not be specifically used except as a strong motivation for the study of martingale theory of point functions in Chapters IV and V.

9. Theorem *Let* $\{(\Omega_\alpha, \Sigma_\alpha, P_\alpha, g_{\alpha\beta})_{\alpha<\beta} : \alpha_0 < \alpha; \alpha_0, \alpha, \beta \text{ in } D\} = K$ *be a projective system of probability spaces. Then the following statements are equivalent*:

(a) *The projective limit of the system K exists.*

(b) *There exists a finite measure* λ_0 *on* $\Sigma = \sigma(\Sigma_0)$, *and for each* $K_1 = \{\tilde{P}_\alpha^*, \alpha > \alpha_0, \alpha \in D\}$ *of probability measures on* Σ *with* $\tilde{P}_\alpha^* | \Sigma_\alpha^* = P_\alpha^*$ *it is true that* K_1 *is terminally uniformly* λ_0*-continuous, where the last term means, for any* $\varepsilon > 0$ *there is a* $\delta_\varepsilon > 0$, *and an* $\alpha_0(\varepsilon) \in D$ *such that* $\alpha > \alpha_0(\varepsilon)$ *implies* $\tilde{P}_\alpha^*(A) < \varepsilon$ *for any* $A \in \Sigma$ *with* $\lambda_0(A) < \delta_\varepsilon$.

(c) *The set* $K_1 = \{\tilde{P}_\alpha^*, \alpha \in D\} \subset ca(\Omega, \Sigma)$ *is relatively weakly sequentially compact, where* $\{\tilde{P}_\alpha^*, \alpha \in D\}$ *is an extension of measures of* P_α^**s as in the above case.*

(d) *There is an isometric lattice isomorphism between* K_1 *of* (c) *and a uniformly integrable positive martingale* $\{f_\alpha, \Sigma_\alpha^*, \alpha \in D\} \subset L^1(\Omega, \Sigma, \lambda)$ *relative* *to some finite positive measure* λ *on* Σ (*or* $f_\alpha = E_\alpha(f)$ *for a unique* $f \in L^1(\Omega, \Sigma, \lambda)$ *in the notation of Theorem* 8).

(e) *There exists a nontrivial continuous symmetric convex function* $\Phi: \mathbb{R}^+ \to \mathbb{R}^+$, $\Phi(0) = 0$, $\Phi(x)/x \to \infty$ *as* $x \to \infty$ (*it is a Young function*) *and a finite measure* λ_1 *on* Σ, $\lambda_1(\Omega) > 0$, *such that each* $\tilde{P}_\alpha^*$ (*of* K_1) *is of uniform, in* α, Φ*-bounded variation relative to* λ_1, *i.e., if* $\pi = \{A_i\}_1^n \subset \Sigma$ *is a partition of* Ω, $\sup \Sigma_\pi \Phi[\tilde{P}_\alpha^*(A_i)/\lambda_1(A_i)]\lambda_1(A_i) \leq c_0 < \infty$, *uniformly in* α *where the supremum is taken over all such* π.

Discussion That (a)$\Leftrightarrow$(d) above leads directly to the general study of (point) martingales. If the measures P_α are, instead of being probabilities, some system of signed measures, then one can prove a corresponding result using a Jordan-type decomposition of set functions. This follows from Proposition 5.2 below. Part (e) above can be stated as K_1 being contained in a bounded set (or ball) of $V^\Phi(\lambda_1)$, an Orlicz space of additive set functions of Φ-bounded variation for λ_1. This line of thought will not be pursued here, but such an identification is useful in a general study of the problem. (Cf. the author's monograph [10, IV.3.14] and paper [14].)

In relation to the abstract projective systems of probability spaces considered in Proposition 1.2 and using certain ideas of Theorem 8, we may present a simple decomposition of a (bounded) projective system into a part that admits a limit and another that does not in such a way that the former is as large as it can ever be. This observation is of some interest in the theory. We use the previous notation.

10. Proposition *Let* $\{(\Omega_\alpha, \Sigma_\alpha, P_\alpha, g_{\alpha\beta})_{\alpha<\beta} : \alpha, \beta \text{ in } D\}$ *be a projective system of probability spaces, and* P *on* (Ω, Σ_0) *be an additive set function given by Proposition* 1.2. *Then the system can be uniquely decomposed into two systems* $\{(\Omega_\alpha, \Sigma_\alpha, P_\alpha^i, g_{\alpha\beta})_{\alpha<\beta} : \alpha, \beta \text{ in } D\}$, $i = 1, 2$, *such that* $0 \leq P_\alpha^i(\Omega_\alpha) \leq 1$ (P_α^i *can be zero or* $P_\alpha^i(\Omega_\alpha) < 1$); *the first system admits the projective limit* (Ω, Σ, P^1), *and the second one is purely finitely additive* (Ω, Σ, P^2) *in the sense that* $0 \leq Q \leq P^2$ *and* Q *is* σ*-additive implies* $Q \equiv 0$. *Moreover, the first system is the largest, admitting the limit such that* $P = P^1 + P^2$. (*Again we are assuming that* $g_\alpha(\Omega) = \Omega_\alpha$, $\alpha \in D$.)

Proof By Proposition 1.2 there is a unique $P: \Sigma_0 \to [0, 1]$, which is finitely additive and which satisfies $P|\Sigma_\alpha^* = P_\alpha^*$, $\alpha \in D$. However, by the classical Yosida–Hewitt [1] decomposition of an additive set function, one has $P = P^1 + P^2$ uniquely, where the P^i have the properties stated in the proposition, $i = 1, 2$. Define $P_\alpha^i = (P^i|\Sigma_\alpha^*) \circ g_\alpha^{-1}$, $i = 1, 2$. Then $P_\alpha = P_\alpha^1 + P_\alpha^2$ are σ-additive $P^i = \lim_{\leftarrow} P_\alpha^i$ and P_α^2s, though σ-additive for $\alpha \in D$, have P^2 as their "limit" given by Proposition 1.2. By the uniqueness of projective limits and that of the Yosida–Hewitt decomposition, the uniqueness assertion is a simple consequence. (See also Dunford–Schwartz [1, III.7.8] for an account of the latter theory.) Everything else is immediate, and since the compatibility of P_α^i is obvious, the result follows.

The quantity $0 \le P^2(\Omega) \le 1$ is sometimes called the *default of σ-additivity* of P. If the system consists of signed measures, then this default is $\text{var}(P^2)$. We remark that it can happen in some situations that $P^1(\Omega) = 0$ so that the given system does not admit a (σ-additive) projective limit at all. The counterexample in Problem I.6.5 possesses this pathological property.

In case the index set D is countable or only has a countable cofinal set, conditions for the existence of projective limits can be somewhat simplified, at least in the topological case. Since this one, in our treatment, does not occur frequently, we shall state the following result and refer the interested reader to Bourbaki [1, p. 54].

11. Theorem *Let $\{(\Omega_\alpha, \Sigma_\alpha, P_\alpha, g_{\alpha\beta})_{\alpha<\beta} : \alpha, \beta \text{ in } D\}$ be a projective system of inner regular (for compact sets in Σ_αs) probability spaces such that D has a countable cofinal set. If g_α: $\Omega(=\lim_{\leftarrow} \Omega_\alpha) \to \Omega_\alpha$ is onto (or g_αs separate points of Ω), $\alpha \in D$, then the system admits a projective limit.*

All the effort in this section has been to find conditions either on the mappings $g_{\alpha\beta}$ or on the topological properties of the measurable spaces $(\Omega_\alpha, \Sigma_\alpha)$, so that the system admits a projective limit. There is another method corresponding to Theorem II.3.20, and to present it, direct limit theory is needed. Since this further illuminates the structure of projective limits, we shall now consider it.

3.3 DIRECT LIMITS AND A GENERALIZED KOLMOGOROV EXISTENCE THEOREM

In this section we present an abstract extension (without topology) of the basic Kolmogorov existence theorem (I.3.1) in the form of a projective limit characterization. Unlike the work in Section 2, this needs a consideration of direct systems and limits of measure spaces. The latter are dual to projective systems, in a certain sense, and we include a relevant portion of that theory here since it also illuminates the previous work. The results of this section and Section 3.4 are not yet in final form and are not essential for the rest of the material in the book. They are of research interest.

To motivate the concept, let us reinterpret again the statement of Theorem I.3.1. If $T \subset \mathbb{R}$, D is the directed (by inclusion) class of all finite subsets of T, $\Omega = \mathbb{R}^T$, $\Omega_\alpha = \mathbb{R}^\alpha$, $\alpha \in D$, and $g_\alpha: \Omega \to \Omega_\alpha$, $g_{\alpha\beta}: \Omega_\beta \to \Omega_\alpha$ $(\alpha < \beta)$ are coordinate projections, then Ω can be identified as $\lim_{\leftarrow}(\Omega_\alpha, g_{\alpha\beta})$. Suppose now we consider the embedding of Ω_α into Ω_β for $\alpha < \beta$ or Ω_α into Ω. Let $i_\alpha: \Omega_\alpha \to \Omega$, $i_{\alpha\beta}: \Omega_\alpha \to \Omega_\beta$ be these embedding mappings. Then we have for $\alpha < \beta < \gamma$ (in D), $i_\alpha = i_\beta \circ i_{\beta\alpha}$, $i_{\gamma\alpha} = i_{\gamma\beta} \circ i_{\beta\alpha}$, and $i_{\alpha\alpha}$ is the identity. Moreover, $\Omega = \bigcup_{\alpha \in D} i_\alpha(\Omega_\alpha)$. These properties of the mappings $\{i_\alpha, i_{\alpha\beta} : \alpha, \beta \text{ in } D\}$ and spaces $\{\Omega_\alpha, \alpha \in D\}$ yield a dual concept (to projective systems) called direct systems. The resulting theory, however, is not a simple analog since, as we see, the direct limit of a system of probability spaces will not generally be a probability space. (Cf. also Vasilach [1].)

Let us start with a definition of this concept.

1. Definition Let $\{(\Omega_\alpha, \Sigma_\alpha, P_\alpha), \alpha \in D\}$ be a family of measure spaces where D is a directed set. Let $h_{\beta\alpha}: \Sigma_\alpha \to \Sigma_\beta$ $(\alpha < \beta)$ be a family of σ-homomorphisms (i.e., $h_{\beta\alpha}(\varnothing) = \varnothing$, $h_{\beta\alpha}(\Omega_\alpha) = \Omega_\beta$, and $h_{\beta\alpha}$ preserves complements and countable unions) such that for $\alpha < \beta < \gamma$ we have $h_{\gamma\beta} \circ h_{\beta\alpha} = h_{\gamma\alpha}$, and $h_{\alpha\alpha}$ is the identity. The family is called a *direct system of measure spaces* if for each $\alpha < \beta$, $P_\alpha(A) = P_\beta(h_{\beta\alpha}(A))$, $A \in \Sigma_\alpha$, and the system is denoted by $\{(\Omega_\alpha, \Sigma_\alpha, P_\alpha, h_{\beta\alpha})_{\alpha<\beta} : \alpha, \beta \text{ in } D\}$.

A connection between the projective and direct systems is provided by the following simple but useful result.

2. Proposition *Let $\{(\Omega_\alpha, \Sigma_\alpha, P_\alpha, h_{\beta\alpha})_{\alpha<\beta} : \alpha, \beta$ in $D\}$ be a direct system of measure spaces. Suppose that for each $\omega \in \Omega_\alpha$, $\{\omega\} \in \Sigma_\alpha$ and that the $h_{\beta\alpha}$ preserve arbitrary unions. Then there exist point mappings $g_{\alpha\beta}: \Omega_\beta \to \Omega_\alpha$ $(\alpha < \beta)$ such that $h_{\beta\alpha} = g_{\alpha\beta}^{-1}$ and that $\{(\Omega_\alpha, \Sigma_\alpha, P_\alpha, g_{\alpha\beta})_{\alpha<\beta} : \alpha, \beta$ in $D\}$ is a projective system of measure spaces.*

Proof For each $\omega \in \Omega_\alpha$, consider $Y_\omega^\beta = h_{\beta\alpha}(\{\omega\}) \in \Sigma_\beta$. Then by hypothesis $\bigcup_\omega Y_\omega^\beta = h_{\beta\alpha}(\bigcup_\omega \{\omega\}) = h_{\beta\alpha}(\Omega_\alpha) = \Omega_\beta$. Since $\{Y_\omega^\beta, \omega \in \Omega_\alpha\}$ is a disjoint collection, we may define $g_{\alpha\beta}$ unambiguously by the equation $g_{\alpha\beta}(y) = \omega$ for all $y \in Y_\omega^\beta$. Then $h_{\beta\alpha} = g_{\alpha\beta}^{-1}$, and $g_{\alpha\alpha}$ is the identity. If $\alpha < \beta < \gamma$, then $g_{\alpha\gamma}^{-1} = (g_{\alpha\beta} \circ g_{\beta\gamma})^{-1}$ by the compatibility of the $h_{\beta\alpha}$; so $\omega = g_{\alpha\beta} \circ g_{\beta\gamma}(y) = g_{\alpha\gamma}(y)$ for all $y \in Y_\omega^\gamma = h_{\gamma\alpha}(\{\omega\})$, $\omega \in \Omega_\alpha$. Since $\bigcup_\omega Y_\omega^\gamma = \Omega_\gamma$, we deduce that $g_{\alpha\beta} \circ g_{\beta\gamma} = g_{\alpha\gamma}$ and then $P_\alpha = P_\beta \circ h_{\beta\alpha} = P_\beta \circ g_{\alpha\beta}^{-1}$. Also $g_{\alpha\beta}$ is $(\Sigma_\beta, \Sigma_\alpha)$-measurable. Hence $\{(\Omega_\alpha, \Sigma_\alpha, P_\alpha, g_{\alpha\beta})_{\alpha<\beta} : \alpha, \beta \text{ in } D\}$ is a projective system, completing the proof.

We observe that if the additional condition on $h_{\beta\alpha}$s is omitted, the conclusion of the above proposition is false as simple counterexamples show. (See, e.g., Royden [1, Chapter 15] on these mappings.) Thus it is *not* always true that a direct system of measures arises from a projective system.

(With a projective system we can obviously associate a direct system by taking $h_{\beta\alpha}$ as $g_{\alpha\beta}^{-1}$.)

It is now desirable to define the concept of a direct limit and relate it to a projective limit, at least in cases where the hypothesis of the above proposition holds. This is not trivial, and the result is useful in completing the proof of the extension theorem contemplated for this section.

Let $\{(\Sigma_\alpha, h_{\beta\alpha})_{\alpha<\beta} : \alpha, \beta \text{ in } D\}$ be a family of spaces where for each $\alpha \in D$, Σ_α is a σ-algebra and $h_{\beta\alpha}: \Sigma_\alpha \to \Sigma_\beta$ $(\alpha < \beta)$ be the compatible σ-homomorphisms as in Definition 1. Let R be a relation defined on this family by the following rule: For $A_\alpha \in \Sigma_\alpha$, $B_\beta \in \Sigma_\beta$, $(A_\alpha, B_\beta) \in R$ (or $A_\alpha R B_\beta$) iff for some $\gamma \in D$ $(\gamma > \alpha, \gamma > \beta)$ we have $h_{\gamma\alpha}(A_\alpha) = h_{\gamma\beta}(B_\beta)$. Let $\tilde{\Sigma} = \bigcup_{\alpha \in D}(\Sigma_\alpha \times \{\alpha\})$ be the *set sum* (or free union) of Σ_αs. It is easy to check that R is an equivalence relation on $\tilde{\Sigma}$ where Σ_α, $\Sigma_\alpha \times \{\alpha\}$ can be and are identified. Let $\Sigma^\infty = \tilde{\Sigma}/R$, the quotient space. For A, B in Σ^∞, define $A < B$ iff there exist $A_\alpha \in \Sigma_\alpha$, $B_\beta \in \Sigma_\beta$ such that $A' R A_\alpha$, $B' R B_\beta$, and a $\gamma \in D$ $(\gamma > \alpha, \gamma > \beta)$ satisfying $h_{\gamma\alpha}(A_\alpha) = h_{\gamma\beta}(B_\beta)$ where $A' \in A$, $B' \in B$ (i.e., A' is any member of the equivalence class of A and similarly B' of B). It is not hard to show that the ordering "$<$" does not depend on the particular A_α, B_β, and $(\tilde{\Sigma}, <)$ becomes a lattice. If $h: \tilde{\Sigma} \to \Sigma^\infty$ is the canonical mapping, which is a lattice homomorphism under the ordering $<$, let $h_\alpha = h | \Sigma_\alpha$ where again Σ_α and $\Sigma_\alpha \times \{\alpha\}$ are identified. Then $\Sigma^\infty = \bigcup_\alpha h_\alpha(\Sigma_\alpha)$ and $\Sigma^\infty$is a Boolean algebra, called the *direct limit* of the system, denoted by $\Sigma^\infty = \lim_\to(\Sigma_\alpha, h_{\alpha\beta})$. This limit is clearly nonempty (unlike the projective limits). In case each Ω_α is a Hausdorff topological space and Σ_α is its Borel (or Baire) σ-algebra (determining the topology of Ω_α), then Σ^∞ can be endowed with the "direct limit" topology by the rule: $A \in \Sigma^\infty$ is open iff $h_\alpha^{-1}(A) \in \Sigma_\alpha$ is open for each $\alpha \in D$. This makes each h_α continuous.

The following result is analogous to Proposition 1.2 and serves a similar purpose in defining the direct limit function.

3. Proposition *Let $\{(\Omega_\alpha, \Sigma_\alpha, P_\alpha, h_{\beta\alpha})_{\alpha<\beta} : \alpha, \beta \text{ in } D\}$ be a direct system of measure spaces and each $h_{\beta\alpha}$ be (for convenience) an isomorphism. [This is automatic if each Σ_α is replaced by its measure algebra determined by P_α.] If $\Sigma^\infty = \lim_\to(\Sigma_\alpha, h_{\beta\alpha})$, as in the preceding paragraph, then there exists uniquely a mapping $P: \Sigma^\infty \to \bar{\mathbb{R}}^+$ such that $P \circ h_\alpha = P_\alpha$, $h_\beta \circ h_{\beta\alpha} = h_\alpha$, $\alpha < \beta$, and P is finitely additive. If $P_\alpha(\Omega_\alpha) = 1$, then $P(I) = 1$ where I is the unit of Σ^∞.*

Proof If $\alpha_1 < \beta_1$, then $P_{\alpha_1} = P_{\beta_1} \circ h_{\beta_1\alpha_1}$ by definition. Let $A \in \Sigma^\infty = \bigcup_\alpha h_\alpha(\Sigma_\alpha)$, so that $A = h_\alpha(A_\alpha)$ for some $\alpha \in D$. If also $A = h_\beta(B_\beta)$ for some β, then there is a $\gamma \in D$ $(\gamma > \alpha, \gamma > \beta)$ such that $h_\alpha = h_\gamma \circ h_{\gamma\alpha}$, $h_\beta = h_\gamma \circ h_{\gamma\beta}$, and

$$h_\gamma(h_{\gamma\alpha}(A_\alpha)) = h_\alpha(A_\alpha) = A = h_\beta(B_\beta) = h_\gamma(h_{\gamma\beta}(B_\beta)). \tag{1}$$

Since h_γ is an isomorphism, we have $h_{\gamma\alpha}(A_\alpha) = h_{\gamma\beta}(B_\beta)$. (With only the homomorphism property, we could have concluded that these two sets are in the same equivalence class of Σ_γ, and this would be sufficient for the following argument.) Define P on Σ^∞ by the equation

$$P(A) = P(h_\alpha(A_\alpha)) = P_\alpha(A_\alpha) \qquad (= P_\beta(B_\beta)). \tag{2}$$

Then (2) implies that P is unambiguously defined and $P_\alpha = P \circ h_\alpha$, $\alpha \in D$. The additivity of P now can be established by a similar (and simple) argument, which is left to the reader. The fact that P is an additive real function and $P(I) = 1$ (since $P_\alpha(\Omega_\alpha) = 1$) is obvious. This completes the proof.

In general, P is not σ-additive on Σ^∞, and P is denoted as $P = \text{w-lim}_\rightarrow P_\alpha$, the weak direct limit of the measures. We introduce the concept precisely as follows, in which reference to Ω_αs will be dropped.

4. Definition Let $\{(\Sigma_\alpha, P_\alpha, h_{\beta\alpha})_{\alpha<\beta} : \alpha, \beta \text{ in } D\}$ be a direct system of probability spaces and $\Sigma^\infty = \lim_\rightarrow(\Sigma_\alpha, h_{\beta\alpha})$, $P = \text{w-lim}_\rightarrow P_\alpha$. Then the pair (Σ^∞, P) is called the *weak direct limit* of the direct system denoted $\text{w-lim}_\rightarrow(\Sigma_\alpha, h_{\beta\alpha})$. If, moreover, P is σ-additive on Σ^∞ (i.e., $A_n \in \Sigma^\infty$, $\bigvee_{n=1}^\infty A_n = A \in \Sigma^\infty$, $A_i \wedge A_j = 0$ for $i \neq j$ implies $P(A) = \sum_{i=1}^\infty P(A_i)$, where $(\vee, \wedge, ')$ are the (join, meet, complement) in Σ^∞ induced by "$<$") and if $\bar{\Sigma}$ is the smallest σ-Boolean algebra containing Σ^∞ (in the sense that there is a subalgebra $\Sigma' \subset \bar{\Sigma}$ that is algebraically isomorphic to Σ^∞, the mapping being a σ-homomorphism and $\bar{\Sigma}$ being the smallest with this property), then denote by $\bar{P}$ the unique extension of P to $\bar{\Sigma}$. (This is a valid and known analog of the Hahn extension theorem.) The pair $(\bar{\Sigma}, \bar{P})$ is called the *strong direct limit* of the direct system denoted $\bar{\Sigma} = \text{s-lim}_\rightarrow(\Sigma_\alpha, h_{\beta\alpha})$ and $\bar{P} = \text{s-lim}_\rightarrow P_\alpha$.

The existence of strong direct limits presents a nontrivial problem. We shall consider this question later. At the moment, the weak direct limit (given by Proposition 3) is sufficient for our next key result. But the distinction between these two limits should be emphasized. We now proceed to the statement and proof of an important representation.

Since in many problems of stochastic analysis probabilities are defined on abstract sets so that the resulting projective system is not topological, it will be important to extract as much of its information as possible. The following representation of such an abstract system plays a fundamental role in the general theory (also compare it with Theorem II.3.20).

5. Theorem *Let $\{(\Omega_\alpha, \Sigma_\alpha, P_\alpha, g_{\alpha\beta})_{\alpha<\beta} : \alpha, \beta \text{ in } D\}$ be a general abstract projective system of probability spaces where each Σ_α is complete for P_α. Then there exists a Hausdorff projective system of (Baire) regular probability spaces*

$\{(S_\alpha, \mathscr{B}_\alpha, \mu_\alpha, \hat{g}_{\alpha\beta})_{\alpha<\beta} : \alpha, \beta \text{ in } D\}$ *admitting a regular projective limit* $(S, \hat{\mathscr{B}}, \mu)$ *having the following structure.*

(i) *For each* $\alpha \in D$ *the space* S_α *is a compact Stone space and* $(S_\alpha, \mathscr{B}_\alpha, \mu_\alpha)$ *is a Baire regular probability triple.*

(ii) *There exist mappings* $u_\alpha : \Omega_\alpha \to S_\alpha$, $\alpha \in D$, *which are* $(\Sigma_\alpha, \mathscr{B}_\alpha)$-*measurable, and mappings* $\tau_\alpha : \Sigma_\alpha \to \mathscr{B}_{0\alpha}$, $\sigma(\mathscr{B}_{0\alpha}) = \mathscr{B}_\alpha$, *where* $\mathscr{B}_{0\alpha}$ *is the algebra of clopen (i.e., closed and open) sets of* S_α, τ_α *being an algebraic isomorphism that preserves finite unions, intersections, and complements (but not a* σ-*homomorphism in general) such that* $\tau_\alpha \circ u_\alpha^{-1}(A) = A$, $A \in \mathscr{B}_{0\alpha}$.

(iii) *The compatible family of continuous mappings* $\{\hat{g}_{\alpha\beta} : \alpha, \beta \text{ in } D\}$ *is related to* $g_{\alpha\beta}$ *for* $\alpha < \beta$ *by* $u_\alpha \circ g_{\alpha\beta} = \hat{g}_{\alpha\beta} \circ u_\beta$ *a.e. Thus outside of a suitable* P_β-*null set* Ω_β^0, *the following diagram is commutative with* $\tilde{\Omega}_\beta = \Omega_\beta - \Omega_\beta^0$:

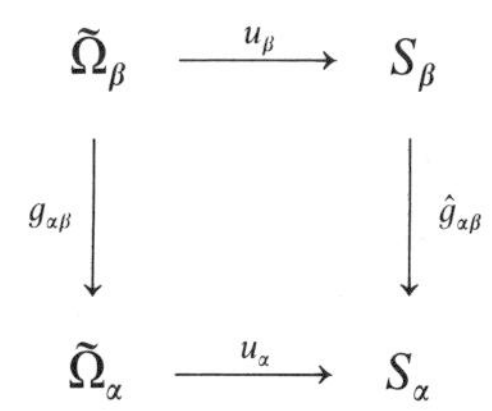

(iv) *The measures* P_α *and* μ_α *are related by the equation* $\mu_\alpha = P_\alpha \circ u_\alpha^{-1}$, *so that* μ_α *is the image measure of* P_α *under* u_α.

(v) *If* $\Omega = \lim_{\leftarrow}(\Omega_\alpha, g_{\alpha\beta})$, $S = \lim_{\leftarrow}(S_\alpha, \hat{g}_{\alpha\beta})$, *and in (iii) the exceptional null sets can be simultaneously ignored, then there is a unique mapping* $u : \Omega \to S$ *(and vacuously so if* $\Omega = \varnothing$*) such that the canonical projections* $g_\alpha : \Omega \to \Omega_\alpha$ *and* $\hat{g}_\alpha : S \to S_\alpha$ *satisfy the relation* $u_\alpha \circ g_\alpha = \hat{g}_\alpha \circ u$, $\alpha \in D$; *whence the following diagram is commutative and* $\mu = P \circ u^{-1}$, $\mu = \lim_{\leftarrow} \mu_\alpha$, *and* μ *is a Baire measure on* $\mathscr{B} = \sigma(\bigcup_\alpha \hat{g}_\alpha^{-1}(\mathscr{B}_\alpha))$.

$$\begin{array}{ccc} \Omega & \xrightarrow{\;u\;} & S \\ {\scriptstyle g_\alpha}\downarrow & & \downarrow{\scriptstyle \hat{g}_\alpha} \\ \Omega_\alpha & \xrightarrow{\;u_\alpha\;} & S_\alpha \end{array}$$

The image system then is unique up to a homeomorphism.

Discussion A family of functions $\{u_\alpha, \alpha \in D\}$ between the projective systems $\{(\Omega_\alpha, g_{\alpha\beta})_{\alpha<\beta} : \alpha, \beta \text{ in } D\}$ and $\{(S_\alpha, \hat{g}_{\alpha\beta})_{\alpha<\beta} : \alpha, \beta \text{ in } D\}$ such that $u_\alpha \circ g_{\alpha\beta} = \hat{g}_{\alpha\beta} \circ u_\beta$ for $\alpha < \beta$ is said to be a *projective system* of mappings of the first set of spaces into the second. When the strengthened hypothesis of

(iii) holds (i.e., the mappings of the diagram can be so chosen that the diagram commutes for *all* points), then part (v) shows the existence of a unique $u: \Omega \to S$, and this u is called the *projective limit* of the family $\{u_\alpha, \alpha \in D\}$ denoted $u = \lim_{\leftarrow} u_\alpha$. However, if we consider merely two projective systems of spaces Ω_αs and S_αs with u_α as connecting mappings, then the following "substitution" need not be true:

$$u(\Omega) = \lim_{\leftarrow} u_\alpha\left(\lim_{\leftarrow} \Omega_\alpha\right) = \lim_{\leftarrow} u_\alpha(\Omega_\alpha), \tag{3}$$

where $\{(u_\alpha(\Omega_\alpha), \hat{g}_{\alpha\beta})_{\alpha<\beta} : \alpha, \beta \text{ in } D\}$ forms a projective system of subspaces of $\{(S_\alpha, \hat{g}_{\alpha\beta})_{\alpha<\beta} : \alpha, \beta \text{ in } D\}$. Only the obvious inclusion $u(\Omega) \subset \lim_{\leftarrow} u_\alpha(\Omega_\alpha)$ holds. That there can be strict inclusion may be seen from the system of Ω_αs of Example B in Section 1. Also, $u(\Omega) \subset S$ need not be $(\Sigma, \mathscr{B})$-measurable. The problem of σ-additivity of P on Σ_0 and the "size" of $u(\Omega)$ in S (as well as the essential equalities in (3)) are interrelated. It may also be observed here that the strengthened hypothesis of (iii) is satisfied if $(\Omega_\alpha, \Sigma_\alpha, P_\alpha)$ is a Radon probability space for each $\alpha \in D$. This can be proved with an application of Theorem 1.5. If $\lim_{\leftarrow}(\Omega_\alpha, \Sigma_\alpha, P_\alpha)$ exists, then such a mapping $u: \Omega \to S$ of (v) can be shown to exist. In any case, when an abstract projective system satisfies the strengthened condition of (iii) (i.e., the equation $u_\alpha \circ g_{\alpha\beta} = \hat{g}_{\alpha\beta} \circ u_\beta$ holds, and so u exists), then the system will be called "*tight*," motivated by the topological case. A related result has been independently established by Schreiber, Sun, and Bharucha-Reid [1], using the theory of algebraic models of probability spaces (cf. Problem 7) of Dinculeanu and Foiaş [1]. The latter representation enables these authors to take the S_αs as compact abelian groups for each $\alpha \in D$, which is then used in their proof of the result. Using another point of view, Scheffer [1] also obtained an analogous representation and related results. On the other hand, the following demonstration leans on the ideas of Choksi [1]. After proving the result, we present the main characterization noted before.

Proof Some facts about the Stone space representation of a Boolean algebra and the preliminary results on direct systems given above will be used in the proof. The argument will be presented in steps for convenience.

I. With the given projective system, consider the associated direct system $\{(\Sigma_\alpha, P_\alpha, h_{\beta\alpha})_{\alpha<\beta} : \alpha, \beta \text{ in } D\}$, where $h_{\beta\alpha} = g_{\alpha\beta}^{-1}$ so that $P_\alpha = P_\beta \circ h_{\beta\alpha}$ and $h_{\beta\alpha}: \Sigma_\alpha \to \Sigma_\beta$ is a σ-isomorphism (since $g_{\alpha\beta}$ is onto). For each α, one can and does regard Σ_α as a Boolean algebra. Let S_α be the Stone space of the algebra Σ_α so that S_α is the space of all homomorphisms of Σ_α into $\{0, 1\}$. It has the following description. If $A_\alpha \in \Sigma_\alpha$, let B_α $(\subset S_\alpha)$ be the set of all homomorphisms taking $A_\alpha(A_\alpha^c)$ into $1(0)$ and let $\mathscr{B}_{0\alpha}$ be the family of all these B_α. Then $\mathscr{B}_{0\alpha}$ is an algebra. Each such B_α may also be regarded as the set of maximal ideals of the Boolean algebra Σ_α containing $A_\alpha(A_\alpha^c)$. Now taking $\mathscr{B}_{0\alpha}$ as an open basis for S_α, one introduces a topology in S_α (i.e., $G \subset S_\alpha$ is open iff G is a union of some sets of $\mathscr{B}_{0\alpha}$). This makes S_α a compact totally disconnected Hausdorff space called the Stone space of Σ_α. If $\tilde{S}_\alpha$ is another representation of Σ_α, then S_α and $\tilde{S}_\alpha$ are homeomorphic. (These facts are standard and are found in many places. See, e.g., Sikorski [1, p. 24].) Let τ_α be the correspondence between the A_α and B_α above.

If $\alpha < \beta$ and S_α and S_β are the corresponding spaces with the algebras $\mathscr{B}_{0\alpha}$ and $\mathscr{B}_{0\beta}$, then the mapping $\tau_\alpha: A_\alpha \mapsto B_\alpha$ in the preceding paragraph defines an isomorphism of Σ_α onto $\mathscr{B}_{0\alpha}$ and we define a mapping $\hat{g}_{\alpha\beta}: S_\beta \to S_\alpha$, which turns out to be onto and continuous. For this, observe that $g_{\alpha\beta}^{-1}(\Sigma_\alpha)$ is a subalgebra of Σ_β and that $g_{\alpha\beta}^{-1}$ is one-to-one. Now each homomorphism of Σ_β into $\{0, 1\}$ induces (by restriction) a corresponding one on $g_{\alpha\beta}^{-1}(\Sigma_\alpha)$, and hence by the one-to-one character of $g_{\alpha\beta}^{-1}$ we can uniquely associate a homomorphism of Σ_α into $\{0, 1\}$. Conversely, with each homomorphism $\Sigma_\alpha \to \{0, 1\}$, we may associate one on $g_{\alpha\beta}^{-1}(\Sigma_\alpha)$ to $\{0, 1\}$ and extend it to Σ_β. Thus each maximal ideal of Σ_α generates an ideal in Σ_β which is then included in a maximal ideal of Σ_β. These maximal ideals (of Σ_β and Σ_α) correspond to the homomorphisms of Σ_β and Σ_α into $\{0, 1\}$ and so to points of S_β and S_α. If $\hat{g}_{\alpha\beta}$ is this correspondence, then the preceding argument implies that $\hat{g}_{\alpha\beta}: S_\beta \to S_\alpha$ is a well-defined onto mapping. Since $\tau_\alpha(\Sigma_\alpha) = \mathscr{B}_{0\alpha}$ and $\tau_\beta(\Sigma_\beta) = \mathscr{B}_{0\beta}$, we deduce that with each $B_\alpha \in \mathscr{B}_{0\alpha}$, $B_\beta = \hat{g}_{\alpha\beta}^{-1}(B_\alpha) \in \mathscr{B}_{0\beta}$, where in fact if B_α is the set of homomorphisms of A_α into 1, then B_β is the set of homomorphisms of $g_{\alpha\beta}^{-1}(A_\alpha)$ $(= h_{\beta\alpha}(A_\alpha) = A_\beta)$ into 1. Thus $\hat{g}_{\alpha\beta}$ is continuous since $\mathscr{B}_{0\alpha}, \mathscr{B}_{0\beta}$ are bases of topologies in S_α and S_β. If $\mu_\alpha = P_\alpha \circ \tau_\alpha^{-1} : \mathscr{B}_{0\alpha} \to [0, 1]$, then μ_α is a probability on $\mathscr{B}_{0\alpha}$, and if $\mathscr{B}_\alpha$ is the σ-algebra generated by $\mathscr{B}_{0\alpha}$, it follows that $(S_\alpha, \mathscr{B}_\alpha, \mu_\alpha)$, $\alpha \in D$, is a regular (Baire) probability space asserted in (i). The preceding identifications also show at once that $\hat{g}_{\alpha\beta} \circ \hat{g}_{\beta\gamma} = \hat{g}_{\alpha\gamma}$, $\hat{g}_{\alpha\alpha}$ is the identity, $\alpha < \beta < \gamma$. In other words, $\{(S_\alpha, \mathscr{B}_\alpha, \mu_\alpha, \hat{g}_{\alpha\beta})_{\alpha<\beta} : \alpha, \beta \text{ in } D\}$ is a projective system of (Baire) regular probability spaces for which Theorem 1.5 applies, and hence this system admits a (Baire) regular limit $(S, \mathscr{B}, \mu)$. [If $\Sigma^\infty = \text{w-lim}_\to(\Sigma_\alpha, h_{\beta\alpha})$, then one can identify (with a nontrivial proof, see, e.g., Haimo [1]) that S represents Σ^∞ and the algebras Σ^∞ and $\mathscr{B}_0 = \bigcup_\alpha \hat{g}_\alpha^{-1}(\mathscr{B}_{0\alpha})$ are isomorphic. This fact is of interest in connecting the weak direct limits and representing projective systems of probability spaces. Haimo's result is given as Proposition 9 below.]

II. We next define $u_\alpha: \Omega_\alpha \to S_\alpha$ and begin checking the properties of the representation asserted in the theorem. Since $(S_\alpha, \mathscr{B}_\alpha, \mu_\alpha)$ is a compact (Baire) regular probability space, there is a unique Borel extension $(S_\alpha, \hat{\mathscr{B}}_\alpha, \hat{\mu}_\alpha)$ where $\hat{\mu}_\alpha$ is the regular Borel measure such that $\hat{\mu}_\alpha | \mathscr{B}_\alpha = \mu_\alpha$ by the standard measure theory, as noted in Theorem I.2.10. So, if $x \in S_\alpha$, then $\{x\} \in \hat{\mathscr{B}}_\alpha$. Moreover, by the relations between the Baire and Borel regularity, there exist a $B_x \in \mathscr{B}_\alpha$ and a set $N \in \hat{\mathscr{B}}_\alpha$, $\hat{\mu}_\alpha(N) = 0$ such that the symmetric difference $B_x \Delta \{x\} \subset N$, i.e., $\hat{\mu}_\alpha(B_x \Delta \{x\}) = 0$. If $A_\alpha \in \Sigma_\alpha$ is the set given by $\tau_\alpha^{-1}(B_x) = A_\alpha$, define $u_\alpha(\omega) = x$ for all $\omega \in A_\alpha$. Then it follows that $u_\alpha(A_\alpha) = B_x$ a.e. $[\hat{\mu}_\alpha]$, and we may identify (ignoring a $\hat{\mu}_\alpha$-null set) $u_\alpha^{-1}(B_\alpha) = \tau_\alpha^{-1}(B_x) = A_\alpha$ so that $\tau_\alpha(u_\alpha^{-1}(B_x)) = B_x$ and for $A \in \Sigma_\alpha$, $u_\alpha^{-1}(\tau_\alpha(A)) = A$ a.e. It follows that the thus-defined $u_\alpha: \Omega_\alpha \to S_\alpha$ is $(\Sigma_\alpha, \hat{\mathscr{B}}_\alpha)$-measurable and $u_\alpha(\Omega_\alpha)$ is dense in S_α, $P_\alpha \circ u_\alpha^{-1} = \mu_\alpha$. In other words, (ii) and (iv) are true. [The μ_α-null sets do not cause any problem in the definition of u_α. Actually, it is possible to select continuously a member of the equivalence class by use of a lifting map whose existence was noted in Theorem I.5.4.]

Using the same notations, consider B_x for $x \in S_\beta$ as above. Let $F_\beta = u_\beta^{-1}(B_x) \in \Sigma_\beta$ so that $u_\beta(F_\beta) = x$. Let $y = \hat{g}_{\alpha\beta}(x) \in S_\alpha$ and let $F_\alpha = u_\alpha^{-1}(B_y)$ so that $u_\beta(F_\alpha) = y$. If $\tilde{F}_\beta = g_{\alpha\beta}^{-1}(F_\alpha)$, then $F_\beta \subset \tilde{F}_\beta$. In fact, $y \in \hat{g}_{\alpha\beta}^{-1}(\{x\})$ and $F_\beta = u_\beta^{-1}(\{x\}) \subset u_\beta^{-1}(\hat{g}_{\alpha\beta}^{-1}(\{y\}))$. Also $u_\alpha^{-1}(\{y\}) = F_\alpha$ a.e., so $\tilde{F}_\beta = g_{\alpha\beta}^{-1}(F_\alpha) = g_{\alpha\beta}^{-1}(u_\alpha^{-1}(\{y\}))$. But $u_\beta^{-1} \circ \hat{g}_{\alpha\beta}^{-1}$ and $g_{\alpha\beta}^{-1} \circ u_\alpha^{-1}$ are one-to-one and take the set $\{y\}$ onto the same set in Ω_β. Hence $\tilde{F}_\beta \supset F_\beta$ follows from the above relations. Moreover, for each $\omega \in F_\beta$ we have $u_\beta(\omega) = x$ a.e., and so

$$\hat{g}_{\alpha\beta} \circ u_\beta(\omega) = y = u_\alpha \circ g_{\alpha\beta}(\omega), \qquad \text{a.a. } (\omega) \text{ in } F_\beta. \tag{4}$$

By the definition of the mappings u_α, u_β, one deduces that this is true for almost all $y \in S_\alpha$ or equivalently for almost all $\omega \in \Omega_\beta$. Since $\alpha < \beta$ is arbitrary, this implies that $\hat{g}_{\alpha\beta} \circ u_\beta = u_\alpha \circ g_{\alpha\beta}$ a.e. It must be noted that even though $u_\alpha: \Omega_\alpha \to S_\alpha$ depends on μ_α (it may change with the null set), $u_\alpha^{-1} = \tau_\alpha^{-1}: \mathscr{B}_{0\alpha} \to \Sigma_\alpha/\mathscr{N}_\alpha$ (the measure algebra of Σ_α, where $\mathscr{N}_\alpha$ is the family of its null sets) is a σ-isomorphism onto, so that this dependence causes no difficulty in the measure relations here. The choice of u_αs then is simply any one from the equivalence class just as in the L^p-theory. For each α, choose and hold fast one such u_α, and with this provision, the property (iii) obtains for each such choice.

III. To prove (v), let $g_\alpha: \Omega \to \Omega_\alpha$, $\hat{g}_\alpha: S \to S_\alpha$ be the canonical mappings from $\Omega = \lim_{\leftarrow} \Omega_\alpha$, and $S = \lim_{\leftarrow} S_\alpha$ as usual. If $u_\alpha: \Omega_\alpha \to S_\alpha$ is as above and (iii) holds *everywhere*, one can define $u: \Omega \to S$ with a set theoretical argument, without reference to the σ-algebras. (The existence of such u is the so-called "universal property" of projective limits of mappings.) Thus the mapping $h_\alpha = u_\alpha \circ g_\alpha: \Omega \to S_\alpha$ is well defined. Since $g_\alpha = g_{\alpha\beta} \circ g_\beta$ and $\hat{g}_\alpha = \hat{g}_{\alpha\beta} \circ \hat{g}_\beta$ for $\alpha < \beta$, we have the relation between h_αs and $\hat{g}_{\alpha\beta}$s as (with the strengthened hypothesis of (iii)):

$$h_\alpha = u_\alpha \circ g_\alpha = u_\alpha \circ g_{\alpha\beta} \circ g_\beta = \hat{g}_{\alpha\beta} \circ u_\beta \circ g_\beta = \hat{g}_{\alpha\beta} \circ h_\beta. \tag{5}$$

To show the existence of $u: \Omega \to S$, we have to verify that $(\hat{g}_\alpha \circ u)(\omega) = h_\alpha(\omega)$, $\omega \in \Omega$, $\alpha \in D$. By definition $u(\omega) = \{\tilde{u}_\alpha(\omega) \in S_\alpha, \alpha \in D\} \in \times_{\alpha \in D} S_\alpha$. If $u(\omega)$ is to be an element of S (as in the paragraph following Definition 1 in Section 1), we have to show that $\tilde{u}_\alpha(\omega) = \hat{g}_{\alpha\beta}(\tilde{u}_\beta(\omega))$. By (5) this is true if $\tilde{u}_\alpha = h_\alpha$. But every element $u(\omega)$ is of this kind for $\omega \in \Omega$. Hence $u(\omega) \in S$, and the desired u is defined. The uniqueness of the mapping is immediate since two such u's can be different only if at least one of the components is distinct, and this is impossible by (5). So (v) follows.

Since the Stone space representation of a Boolean algebra is unique to within a homeomorphism, it follows that the representing family $\{(S_\alpha, \mathscr{B}_\alpha, \mu_\alpha, \hat{g}_{\alpha\beta})_{\alpha<\beta} : \alpha, \beta \text{ in } D\}$ of the given system is unique to within a measure preserving homeomorphism. Since such a system admits a Baire regular limit $(S, \mathscr{B}, \mu)$ by Theorem 1.5, it admits a unique Borel regular (or Radon) extension. Thus the proof of the theorem is complete.

6. Remark The existence of $u_\alpha: \Omega_\alpha \to S_\alpha$ can also be proved by using the fact that $L^\infty(\Omega_\alpha, \Sigma_\alpha, P_\alpha) \cong C(\hat{S}_\alpha)$, where $\hat{S}_\alpha$ is a Stone space which is homeomorphic to S_α. Here $\hat{S}_\alpha$ is the set of multiplicative linear functionals on $L^\infty(P_\alpha)$, which is a compact set in the weak* topology of $(L^\infty(P_\alpha))^*$, the adjoint space of $L^\infty(P_\alpha)$. For each $\omega \in \Omega_\alpha$, if the evaluation functional is $x_\omega^* \in \hat{S}_\alpha$, i.e., $x_\omega^*(f) = f(\omega)$ for $f \in L^\infty(P_\alpha)$, then $u_\alpha: \omega \mapsto x_\omega^*$ is the desired mapping. It is easily shown that $u_\alpha(\Omega_\alpha)$ is a dense subspace of $\hat{S}_\alpha$, and if $v_\alpha: \hat{S}_\alpha \to S_\alpha$ is the above homeomorphism, then $\tilde{u}_\alpha = v_\alpha \circ u_\alpha: \Omega_\alpha \to S_\alpha$ gives the mapping we constructed in the above theorem with $\tau_\alpha \circ \tilde{u}_\alpha^{-1}$ is the identity on $\mathscr{B}_{0\alpha}$.

We can now present the previously announced characterization.

7. Theorem *Let* $\{(\Omega_\alpha, \Sigma_\alpha, P_\alpha, g_{\alpha\beta})_{\alpha<\beta} : \alpha, \beta \text{ in } D\}$ *be a "tight" projective system of probability spaces satisfying the s.m. condition (or more generally that* $g_\alpha(\Omega) = \Omega_\alpha$, $\alpha \in D$, *where* $\Omega = \lim_\leftarrow(\Omega_\alpha, g_{\alpha\beta})$). *Let* $\{(S_\alpha, \mathscr{B}_\alpha, \mu_\alpha, \hat{g}_{\alpha\beta})_{\alpha<\beta} : \alpha, \beta \text{ in } D\}$ *be the (compact) representing projective system of the above family guaranteed by Theorem* 5 *and let* $u_\alpha: \Omega_\alpha \to S_\alpha$, $u = \lim_\leftarrow u_\alpha$, $\mu = \lim_\leftarrow \mu_\alpha$ *(which exists and is regular by Theorem* 1.5) *be the connecting mappings between these systems and the limit measure of the latter system. Then* $P = \lim_\leftarrow P_\alpha$ *exists (i.e.,* P *of Proposition* 1.2 *is* σ*-additive) iff* $u(\Omega)$ *is* μ*-full or thick in* S *(i.e.,* $\mu^*(u(\Omega)) = \mu(S) = 1$, *where* μ^* *is the outer measure generated by* $(\mathscr{B}, \mu)$ *in* S). *Moreover,* $\mu = P \circ u^{-1}$ *when the last condition holds.*

Proof For the sufficiency, suppose that $u(\Omega)$ is μ-full in S. Consider the trace σ-algebra $\tilde{\mathscr{B}} = \{B \cap u(\Omega) : B \in \mathscr{B}\}$ where $(S, \mathscr{B}, \mu)$ is the (regular) limit of the representing system. Define $\tilde{\mu}: \tilde{\mathscr{B}} \to [0,1]$ by the equation $\tilde{\mu}(B \cap C) = \mu(B)$ where $C = u(\Omega)$. The fact that C is μ-full (or thick) and μ is σ-additive implies that $\tilde{\mu}$ is uniquely defined and is a probability on $\tilde{\mathscr{B}}$ (though $\tilde{\mu}$ is no longer "regular"). This is a standard (but not entirely obvious) fact. [We have discussed this in more detail after the proof of Corollary I.3.4. The specific result on "thick" sets, which is applicable here, is precisely that given in Theorem I.3.5.] Let $\tilde{P}$ be a function on $\tilde{\Sigma} = u^{-1}(\mathscr{B})$ defined by the equation $\tilde{P}(u^{-1}(B)) = \tilde{\mu}(B \cap C)$ $(=\mu(B))$. Then $\tilde{P}$ is a probability measure on $\tilde{\Sigma}$. We assert that $P = \tilde{P}$ on Σ_0 so that the P of Proposition 1.2 is σ-additive and hence $P = \lim_{\leftarrow} P_\alpha$ exists.

To prove this, let $A \in \Sigma_0 = \bigcup_\alpha g_\alpha^{-1}(\Sigma_\alpha)$. So $A = g_\alpha^{-1}(A_\alpha)$ for some $A_\alpha \in \Sigma_\alpha$, $\alpha \in D$, and $P(A) = P_\alpha(A_\alpha)$. By Theorem 5(ii) there exists an isomorphism $\tau_\alpha: \Sigma_\alpha \to \mathscr{B}_{0\alpha}$ (using the notation introduced there) such that if $B_\alpha = \tau_\alpha(A_\alpha)$ so that $B_\alpha = \tau_\alpha \circ u_\alpha^{-1}(B_\alpha)$, where $u_\alpha: \Omega_\alpha \to S_\alpha$ is the measurable mapping of the spaces shown, then $\hat{g}_\alpha^{-1}(B_\alpha) \in \mathscr{B}$ and $u^{-1} \circ \hat{g}_\alpha^{-1}(B_\alpha) \in \tilde{\Sigma}$. Moreover, using the relation $u_\alpha \circ g_\alpha = \hat{g}_\alpha \circ u$ of Theorem 5(iii) (invoking the "tightness" hypothesis now), we also have

$$\begin{aligned} u^{-1} \circ \hat{g}_\alpha^{-1}(B_\alpha) &= (u_\alpha \circ g_\alpha)^{-1}(B_\alpha) = g_\alpha^{-1} \circ u_\alpha^{-1}(B_\alpha) \\ &= g_\alpha^{-1}(\tau_\alpha^{-1}(B_\alpha)) = g_\alpha^{-1}(A_\alpha) = A. \end{aligned} \tag{6}$$

Since $\hat{g}_\alpha^{-1}(B_\alpha) \in \mathscr{B}$, we deduce that $A \in u^{-1}(\mathscr{B}) = \tilde{\Sigma}$, or that $\Sigma_0 \subset \tilde{\Sigma}$. Moreover,

$$\begin{aligned} P(A) = P_\alpha(A_\alpha) &= \mu_\alpha(B_\alpha) \qquad \text{by the definition of } \tau_\alpha \\ &= \mu(\hat{g}_\alpha^{-1}(B_\alpha)) = \tilde{\mu}(\hat{g}_\alpha^{-1}(B_\alpha) \cap C) \\ &= \tilde{P}(u^{-1} \circ \hat{g}_\alpha^{-1}(B_\alpha)) = \tilde{P}(A). \end{aligned} \tag{7}$$

It follows that P and $\tilde{P}$ agree on Σ_0, $\mu = \tilde{P} \circ u^{-1} = P \circ u^{-1}$ on Σ_0 and hence on $\Sigma = \sigma(\Sigma_0) \subset \tilde{\Sigma}$. (Observe that $u^{-1}: \tilde{\mathscr{B}} \to \Sigma$ is one-to-one also.) Denoting the unique extension of P to Σ by the same symbol, we deduce that $P = \lim_{\leftarrow} P_\alpha$ and the original system admits the projective limit, and then $\mu = P \circ u^{-1}$ on $\mathscr{B}$.

To prove the converse, again let $C = u(\Omega) \subset S$ and define $\tilde{\mathscr{B}}_0 = \{C \cap B : B \in \mathscr{B}_0\}$, the trace of $\mathscr{B}_0$ on C. Then $\tilde{\mathscr{B}}_0$ is an algebra on C. Define a function $\tilde{\mu}: \tilde{\mathscr{B}}_0 \to \mathbb{R}^+$ in terms of P $(=\lim_{\leftarrow} P_\alpha)$ by the equation $\tilde{\mu}(C \cap B) = P(u^{-1}(B))$. Clearly $\tilde{\mu}$ will be well defined iff $C \cap B = \varnothing$ implies $P(u^{-1}(B)) = 0$. To see that this is indeed the case, apply u^{-1} to both sides of this equation. Thus

$$\varnothing = u^{-1}(C \cap B) = u^{-1}(C) \cap u^{-1}(B) = \Omega \cap u^{-1}(B) = u^{-1}(B) \tag{8}$$

since $u: \Omega \to C$ is onto. Hence $\tilde{\mu}$ is well defined, and since P is σ-additive, $\tilde{\mu}$ must also be σ-additive on $\tilde{\mathscr{B}}_0$. We show that $\tilde{\mu}$ agrees with the outer measure generated by μ and $\mathscr{B}_0$, which will complete the argument.

Since $B \in \mathscr{B}_0$ implies $B = \hat{g}_\alpha^{-1}(B_\alpha)$, $B_\alpha \in \mathscr{B}_\alpha$ for some $\alpha \in D$, we have for this α

$$\begin{aligned} \mu_\alpha(B_\alpha) &= P_\alpha(u_\alpha^{-1}(B_\alpha)) = P(g_\alpha^{-1}(u_\alpha^{-1}(B_\alpha))) \\ &= P(u^{-1} \circ \hat{g}_\alpha^{-1}(B_\alpha)) \qquad \text{as in} \quad (6) \\ &= P(u^{-1}(B)) = \tilde{\mu}(C \cap B) = \tilde{\mu}(C \cap \hat{g}_\alpha^{-1}(B_\alpha)). \end{aligned} \tag{9}$$

It follows that if μ^* is the outer measure generated by μ (cf. Proposition 1.2 with μ_α) and $\mathscr{B}_0$, $\mu^*(C \cap \hat{g}_\alpha^{-1}(B_\alpha)) = \tilde{\mu}(C \cap \hat{g}_\alpha^{-1}(B_\alpha))$ and $\tilde{\mathscr{B}}_0$ is a subring of the σ-algebra of μ^*-measurable sets $\mathscr{B}_{\mu^*}$ (since $\tilde{\mu}$ is σ-additive) by the fundamental Theorem I.2.1. Thus $\tilde{\mu} = \mu^* | \mathscr{B}_0$. Taking $B = S$, we deduce that $\mu^*(C) = \mu_\alpha(S_\alpha) = 1$ by (9). Then with the earlier argument $\mu = P \circ u^{-1}$. This completes the proof.

Remark In case $u(\Omega) \in \mathscr{B}$, we can state that $P = \lim_{\leftarrow} P_\alpha$ exists iff $S - u(\Omega)$ is μ-null. However, $u(\Omega)$ is not necessarily μ-measurable. With the algebraic model representation, and under a corresponding assumption, a result related to the above theorem has been independently obtained by Schreiber, Sun, and Bharucha–Reid [1]. (The two results were obtained in 1969 by the author.) It is likely that this theorem is true if the "tightness" hypothesis is further weakened.

Let us now present a sufficient condition for the existence of a strong direct limit of a direct system of probability spaces as given in Definition 4. This will illuminate the preceding considerations and explain the fact that this subject needs certain tools other than those used in the projective limit theory.

We first recall some facts and other concepts for this purpose. Observe that from Step I of the proof of Theorem 5 every Boolean algebra $\mathscr{B}$ has a representation $(S, \mathscr{A}_t)$, where S is a point set, $\mathscr{A}_t$ an algebra of subsets of S, and the mapping $t: \mathscr{B} \to \mathscr{A}_t$ a *faithful representation*, i.e., t is an algebraic isomorphism. If S is topologized with the sets of $\mathscr{A}_t$ as a base, then S becomes a compact Stone space. In this case t is called a *perfect representation*. In general, t does *not* preserve infinite operations. If $\mu: \mathscr{B} \to \mathbb{R}^+$ is a bounded additive function, then $\mu \circ t^{-1}: \mathscr{A}_t \to \mathbb{R}^+$ is also a bounded additive function, and it becomes σ-additive if t is a perfect representation because of the extremal disconnectedness of the topology of S. (If $A_n \downarrow \varnothing$ in $\mathscr{A}_t$, then $A_n = \varnothing$ for $n \geq n_0 \geq 1$, for some n_0.) The alternative method indicated in Remark 6 shows that $\mathscr{B}$ admits infinitely many representations unless it is a finite algebra. In fact, this is a consequence of the result that such a representation space is obtained from the set of all maximal ideals of $\mathscr{B}$ and since every proper ideal of $\mathscr{B}$ is contained in a maximal ideal, which is not unique if $\mathscr{B}$ is not finite. The following proposition then contains two conditions for μ to be σ-additive on $\mathscr{B}$, based on its behavior on $\mathscr{A}_t$ when there is topology on S and when there is none. This result is a consequence of the works of Yosida–Hewitt [1] and Hewitt [1].

8. Proposition *Let $\{(\Sigma_\alpha, \mu_\alpha, g_{\beta\alpha})_{\alpha<\beta}: \alpha, \beta \text{ in } D\}$ be a direct system of probability (or bounded measure) spaces and (Σ^∞, μ) be their weak direct limit. Then μ on Σ^∞ is σ-additive (and hence the system admits a strong direct limit) iff one of the two following conditions holds:*

(a) *for every faithful representation τ of Σ^∞, $\mu \circ \tau: \mathscr{A}_\tau \to \mathbb{R}$ is σ-additive;*

(b) *if S is a compact Stone space and $\mathscr{A}_\tau$ is the (Baire) algebra of its clopen sets (i.e., $(S, \mathscr{A}_\tau)$ is a perfect representation) and $\hat{\mu} = \mu \circ \tau$, then every closed G_δ set $\hat{A} \subset S$ with empty interior (i.e., $\hat{A}$ is a nowhere dense closed G_δ-set) satisfies $\hat{\mu}(\hat{A}) = 0$.*

If μ is σ-additive on Σ^∞, then there is a unique σ-additive extension of μ to the σ-Boolean algebra $\bar{\Sigma}$ of Σ^∞ analogous to the Hahn extension theorem, as noted in Definition 4 (cf. also Kappos [1, Section II.2]). The details of the above statements will be omitted here, referring the reader to the cited works. We present a sufficient condition to accompany Proposition 8.

On the structure of Boolean algebras and their direct and projective limits, the following technical result from Haimo [1] is also of interest. It is stated here for use in the next proposition.

9. Proposition *Let $\{(\Sigma_\alpha, g_{\beta\alpha})_{\alpha<\beta} : \alpha, \beta$ in $D\}$ be a direct system of Boolean algebras and $\Sigma^\infty = \lim_{\to}(\Sigma_\alpha, g_\alpha)$ be its direct limit in the sense described above in Proposition 3. Let $\{(S_\alpha, f_{\alpha\beta})_{\alpha<\beta} : \alpha, \beta$ in $D\}$ be the corresponding Stone representation spaces forming a projective system, i.e., S_α is a compact Stone space whose algebra of clopen sets represents Σ_α and $f_{\alpha\beta}: S_\beta \to S_\alpha$ is the induced (by $g_{\beta\alpha}$) point mapping defined by $f_{\alpha\beta}(\mathscr{I}_\beta) = g_{\alpha\beta}^{-1}(\mathscr{I}_\beta)$ for each principal ideal $\mathscr{I}_\beta$ in Σ_β (which is a point of S_β by definition). If $S = \lim_{\leftarrow}(S_\alpha, f_{\alpha\beta})$ and T is the Stone space of Σ^∞, then S and T are homeomorphic.*

With the above two results, and using Theorem 7, it is possible to present sufficient conditions for the existence of (strong) direct limits in a special case containing the second interpretation of Kolmogorov's theorem discussed at the beginning of this section. We give this in the following form, and sketch its proof.

10. Theorem *Let* $\{(\Omega_\alpha, \Sigma_\alpha, P_\alpha, g_{\beta\alpha})_{\alpha<\beta} : \alpha, \beta$ in $D\}$ *be a direct system of probability spaces such that (i) each* Ω_α *is compact, (ii)* $(\Omega_\alpha, \Sigma_\alpha, P_\alpha)$ *is a Baire probability space,* $\alpha \in D$, *and (iii) the* $g_{\beta\alpha}$ *satisfy the conditions of Proposition 2 (i.e., for* $\omega \in \Omega_\alpha$, $\{\omega\} \in \Sigma_\alpha$, *and* $g_{\beta\alpha}$ *is closed under arbitrary unions). Then the strong direct limit* $(\bar{\Sigma}, \bar{P})$ *of the given system exists.*

Proof The hypothesis on $g_{\beta\alpha}$ implies by Proposition 2 that there exist $f_{\alpha\beta}: \Omega_\beta \to \Omega_\alpha$ such that $\{(\Omega_\alpha, \Sigma_\alpha, P_\alpha, f_{\alpha\beta})_{\alpha<\beta} : \alpha, \beta$ in $D\}$ is a projective system of (compact) Baire probability spaces and $g_{\beta\alpha} = f_{\alpha\beta}^{-1}$. Then by (the last part of) Theorem 1.5 this system admits a (unique) projective limit (Ω, Σ, P) where $\Omega = \lim_{\leftarrow}(\Omega_\alpha, f_\alpha)$, $\Sigma = \sigma(\Sigma_0)$, $\Sigma_0 = \bigcup_{\alpha\in D} f_\alpha^{-1}(\Sigma_\alpha)$, and $P \circ f_\alpha^{-1} = P_\alpha$, $\alpha \in D$. But by Proposition 3, for the given system, a (weak) direct limit (Σ^∞, μ) exists with $\mu = \lim_{\rightarrow} P_\alpha$, and μ is a finitely additive real function on the Boolean algebra Σ^∞. Such a pair (Σ^∞, μ) is sometimes called a "quasi-probability algebra" (cf. Kappos [1]). In general, Σ^∞ is not an algebra of sets. We now show that μ is σ-additive on Σ^∞, so that it has a (Hahn-type) unique extension to $(\bar{\Sigma}, \bar{P})$ as already noted.

Let T be the Stone space of Σ^∞ (i.e., (T, τ) is a perfect representation) and (S_α, τ_α) be likewise a representation of Σ_α. By Theorem 5 there exists $\hat{f}_{\alpha\beta}: S_\beta \to S_\alpha$ such that if $\mathscr{B}_\alpha$ is the algebra of clopen sets of S_α (so $\tau_\alpha: \Sigma_\alpha \to \mathscr{B}_\alpha$ is faithful), then $\{(S_\alpha, \mathscr{B}_\alpha, \hat{P}_\alpha, \hat{f}_{\alpha\beta})_{\alpha<\beta} : \alpha, \beta$ in $D\}$ is a projective system of Baire probability spaces where $\hat{P}_\alpha = P_\alpha \circ \tau_\alpha$. Let $S = \lim_{\leftarrow}(S_\alpha, \hat{f}_\alpha)$. Since $\{(\Omega_\alpha, \Sigma_\alpha, P_\alpha, f_{\alpha\beta})_{\alpha<\beta} : \alpha, \beta$ in $D\}$ admits the projective limit (Ω, Σ, P), one *can* assume that the system is "tight." So by Theorems 5 and 7 there is a mapping $u: \Omega \to S$ such that $u(\Omega)$ is $\hat{P}$ $(= \lim_{\leftarrow} \hat{P}_\alpha)$-thick in S and then $P \circ u^{-1} = \hat{P}$. If $\lambda: S \to T$ is the homeomorphism given by Proposition 9, then (T being a Stone space) we may give S also the topology of a Stone space. In fact, $S = \lim_{\leftarrow} S_\alpha$ as a set, and we may regard λ as a bijective mapping between T and S, so that S can be given the topology derived from T, i.e., a set $A \subset S$ is open iff $\lambda(A)$ is open in T. Thus S becomes a Stone space. But the mapping u induces a homomorphism $t: \Sigma_0 \to \mathscr{B}_0 = \bigcup_{\alpha\in D} \hat{f}_\alpha^{-1}(\mathscr{B}_\alpha)$, where $\mathscr{B}_0$ is an algebra of sets of S and t is an (algebraic) isomorphism. By the essential uniqueness of the Stone representation, $(S, \mathscr{B}_0)$ can be identified as a faithful (by t) perfect representation of Σ_0. Hence by Proposition 8, and the fact that P on Σ_0 is σ-additive, we conclude that $P \circ t$ gives measure zero for each closed nowhere dense G_δ subset of S. Thus we may identify the representation spaces $(T, \mathscr{A}_\tau)$ and $(S, \mathscr{B}_0)$ of the algebras Σ^∞ and Σ_0, respectively. But since Σ_0 is an algebra of sets and the mappings $\tau: \Sigma^\infty \to \mathscr{A}_\tau$ and (by identification through λ) $t: \Sigma_0 \to \mathscr{A}_\tau$ are isomorphisms, the mapping

$h_\alpha: \Sigma_\alpha \to \mathscr{A}_\tau$ defined by the composition $h_\alpha = \tau \circ g_\alpha = t \circ f_\alpha^{-1}$ is an isomorphism for each $\alpha \in D$, i.e., the diagram

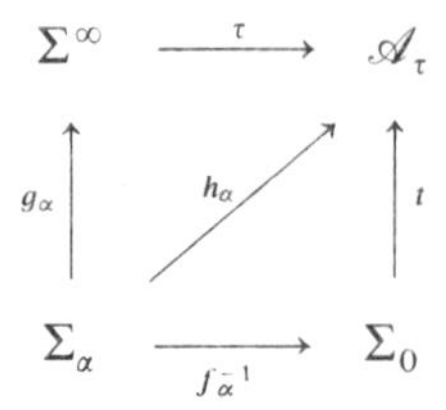

is commutative with g_α as the displayed canonical mapping.

The preceding reduction allows us to conclude the desired result. Thus let $\hat{A} \in \mathscr{A}_\tau$ be a nowhere dense closed G_δ set. Then there exists for some $\alpha \in D$ an $A_\alpha \in \Sigma_\alpha$ such that $\hat{A} = h_\alpha(A_\alpha) = \tau(g_\alpha(A_\alpha)) = t(f_\alpha^{-1}(A_\alpha))$. Also $P = \hat{P} \circ t$. Consequently,

$$\hat{\mu}(\hat{A}) = \mu(\tau^{-1}(\hat{A})) = P_\alpha(A_\alpha) = P \circ f_\alpha^{-1}(A_\alpha) = \hat{P}(t(f_\alpha^{-1}(A_\alpha))) = \hat{P}(\hat{A}), \tag{10}$$

and as seen in (1), the equation is unaltered if A_α is replaced by A_β with $\hat{A} = h_\beta(A_\beta)$ since they will be in the same equivalence class. But P being σ-additive, $\hat{P}(\hat{A}) = 0$, and so $\hat{\mu}(\hat{A}) = 0$. By Proposition 8(b) we then conclude that μ on Σ^∞ must be σ-additive. This completes the proof of the theorem.

Note Under the hypothesis of the theorem, the proof shows that (Ω, Σ_0) is a representation space of the Boolean algebra Σ^∞, which is faithful but *not* necessarily perfect.

In the case of Kolmogorov's theorem, $\Omega_\alpha = \bar{\mathbb{R}}^\alpha$ and μ_α are Lebesgue–Stieltjes measures where $\bar{\mathbb{R}}$ is the compactified line. So this is again covered here. The procedure used the compactness hypothesis of Ω_αs only to deduce the existence of the projective limit of the "associated" projective system. However, the assumption that the $g_{\alpha\beta}$ be adjoint to point mappings is restrictive. A general study of the strict direct limit and related problems is clearly desirable.

3.4 INFINITE PRODUCT CONDITIONAL PROBABILITY MEASURES

The work of Section 3.3 enables us to present an extension of a theorem of Tulcea [1] on infinite product regular conditional probabilities to projective systems. If the system is not necessarily regular, then one uses the

Dunford–Schwartz (or order) integral and the corresponding theory of Section II.3. The result will be seen ultimately as another application of Theorem 3.7.

It was noted in Section II.3 (just prior to Proposition II.3.19) that a conditional probability can also be defined (alternative to Definition II.1.1) as follows. If $(\Omega_i, \Sigma_i, P_i)$, $i = 1, 2$, are probability spaces such that $P_2 = P_1 \circ h^{-1}$ where $h: \Omega_1 \to \Omega_2$ is a measurable mapping, then there is a function q_{21}, which is P_2-uniquely defined by the equation (cf. also page 80)

$$P_1(A \cap h^{-1}(B)) = \int_B q_{21}(A, \omega)\, dP_2(\omega), \qquad A \in \Sigma_1, \quad B \in \Sigma_2. \tag{1}$$

The existence of $q_{21}: \Sigma_1 \times \Omega_2 \to \mathbb{R}^+$ is again a consequence of the Radon–Nikodým theorem. Thus $q_{21}: \Sigma_1 \to L^\infty(\Sigma_2)$ is a vector valued measure. Its (total) variation measure is P_1 when q_{21} is regarded as a measure into $L^1(\Sigma_2)$ since for $A \in \Sigma_1$

$$\begin{aligned} |q_{21}|(A) &= \sup\left\{ \sum_{i=1}^n \|q_{21}(A_i)\|_1 : A_i \subset A,\ A_i \in \Sigma_1, \text{ disjoint} \right\} \\ &= \sup\left\{ \int_{\Omega_2} q_{21}\left(\bigcup_{i=1}^n A_i \right) dP_2 : \bigcup_{i=1}^n A_i \subset A \right\} = P_1(A) \qquad \text{by (1).} \end{aligned}$$

Even though $0 \le q_{21}(A, \omega) \le 1$ a.e. (P_2), $q_{21}(\Omega_1, \omega) = 1$ a.e. and is pointwise σ-additive, $q_{21}(\cdot, \omega_2)$ is not necessarily a measure. So using the Dunford–Schwartz (or order) integral, we shall present a result on projective limits of probability spaces by postulating a condition on such "vectorial" probabilities. This will add to the characterizations obtained in Section 3.

If $\{(\Omega_\alpha, \Sigma_\alpha, P_\alpha, g_{\alpha\beta})_{\alpha < \beta} : \alpha, \beta \text{ in } D\}$ is a projective system of probability spaces, then for $\alpha < \beta$, $P_\alpha = P_\beta \circ g_{\alpha\beta}^{-1}$ so that we have, with $h = g_{\alpha\beta}: \Omega_\beta \to \Omega_\alpha$ in (1), a conditional probability function $q_{\alpha\beta}: \Sigma_\beta \to L^\infty(\Sigma_\alpha)$. Let $\alpha_1 < \alpha_2 < \cdots < \alpha_{n+1}$ be a finite set from D and define $\mathscr{B}_i = g_{\alpha_i\alpha_{n+1}}^{-1}(\Sigma_{\alpha_i})$, so that $\mathscr{B}_1 \subset \mathscr{B}_2 \subset \cdots \subset \mathscr{B}_n \subset \Sigma_{n+1}$. As a consequence of Corollary II.1.3, writing $E^{\mathscr{B}_i}(f)$ as $E_{\alpha_i}(f \mid g_{\alpha_i\alpha_{i+1}})$, we get

$$\int_{\Omega_{\alpha_{n+1}}} f\, dP_{\alpha_{n+1}} = \int_{\Omega_{\alpha_1}} E_{\alpha_1}(\cdots(E_{\alpha_n}(f \mid g_{\alpha_n\alpha_{n+1}}) \mid g_{\alpha_{n-1}\alpha_n}) \cdots \mid g_{\alpha_1\alpha_2})\, dP_{\alpha_1}. \tag{2}$$

Using the fact that $g_{\alpha_1\alpha_{n+1}} = g_{\alpha_1\alpha_2} \circ g_{\alpha_2\alpha_3} \cdots g_{\alpha_n\alpha_{n+1}}$ and with formula (1) (or by induction with (1)), (2) can be expressed as

$$\begin{aligned} \int_{\Omega_{\alpha_{n+1}}} f\, dP_{\alpha_{n+1}} = \int_{\Omega_{\alpha_1}} dP_{\alpha_1}(\omega_1) \int_{\Omega_{\alpha_2}} q_{\alpha_1\alpha_2}(d\omega_2, \omega_1) \cdots \\ \times \int_{\Omega_{\alpha_{n+1}}} f(\omega_{n+1}) q_{\alpha_n\alpha_{n+1}}(d\omega_{n+1}, \omega_n), \end{aligned} \tag{3}$$

where all the integrals after the first on the right are the Dunford–Schwartz integrals, the others being in Lebesgue's sense. Writing $f = \chi_A$, $A \in \Sigma_{\alpha_{n+1}}$ and using (1) for P_{α_1} and $P_{\alpha_{n+1}}$, we get from (3):

$$q_{\alpha_1\alpha_{n+1}}(A, \omega_1) = \int_{\Omega_{\alpha_2}} q_{\alpha_1\alpha_2}(d\omega_2, \omega_1) \int_{\Omega_{\alpha_3}} q_{\alpha_2\alpha_3}(d\omega_3, \omega_2) \cdots \times \int_{\Omega_{\alpha_{n+1}}} \chi_A q_{\alpha_n\alpha_{n+1}}(d\omega_{n+1}, \omega_n). \tag{4}$$

This holds a.e. $[P_{\alpha_1}]$, and these are again Dunford–Schwartz integrals. Now the projective system implies (4), but integrating (4) relative to P_{α_1}, we get (3) and hence can recover the system. Thus $q_{\alpha\beta}$s are just as important as the projective system, and we present a characterization in terms of these functions. Later the result will be specialized to get an important theorem due to Tulcea [1] on product spaces. We use the above notations without comment.

The following is the desired characterization, the necessity part of which, however, uses a result on disintegration of measures and ultimately Theorem 3.7 in both directions. So it cannot be presented with adequate details. Thus we omit its proof. Except for illustration, this result is not essential for the material in this book. The purpose of its inclusion is to show the progression of the abstract ideas in this area and to indicate a direction for future research.

1. Theorem *Let $\{(\Omega_\alpha, \Sigma_\alpha, P_\alpha, g_{\alpha\beta})_{\alpha<\beta} : \alpha, \beta \text{ in } D\}$ be a projective system of (complete) probability spaces satisfying the s.m. and "tightness" conditions (as in Theorem 3.7). Then the system admits a unique projective limit if for each $\alpha < \beta$ the support of $q_{\alpha\beta}(\cdot)$ intersects the set $g_{\alpha\beta}^{-1}(\{\omega_\alpha\})$ for a.a. $\omega_\alpha \in \Omega_\alpha$. In other words,*

$$A_\beta \in \Sigma_\beta,\ q_{\alpha\beta}(A_\beta, \omega_\alpha) > 0, \quad \Rightarrow A_\beta \cap g_{\alpha\beta}^{-1}(\{\omega_\alpha\}) \neq \varnothing, \quad \text{a.a.} \quad \omega_\alpha \text{ in } \Omega_\alpha. \tag{5}$$

Conversely, if the system admits a projective limit, then for each pair $\alpha < \beta$, $q_{\alpha\beta}$ can be chosen such that (5) is true.

2. Remark The sufficiency of this theorem was already established by Choksi [1] under the restriction that each $q_{\alpha\beta}$ is *regular* but the system is not assumed "tight." His result then implies, however, that the system is indeed "tight" in our sense. Also condition (5) was discovered by him. There are extensions (sufficient conditions) of Tulcea's result [1] by others (cf., e.g., Raoult [1]), but these results seem to have more involved conditions.

The following observation of Choksi [1] enables us to obtain Tulcea's theorem from the above result.

3. Lemma *Let $(\Omega, \Sigma, P) = \times_{i=1}^{2}(\Omega_i, \Sigma_i, P_i)$ be the product space of two probability spaces. If $\pi_i: \Omega \to \Omega_i$ is the coordinate projection for $i = 1, 2$ and if*

$p_{01}: \Sigma \times \Omega_1 \to [0,1]$ is a conditional probability for P and P_1 (and π_1 for h) given by (1), *then there is a modification* (*or version*) *q_{01} of p_{01}* (*which satisfies* (1)) *such that for each $A \in \Sigma$, $\omega_1 \in \Omega_1$ we have*

$$q_{01}(A, \omega_1) > 0 \Rightarrow A \bigcap \pi_1^{-1}(\{\omega_1\}) \neq \varnothing, \tag{6}$$

so that (5) *holds.*

Proof For $A \in \Sigma$, consider the ω_1-section $A_{\omega_1} = \{\omega_2 : (\omega_1, \omega_2) \in A\}$. Let q_{01} be the function defined by $q_{01}(A, \omega_1) = p_{01}(\Omega_1 \times A_{\omega_1}, \omega_1)$, where p_{01} is a conditional probability satisfying the equation

$$P(C \cap \pi_1^{-1}(D)) = \int_D p_{01}(C, \omega_1)\, dP_1(\omega), \qquad C \in \Sigma, \quad D \in \Sigma_1. \tag{7}$$

To see that the q_{01} satisfies (6), we only need to check the result for all measurable rectangles $A \times B = C$. Note that if $\omega_1 \notin A$, $C = A \times B$, then $p_{01}(\Omega_1 \times C_{\omega_1}, \omega_1) = 0$. Hence

$$\int_{\Omega 1} q_{01}(A \times B, \omega_1)\, dP_1(\omega_1)$$

$$= \int_A p_{01}(\Omega_1 \times B, \omega_1)\, dP_1(\omega_1) = P((\Omega_1 \times B) \cap \pi_1^{-1}(A)) \qquad \text{by} \quad (7)$$

$$= P((\Omega_1 \times B) \cap (A \times \Omega_2)) = P(A \times B) = \int_{\Omega_1} p_{01}(A \times B, \omega_1)\, dP_1(\omega_1). \tag{8}$$

Thus by (8), both q_{01} and p_{01} are the appropriate Radon–Nikodým derivatives of P relative to P_1, which agree on all the measurable rectangles and hence on Σ itself by the product measure theory, i.e., $q_{01}(E) = p_{01}(E)$ a.e. $[P_1]$ for $E \in \Sigma$. Now, if $q_{01}(E, \omega_1) > 0$, then E_{ω_1} cannot be empty. So if $E = \Omega_1 \times B$, $\varnothing \neq B \in \Sigma_2$, then $E \cap \pi_1^{-1}(\{\omega_1\}) = (\Omega_1 \times B) \cap (\{\omega_1\} \times \Omega_2) = \{\omega_1\} \times B \neq \varnothing$. Thus (6) holds.

With this preparation we can present, as a consequence of the above result, the following useful theorem due to Tulcea [1]. He proved it with suitable modifications of Kolmogorov's arguments in [1, §III.4].

4. Theorem *Let $\{(R_n, \mathcal{R}_n), n \geq 1\}$ be a family of measurable spaces. Let $\{p_n, n \geq 1\}$ be regular conditional probabilities in the sense that $p_{n+1}: \mathcal{R}_{n+1} \times \Omega_n \to [0,1]$ satisfies (i) $p_{n+1}(\cdot, \omega_n)$ is a probability on $\mathcal{R}_{n+1}$ for each $\omega_n \in \Omega_n = \times_{i=1}^n R_i$ and (ii) $p_n(A, \cdot)$ is measurable relative to $\Sigma_n = \bigotimes_{i=1}^n \mathcal{R}_i$, for each $A \in \mathcal{R}_{n+1}$, $n \geq 1$. If p_{mn} is defined for $1 \leq m < n$ by*

$$p_{mn}(A^{(n)}, \omega_m) = \chi_{A^{(n)}} \int_{A_{m+1}} p_{m+1}(dr_{m+1}, \omega_m) \cdots \int_{A_n} p_n(dr_n, \omega_{n-1}), \tag{9}$$

where $A^{(n)} = A_1 \times \cdots \times A_n \in \Sigma_n$, *and* $\omega_k = (r_1, \ldots, r_k)$, $k \geq 1$, *then with the measures* $P_m: \Sigma_n \to [0, 1]$ *defined by* $P_m(A) = \int_{\Omega_1} p_{1m}(A, \omega_1)\, dP_1(\omega)$ *for any initial probability* P_1 *on* $\mathscr{R}_1 \to [0, 1]$, *we have* $\{(\Omega_n, \Sigma_n, P_n, \pi_{mn}) : n \geq m \geq 1\}$ *as a "tight" projective system of probability measures which then admits a limit* (Ω, Σ, P) *such that*

$$P(\pi_n^{-1}(A^{(n)})) = \int_{\Omega_m} p_{mn}(A^{(n)}, \omega_m)\, dP_m(\omega_m)$$

$$= \int_{A_1} dP_1(\omega_1) \int_{A_2} p_2(dr_2, \omega_1) \cdots \int_{A_n} p_n(dr_n, \omega_{n-1}). \tag{10}$$

Here $\pi_{mn}: \Omega_n \to \Omega_m$, $\pi_n: \Omega \to \Omega_n$ *are the coordinate projections.*

Proof The definition of p_{mn} by (9) (with Lebesgue integrals) and Formula (4) show that the resulting $\{P_n, n \geq 1\}$ form a consistent family with natural projections. Hence $\{(\Omega_n, \Sigma_n, P_n, \pi_{mn}) : n \geq m \geq 1\}$ is a projective system of probability spaces. Since the given p_n, $n \geq 1$ are regular, it follows by Choksi's result noted in Remark 2 after the statement of Theorem 1 that the system is "tight." By Lemma 3 then condition (5) is valid for this product system. So the conclusion is a consequence of Theorem 1. The formula of Eq. (10) is then immediate.

We observe that the regularity implies, by Choksi's theorem [1], the system admits a limit. Then it is easily shown that there exists a mapping $u: \Omega \to S$ such that $u(\Omega)$ is μ ($= \lim_{\leftarrow} \mu_n$) thick, as required in Theorem 3.7. Thus the point of the above argument is that if p_n are not necessarily regular so that the integrals of (9) are in the Dunford–Schwartz sense, then again we have a projective system $\{(\Omega_n, \Sigma_n, P_n, \pi_{mn}) : n \geq m \geq 1\}$ satisfying condition (5). To invoke Theorem 3.7, however, we need some additional hypothesis to ensure the "tightness" condition. Thus any other (mild) restriction satisfying this condition on the system will yield a different generalization of the above result.

3.5 A MULTIDIMENSIONAL EXTENSION

Thus far only the projective limits of probability spaces are considered. But the set martingale problem shows that one needs to treat the results for signed measures, and perhaps for vector measures also (since the processes can be vector valued). These extensions, however, can be reduced to the cases of the theory already given in the preceding sections, and we show here how this might be done by presenting a vector analog of Theorem 2.2. Such extensions have been discussed by Métivier [1].

First observe that the definition of a projective system of vector measures is the same as that of scalar (or even probability) measures, and the projective limit concept is also the same. This is because the argument of the scalar case holds verbatim here. Also the Hahn extension (cf. Theorem I.2.2) is available for vector measures. Recall that if $\mathscr{A}$ is a σ-algebra of a set Ω, then a vector measure $\mu: \mathscr{A} \to \mathscr{X}$ (a Banach space) has the approximation property relative to a class $\mathscr{C} \subset \mathscr{A}$ if for each $\varepsilon > 0$, $A \in \mathscr{A}$ there is a $C \in \mathscr{C}$, $C \subset A$ such that $\|\mu(B)\| < \varepsilon$ for all $B \in \mathscr{A}$, $B \subset A - C$. Also $\mathscr{C}$ is a compact class if $\{C_n\}_1^\infty \subset \mathscr{C}$, $\bigcap_{n=1}^\infty C_n = \varnothing$, implies the existence of $1 \le n_0 < \infty$ such that $\bigcap_{k=1}^{n_0} C_k = \varnothing$. By analogy with the scalar case, we say that μ is a *compact* $\mathscr{X}$-valued set function relative to $\mathscr{C}$ if μ has the approximation property with respect to a compact class $\mathscr{C}$. When $\mathscr{X} \neq \mathbb{R}$, we often exhibit its presence in the notation of a projective system of these vector measures.

The following result is a vector analog of Theorem 2.2 and indicates how other results may similarly be extended. It is a typical representative and not the most general result possible in this context.

1. Theorem *Let* $\{(\Omega_\alpha, \Sigma_\alpha, \mu_\alpha, f_{\alpha\beta}, \mathscr{X})_{\alpha<\beta} : \alpha, \beta \text{ in } D\}$ *be a projective system of (vector or)* $\mathscr{X}$*-valued measure spaces where* $\mathscr{X}$ *is a (real) reflexive Banach space and* $\{(\Omega_\alpha, f_{\alpha\beta}) : \alpha, \beta \text{ in } D\}$ *satisfies the sequential maximality condition. Suppose that* $\sup_\alpha |\mu_\alpha|(\Omega_\alpha) < \infty$, *i.e., the* μ_α *are uniformly of finite variation (or if* $\mu = \lim_{\leftarrow} \mu_\alpha$, *then* $|\mu|(\Omega) < \infty$), *and that the conditions (i)–(iii) of Theorem 2.2 hold: Namely,*

(i) $\mathscr{C}_\alpha \subset \Sigma_\alpha$ *is a compact class for each* α, *and* $f_{\alpha\beta}(\mathscr{C}_\beta) \subset \mathscr{C}_\alpha$ *for* $\alpha < \beta$;

(ii) $f_{\alpha\beta}^{-1}(\{\omega_\alpha\}) \cap \mathscr{C}_\beta$ *is a compact class for each* $\omega_\alpha \in \Omega_\alpha$, *and for each* $A_\alpha \in \mathscr{C}_\alpha$, $B_\beta \in \mathscr{C}_\beta$ *there is a* $\gamma \in D$, $C_\gamma \in \mathscr{C}_\gamma$ *such that* $\gamma > \beta > \alpha$ *and* $C_\gamma = f_{\alpha\gamma}^{-1}(C_\alpha) \cap f_{\beta\gamma}^{-1}(C_\beta)$.

Suppose that each μ_α *has the approximation property for* $\mathscr{C}_\alpha$. *Then there exists a vector measure* $\mu: \Sigma\,(=\sigma(\Sigma_0)) \to \mathscr{X}$ *[i.e.,* $(\Omega, \Sigma, \mu, \mathscr{X})$ *is the projective limit of the system] such that* $\mu_\alpha = \mu \circ f_\alpha^{-1}$. *Moreover,* μ *has the approximation property on* Σ *relative to* $\mathscr{C}_{\delta\sigma}^* = (\bigcup_{\alpha \in D} f_\alpha^{-1}(\mathscr{C}_\alpha))_{\delta\alpha}$ *if also* $\mathscr{X}$ *is a separable space. Thus for any* $A \in \Sigma$ *and* $\varepsilon > 0$ *there exists a* $C_\varepsilon \in \mathscr{C}_{\delta\sigma}^*$ *with* $|\mu|(A - C_\varepsilon) \le \varepsilon$.

The idea of proof is as follows. We consider the scalar measures $x^* \circ \mu_\alpha: \Sigma_\alpha \to \mathbb{R}$ for any fixed but arbitrary $x^* \in \mathscr{X}^*$. Then $\{(\Omega_\alpha, \Sigma_\alpha, v_\alpha^{x^*}, f_{\alpha\beta})_{\alpha<\beta} : \alpha, \beta \text{ in } D\}$ is a projective system of signed measures, where $v_\alpha^{x^*} = x^* \circ \mu_\alpha$. Next we show under the given hypothesis that this system admits a limit $(\Omega, \Sigma, v^{x^*})$ on reducing it to the nonnegative case and applying Theorem 2.2. Then $v^{x^*}: \Sigma \to \mathbb{R}$ is a signed measure and defines a unique set function $\mu: \Sigma \to \mathscr{X}^{**} = \mathscr{X}$ such that $x^* \circ \mu = v^{x^*}$. This means that μ is weakly σ-additive on Σ and hence by a fundamental result of Pettis (cf. Theorem I.4.13) it is σ-additive in the norm topology of $\mathscr{X}$ proving the result. Let us fill in the details now.

As the above outline indicates, the connecting link between the vector case and the positive system is the analysis of signed measures. We simply write $v_\alpha: \Sigma_\alpha \to \mathbb{R}$ for $v_\alpha^{x^*}$ in the treatment of the signed case and recall that by

Definition 1.6, the resulting system $\{(\Omega_\alpha, \Sigma_\alpha, \nu_\alpha, f_{\alpha\beta})_{\alpha<\beta} : \alpha, \beta \text{ in } D\}$ determines a (σ-additive) set martingale $\{(\nu_\alpha^*, \Sigma_\alpha^*)_{\alpha\in D}\}$, where $\nu_\alpha^* \circ f_\alpha^{-1} = \nu_\alpha$, or $\nu_\alpha^* = \nu|\Sigma_\alpha^*$, the restriction. Let us first establish some technical preliminaries based somewhat on the works of Krickeberg [2] and Métivier [1].

2. Proposition *Let $\{(\Omega_\alpha, \Sigma_\alpha, \nu_\alpha, f_{\alpha\beta})_{\alpha<\beta} : \alpha, \beta \text{ in } D\}$ be a projective system of signed measures such that for each $\alpha \in D$, ν_α is a compact measure for a compact class $\mathscr{C}_\alpha \subset \Sigma$, i.e., for each $A \in \Sigma_\alpha$, $\varepsilon > 0$, there is a $C \in \mathscr{C}_\alpha$ such that $C \subset A$ and $|\nu|(A - C) < \varepsilon$. If $|\nu_\alpha|(\Omega_\alpha) \le k_0 < \infty$, $\alpha \in D$ (so $|\nu|(\Omega) < \infty$), $f_{\alpha\beta}(\mathscr{C}_\beta) \subset \mathscr{C}_\alpha$ for $\alpha < \beta$, and if $\tilde{\nu}_\alpha^+ = \nu^+ \circ f_\alpha^{-1}$, $\tilde{\nu}_\alpha^- = \nu^- \circ f_\alpha^{-1}$, and $|\tilde{\nu}_\alpha| = |\nu| \circ f_\alpha^{-1}$, then $\tilde{\nu}_\alpha^+$, $\tilde{\nu}_\alpha^-$, and $|\tilde{\nu}_\alpha|(\cdot)$ are compact (positive) measures on Σ_α for $\mathscr{C}_\alpha$, $\alpha \in D$, where ν^+, ν^-, and $|\nu|$ are the positive, negative, and total variations of ν on Σ_0.*

Proof Since ν_α is σ-additive by the classical Jordan decomposition of signed measures, ν_α^+, ν_α^-, $|\nu_\alpha|$ are σ-additive for each $\alpha \in D$. In general, however, $\tilde{\nu}_\alpha^+$ and ν_α^+ are unequal, and $\nu_\alpha^+ \le \tilde{\nu}_\alpha^+$. To see this, for any $A_\alpha \in \Sigma_\alpha$, by definition of ν_α^+ and using $A_\alpha^* = f_\alpha^{-1}(A_\alpha) \in \Sigma_\alpha^*$, $A_\beta = f_{\alpha\beta}^{-1}(A_\alpha) \in \Sigma_\beta$, we have

$$\begin{aligned} \nu_\alpha^+(A_\alpha) &= \sup\{\nu_\alpha(B) : B \subset A_\alpha, B \in \Sigma_\alpha\} \\ &\le \sup\{\nu_\beta(B) : B \subset f_{\alpha\beta}^{-1}(A_\alpha), \alpha < \beta, B \in \Sigma_\beta\}. \quad (1) \\ \nu_\alpha^+(A_\alpha) &= \sup\{\nu \circ f_\beta^{-1}(B) : f_\beta^{-1}(B) \subset f_\beta^{-1} \circ f_{\alpha\beta}^{-1}(A_\alpha) = f_\alpha^{-1}(A_\alpha), B \in \Sigma_\beta, \alpha < \beta\} \\ &= \sup\{\nu(B_\beta^*) : B_\beta^* \in \Sigma_\beta^*, B_\beta^* \subset A_\alpha^*, \alpha < \beta\} \\ &= \nu^+(A_\alpha^*) = \nu^+ \circ f_\alpha^{-1}(A_\alpha) = \tilde{\nu}_\alpha^+(A_\alpha). \quad (2) \end{aligned}$$

If we let $(\tilde{\nu}_\alpha^*)^+ = \nu^+|\Sigma_\alpha^*$ (and similarly $(\tilde{\nu}_\alpha^*)^-$ and $|\tilde{\nu}_\alpha^*|$), then (2) implies the useful relations:

$$(\tilde{\nu}_\alpha^*)^+(A_\alpha^*) = \tilde{\nu}_\alpha^+(A_\alpha), \quad A_\alpha \in \Sigma_\alpha, \quad (3a)$$

$$\nu_\alpha^+ \le \tilde{\nu}_\alpha^+. \quad (3b)$$

Since we need to conclude the compactness of $\tilde{\nu}_\alpha^+$, the last inequality shows that the σ-additivity of $\tilde{\nu}_\alpha^+$ cannot be directly deduced from that of ν_α^+. But that property is essential for the conclusion. We deduce it using (3a) and proving the σ-additivity of $(\tilde{\nu}_\alpha^*)^+$ with the key property that $\{\nu_\alpha^*, \Sigma_\alpha^*, \alpha \in D\}$ is a (bounded) set martingale (and ν_α^* is σ-additive). This fact is nontrivial, and a proof will be given. We present the result separately since it can be used again in Chapter IV.

3. Proposition (Set martingale decomposition) *Let $\{X, \mathscr{B}_\alpha, \lambda_\alpha, \alpha \in D\}$ be a bounded set martingale, i.e., $\lambda_\alpha = \lambda_\beta|\mathscr{B}_\alpha$* for $\alpha < \beta$, $X \in \mathscr{B}_\alpha$, $\sup_\alpha|\lambda_\alpha|(X) < \infty$, *and $\mathscr{B}_\alpha \subset \mathscr{B}_\beta$. Then there exist nonnegative set martingales $\{\xi_\alpha, \eta_\alpha$ on $\mathscr{B}_\alpha, \alpha \in D\}$ such that $\lambda_\alpha = \xi_\alpha - \eta_\alpha$, $\alpha \in D$, and $\sup_\alpha|\lambda_\alpha|(X) = \xi_\alpha(X) + \eta_\alpha(X)$. Such a decomposition is unique.*

Proof By definition of positive variation (cf. (1) above), $\lambda_\alpha^+(\cdot)$ is a σ-additive function such that for each $A_\alpha \in \mathscr{B}_\beta$, $\lambda_\alpha^+(A_\alpha) \le \lambda_\beta^+(A_\alpha)$ if $\alpha < \beta$, and in general there is inequality when $\{\lambda_\alpha\}_{\alpha\in D}$ itself is not a positive set martingale. Define ξ_α on $\mathscr{B}_\alpha$ by the equation

$$\xi_\alpha(A) = \sup\{\lambda_\beta^+(A) : \beta > \alpha\} = \lim_{\beta > \alpha} \lambda_\beta^+(A), \qquad A \in \mathscr{B}_\alpha, \tag{4}$$

where we can replace sup by lim because of the monotonicity of λ_β^+s. Also $\xi_\alpha(X) = \lim_{\beta>\alpha} \lambda_\beta^+(X) \le \sup_\alpha|\lambda_\alpha|(X) = c_0 < \infty$. It is clear from (4) that for $\alpha_1 < \alpha_2$, $\xi_{\alpha_1}(A) = \xi_{\alpha_2}(A)$ for any $A \in \mathscr{B}_{\alpha_1}$, and that ξ_α is (positive and) additive since each λ_β^+ has this property. We assert that ξ_α is actually σ-additive. For if this is false, then there would exist a decreasing sequence $\{A_n\}_1^\infty \subset \mathscr{B}_\alpha$ with empty intersection, a sequence $\alpha < \beta_1 < \beta_2 < \cdots$ in D, and an $\varepsilon > 0$ such that

$$\lambda_{\beta_n}^+(A_n) > \varepsilon, \qquad n \ge 1. \tag{5}$$

Since each $\lambda_{\beta_n}^+$ is σ-additive and bounded, it is absolutely continuous relative to the bounded measure ζ defined by

$$\zeta(A) = \sum_{n=1}^\infty \frac{1}{2^n} \frac{\lambda_{\beta_n}^+(A)}{\lambda_{\beta_n}^+(X)}, \qquad A \in \mathscr{B}_\alpha. \tag{6}$$

Hence by the Radon–Nikodým theorem there exist (ζ-unique) $\mathscr{B}_\alpha$-measurable $f_n = d\lambda_{\beta_n}^+/d\zeta$ such that $0 \le f_n \le f_{n+1}$ a.e. (ζ). If $f = \lim_n f_n$, then by the monotone convergence theorem $\int_X f\,d\zeta = \lim_n \int_X f_n\,d\zeta = \lim_n \lambda_{\beta_n}^+(X) \le c_0 < \infty$. Hence the set $F = \{f_n, n \ge 1\}$ dominated by the integrable f is uniformly integrable so that $\int_{(\cdot)} f_n\,d\zeta = \lambda_{\beta_n}^+(\cdot)$ is uniformly countably additive. It follows that $\lim_{n\to\infty} \int_{A_n} f_i\,d\lambda = 0$ uniformly in $f_i \in F$ and in particular $\lim_{n\to\infty} \int_{A_n} f_n\,d\lambda = 0 = \lim \lambda_{\beta_n}^+(A_n)$ contradicting (5). So we conclude that $\{X, \mathscr{B}_\alpha, \xi_\alpha, \alpha \in D\}$ is a positive set martingale such that

$$\xi_\alpha(A) = \lim_{\beta>\alpha} \lambda_\beta^+(A) \ge \lambda_\alpha^+(A) \ge \lambda_\alpha(A), \quad A \in \mathscr{B}_\alpha, \qquad \lim_\alpha \xi_\alpha(X) = \lim_\alpha \lambda_\alpha^+(X). \tag{7}$$

If $\eta_\alpha = \xi_\alpha - \lambda_\alpha \ge 0$, then η_α, being the difference of two set martingales on the same stochastic base $\{\mathscr{B}_\alpha, \alpha \in D\}$, is also a set martingale. Since $|\lambda_\alpha| = \lambda_\alpha^+ + \lambda_\alpha^-$, it follows by (7) that $\eta_\alpha(X) = \lim_\alpha \lambda_\alpha^-(X)$ and hence $\sup_\alpha|\lambda_\alpha|(X) = \xi_\alpha(X) + \eta_\alpha(X)$, the right side being independent of α by the martingale property. Thus $\lambda_\alpha = \xi_\alpha - \eta_\alpha$, $\alpha \in D$ is a decomposition.

For the uniqueness, let $\lambda_\alpha = \xi'_\alpha - \eta'_\alpha$ having the same properties as ξ_α and η_α. Then $\xi'_\alpha \ge \lambda_\alpha$, $\xi'_\alpha \ge 0$, implies that $\xi'_\alpha \ge \lambda_\alpha^+$. Consequently, using the definition (4) and the fact that for positive set martingales the definition reproduces the given family (i.e., $\xi'_\alpha(A) = \lim_{\beta>\alpha} \xi'_\beta(A)$, $A \in \mathscr{B}_\alpha$), we deduce that $\xi'_\alpha \ge \xi_\alpha$, $\alpha \in D$. Considering the set martingale $\{-\lambda_\alpha, \mathscr{B}_\alpha, \alpha \in D\}$ and noting that $(-\lambda_\alpha)^+ = \lambda_\alpha^-$, $-\lambda_\alpha = \eta'_\alpha - \xi'_\alpha$, we get from the above sentence that $\eta'_\alpha \ge \eta_\alpha$. But $\sup_\alpha|\lambda_\alpha|(X) = \xi'_\alpha(X) + \eta'_\alpha(X) = \xi_\alpha(X) + \eta_\alpha(X)$. This and the preceding inequalities imply $\xi'_\alpha = \xi_\alpha$ and $\eta'_\alpha = \eta_\alpha$, $\alpha \in D$, completing the proof.

Proof of Proposition 2 (Completed) Since $\{\Omega, \Sigma_\alpha^*, \nu_\alpha^*, \alpha \in D\}$ is a bounded set martingale where $\Omega = \lim_{\leftarrow}(\Omega_\alpha, f_\alpha)$, it follows by the above proposition that $(\tilde{\nu}_\alpha^*)^+$ is precisely the positive martingale corresponding to ξ_α (cf. the definition of $(\tilde{\nu}_\alpha^*)^+$ in (2)). Hence it is σ-additive for each $\alpha \in D$. By (3a) this implies that $\tilde{\nu}_\alpha^*$ is σ-additive (and bounded). Hence for any $\varepsilon > 0$ and any $A_\alpha \in \Sigma_\alpha$ there exists a $\beta > \alpha$ and $B \in \Sigma_\beta$, $B \subset f_{\alpha\beta}^{-1}(A_\alpha)$, such that (cf. (2))

$$\tilde{\nu}_\alpha^+(A_\alpha) \leq \nu_\beta(B) + \varepsilon/2. \tag{8}$$

But by hypothesis ν_β has the approximation property relative to the compact class $\mathscr{C}_\beta \subset \Sigma_\beta$. So there is a $C_\beta \in \mathscr{C}_\beta$, $C_\beta \subset B$ such that

$$|\nu_\beta|(B - C_\beta) < \varepsilon/2. \tag{9}$$

The boundedness of $|\nu_\beta|(\Omega_\beta) \leq |\tilde{\nu}_\beta^*|(f_\beta^{-1}(\Omega_\beta)) \leq |\nu|(\Omega) < \infty$ is used here.

On the other hand, $f_{\alpha\beta}(\mathscr{C}_\beta) \subset \mathscr{C}_\alpha$ so that $C_\alpha = f_{\alpha\beta}(C_\beta) \in \mathscr{C}_\alpha \subset \Sigma_\alpha$. Consequently, (8) and (9) imply since $C_\alpha \subset f_{\alpha\beta}(B) \subset f_{\alpha\beta} \circ f_{\alpha\beta}^{-1}(A_\alpha) = A_\alpha$ ($f_{\alpha\beta}$ being onto by the s.m. condition and the fact that for any mapping $u: X \to Y$, $u(u^{-1}(A)) = A$ for all $A \subset X$ iff u is onto):

$$\begin{aligned}
\tilde{\nu}_\alpha^+(A_\alpha - C_\alpha) &= \tilde{\nu}_\alpha^+(A_\alpha) - \tilde{\nu}_\alpha^+(C_\alpha) \leq \nu_\beta(B) + \varepsilon/2 - \tilde{\nu}_\alpha^+(C_\alpha) \\
&= \nu_\beta(B) + \varepsilon/2 - \nu^+ \circ f_\beta^{-1} \circ f_{\alpha\beta}^{-1}(C_\alpha) \qquad (\text{since } f_\alpha = f_{\alpha\beta} \circ f_\beta \text{ and using (2)}) \\
&\leq \nu_\beta(B) + \varepsilon/2 - (\nu^+ \circ f_\beta^{-1})(C_\beta) \qquad (\text{since } C_\beta \subset f_{\alpha\beta}^{-1}(C_\alpha)) \\
&\leq \nu_\beta(B) + \varepsilon/2 - (\nu \circ f_\beta^{-1})(C_\beta) \qquad (\text{since } \nu \leq \nu^+) \\
&= \nu_\beta(B - C_\beta) + \varepsilon/2 \qquad (\text{by the additivity of } \nu_\beta = \nu \circ f_\beta^{-1}) \\
&\leq |\nu_\beta|(B - C_\beta) + \varepsilon/2 \leq \varepsilon/2 + \varepsilon/2 = \varepsilon \qquad \text{by (9).}
\end{aligned}$$

Hence $\tilde{\nu}_\alpha^+$ has the approximation property on Σ_α relative to $\mathscr{C}_\alpha$. If we consider $\{-\nu_\alpha, \alpha \in D\}$ in the above, the result holds for $\tilde{\nu}_\beta^-$ and then by addition for $|\tilde{\nu}_\alpha| = \tilde{\nu}_\alpha^+ + \tilde{\nu}_\alpha^-$. This completes the proof of Proposition 2.

Remark By construction, $\{(\Omega_\alpha, \Sigma_\alpha, \tilde{\nu}_\alpha^\pm, f_{\alpha\beta})_{\alpha<\beta} : \alpha, \beta \text{ in } D\}$ are projective systems and under the hypothesis of this proposition they are compact systems, and so they admit limits by Theorem 2.2.

Proof of Theorem 1 Let $x^* \in \mathscr{X}^*$ be a fixed but arbitrary element and consider the signed measures $\nu_\alpha = x^* \circ \mu_\alpha$. (Recall that $\mathscr{X}^*$ is the space of all continuous linear functionals on $\mathscr{X}$.) Then $\{(\Omega_\alpha, \Sigma_\alpha, \nu_\alpha, f_{\alpha\beta})_{\alpha<\beta} : \alpha, \beta \text{ in } D\}$ is a projective system of signed measures. The system is bounded since $|\nu_\alpha|(\Omega_\alpha) \leq \|x^*\| \, |\mu_\alpha|(\Omega_\alpha) \leq \|x^*\| \sup_n |\mu_\alpha|(\Omega_\alpha) < \infty$ by hypothesis, where we used the fact that

$$\begin{aligned}
|x^* \circ \mu_\alpha|(\Omega_\alpha) &= \sup\left\{\sum_{k=1}^{n} |x^* \circ \mu_\alpha(A_k)| : \{A_i\}_1^n \subset \Sigma_\alpha, A_i \text{ disjoint}\right\} \\
&\leq \sup\left\{\|x^*\| \sum_{k=1}^{n} \|\mu_\alpha(A_k)\| : \{A_i\}_1^n \subset \Sigma_\alpha, A_i \text{ disjoint}\right\} \\
&= \|x^*\| \, |\mu_\alpha|(\Omega_\alpha).
\end{aligned} \tag{10}$$

This inequality also shows, on replacing Ω_α by $A_\alpha \in \Sigma_\alpha$, that the approximation property of μ_α implies that of v_α for $\mathscr{C}_\alpha$, $\alpha \in D$. But then by Proposition 2, we can deduce that $\{(\Omega_\alpha, \Sigma_\alpha, \tilde{v}_\alpha^{\pm}, f_{\alpha\beta})_{\alpha<\beta}: \alpha, \beta \text{ in } D\}$ are two projective systems of positive uniformly bounded measure spaces satisfying the approximation conditions relative to the compact classes $\mathscr{C}_\alpha \subset \Sigma_\alpha$, $\alpha \in D$. By Theorem 2.2 applied to these systems, we conclude that they admit projective limits $(\Omega, \Sigma, v^{\pm})$ and that $v^{\pm}$ have the approximation property relative to $(\mathscr{C}^*)_\delta$. This implies, by addition, that $|v|$ is a measure on Σ with the same approximation property.

Since $v_\alpha = v \circ f_\alpha^{-1} = x^* \circ \mu \circ f_\alpha^{-1} = x^* \circ \mu_\alpha$, and $x^* \in \mathscr{X}^*$ is arbitrary, we have proved that $x^* \circ \mu: \Sigma \to \mathbb{R}$ is σ-additive. Now the mapping $x^* \mapsto x^* \circ \mu(A)$ for each $A \in \Sigma$ is a continuous linear operation on $\mathscr{X}^*$ to $\mathbb{R}$ and hence is an element of its adjoint $(\mathscr{X}^*)^*$ which by the reflexivity hypothesis is identifiable with $\mathscr{X}$. We denote this element by $\mu(A) \in \mathscr{X}$. Thus $(x^* \circ \mu)(A) = x^*(\mu(A))$ and so μ is weakly σ-additive on Σ with values in $\mathscr{X}$. By the theorem of Pettis, referred to earlier, we conclude that μ is strongly (i.e., in the norm topology) σ-additive and $\mu_\alpha = \mu \circ f_\alpha^{-1}$.

It remains to prove the stated approximation property, using (for the first time) the extra hypothesis of $\mathscr{X}$. We have to use again some results from linear analysis (with references to the sources). The separability of $\mathscr{X}$ implies the existence of a denumerable collection $\{x_i^*\}_1^\infty \subset \mathscr{X}^*$, and $\|x_i^*\| = 1$, which are norm determining for $\mathscr{X}$, in that $\|x\| = \sup_i |x_i^*(x)|$ for each $x \in \mathscr{X}$. Thus if $A \in \Sigma$ and $\varepsilon > 0$ are given, consider $x_i^* \circ \mu$. By the preceding proof (or Theorem 2.2) there exists $C_i = C(x_i^*, \varepsilon) \in \mathscr{C}_\delta^*$ such that $C_i \subset A$ and $|x_i^* \circ \mu|(A - C_i) < \varepsilon$. Let $C_\varepsilon = \bigcup_{i=1}^\infty C_i \in \mathscr{C}_{\delta\sigma}^* \subset \Sigma$. Then $A \supset C_\varepsilon$ and hence

$$|x_n^* \circ \mu|(A - C_\varepsilon) < \varepsilon, \qquad n \geq 1. \tag{11}$$

Now taking the supremum over $n \geq 1$ on the left, we get, by using a standard result (which says that in the case under consideration the variation and semivariation are equal, cf. Dinculeanu [1, Propositions 4 and 5, pp. 54–55]), that

$$|\mu|(A - C_i) = \sup_i |x_i^* \circ \mu|(A - C_i) \leq \varepsilon. \tag{12}$$

This completes the proof of the theorem.

We present some complements to the theory of the preceding sections as problems in the last section.

Complements and Problems

1. A probability space (Ω, Σ, μ) is said to be (i) *perfect* if for each real measurable f on Ω the image measure $v = \mu \circ f^{-1}$ on $\mathbb{R}$ is a Lebesgue–Stieltjes measure, or equivalently for any $A \subset \mathbb{R}$ for which $f^{-1}(A) \in \Sigma$ there is a Borel set $B \subset \mathbb{R}$ such that $\mu(f^{-1}(A)) = \mu(f^{-1}(B))$ (the concept is due to

Gnedenko and Kolmogorov) and (ii) *quasi-compact* if for each sequence $\{E_n, n \geq 1\} \subset \Sigma$, and $\varepsilon > 0$, there exists a set $E_\varepsilon \in \Sigma$ with $\mu(E_\varepsilon) > 1 - \varepsilon$ such that $\{E_\varepsilon \cap E_n, n \geq 1\}$ is compact in the sense of Definition 2.1. This concept and the following implication are due to Ryll–Nardzewski [1]: μ is compact $\Rightarrow \mu$ is quasi-compact $\Leftrightarrow \mu$ is perfect. Let $(X, \mathscr{A})$ be a measurable space and I be an index set. If $X_i = X$, $\mathscr{A}_i = \mathscr{A}$, and D is the directed (by inclusion) set of all finite subsets of I, let $\Omega_\alpha = \times_{i \in \alpha} X_i = X^\alpha$, $\Sigma_\alpha = \bigotimes_{i \in \alpha} \mathscr{A}_i = \mathscr{A}^\alpha$, and $P_\alpha: \Sigma_\alpha \to [0, 1]$ be a probability such that $(\Omega_\alpha, \Sigma_\alpha, P_\alpha)$ is perfect for each $\alpha \in D$. Show that $\{(\Omega_\alpha, \Sigma_\alpha, P_\alpha, \pi_{\alpha\beta})_{\alpha<\beta} : \alpha, \beta \text{ in } D\}$ is a projective system admitting a limit (Ω, Σ, P) which is perfect iff $P_\alpha = P_\beta \circ \pi_{\alpha\beta}^{-1}$ for each $\alpha < \beta$ in D where $\pi_{\alpha\beta}: \Omega_\beta \to \Omega_\alpha$ is the coordinate projection. [*Hints*: Either use the above implications with compactness and reduce the result to Theorem 2.2 or verify by direct computation that the measure P on $\Sigma_0 = \bigcup_\alpha \pi_\alpha^{-1}(\Sigma_\alpha)$ given by Proposition 1.2 must satisfy $P(E_n) \to 0$ for any $E_n \downarrow \varnothing$, $E_n \in \Sigma_0$ by supposing the contrary. Then check for quasi-compactness of P. It is also of interest to note that the existence result is deducible from the sufficiency of Theorem 3.7 on noting that for any $f: X \to \mathbb{R}$ in (i) we may take $f(X) = B$ to be measurable in $(\mathbb{R}, \mathscr{B})$, the Borel line, so that $S_\alpha = \times_{i \in \alpha} B_i$, $\mathscr{B}_\alpha = \bigotimes_{i \in \alpha} \mathscr{B}(B_i)$ and $u_\alpha = \times_{i \in \alpha} f_i = (f_{i_1}, \ldots, f_{i_n})$ if $\alpha = (i_1, \ldots, i_n)$, where $B_i = B$, $f_i = f$ in that theorem. One can then show that $u = \lim_{\leftarrow} u_\alpha$ gives $u(\Omega)$ to be full. An extended account of perfect measures is given by Sazonov [1] to whom the above result is due.]

2. Let $(X, \mathscr{A})$ be a measurable space and $\mu: \mathscr{A} \to \mathbb{R}^+$ be additive. Then a subring $\mathscr{A}_0(\mu) \subset \mathscr{A}$, depending on μ, is called μ-pure if the following is true: (i) $\{A_n, n \geq 1\} \subset \mathscr{A}_0(\mu)$, A_n disjoint, and $\bigcup_{n=1}^{\infty} A_n \in \mathscr{A} \Rightarrow \mu(\bigcup_{n \geq n_0} A_n) = 0$ for some $1 < n_0 < \infty$, and (ii) the Carathéodory generated measure μ^*, with $\mathscr{A}_\mu$ as μ^*-measurable sets, by the pair $(\mu, \mathscr{A})$ is the same as that generated by the pair $(\mu, \mathscr{A}_0(\mu))$ (cf. Theorem I.2.1 on Carathéodory generated measures). An additive set function $\mu: \mathscr{A} \to \mathbb{R}^+$ is called *pure* if there is a μ-pure ring $\mathscr{A}_0(\mu) \subset \mathscr{A}$. It may be verified that a pure additive μ on $\mathscr{A}$ is σ-additive. Comparing this with Definition 3.1, it is seen that if μ is compact on $(X, \mathscr{A})$, then it is pure. If $(X, \mathscr{A})$ is a measurable space and I is an index set, let $(\Omega_\alpha, \Sigma_\alpha)_{\alpha \in D}$ be product-measurable spaces of $(X, \mathscr{A})$ as in Problem 1. If $P_\alpha: \Sigma_\alpha \to [0, 1]$ is a probability such that for each $\alpha \in D$ it is pure, show that $\{(\Omega_\alpha, \Sigma_\alpha, P_\alpha, \pi_{\alpha\beta})_{\alpha<\beta} : \alpha, \beta \text{ in } D\}$ is a projective system of pure probability spaces admitting a pure limit (Ω, Σ, P) iff for each $\alpha < \beta$ in D, $P_\alpha = P_\beta \circ \pi_{\alpha\beta}^{-1}$, where $\pi_{\alpha\beta}$ are coordinate projections. State and prove an analog of Theorem 2.2 in place of the product spaces here. (*Remarks*: Proof is similar to the above problem, or the original Kolmogorov's Theorem I.3.1. Note that a pure measure is complete by its definition. Hence if we start with a σ-additive complete measure, then (i) may be replaced (by considering measure algebras if necessary) by a "compact system." With Theorem I.5.4, one may use a lifting map and select actually a compact class from $\mathscr{A}_0(\mu)$ and hence if the completed μ is pure, then it will also be compact. Thus these two concepts are the same. In some ways the concept of pure measures is more convenient. Some elementary consequences and related extensions, such as "purely $\aleph_0$-compact" measures, were reported by Frolik and Pachl [1].)

3. Let X be a completely regular Hausdorff space and $\mathscr{A}$ be the Baire σ-algebra of X (i.e., the smallest σ-algebra relative to which each real continuous function on X is measurable). A function $\mu: \mathscr{A} \to \mathbb{R}^+$ is *net-additive* if for each net $\{A_t, t \in I\} \subset \mathscr{A}$, $A_t \downarrow \varnothing$ one has $\mu(A_t) \downarrow 0$. The space X is *B-compact* (or boundedly compact) if every Baire regular $\mu: \mathscr{A} \to \mathbb{R}^+$ is net-additive. Clearly every compact space X is B-compact and more generally one can show that each Lindelöf space (i.e., each open covering has a countable subcovering so that each $\mathbb{R}^n$) is B-compact. The cartesian product of a B-compact and a compact space is B-compact though such a product of two arbitrary B-compact spaces need not be B-compact. Let $(X, \mathscr{A})$ be a Baire measurable space and T an index set. If D is the directed set of all finite subcollections of T and $\Omega_\alpha = X^\alpha$, $\Sigma_\alpha = \mathscr{A}^\alpha$, the product spaces as in Problem 1 with $\pi_{\alpha\beta}: \Omega_\beta \to \Omega_\alpha$ as coordinate projections, suppose that $\{(\Omega_\alpha, \Sigma_\alpha, P_\alpha, \pi_{\alpha\beta})_{\alpha<\beta} : \alpha, \beta \text{ in } D\}$ is a projective system of Radon probability spaces where for each α, Ω_α is a locally compact B-compact space. Show that the system admits a limit (Ω, Σ, P) (this much is a special case of Bochner's earlier result, cf. Theorem 1.5) and that if each P_α is net-additive, P has a net-additive Borel extension to Ω $(=\lim_{\leftarrow}(\Omega_\alpha, \pi_{\alpha\beta}))$. This result and the concept of B-compactness have been given by Kirk [1]. How does this compare with Theorem 2.3?

4. Show that the proof of Theorem 2.3 can be simplified if the spaces there are assumed to be completely regular, by the following reduction. Use Stone–Čech compactification and extend the given measures from the given spaces Ω_α to their compactifications $\hat{\Omega}_\alpha$ since the mappings $g_{\alpha\beta}$ then extend to $\hat{g}_{\alpha\beta}$ uniquely. If i_α is the injection and $\hat{P}_\alpha = P_\alpha \circ i_\alpha^{-1}$, $\alpha \in D$, $\Sigma_\alpha = i_\alpha^{-1}(\hat{\Sigma}_\alpha)$, then $\{\hat{\Omega}_\alpha, \hat{\Sigma}_\alpha, \hat{P}_\alpha, \hat{g}_{\alpha\beta})_{\alpha<\beta} : \alpha, \beta \text{ in } D\}$ becomes a projective system of Radon probabilities on *compact* spaces. Then invoke the last part of Theorem 1.5 (i.e., Choksi's [1] result applies). [This shows the difference between these two results.]

5. This and the next two problems contain important applications of the projective limits to entropy and information theory. So extended sketches of arguments are included. Let us recall some notation and terminology.

If $(\Omega_i, \Sigma_i, P_i)$, $i = 1, 2$, are probability spaces, let $T: \Omega_1 \to \Omega_2$ be a measurable and measure-preserving mapping so that $P_1 = P_2 \circ T^{-1}$. The transformation T is called an *endomorphism* if $\Omega_1 = \Omega_2$ and the measure spaces are identical. If also T is one-to-one and its inverse is measure preserving, it is an *isomorphism* between the spaces; and if T is an endomorphism, the latter is called an *automorphism*. Thus $T_i: \Omega_i \to \Omega_i$, $i = 1, 2$, are *isomorphic* if there is an isomorphism $\tau: \Omega_1 \to \Omega_2$ such that $T_2 = \tau \circ T_1 \circ \tau^{-1}$. If τ is not invertible, but measure preserving such that $\tau \circ T_1 = T_2 \circ \tau$, then T_2 is a *factor* of T_1. If T_1 is a factor of T_2 and T_2 is a factor of T_1, then T_1 and T_2 are *weakly isomorphic*. In the equations $\tau \circ T_1 = T_2 \circ \tau$, etc., it suffices to assume these relations to be valid, outside of P_2 (hence also of $P_1 = P_2 \circ \tau^{-1}$) null sets. Since an isomorphism preserves, in general, only finite set operations and since a complete probability space (Ω, Σ, P) is isomorphic to $(S, \mathscr{A}, \hat{P})$, which is a regular compact measure space, as seen in the proof of Theorem 3.5, it follows that every automorphism $T: \Omega \to \Omega$ induces a set mapping $\tilde{T} = \bar{\tau} \circ T^{-1} \circ \bar{\tau}^{-1}: \mathscr{A} \to \mathscr{A}$ which is measurable and measure preserving where $\bar{\tau}: \Sigma \to \mathscr{A}$ is an isomorphism. [We may even assume that $\tilde{T}$ is induced by a (continuous) point mapping on S by replacing (if necessary) the measure algebra $(\mathscr{A}, \hat{P})$ with another isomorphic mapping (cf. Proof of Theorem 3.5).] Such a possibility of replacing a "bad" space (Ω, Σ, P) with a "nice" $(S, \mathscr{A}, \hat{P})$ will be useful in this problem. If $Þ = \{A_1, \ldots, A_n\} \subset \Sigma$ is a partition of Ω (i.e., A_i are disjoint with union Ω except for a null set), we define the *information* of Þ relative to P as the quantity $H(Þ) = -\sum_{i=1}^{n} P(A_i) \log P(A_i)$, where $f(x) = x \log x$ such that $f(0) = 0$. If T is an endomorphism, then the information in T, relative to a (partition) Þ and P, is defined as the "long-time average" information. This is given by $h(T, Þ) = \lim_n \sup (1/n) H(\bigvee_{k=0}^{n-1} T^{-k}(Þ))$, where $\bigvee_{k=0}^{n-1} T^{-k}(Þ) = \sigma(\bigcup_{k=0}^{n-1} T^{-k}(Þ))$ as usual. The *entropy* of T is then defined as $h(T) = \sup\{h(T, Þ): Þ \subset \Sigma,\ Þ \text{ a partition}\}$. Observe or prove, with the convexity properties of $f(\cdot)$ (Jensen's inequality, etc.), that the "lim sup" may be replaced by "lim" and that $H(\cdot)$, $h(T, \cdot)$ are monotone; $h(T)$ has the same value if Σ is replaced by an algebra Σ_0 such that $\Sigma = \sigma(\Sigma_0)$. Also if $k \in \mathbb{N}$, $h(T^k) = kh(T)$, and if T is an automorphism, then $h(T^k) = |k| h(T)$ for all integers k. (Cf., e.g., Billingsley [1] for details.) It follows immediately from definition that two isomorphic endomorphisms have the same entropy. Consequently, by the above discussion, for the problems involving entropy of endomorphisms, *we may replace the given (complete) probability space (Ω, Σ, P) by a topological space $(S, \mathscr{A}, \hat{P})$ where S is a compact space, $\mathscr{A}$ is a Baire (or even Borel) σ-algebra, and $\hat{P}$ is a Radon probability*. We do this in what follows without comment. Then the endomorphisms are also continuous. It is less obvious but true that two endomorphisms that are weakly isomorphic have the same entropy (cf. Rohlin [1]).

(a) Let $\{(\Omega_\alpha, \Sigma_\alpha, P_\alpha, f_{\alpha\beta})_{\alpha<\beta}: \alpha, \beta \text{ in } D\}$ be a (Radon) projective system of compact probability spaces admitting a (Baire) limit (Ω, Σ, P) (cf. Theorem 1.5). If T_α is an endomorphism on $(\Omega_\alpha, \Sigma_\alpha, P_\alpha)$, $\alpha \in D$, assume that $\{(T_\alpha, f_{\alpha\beta}): \alpha, \beta \text{ in } D\}$ is *consistent*, i.e., $T_\alpha \circ f_{\alpha\beta} = f_{\alpha\beta} \circ T_\beta$ for $\alpha < \beta$ so that

$P_\alpha \circ T_\alpha^{-1} = P_\beta \circ f_{\alpha\beta}^{-1} \circ T_\alpha^{-1}$. Let $T = \lim_\leftarrow(T_\alpha, f_{\alpha\beta})$ that exists (cf. Theorem 3.5(v)). Show that T is an endomorphism of (Ω, Σ, P). If $h(T_\alpha)$ is the entropy of T_α and $h(T)$ that of T, show that $h(T) = \lim_\alpha h(T_\alpha)$. [*Hints*: Observe that $h(T_\alpha) \leq h(T)$. On the other hand, since $\Sigma_0 = \bigcup_{\alpha \in D} f_\alpha^{-1}(\Sigma_\alpha) = \bigcup_{\alpha \in D} \Sigma_\alpha^*$, $\Sigma = \sigma(\Sigma_0)$, consider a (finite) partition $\text{Þ} \subset \Sigma_0$. Then $\text{Þ} \subset \Sigma_\alpha^*$ for some $\alpha \in D$, so that $h(T, \text{Þ}) \leq h(T, \Sigma_\alpha^*) = h(T, \Sigma_\alpha) = h(T_\alpha)$ where we used the facts that $P \circ f_\alpha^{-1} = P_\alpha$ and $T_\alpha \circ f_\alpha = f_\alpha \circ T$. Hence $h(T) \leq \lim_\alpha \inf h(T_\alpha)$.] Deduce that for the independent case of the classical Fubini–Jessen system, where $\Omega = \times_{\alpha \in D} \Omega_\alpha$, $\Sigma_\alpha = \bigotimes_{i \in D} \Sigma_i$, $P_\alpha = \bigotimes_{i \in D} P_i$, and $(T_\alpha \times T_\beta)(\omega_\alpha, \omega_\beta) = (T_\alpha \omega_\alpha, T_\beta \omega_\beta)$, we get $h(T) = \sum_{\alpha \in D} h(T_\alpha)$. [Here one uses the fact that $h(T_\alpha \times T_\beta) = h(T_\alpha) + h(T_\beta)$.]

(b) Let T be an endomorphism on a (compact) space (Ω, Σ, P). Then T is said to have a *natural extension* $\hat{T}$ on some (compact) space $(\hat{\Omega}, \hat{\Sigma}, \hat{P})$ if $\hat{T}$ is an automorphism such that there exists an algebra $\hat{\text{Þ}} \subset \hat{\Sigma}$ ($\hat{\text{Þ}}$ can be taken finite or countable if $\hat{P}$ is separable) such that (i) $\hat{T}$ is invariant for $\hat{\text{Þ}}$ (i.e., $\hat{T}(\hat{\text{Þ}}) \supset \hat{\text{Þ}}$), (ii) $\{\hat{T}^k(\hat{\text{Þ}}),\ k \in \mathbb{N}\}$ generates Σ, and (iii) T and $\hat{T} | \sigma(\hat{\text{Þ}})$ are isomorphic. Show that every endomorphism has a natural extension $\hat{T}$ which may be described as follows: Let $f_{mn} = T^{n-m}: \Omega \to \Omega$, $n \geq m \geq 1$, $\mathscr{B}_n = T^{-1}(\mathscr{B}_{n-1})$, $\mathscr{B}_0 = \Sigma$, and $P_n = P | \mathscr{B}_n$. Then $\{(\Omega, \mathscr{B}_n, P_n, f_{mn}): n \geq m \geq 1\}$ is a (Radon) projective system of probability spaces admitting the limit $(\hat{\Omega}, \hat{\mathscr{B}}, \hat{P})$, where $\hat{\Omega} = \lim_\leftarrow(\Omega, f_{mn})$, $\hat{P} = \lim_\leftarrow(P_n, f_{mn})$, and $\hat{\mathscr{B}} = \sigma(\hat{\text{Þ}})$, $\hat{\text{Þ}} = \bigcup_{n \geq 1} f_n^{-1}(\mathscr{B}_n)$. If $T_n = T$ on $(\Omega, \mathscr{B}_n, P_n)$, then $\{(T_n, f_{mn})\}$ is a consistent family and $\hat{T} = \lim_\leftarrow(T_n, f_{mn})$ is a natural extension of T. [*Hints*: Note that $\hat{\Omega} \subset \Omega^{\mathbb{N}}$ is a compact set consisting of all threads $\hat{\omega} = \{(\omega_i, i \geq 1): \omega_m = f_{mn}(\omega_n) = T^{n-m}(\omega_n),\ \omega_n \in \Omega_n = \Omega\} = \{(\omega_i)_1^\infty : \omega_n = T(\omega_{n+1}),\ n \geq 1,\ \omega_n \in \Omega\}$ and $\hat{T}(\hat{\omega}) = \{T(\omega_n),\ n \geq 1\} = \{T\omega_1, \omega_1, \omega_2, \ldots\}$. Since $\hat{T}$ is clearly invertible, and by (a) it is an endomorphism, one has to show that $\hat{T}^{-1}$ is measure preserving. But $\hat{T}$ is measurable for $\hat{\mathscr{B}}$, i.e., $\hat{T}^{-1}(\hat{\text{Þ}}) \subset \hat{\text{Þ}}$ and $\hat{P} \circ \hat{T}^{-1} = \hat{P}$. So $\hat{P} \circ \hat{T} = (\hat{P} \circ \hat{T}^{-1}) \circ \hat{T} = \hat{P}$, and hence $\hat{T}$ is an automorphism. If one takes $T^{-1}(\Sigma)$ as $\hat{\text{Þ}}_0$, then on noting $f_1 \circ \hat{T} = T_1 \circ f_1$ and that $f_1(\hat{\Omega}) = T(\Omega)$, we conclude that $\hat{T}(\hat{\text{Þ}}_0) \supset \hat{\text{Þ}}_0$ and $\bigcup_{k=1}^\infty \hat{T}^k(\hat{\text{Þ}}_0) \supset \bigcup_{n \geq 1} f_n^{-1}(\mathscr{B}_n)$; so $\{\hat{T}^k(\hat{\text{Þ}}_0)\}_1^\infty$ generates $\hat{\mathscr{B}}$. Since $\hat{T}$ is the projective limit of T_n $(= T)$ by definition, T_n is a factor of $\hat{T}$ and hence $\hat{T}$ is a natural extension of T. It may be shown that a natural extension is unique up to isomorphism and hence if T itself is an automorphism, then T and $\hat{T}$ are isomorphic.]

(c) Show that if T and $\hat{T}$ are as in (b), then $h(\hat{T}) = h(T)$. [*Hint*: Note that $h(\hat{T}) \geq h(T) \geq h(T, \mathscr{B}_n) \geq h(T_n, \mathscr{B}_n) \to h(\hat{T})$ as $n \to \infty$ by (a). It may be shown that if $h(T) = 0$, then T must be an automorphism.]

(d) Let T be an endomorphism and Þ any nontrivial partition in (Ω, Σ, P). Then T is said to have *completely positive entropy* if $h(T, \text{Þ}) > 0$. Consider now $\{(\Omega_\alpha, \Sigma_\alpha, P_\alpha, f_{\alpha\beta})_{\alpha < \beta} : \alpha, \beta \text{ in } D\}$, a projective system of probability spaces with (Ω, Σ, P) as their limit. As noted at the beginning, the existence may be assumed here. If $\{T_\alpha: \Omega_\alpha \to \Omega_\alpha\}$ is a consistent family of endomorphisms with completely positive entropies and if $T = \lim_{\leftarrow}(T_\alpha, f_{\alpha\beta})$, show that $T: \Omega \to \Omega$ has a completely positive entropy. (If the spaces Ω_α are restricted to be abelian groups also, then a converse statement holds. Cf. 7(f) below for details.) [*Hints*: This result holds if $D = \mathbb{N}$ by the work of Rohlin [1, §13.14], whose result, given for separable Radon spaces, is seen to be true in the present case. Hence it is also true by Theorem 2.11 (or Bourbaki [1, p. 54]) if D has merely a countable cofinal set. Suppose the result is false in the general case so that $h(T) = 0$. Then for each $n \geq 1$ there exists $\alpha_n \in D$ such that $\alpha > \alpha_n$ implies $h(T_\alpha) < 1/n$. Consequently, the sequence $\{h(T_{\alpha_n})\}_1^\infty$ has also the limit equal to zero. Let $J \subset D$ be a (directed) subset for which $\{\alpha_n, n \geq 1\}$ is cofinal, e.g., $J = \bigcup_{n=1}^\infty \{\alpha \in D : \alpha \leq \alpha_n\}$ will suffice. Then the subsystem $\{(\Omega_\alpha, \Sigma_\alpha, P_\alpha, f_{\alpha\beta}, T_\alpha)_{\alpha<\beta} : \alpha, \beta \text{ in } J\}$ has completely positive entropy for each $\alpha \in J$. If $\Omega = \lim_{\bar{D}}(\Omega_\alpha, f_{\alpha\beta})$ and $\Omega_J = \lim_{\bar{J}}(\Omega_\alpha, f_{\alpha\beta})$, let $f_\alpha: \Omega \to \Omega_\alpha, f_{J\alpha}: \Omega_J \to \Omega_\alpha$, and $f_J: \Omega \to \Omega_J$ be the canonical mappings. Then $f_\alpha = f_{J\alpha} \circ f_J$ by the projective limit theory of Section 2. We then have $P_J = P \circ f_J^{-1} = \lim_{\alpha \bar{\in} J} P_\alpha$, and $(\Omega_J, \Sigma_J, P_J)$ is the projective limit of the subsystem where $\Sigma_J = \sigma(\bigcup_{\alpha \in J} f_{J\alpha}^{-1}(\Sigma_\alpha))$. If $T_J = \lim_{\bar{J}}(T_\alpha, f_{\alpha\beta})$, then by (a) and Rohlin's result noted above, we must have $h(T_J) = \lim_n h(T_{\alpha_n}) = \lim_{\alpha \in J} h(T_\alpha) > 0$, contradicting the choice giving zero to this limit.]

(e) It was noted that two endomorphisms T_1, T_2 have the same entropy if both are isomorphic or even weakly isomorphic. By (b) and (c), T_is have natural extensions $\hat{T}_i$ such that $h(\hat{T}_i) = h(T_i)$, $i = 1, 2$. Thus if entropy determines the isomorphism of the endomorphisms T_1, T_2 then *a fortiori* it must determine the same for any two automorphisms. However, recently Ornstein and his associates have shown the existence (in plentitude) of automorphisms, called the K-automorphisms (which are not Bernoulli—the result is true for the latter; essentially the Bernoulli automorphisms are determined by the shift transformations on projective limits of products of k-point spaces), which have the same entropy but no two of them are isomorphic. In fact, they are not even isomorphic to their inverses. Thus the concept of weak isomorphism in this context appears relevant. With this background, we can assert the following: If T_1, T_2 are weakly isomorphic endomorphisms on (Ω, Σ, P), which we may and do assume to be a Radon space with Ω compact, then their natural extensions $\hat{T}_1$, $\hat{T}_2$ have the same entropy. However, they need not be isomorphic.† [*Hints*: If τ_1, τ_2 are endomorphisms such that (by the weak isomorphism hypothesis) $\tau_1 \circ T_1 = T_2 \circ \tau_1$, $\tau_2 \circ T_2 = T_1 \circ \tau_2$ a.e., then consider the projective system $\{(\Omega, \Sigma_n, P_n, f_{mn}) : n \geq m \geq 1\}$ defined by $f_{mn} = \tau_1$ if m is even $n = m + 1$, and $f_{mn} = \tau_2$ if n is odd. Then $f_{mn} = (\tau_1 \circ \tau_2)^{(1/2)(n-m)}$ if m and n are even,

† D. S. Ornstein has kindly informed me that there are counterexamples showing that the natural extensions $\hat{T}_1$ and $\hat{T}_2$ of T_1, T_2 are *not* isomorphic.

$f_{mn} = (\tau_2 \circ \tau_1)^{(1/2)(n-m)}$ if both are odd and are suitably (left or right) multiplied by τ_1 or τ_2 in the mixed case. Similarly, let $\Sigma_n = f_{mn}^{-1}(\Sigma_m)$ with $\Sigma_0 = \Sigma$ and $P_n = P|\Sigma_n$. As in (a), we see that the system admits a projective limit $(\hat{\Omega}, \hat{\Sigma}, \hat{P})$. Let $T_n = T_1$ if n is odd, and $T_n = T_2$ if n is even. Then $\hat{T} = \lim_{\leftarrow} T_n$ is a natural extension of the sequence $\{T_n, n \geq 1\}$. Since $J_i \subseteq \mathbb{N}$, $i = 1, 2$, and since the odd and even integers form cofinal sets of $\mathbb{N}$, one concludes that these two subsystems admit projective limits. Even though there is isomorphism between the systems at each stage, this does not imply the same for their limits. If $\hat{T}_1 = \lim_{J_1^{\leftarrow}} T_n$, $\hat{T}_2 = \lim_{J_2^{\leftarrow}} T_n$, then from the fact that $h(T_1) = h(T_2)$, by the weak isomorphism, we get $h(\hat{T}_1) = h(\hat{T}_2)$ by (c). Regarding this result, see J. R. Brown [1], and (d) verifies a conjuncture of his. For other results referred to here, see the memoirs of Rohlin [1] and Ornstein [1].]

6. (a) The following formula is useful in some calculations of entropy (as defined above) and will be used in Section IV.5 to give a proof of the Kolmogorov–Sinaĭ theorem with martingale convergence. Let $\mathscr{A}, \mathscr{B}$ be finite algebras and $\mathscr{C}$ any σ-algebra in Σ where (Ω, Σ, P) is a probability space. Let $\{A_i\}_1^m$ and $\{B_i\}_1^n$ be the partitions of Ω generating $\mathscr{A}$ and $\mathscr{B}$. Show that

$$P^{\mathfrak{F}}(E) = \sum_{i=1}^{m} \chi_{A_i} P^{\mathscr{C}}(E \cap A_i)[P^{\mathscr{C}}(A_i)]^{-1} \quad \text{a.e.,} \qquad E \in \Sigma. \tag{$*$}$$

Here $\mathfrak{F} = \mathscr{A} \vee \mathscr{C} = \sigma(\mathscr{A} \cup \mathscr{C})$ and $P^{\mathscr{B}}$ is a conditional probability. [*Hints*: Integrating both sides of (*) on a set $F_i = A_i \cap C$, we find that the equation is valid. Since finite disjoint unions of such F_i generate $\mathfrak{F}$ and both sides are $\mathfrak{F}$-measurable, they can be identified.]

(b) Analogous to entropy, we define the conditional entropy $H(\mathscr{A}|\mathscr{B})$ as $H(\mathscr{A}|\mathscr{B}) = -\sum_{i=1}^m \sum_{j=1}^n P(A_i \cap B_j) \log P(A_i|B_j)$, where $P(A|B)$ is the conditional probability of A, given B. Show that $H(\mathscr{A}|\mathscr{B}) = E(\sum_{i=1}^m \varphi(P^{\mathscr{B}}(A_i)))$, where $\varphi(t) = t \log t$ for $t > 0$ and $\varphi(0) = 0$. This definition is meaningful even if $\mathscr{B}$ is not finite. Using this alternate definition of $H(\cdot|\cdot)$, show that for $\mathscr{A}, \mathscr{B}, \mathscr{C}$ given above we have (i) $H(\mathscr{A} \vee \mathscr{B}|\mathscr{C}) = H(\mathscr{A}|\mathscr{C}) + H(\mathscr{B}|\mathscr{A} \vee \mathscr{C})$, (ii) $H(\cdot|\mathscr{C})$ is increasing, (iii) $H(\mathscr{A}|\cdot)$ is decreasing, and (iv) invariant under an endomorphism $T: \Omega \to \Omega$, i.e., if $T^{-1}(\Sigma) \subset \Sigma$ and $P \circ T^{-1} = P$, then $H(T^{-1}(\mathscr{A})|T^{-1}(\mathscr{C})) = H(\mathscr{A}|\mathscr{C})$. [*Hints*: In the formula ($*$), since the sum has only one nonzero term for each $\omega \in \Omega$, we can take "logs" on both sides and note that

$$\sum_{i=1}^{m} \sum_{j=1}^{n} \chi_{A_i \cap B_j} \log P^{\mathscr{C}}(A_i \cap B_j) = \sum_{i=1}^{m} \chi_{A_i} \log P^{\mathscr{C}}(A_i) + \sum_{j=1}^{n} \chi_{B_j} \log P^{\mathfrak{F}}(B_j).$$

Taking expectations on both sides and noting that it can be rearranged and simplified, we get (i). By using conditional Jensen's inequality (Theorem II.1.6 and the following remark), we get (iii). Using the fact that $P^{T^{-1}(\mathscr{C})}(T^{-1}(A))(\omega) = P^{\mathscr{C}}(A)(T\omega)$ a.e., we get (iv) by a simple computation. Others are similar. See Billingsley [1] for such results.]

7. Let T_1 and T_2 be endomorphisms on the probability spaces $(\Omega_i, \Sigma_i, P_i)$, $i = 1, 2$, respectively. One says that T_1 and T_2 are *conjugate* if the operators $U_{T_i}: f \mapsto f \circ T_i$, $i = 1, 2$, $f \in L^2(\Omega_i, \Sigma_i, P_i)$ (which are clearly isometries) are connected by $VU_{T_1} = U_{T_2}V$, where $V: L^2 \cap L^\infty(\Omega_1, \Sigma_1, P_1) \to L^2 \cap L^\infty(\Omega_2, \Sigma_2, P_2)$ is an isometry on L^2 and is a multiplicative linear operator. If for the endomorphisms $T_i: \Omega_i \to \Omega_i$, $i = 1, 2$, T_2 is a factor of T_1, then the associated U_{T_i} are conjugate. If the identity operators are conjugate, then the measure spaces are called conjugate so that, in this case, V induces (and is induced by) a set mapping $\tau: \Sigma_1 \to \Sigma_2$, which is measure preserving. An *algebraic triple* $\Phi = (\Gamma, U, \varphi)$ consists of an abelian group Γ, an injective homomorphism $U: \Gamma \to \Gamma$, and a positive definite function $\varphi: \Gamma \times \Gamma \to \mathbb{R}$ such that $\varphi(\gamma_1, \gamma_2) = \varphi(\gamma_1 - \gamma_2)(= \varphi(\gamma_1\gamma_2^{-1}))$, and $\varphi(U\gamma) = \varphi(\gamma)$, $\gamma \in \Gamma$, with $\varphi(\gamma) = 1$ iff $\gamma =$ identity. An algebraic triple Φ is an *algebraic model* for a probability space (Ω, Σ, P) with an endomorphism $T: \Omega \to \Omega$ if there exists an injective homomorphism $J: \Gamma \to \Gamma_P$ such that $J(\Gamma)$ generates $L^2(\Omega, \Sigma, P)$ and $\varphi(\gamma) = \varphi_P(J\gamma)$, $\gamma \in \Gamma$, where Γ_P is the group of (equivalence classes of) functions f of $L^2(\Omega, \Sigma, P)$ such that $|f| = 1$ a.e. and $\varphi_P(f) = \int_\Omega f\,dP$, $f \in \Gamma_P$. So φ_P is positive definite. Note that $\Phi_T = (\Gamma_P, U_T, \varphi_P)$ is an algebraic triple for the endomorphism T. If J is an isomorphism, then Φ and Φ_T are isomorphic algebraic triples. Also a projective system $\{(\Omega_\alpha, \Sigma_\alpha, P_\alpha, f_{\alpha\beta})_{\alpha<\beta} : \alpha, \beta \text{ in } D\}$ of probability spaces with a consistent family of endomorphisms $\{(T_\alpha, f_{\alpha\beta}) : \alpha, \beta \text{ in } D\}$ determines an algebraic triple system $\{(\Phi_{T_\alpha}, \tilde{f}_{\beta\alpha})_{\alpha<\beta} : \alpha, \beta \text{ in } D\}$ where $\tilde{f}_{\beta\alpha} = U_{f_{\alpha\beta}}$ (so that $\tilde{f}_{\gamma\beta} \circ \tilde{f}_{\beta\alpha} = \tilde{f}_{\gamma\alpha}$ for $\alpha < \beta < \gamma$) and thus is a directed family. In relation to Theorems 3.5, 3.7 (or 2.3), we have the following results complementing the former in certain respects. Here we use the terminology and some of the results of Problem 5 above.

(a) Two projective systems of probability spaces are conjugate iff they have isomorphic algebraic model systems (which are necessarily directed systems).

(b) Every directed system of algebraic triples is an algebraic model for some projective system of probability spaces admitting a projective limit and having a consistent family of endomorphisms on it.

(c) Deduce from (a) and (b) that every projective system of probability spaces with a consistent family of endomorphisms is conjugate to the one that admits a projective limit. [This and part (b) may be compared with the

Stone space representation given in Theorem 3.5. In the present case the Stone spaces may also be chosen to be compact abelian groups. This finer structure allows us to take the isomorphic images to have the most well-behaved measures on them. These can be taken to be Haar measures when (and only when) Γ_P has a subgroup that forms an orthonormal basis for $L^2(\Omega, \Sigma, P)$, as in the case of Walsh systems (cf. Chapter V), if (Ω, Σ, P) is the Lebesgue unit interval.]

(d) If $\{(\Omega_\alpha^i, \Sigma_\alpha^i, P_\alpha^i, f_{\alpha\beta}^i)_{\alpha<\beta} : \alpha, \beta \text{ in } D\}$, $i = 1, 2$, are projective systems of probability spaces admitting limits $(\Omega^i, \Sigma^i, P^i)$, $i = 1, 2$, and $\{(T_\alpha^i, f_{\alpha\beta}^i) : \alpha, \beta \text{ in } D\}$, $i = 1, 2$, are respective systems of endomorphisms with limits T^i, $i = 1, 2$, the conjugacy of the given systems implies the same property for their limits.

(e) If there is only one system in (d), where Ω_α is a compact abelian group, Σ_α is the Borel σ-algebra, and P_α is the Haar measure (then the limit (Ω, Σ, P) exists and is Baire regular by Theorem 1.5), let $T = \lim_{\leftarrow} T_\alpha$, for the associated system of continuous endomorphisms. [Thus each Γ_{P_α} must have a subgroup forming an orthonormal basis of $L^2(\Omega_\alpha, \Sigma_\alpha, P_\alpha)$ by the remarks in (c).] If each $T_\alpha = \tau_\alpha \circ \rho_\alpha$ where τ_α is a continuous endomorphism on Ω_α and ρ_α is a rotation (i.e., $\rho_\alpha(x) = cx$ for some $c \in \Omega_\alpha$ and all $x \in \Omega_\alpha$), then (*) $T = \lim_{\leftarrow} T_\alpha = (\lim_{\leftarrow} \tau_\alpha) \circ (\lim_{\leftarrow} \rho_\alpha) = \tau \circ \rho$ and ρ is a rotation on the group Ω. Conversely, if $T = \tau \circ \rho$, then each T_α factors, as above, and satisfies the equation (*).

(f) If $\{(\Omega_\alpha, \Sigma_\alpha, P_\alpha, f_{\alpha\beta})_{\alpha<\beta} : \alpha, \beta \text{ in } D\}$ is a projective system as in (e) where Ω_α is also separable and $\{(T_\alpha, f_{\alpha\beta}) : \alpha, \beta \text{ in } D\}$ is the associated consistent family of endomorphisms with $T = \lim_{\leftarrow} T_\alpha$, then T has completely positive entropy iff each T_α, $\alpha \in D$, has the same property. [In one direction the result follows from Problem 5(d). Now use all the hypothesis for the converse, to invoke a result of Rohlin [1] and infer that an endomorphism T on such a compact group is ergodic (i.e., $P \circ T^{-1} = P$ only on the trivial σ-algebra of null sets and their complements) iff it has completely positive entropy. Then verify that for the consistent family $\{(T_\alpha, f_{\alpha\beta})_{\alpha<\beta} : \alpha, \beta \text{ in } D\}$, $T = \lim_{\leftarrow} T_\alpha$, T is ergodic iff each T_α is. This implies the statement.]

(The other parts are similarly involved. The basic theory of algebraic models was established by Dinculeanu and Foiaş [1] and (b)–(d) are then deduced by Chi and Dinculeanu [1]. These papers have other related results also. In case Ω_αs above are separable metric spaces, we note that (cf. Billingsley [1, p. 69]) conjugacy is equivalent to isomorphism for a pair of measure-preserving transformations.)

CHAPTER

IV

Martingales and Likelihood Ratios

The fundamental inequalities and the basic convergence theory of discrete (sub)martingales are developed in this chapter. As an important application the Gaussian dichotomy theorem is established and then various likelihood ratios for Gaussian processes are given. Some other extensions (e.g., asymptotic martingales) and consequences are included in the text and as complements.

4.1 DEFINITION AND FUNDAMENTAL INEQUALITIES

In discussing the relation between projective systems and set martingales, we have noted the concept of a martingale of point functions at the end of Section III.1. Let us introduce the concept abstractly and then explain its significance.

1. Definition Let $\{X_i, i \in I\}$ be a family of integrable random variables on a probability space (Ω, Σ, P), where I is partially ordered. If $i_1 < i_2 < \cdots < i_n < i_{n+1}$ are any $n+1$ points from I and $\mathscr{B}_n = \sigma(X_{i_j}, 1 \leq j \leq n)$ is the σ-algebra generated by $X_{i_1}, \ldots, X_{i_n}$, then the family (or process) is called a *martingale* if the following equation holds with probability one:

$$E(X_{i_{n+1}} | X_{i_1}, \ldots, X_{i_n}) = E^{\mathscr{B}_n}(X_{i_{n+1}}) = X_{i_n}, \qquad n \geq 1, \tag{1}$$

where $E^{\mathscr{B}_n}$ (or $E(\cdot | X_{i_1}, \ldots, X_{i_n})$) is the conditional expectation operator.

If one regards the process as the successive fortunes of a gambler in a certain game, then (1) says that the player's expected fortune on the i_{n+1}th game, having known the preceding fortunes at the i_1th, i_2th, ..., i_nth plays, is simply the latest (present) one. Such games are considered *fair*. If this expected fortune is at least as much as (and possibly higher than) the present one, then the game is regarded as *favorable* (to the player), and it is called a *submartingale*; and if it is *unfavorable*, it is called a *supermartingale*. Thus one may state this more general concept abstractly as follows:

2. Definition If $\{X_i, i \in I\}$ is an integrable real process on (Ω, Σ, P) and $i_1 < i_2 < \cdots < i_n < i_{n+1}$, then it is called a *submartingale* if

$$E^{\mathscr{B}_n}(X_{i_{n+1}}) \geq X_{i_n} \quad \text{a.e.,} \qquad n \geq 1, \tag{2}$$

and it is a *supermartingale* if the inequality in (2) is reversed.

From these definitions it is clear that the process $\{X_i, i \in I\}$ is a martingale iff it is both a submartingale and a supermartingale. Note that one must have an ordered index for these concepts to be meaningful. Also the process can be vector valued for martingales, but for the submartingale and supermartingales the range of X_is must be an ordered space, and thus the generalization involved in Definition 2 is in the real (or ordered) case. In what follows, if $\{\mathfrak{F}_i, i \in I\}$ is a net of (increasing) σ-algebras in Σ, and X_i is $\mathfrak{F}_i$-measurable, we say that X_i is $\mathfrak{F}_i$*-adapted*, or simply $\{X_i, \mathfrak{F}_i, i \in I\}$ a (an adapted) process. Also if X_is form a submartingale, then the $-X_i$ form a supermartingale, so only one of these two processes may be considered in the following developments.

Some immediate consequences of these concepts can be given. If $\{X_t, \mathfrak{F}_t, t \in T\}$ is a martingale, where $T \subset \mathbb{R}$, then $\{|X_t|, \mathfrak{F}_t, t \in T\}$ is a submartingale. More generally we have the following result.

3. Proposition *If* $\{X_t, \mathfrak{F}_t, t \in T\}$ *is a (super)martingale on* (Ω, Σ, P) *and* $\varphi: \mathbb{R} \to \mathbb{R}$ *is a (nonincreasing) concave function such that* $E(|\varphi(X_{t_0})|) < \infty$ *for some* $t_0 \in T$, *then* $\{\varphi(X_t), \mathfrak{F}_t, t \in T, t \leq t_0\}$ *is a supermartingale. If "(super)" is replaced by "(sub)" and* φ *is a (nondecreasing) convex function, then under the same conditions* $\{\varphi(X_t), \mathfrak{F}_t, t \leq t_0, t \in T\}$ *is a submartingale.*

Proof By (2), for any $t < t_0$ we have

$$E^{\mathfrak{F}_t}(X_{t_0}) \leq X_t \quad \text{a.e.,} \tag{3}$$

with equality in the martingale case and a reverse inequality for submartingales. So the conditional Jensen inequality (applying Theorem II.1.6 for $-\varphi$) yields

$$\varphi(X_t) \geq \varphi(E^{\mathfrak{F}_t}(X_{t_0})) \geq E^{\mathfrak{F}_t}(\varphi(X_{t_0})) \quad \text{a.e.,} \tag{4}$$

provided $\varphi(X_t)$ is integrable. By the concavity of φ, there is a line touching φ at each point such that $ax + b \geq \varphi(x)$, $x \in \mathbb{R}$ for some real a, b. Hence $aX_t + b \geq \varphi(X_t)$ so that the integrability of X_t implies that $\int_\Omega \varphi(X_t)\, dP < \infty$. This, however, is sufficient to apply Theorem II.1.6, and (4) is true so that $E(\varphi(X_t)) \geq E(\varphi(X_{t_0})) > -\infty$. Thus $\varphi(X_t)$ is integrable. Replacing (t, t_0) by (s, t) for $s < t$, $s \in T$, we deduce that $\{\varphi(X_t), \mathfrak{F}_t, t \leq t_0, t \in T\}$ is a supermartingale. If φ is convex, then the inequality (4) is reversed, and then the last statement follows by the same argument.

Taking $\varphi(x) = e^{-x}$ and $\varphi(x) = \max(x, 0) = x^+$, one gets the following useful specialization.

4. Corollary *If $\{X_t, \mathfrak{F}_t, t \in T\}$ is a positive martingale, then $\{e^{-X_t}, \mathfrak{F}_t, t \in T\}$ is a uniformly bounded (by 1) positive submartingale. If the process is any submartingale, then $\{X_t^+, \mathfrak{F}_t, t \in T\}$ is a submartingale.*

5. Remark If $\{X_n, \mathfrak{F}_n, n \geq 1\} \subset L^1(P)$ is an adapted sequence and if $\xi_n = X_n - X_{n-1}$, then it is a martingale iff $E^{\mathfrak{F}_n}(\xi_{n+1}) = 0$, $n \geq 1$. In fact, if it is a martingale, then $0 = E^{\mathfrak{F}_n}(X_{n+1} - X_n) = E^{\mathfrak{F}_n}(\xi_{n+1})$, $n \geq 1$, a.e. Conversely, if this holds, since $X_n = \sum_{k=1}^n \xi_k$, then $E^{\mathfrak{F}_n}(X_{n+1}) = E^{\mathfrak{F}_n}(\xi_{n+1} + X_n) = X_n$ a.e. In case the X_n are in $L^2(P)$ forming a martingale, then $E(\xi_n \xi_m) = 0$ for $n \neq m$. Indeed, if $n > m$, $E(\xi_n \xi_m) = E(E^{\mathfrak{F}_m}(\xi_n \xi_m)) = E(\xi_m E^{\mathfrak{F}_n}(\xi_n)) = 0$ (cf. Proposition II.1.20). So $\{\xi_n, n \geq 1\}$ is an orthogonal sequence.

We shall establish the fundamental inequalities for (sub)martingales with elementary "stopping time" techniques. These concepts will be useful for some generalizations called (semimartingales and) asymptotic martingales, or "amarts."

Let $\{\mathfrak{F}_i, i \in I\}$ be a right filtering net of σ-algebras from (Ω, Σ, P), where I is directed. A mapping $\tau\colon \Omega \to I$ is said to be an *elementary* (*simple*) *stopping time* of $\{\mathfrak{F}_i, i \in I\}$ if τ takes at most countably (finitely) many values such that $\{\omega : \tau(\omega) \leq i\} \in \mathfrak{F}_i$ and $\{\omega : \tau(\omega) \geq i\} \in \mathfrak{F}$, $i \in I$. The words "elementary" or "simple" will be dropped when there is no confusion. If I is linearly ordered, then both these may be replaced by the single condition $\{\omega : \tau(\omega) = i\} \in \mathfrak{F}$, $i \in I$. The stopping time operations are very effective, if $I \subset \mathbb{R}$, in classifying martingales depending on their sample path behavior. If I is only directed (e.g., $I \subset \mathbb{R}^2$), then they have not yet been proved of equal importance. Going in the other direction (of generalizing the processes), they are not only useful, but the very definition of some of these extensions will be based on such transformations.

If $\tau_1, \ldots, \tau_n$ are stopping times of $\{\mathfrak{F}_i, i \in I\}$, define $Y_j = X \circ \tau_j$, where this means $(X \circ \tau_j)(\omega) = X_{\tau_j(\omega)}(\omega) = Y_j(\omega)$, $\omega \in \Omega$. Since clearly each τ_j is measurable relative to $\sigma(\bigcup_i \mathfrak{F}_i) \subset \Sigma$, and X_i is $\mathfrak{F}_i$-adapted, it is clear that the composition function Y_j is measurable for Σ but not necessarily so for $\mathfrak{F}_i$, $i \in I$. So a natural question is to find conditions (and suitable σ-algebras) for the transformed process $\{Y_j, j \geq 1\}$ of the (sub) martingale $\{X_i, \mathfrak{F}_i, i \in I\}$ to be of the same kind when $\tau_j \leq \tau_{j+1}$, $j \geq 1$. Such an increasing sequence is called a *stopping time process*. The following result, due to Doob [1] and Chow [1], gives a solution to this problem.

6. Theorem (a) *If $\{X_i, \mathfrak{F}_i, i \in I\}$ is an adapted process and $\{\tau_j, j \geq 1\}$ is a stopping time process, then $\{Y_j, \mathcal{G}_j, j \geq 1\}$ is also an adapted process where $Y_j = X \circ \tau_j$, and*

$$\mathcal{G}_j = \sigma\{A \cap [\tau_j \geq i] : A \in \mathfrak{F}_i, i \in I\}, \qquad j \geq 1. \tag{5}$$

(b) *If the given X_i-process is a (sub)martingale, each τ_j is finitely valued for each j, and I is linearly ordered, then $\{Y_j, \mathcal{G}_j, j \geq 1\}$ is also a (sub)martingale.*

Proof (a) If $x \in \mathbb{R}$, consider for any j

$$[Y_j < x] \cap [\tau_j = i] = [X_i < x] \cap [\tau_j = i] \in \mathfrak{F}_i, \qquad i \in I.$$

But $[Y_j < x] \cap [\tau_j \geq i] = [X_i < x] \cap [\tau_j = i] \cap [\tau_j \geq i] \in \mathcal{G}_j$ for all $i \in I$, $x \in \mathbb{R}$. Hence

$$[Y_j < x] = \bigcup_{i \in I} [Y_j < x] \cap [\tau_j \geq i] \in \mathcal{G}_j \tag{6}$$

since the right-hand side union is at most countable. So Y_j is $\mathcal{G}_j$-adapted.

To see that $\mathcal{G}_j \subset \mathcal{G}_{j+1}$, observe that for each $A \in \mathfrak{F}_i$, $A \cap [\tau_j \geq i] \in \mathfrak{F}_i$ and is a generator of $\mathcal{G}_j$ so that, $\tau_j < \tau_{j+1}$ being true, $A \cap [\tau_j \geq i] = A \cap [\tau_j \geq i] \cap [\tau_{j+1} \geq i] \in \mathfrak{F}_i$ and is a generator of $\mathcal{G}_{j+1}$. Hence $\mathcal{G}_j \subset \mathcal{G}_{j+1}$.

(b) Now let the X-process be a submartingale (so each $X_i \in L^1(P)$). We have to show that $E^{\mathcal{G}_j}(Y_{j+1}) \geq Y_j$ a.e. with equality for martingales since it is seen that $Y_j \in L^1(P)$. In fact, $Y_j = \sum_{r=1}^{n_j} X_{i_r} \chi_{[\tau_j = i_r]}$, where $i_1 < i_2 < \cdots < i_{n_j}$ are the values of τ_j on using the *linear* ordering of I. Since each $X_{i_r} \in L^1(P)$, it follows that $Y_j \in L^1(P)$ because $n_j < \infty$. Let $A \in \mathcal{G}_1$ and if $i_1 < i_2 < \cdots < i_n$ are the values of τ_j $(< \tau_{j+1})$, let $A_k = A \cap [\tau_j = i_k]$, $1 \leq k \leq n$. Then it suffices to show that since

$$\int_A Y_{j+1}\, dP \geq \int_A Y_j\, dP \Leftrightarrow \int_{A_{k_0}} Y_{j+1}\, dP \geq \int_{A_{k_0}} Y_j\, dP, \qquad 1 \leq k_0 \leq n, \tag{7}$$

the second inequality holds for an arbitrary $k_0 \in \{1, \ldots, n\}$. This is proved using the following elegant argument due to Hunt [1].

Denote by $A_i = A_{k_0} \cap [\tau_{j+1} = i]$, $A^i = A_{k_0} \cap [\tau_{j+1} > i]$, $i \in I$, so that $A_{i_k} \cup A^{i_k} = A^{i_{k-1}}$, $k > 1$, a disjoint union. Hence

$$\begin{aligned}
\int_{A_{k_0}} Y_j \, dP &= \int_{A_{k_0}} X_{i_{k_0}} \, dP \qquad \text{since } \tau_j = i_{k_0} \quad \text{on} \quad A_{k_0} \\
&= \int_{A_{i_{k_0}}} Y_{j+1} \, dP + \int_{A^{i_{k_0}}} X_{i_{k_0}} \, dP \qquad \text{by definition} \\
&\leq \int_{A_{i_{k_0}}} Y_{j+1} \, dP + \int_{A^{i_{k_0}}} X_{i_{k_0+1}} \, dP \qquad \text{(the submartingale hypothesis)} \\
&= \int_{A_{i_{k_0}}} Y_{j+1} \, dP + \int_{A_{i_{k_0}+1}} Y_{j+1} \, dP + \int_{A^{i_{k_0}+1}} X_{i_{k_0+1}} \, dP \\
&\;\;\vdots \\
&\leq \int_{\bigcup_{k \geq k_0} A_{i_k}} Y_{j+1} \, dP = \int_{A_{k_0}} Y_{j+1} \, dP. \qquad (8)
\end{aligned}$$

There is equality throughout in the martingale case. This completes the proof.

Remarks Part (b) of the theorem is not valid if I is not linearly ordered. On the other hand, it is not hard to show, in the case that I is linearly ordered, that $\mathscr{G}_j$ of (5) is given by $\{A \in \tilde{\mathfrak{F}} : A \cap [\tau_j \leq i] \in \mathfrak{F}_i, i \in I\}$, where $\tilde{\mathfrak{F}} = \sigma(\bigcup_i \mathfrak{F}_i)$, i.e., this collection is a σ-algebra, so that $\mathscr{G}_j$ is called the *σ-algebra of events prior to* τ_j. Since τ_j takes only a countable set of values, it may be shown that both here and in (5) one can use $[\tau_j = i]$ instead of the weaker inequality. Thus both $\mathscr{G}_j$ and $\mathfrak{F}_i$ have the same trace algebras on $[\tau_j = i]$, $i \in I$.

We are now ready to establish the maximal inequalities due to Doob [1], which are extensions of Kolmogorov's classical inequality (see below).

7. Theorem *Let* $\{X_k, \mathfrak{F}_k, 1 \leq k \leq n\}$ *be a submartingale and* $\lambda \in \mathbb{R}$. *If* $A_\lambda = \{\omega : \max_{k \leq n} X_k(\omega) > \lambda\}$, $B_\lambda = \{\omega : \min_{k \leq n} X_k(\omega) < \lambda\}$, *then*

$$\lambda P(A_\lambda) \leq \int_{A_\lambda} X_n \, dP, \qquad (9)$$

$$\lambda P(B_\lambda) \geq \int_{\Omega} X_1 \, dP - \int_{B_\lambda^c} X_n \, dP. \qquad (10)$$

Proof Let $\tau_1(\omega) = \inf\{k : X_k(\omega) > \lambda\}$, with $\tau_1(\omega) = n$ if $\{\ \} = \varnothing$. Then $[\tau_1 = r] = \bigcup_{k=1}^r [X_j \le \lambda, 1 \le j \le k-1, X_k > \lambda] = \bigcup_{k=1}^r A_k$ (say), where for $k = 1$ we take $A_1 = [X_1 > \lambda]$. Hence $[\tau_1 = r] \in \mathfrak{F}_r$, $1 \le r \le n$, so that it is a stopping time (by the above remark). Let $\tau_2 = n$ on Ω. Hence $\tau_1 \le \tau_2$ and $\mathscr{G}_2 = \mathfrak{F}_n$ so that τ_1, τ_2 are stopping times of $\{\mathfrak{F}_k\}_1^n$ and by Theorem 5 $\{Y_j, \mathscr{G}_j\}_1^2$ is a submartingale, where $Y_j = X(\tau_j)$. Since $A_\lambda = [Y_1 \ge \lambda] \in \mathscr{G}_1$, we get, from $E^{\mathscr{G}_1}(Y_2) \ge Y_1$ a.e., the following string of inequalities:

$$\int_{A_\lambda} X_n\, dP = \int_{A_\lambda} Y_2\, dP \ge \int_{A_\lambda} Y_1\, dP = \sum_{j=1}^n \int_{A_j} Y_1\, dP \qquad \text{since} \quad A_j \cap A_{j+1} = \varnothing$$

$$= \sum_{j=1}^n \int_{A_\lambda} X_j\, dP \ge \lambda P\left(\bigcup_{j=1}^n A_j\right) = \lambda P(A_\lambda).$$

This proves (9).

Similarly, let $\tau = \inf\{k\colon X_k < \lambda\}$, and $= n$ if $\{\ \} = \varnothing$. Then taking $\tau_0 = 1$, we get $\tau_0 \le \tau$, and both are stopping times of $\{\mathfrak{F}_k\}_1^n$. So $E(Y_0) \le E(Y)$, where $Y_0 = X_1$ and $Y = X(\tau)$. Thus noting that $\tau = n$ on B_λ^c and letting $B_j = [X_i \ge \lambda, i \le j-1, X_j < \lambda]$, we get $B_\lambda = \bigcup_{j=1}^n B_j$, disjoint union, and

$$E(X_1) \le \int_{B_\lambda} Y\, dP + \int_{B_\lambda^c} Y\, dP = \int_{\bigcup_{j=1}^n B_j} Y\, dP + \int_{B_\lambda^c} X_n\, dP$$

$$\le \sum_{j=1}^n \lambda \int_{B_j} dP + \int_{B_\lambda^c} X_n\, dP = \lambda P(B_\lambda) + \int_{B_\lambda^c} X_n\, dP.$$

This is (10), and the proof is complete.

It should be noted that this result can be proved without using stopping times by decomposing A_λ, B_λ as in the above work. This is how the classical inequalities were originally established.

If $\xi_1, \xi_2, \ldots$ are independent random variables with means zero and variances $\sigma_1^2, \sigma_2^2, \ldots$, let $S_n = \sum_{i=1}^n \xi_i$. Then $\{S_n, \mathfrak{F}_n, n \ge 1\}$ is clearly a martingale, where $\mathfrak{F}_n = \sigma\{\xi_1, \ldots, \xi_n\}$. Taking $\varphi(x) = x^2$ in Proposition 3, we get $\{S_n^2, \mathfrak{F}_n, n \ge 1\}$ to be a submartingale. Applying (9) to this sequence, one obtains:

8. Corollary (Kolmogorov's inequality) *If* $S_n = \xi_1 + \cdots + \xi_n$ *where* ξ_ns *are mutually independent with means zero and variances* $\sigma_1^2, \sigma_2^2, \ldots$, *then for each* $\varepsilon > 0$

$$P\left[\sup_{1 \le k \le n} |S_n| \ge \varepsilon\right] \le \frac{1}{\varepsilon^2} \sum_{i=1}^n \sigma_i^2. \tag{11}$$

(If $n = 1$, this is the *Čebyšev inequality*.)

To illustrate the techniques involved, we include the following inequality, due to Chacón [1], which will be of interest in the asymptotic martingale theory later. His proof is different. We base it on an earlier result by Sudderth [1].

9. Theorem *Let $\{X_n, n \geq 1\}$ be a sequence of random variables lying in some ball of $L^1(\Omega, \Sigma, P)$. If $\mathfrak{F}_n = \sigma(X_i, 1 \leq i \leq n)$ and $\mathscr{S}$ is the set of all simple stopping times of $\{\mathfrak{F}_n, n \geq 1\}$, then*

$$0 \leq \int_\Omega \left(\overline{\lim_n}\, X_n - \underline{\lim_n}\, X_n\right) dP \leq \overline{\lim_{\substack{\tau_1,\tau_2\\ \tau_i \in \mathscr{S}}}} \int_\Omega (X(\tau_1) - X(\tau_2))\, dP, \tag{12}$$

where the $\{X(\tau_i)\}_{i \geq 1}$ is the transformed process of $\{X_n, \mathfrak{F}_n, n \geq 1\}$.

Proof Since it is clear that with $\tau_i \in \mathscr{S}$, $i = 1, 2$, there is a simple stopping time $\tau \geq \tau_i$, $\mathscr{S}$ is directed. First suppose that $|X_n| \leq c < \infty$ a.e. Hence $X^* = \lim_n \sup X_n$, $X_* = \lim_n \inf X_n$ are both bounded. If $V_n = E^{\mathfrak{F}_n}(X^*)$, then $\{V_n, \mathfrak{F}_n, n \geq 1\}$ is a bounded (hence in $L^2(P)$) martingale. Let $\mathfrak{F}_\infty = \sigma(\bigcup_n \mathfrak{F}_n)$. By Remark 5, if $\xi_n = V_n - V_{n-1}$ ($V_0 = 0$), then $\{\xi_k, k \geq 1\} \subset L^2(\Omega, \mathfrak{F}_\infty, P)$ is an orthogonal sequence, $V_n = \sum_{k=1}^n \xi_k$, and $E(X^*\xi_n) = E(E^{\mathfrak{F}_n}(X^*)\xi_n) = E(V_n\xi_n) = E(\xi_n^2) = a_n$ (say). Hence if $\tilde{\xi}_n = \xi_n a_n^{-1/2}$, then $E(X^*\tilde{\xi}_n) = a_n^{1/2}$, and by Bessel's inequality, $\sum_{n=1}^\infty a_n \leq E((X^*)^2) < \infty$. This implies $E[(V_n - V_m)^2] = \sum_{k=m+1}^n a_k \to 0$ as $n, m \to \infty$. Thus $V_n \to V$ in L^2 and V is $\mathfrak{F}_\infty$-measurable. Also by the L^2-continuity of $E^{\mathfrak{F}}$ (cf. Theorem II.1.7), we deduce that $E^{\mathfrak{F}_n}(V) = V_n$ a.e. Thus

$$\nu(A) = \int_A V\, dP = \int_A V_n\, dP = \int_A X^*\, dP, \qquad A \in \mathfrak{F}_n, \quad n \geq 1. \tag{13}$$

So ν being σ-additive on $\bigcup_n \mathfrak{F}_n$, it follows that $\int_A V\, dP = \int_A X^*\, dP$, $A \in \mathfrak{F}_\infty$ and hence that $V = X^*$ a.e. Thus $V_n \to X^*$ in L^2 and so pointwise a.e. for a subsequence. [Actually the whole sequence converges a.e. We shall prove this later in the next section, with some general results, by a slightly more detailed argument. But this weaker statement is sufficient here.] With this preparation we assert that

$$E(X^*) \leq \limsup_{\tau \in \mathscr{S}} E(X \circ \tau). \tag{14}$$

In fact, let N be a large fixed integer. If $\tau_0 \in \mathscr{S}$, $\tau_0 < N$, and $\varepsilon > 0$ are arbitrary, define

$$\tau_1 = \begin{cases} \inf\{n \geq \tau_0, V_n < X_n + \varepsilon\} \\ N \qquad \text{if} \quad \{\ \} = \varnothing. \end{cases} \tag{15}$$

Because $V_{n_i} \to X^*$ a.e., by the above, $\{\ \} \neq \emptyset$ for any such τ_0 if N is large and $\tau_1 = n$ on sets of positive measure for some $n < N$. Thus

$$E(X(\tau_1)) \geq \sum_{n=1}^{N} \int_{[\tau_1 = n]} V_n \, dP - \varepsilon = \int_{\Omega} \chi_{[\tau_1 \leq N]} X^* \, dP - \varepsilon \tag{16}$$

since $[\tau_1 = n] \in \mathfrak{F}_n$. But $E(X(\tau_1)) \leq \sup\{E(X(\tau)) : \tau \in \mathscr{S}, \tau \geq \tau_0\}$, so (16) implies (14) on letting $N \to \infty$ by monotone convergence, $\varepsilon > 0$ and τ_0 being arbitrary. Considering $-X_n$ in (14) and recalling $X_* = -\limsup(-X_n)$, one has from (14)

$$E(X_*) \geq \liminf_{\tau \in \mathscr{S}} E(X(\tau)). \tag{17}$$

But now by the classical Fatou's lemma and dominated convergence, if $u_n = \sup_{k \geq n} X_k$ so that $u_n \downarrow X^*$, and $X(\tau) \leq u_n$ for $n \leq \tau$, we have

$$\limsup_{\tau \in \mathscr{S}} E(X(\tau)) \leq \lim_n \left(\sup_{n \leq \tau \in \mathscr{S}} E(X_\tau) \right) \leq \lim_n E(u_n) = E(X^*). \tag{18}$$

Consequently, (14) and (18) yield

$$E(X^*) = \limsup_{\tau \in \mathscr{S}} E(X(\tau)). \tag{19}$$

A similar reasoning with (17) gives

$$E(X_*) = \liminf_{\tau \in \mathscr{S}} E(X(\tau)). \tag{20}$$

Since all these quantities are finite, there exists a pair of sequences $\{\tau_n, \tau'_n\}_1^\infty \subset \mathscr{S}$ such that "lim sup" and "lim inf" can be replaced by limits and then

$$0 \leq E(X^* - X_*) = \lim_{\tau_n \to \infty} E(X(\tau_n)) - \lim_{\tau'_n \to \infty} E(X(\tau'_n)) \tag{21}$$

$$\leq \limsup_{\{\tau_1, \tau_2\} \subset \mathscr{S}} E(X(\tau_1) - X(\tau_2)). \tag{22}$$

This is (12) if X_ns are uniformly bounded.

If the X_ns are as given, let $Y_n^c = X_n$ if $|X_n| \leq c$, $= c$ if $X_n > c$ and $= -c$ if $X_n < -c$. Then $\{Y_n^c, \mathfrak{F}_n, n \geq 1\}$ satisfies the hypothesis of the first part, and so (21) holds for the Y_n^c-sequence for each $c > 0$. Now let $c \to \infty$ through a sequence after replacing the right side with the $X(\tau)$ and noting that $(Y^c)^* - (Y^c)_*$ increases to $X^* - X_*$. If this is not in $L^1(P)$, then the right side of (21) is infinite and $\lim_n E(X(\tau_n)) = E(X^*) = \infty$ (or the second $= -\infty$). Since $\int_\Omega |X_n| \, dP \leq k < \infty$, let $\tau \in \mathscr{S}$, and set $\tau' = \tau$ on $[X(\tau) > 0]$, $= N$ on its complement. Then $\int_\Omega (X(\tau') - X_N) \, dP \geq \int_\Omega X(\tau) \, dP - k$. Taking suprema on $\tau \in \mathscr{S}$, we get this $= \infty$. Similarly, if $E(X_*) = -\infty$. So (21) holds again. This completes the proof.

The following result, which is extended later, is of interest.

10. Proposition *Let* $\{X_n, \mathfrak{F}_n, n \geq 1\}$ *be a martingale,* $|X_n| \leq Y$, $Y \in L^1(P)$. *Then* $X_n \to X$ *a.e. and in* $L^1(P)$.

Proof Let $Z_n = X(\tau_n)$, $\tau_n \in \mathscr{S}$. If $\{\tau_n, n \geq 1\} \subset \mathscr{S}$ is a stopping time process and $\mathscr{G}_n$ is the σ-algebra of events prior to τ_n, then by Theorem 6 $\{Z_n, \mathscr{G}_n, n \geq 1\}$ is a martingale. Hence $E^{\mathscr{G}_n}(Z_{n+1}) = Z_n$ a.e., so that $E(Z_n) = E(Z_{n+1}) =$ a constant. Since X_ns are dominated by an integrable random variable Y, Eq. (21) is valid. But $E(X(\tau_n)) = a = E(X(\tau_n'))$; since for both sequences we may take the same initial element $\tau_0 = \tau_0'$, so $E(X(\tau_n)) = E(X(\tau_0)) = a = E(X(\tau_n'))$. Hence $E(X^* - X_*) = 0$ and $X^* = X_*$ a.e. Thus $X_n \to X^* = X_* = X_\infty$ a.e. Since the sequence is uniformly integrable, the L^1-convergence is a consequence of the dominated convergence criterion. This completes the proof.

11. Remark In the above proof we used the property of constant expected values for a martingale. This property characterizes martingales from sub- or supermartingales, i.e., a sub- (or super-) martingale is a martingale iff its expectation is a constant. In fact, let $\{X_n, \mathfrak{F}_n, n \geq 1\}$ be a submartingale. Then $E^{\mathfrak{F}_n}(X_{n+1}) \geq X_n$ a.e., $n \geq 1$. Hence $a_n = E(X_n) \leq E(X_{n+1}) = a_{n+1}$, and the $a_n \uparrow$. If the process is a martingale, then $a_n = a_{n+1}$, all $n \geq 1$, so that they are all equal to a constant $a_n = a$, n ≥ 1. Conversely, if the latter is true, then

$$0 \leq E(E^{\mathfrak{F}_n}(X_{n+1}) - X_n) = a_{n+1} - a_n = 0, \qquad n \geq 1.$$

But since the integrand is nonnegative a.e., we must have $E^{\mathfrak{F}_n}(X_{n+1}) = X_n$ a.e., $n \geq 1$, so that the process must be a martingale. The argument applied to $-X_n$s proves the supermartingale case. Note also that when the stochastic *base* $\{\mathfrak{F}_n, n \geq 1\}$ is fixed, the class of all martingales with the same base is a linear space, while the class of all positive linear combinations of submartingales forms a cone.

The following is another interesting inequality, also due to Doob [1]. For an analogous classical result, see Zygmund [1, XII.4.6]:

12. Theorem *If* $\{X_k, \mathfrak{F}_k, 1 \leq k \leq n\}$ *is a positive submartingale in* $L^p(P)$, *and* $X^* = \max_k X_k$, *then the following inequalities hold*:

$$E((X^*)^p) \leq \begin{cases} q^p E(X_n^p) & \text{if} \quad p > 1 \quad \text{so that} \quad q = p/(p-1) \\[2ex] \dfrac{e}{e-1}[1 + E(X_n \log^+ X_n)] & \text{if} \quad p = 1. \end{cases} \tag{23}$$

Proof By Theorem 7, we have for $\lambda > 0$,

$$\lambda P([X^* > \lambda]) \leq \int_{[X^* > \lambda]} X_n \, dP. \tag{24}$$

Consider the relations between P and its image $Q = P \circ X^{*-1}$ on $\mathbb{R}$:

$$\begin{aligned}
\int_\Omega (X^*)^p \, dP &= \int_{\mathbb{R}} x^p \, dQ(x) = p \int_{\mathbb{R}^+} Q([x, \infty))x^{p-1} \, dx \qquad \text{(integrating by parts)} \\
&\leq p \int_{\mathbb{R}^+} x^{p-2} \, dx \int_{[X^* \geq x]} X_n \, dP \qquad \text{(by (24))} \qquad (25) \\
&= p \int_\Omega X_n \, dP \int_0^{X^*} x^{p-2} \, dx \qquad \text{(by Tonelli's theorem)} \\
&= (p/(p-1)) \int_\Omega X_n (X^*)^{p-1} \, dP \qquad (\text{if } \; p > 1) \\
&\leq q \|X_n\|_p \|(X^*)^{p-1}\|_q \qquad \text{(by Hölder's inequality)} \\
&= q \|X_n\|_p (\|X^*\|_p)^{p/q} \qquad (p > 1). \qquad (26)
\end{aligned}$$

If $\|X^*\|_p = 0$, then (23) is true. Suppose $0 < \|X^*\|_p < \infty$. Then dividing (26) by $\|X^*\|_p^{p/q}$, we get (23). Since L^p is a vector lattice and $n < \infty$, $\|X^*\| < \infty$ always. It remains to consider $p = 1$. In this case (25) becomes

$$\int_{[X^* \geq 1]} X^* \, dP \leq \int_{\mathbb{R}^+} \frac{dx}{x} \int_{[X^* \geq x \geq 1]} X_n \, dP = \int_\Omega X_n \log^+ X^* \, dP. \tag{27}$$

However for $a \geq 0$, $b > 0$, $a \log b \leq a \log^+ a + \log(b/a) \leq a \log^+ a + (b/e)$ since $\log(b/a)$ has a maximum at $a = b/e$ for each fixed b. Hence (27) becomes

$$\int_\Omega X^* \, dP - 1 \leq \int_{[X^* \geq 1]} X^* \, dP \leq \int_\Omega X_n \log^+ X_n \, dP + \frac{1}{e} \int_\Omega X^* \, dP. \tag{28}$$

Transposing the quantities in (28), we get the second part of (23), completing the proof.

The argument for inequalities in (25) and (27) can be stated as follows, the proof being the same as above.

13. Proposition *If X, Y are two positive random variables such that $\lambda P(X \geq \lambda) \leq \int_{[X \geq \lambda]} dP$, then for any $\varphi: \mathbb{R}^+ \to \mathbb{R}^+$, increasing and $\varphi(0) = 0$, we have*

$$\int_\Omega \varphi(X) \, dP \leq \int_\Omega Y \, dP \int_0^X \frac{d\varphi(x)}{x}.$$

As a consequence of Theorem 12 and Proposition 10, one has:

14. Proposition *Every $L^p(P)$-bounded martingale $\{X_n, \mathfrak{F}_n, n \geq 1\}$ converges a.e. and in $L^p(P)$, $p > 1$.*

Proof Let $X_n^* = \max_{k \geq n}|X_k|$. Since $\{|X_k|, \mathfrak{F}_k, 1 \leq k \leq n\}$ is a submartingale, (23) implies if $X^* = \lim_n X_n^*$ $(=\sup_{k \geq 1}|X_k|)$,

$$E((X^*)^p) = \lim_n E((X_n^*)^p) \leq q^p \lim_n E(|X_n|^p) = q^p \sup_n E(|X_n|^p) < \infty \tag{29}$$

by the L^p-boundedness of X_ns. Hence $X^* \in L^p(P) \subset L^1(P)$ and $|X_n| \leq X^*$ a.e. So by Proposition 10, $X_n \to X_\infty$ a.e., and $|X_\infty| \leq X^*$ a.e. Hence $X_\infty \in L^p(P)$. Since $|X_n - X_\infty|^p \leq (2X^*)^p$, $n \geq 1$, it follows by the dominated convergence theorem that $X_n \to X_\infty$ in $L^p(P)$ also, completing the proof.

4.2 CONVERGENCE THEORY

A basic interest in the (sub)martingale theory stems from its convergence properties. Therefore we present different versions and proofs of the fundamental convergence theorems due to Doob [1] and in a different (but equivalent) formulation to Andersen and Jessen [1]. It will be convenient for the convergence (as well as the structure) theory if certain decompositions are first established.

The following result, known as *Krickeberg's decomposition* of an L^1-bounded martingale, is an immediate consequence of what has already been established as Proposition III.5.3. Let us restate the special case in terms of point functions for reference.

1. Proposition *Let $\{X_n, \mathfrak{F}_n, n \geq 1\}$ be a martingale lying in a ball of $L^1(P)$. Then there exist two positive martingales $\{X_n^{(i)}, \mathfrak{F}_n, n \geq 1\}$, $i = 1, 2$, such that $X_n = X_n^{(1)} - X_n^{(2)}$, a.e., $n \geq 1$, and $\sup_n E(|X_n|) = E(X_n^{(1)}) + E(X_n^{(2)})$ and conversely. The decomposition is unique.*

If $v_n: A \mapsto \int_A X_n \, dP$, $A \in \mathfrak{F}_n$, then $\{v_n, \mathfrak{F}_n, n \geq 1\}$ is a bounded set martingale, each v_n being P-continuous; Proposition III.5.3 implies the existence of nonnegative set martingales $\{\lambda_n^{(i)}, \mathfrak{F}_n, n \geq 1\}$, $i = 1, 2$, with $v_n = \lambda_n^{(1)} - \lambda_n^{(2)}$, and (as the proof there shows) $\lambda_n^{(i)}$ is P-continuous on $\mathfrak{F}_n$. If $X_n^{(i)} = d\lambda_n^{(i)}/dP$ on $\mathfrak{F}_n$, then the result is a restatement of the earlier one. The converse is obvious. Uniqueness part is the same as before.

The following result, known as *Doob's decomposition*, is useful to reduce the convergence study of submartingales to martingales, as well as for other applications.

2. Proposition *Let $\{X_n, \mathfrak{F}_n, n \geq 1\}$ be a submartingale. Then there is a martingale $\{X'_n, \mathfrak{F}_n, n \geq 1\}$ and an increasing process $\{A_n, \mathfrak{F}_{n-1}, n \geq 1\}$ such that $X_n = X'_n + A_n$, $n \geq 1$, $A_1 = 0$ a.e., and conversely. The decomposition is unique and in fact A_n can be defined by the equation*

$$A_n = \sum_{j=1}^{n} (E^{\mathfrak{F}_{j-1}}(X_j) - X_{j-1}), \qquad n \geq 2 \quad \text{a.e.} \tag{1}$$

Proof If $\{X_n, \mathfrak{F}_n, n \geq 1\}$ is a submartingale and A_n is given by (1), then $A_n \leq A_{n+1}$, and if $X'_n = X_n - A_n$, we have for $n \geq 1$

$$E^{\mathfrak{F}_n}(X'_{n+1}) = E^{\mathfrak{F}_n}(X_{n+1}) - (E^{\mathfrak{F}_n}(X_{n+1})) - \sum_{j=1}^{n} [E^{\mathfrak{F}_{j-1}}(X_j) - X_{j-1}] + X_n$$

$$= X_n - A_n = X'_n \quad \text{a.e.} \tag{2}$$

So $\{X'_n, \mathfrak{F}_n, n \geq 1\}$ is a martingale, and $\{A_n, \mathfrak{F}_{n-1}, n \geq 1\}$ is an increasing process.

Conversely, if $X_n = X'_n + A_n$, where $A_n = \sum_{j=1}^{n} a_j$, $a_j = A_j - A_{j-1}$, $a_1 = 0$, $a_j \geq 0$, and $\{X'_n, \mathfrak{F}_n, n \geq 1\}$ is a martingale, then

$$E^{\mathfrak{F}_n}(X_{n+1}) = X'_n + \sum_{j=1}^{n} a_j + a_{n+1} \geq X'_n + A_n = X_n \quad \text{a.e.}$$

since $a_{n+1} \geq 0$. So the process $\{X_n, \mathfrak{F}_n, n \geq 1\}$ is a submartingale. If $X_n = X'_n + A_n = Y'_n + B_n$ are two representations, then $X'_n - Y'_n = B_n - A_n$ is a martingale. Hence $E(B_n - A_n) =$ a constant (cf. Remark 1.11). But $B_1 = A_1 = 0$ so that this constant is zero. Also $B_n - A_n = E^{\mathfrak{F}_n}(B_{n+1} - A_{n+1}) = B_{n+1} - A_{n+1}$ since both A_n and B_n are $\mathfrak{F}_{n-1}$-adapted. Hence $B_1 - A_1 = B_n - A_n = 0$, for all n. Thus $A_n = B_n$ and then $Y'_n = X'_n$ a.e. This completes the proof.

Let us now turn to the convergence statements. We first prove the following key special result. (Compare with Proposition 1.14.)

3. Theorem *Every positive $L^2(P)$-bounded submartingale converges both in L^2-mean and a.e.*

Proof Since the process $\{X_n, \mathfrak{F}_n, n \geq 1\} \subset L^2(P)$ is nonnegative, it follows by Proposition 1.3 that $\{X_n^2, \mathfrak{F}_n, n \geq 1\} \subset L^1(P)$ is a submartingale and that $E(X_n^2) \uparrow a < \infty$, as $n \to \infty$. Hence for $m < n$

$$0 \leq E(X_n^2 - X_m^2) = E((X_n - X_m)^2) + 2E(X_m(X_n - X_m)). \tag{3}$$

Since $E(X_m(X_n - X_m)) = E(X_m E^{\mathfrak{F}_m}(X_n - X_m)) \geq 0$ by the submartingale property and the left side of (3) tends to zero, both the right-side (positive) terms must tend to zero as $m, n \to \infty$. Hence $X_n \to X$ in $L^2(P)$. We now show that the a.e. convergence holds by using the maximal inequality.

For any $m \geq 1$, the sequence $\{X_k - X_m, \mathfrak{F}_k, m < k \leq n\}$ is clearly a submartingale. Hence by the maximal inequalities (cf. Theorem 1.7) for any $\varepsilon > 0$,

$$\varepsilon P\left[\max_{m<k\leq n} (X_k - X_m) \geq \varepsilon\right] \leq E(|X_n - X_m|),$$

$$-\varepsilon P\left[\min_{m<k\leq n} (X_k - X_m) \geq -\varepsilon\right] \geq E((X_{m+1} - X_m)) - E(|X_n - X_m|). \quad (4)$$

Letting $n \to \infty$ and noting that $L^2(P)$-convergence implies $L^1(P)$-convergence, we get from (4)

$$\varepsilon\left(P\left[\sup_{k>m}(X_k - X) \geq \varepsilon\right] + P\left[\inf_{k>m}(X_k - X_m) \geq -\varepsilon\right]\right)$$
$$\leq E(|X_{m+1} - X|) + 2E(|X_m - X|). \quad (5)$$

If now $m \to \infty$, one gets from (5)

$$\lim_n P\left[\sup_{k\geq n}|X_k - X_n| \geq \varepsilon\right] = 0. \quad (6)$$

Consequently, $\limsup_n X_n = X^*$, $\liminf_n X_n = X_*$ are finite and

$$P[X^* - X_* \geq 2\varepsilon] \leq 2\lim_n P\left[\sup_{k\geq n}|X_k - X_m| \geq \varepsilon\right] = 0. \quad (7)$$

Thus $X^* = X_*$ a.e. and $X_n \to X = X_* = X^*$ a.e., as asserted.

It is now possible to prove the following fundamental convergence theorem due to Doob [1]. Two proofs will be given based on the recent arguments of Isaac [1] and Lamb [1] (cf. also Meyer [2]). They have independent interest, and the work will be used later.

4. Theorem *If $\{X_n, \mathfrak{F}_n, n \geq 1\}$ is an L^1-bounded martingale, then $X_n \to X$ a.e. and $E(|X|) \leq \liminf_n E(|X_n|)$.*

First proof By Proposition 1, $X_n = X_n^{(1)} - X_n^{(2)}$, where $\{X_n^{(i)}, \mathfrak{F}_n, n \geq 1\}$ is a positive martingale. So it suffices to show that every positive martingale converges a.e. Let $\{Y_n, \mathfrak{F}_n, n \geq 1\}$ be such a process. Then $Z_n = e^{-Y_n}$ defines $\{Z_n, \mathfrak{F}_n, n \geq 1\}$ to be a strictly positive and bounded submartingale by Corollary 1.4, and so it suffices to show that each positive bounded submartingale converges a.e. since then $Y_n = \log(1/Z_n) \to \log(1/Z) = Y$ a.e., by the continuity of the "log" function. By Theorem 3, this is true because

$L^\infty \subset L^2$. Since $E(Y) \leq \underline{\lim}_n E(Y_n) < \infty$ by Fatou's inequality, $0 < Z < \infty$ a.e. Thus $X_n \to X$ a.e., and then again the Fatou theorem implies $E(|X|) \leq \underline{\lim}_n E(|X_n|)$, proving the result.

Second proof The argument here, except for using the maximal inequality, is independent of other results used above. It consists of two steps: (i) every $L^1(P)$-bounded martingale can be approximated by (or it coincides with) a uniformly integrable martingale on a set of arbitrarily large probability and (ii) each uniformly integrable martingale converges a.e. (and in $L^1(P)$ also). Let us fill in the details.

Step (i). If $r > 0$ is any number, let $\tau = \inf\{n : |X_n| > r\}$, with $\inf(\phi) = +\infty$. It is clear that $\tau : \Omega \to \overline{\mathbb{R}}^+$ is an elementary stopping time of $\{\mathfrak{F}_n, n \geq 1\}$. Let $\tau_n = \min(\tau, n)$, so that τ_n is a simple stopping time, and by Proposition 1.6 $\{X(\tau_n), \mathscr{G}_n, n \geq 1\}$ is a martingale. It is plain that $|X(\tau_n)| \leq |X(\tau)|$. The uniform integrability of $\{X(\tau_n), \mathscr{G}_n, n \geq 1\}$ follows if it is shown that $X(\tau) \in L^1(P)$. In fact, it is seen that $\mathscr{G}_n \subset \mathfrak{F}_n$ in this case ($\tau_n = \tau \wedge n$), and one has

$$\begin{aligned}\int_\Omega |X(\tau)|\, dP &= \int_{[\tau = \infty]} |X(\tau)|\, dP + \int_{[\tau < \infty]} |X(\tau)|\, dP \leq r + \int_{[\tau < \infty]} \lim_n |X(\tau_n)|\, dP \\ &\leq r + \liminf_n \int_{[\tau < \infty]} |X(\tau_n)|\, dP \qquad \text{by Fatou's lemma} \\ &\leq r + \sup_n E(|X(\tau_n)|) \leq r + \sup_n E(|X_n|) = r + c_0 < \infty. \end{aligned} \tag{8}$$

But by the maximal inequality,

$$P\left[\sup_n |X_n| > r\right] \leq (1/r) \sup_n E(|X_n|) \leq c_0/r,$$

which can be made arbitrarily small, i.e., given $\varepsilon > 0$, choose $r_\varepsilon = c_0/\varepsilon$. Then if $A_{r_\varepsilon} = [\sup_n |X_n| > r_\varepsilon]$, $P(A_{r_\varepsilon}) \leq \varepsilon$. On $A^c_{r_\varepsilon}$, $X(\tau_n) = X_n$, $n \geq 1$. If we show that $X(\tau_n) \to \tilde{X}$ a.e., then so does X_n on $A^c_{r_\varepsilon}$ for each $\varepsilon > 0$, and this implies the result.

Step (ii). Each uniformly integrable martingale $\{Y_n, \mathfrak{F}_n, n \geq 1\}$ converges a.e. and in $L^1(P)$. To see this, let $\mathfrak{F}_\infty = \sigma(\bigcup_n \mathfrak{F}_n)$. Observe that $\bigcup_n L^1(\mathfrak{F}_n)$ is a norm dense subspace of $L^1(\mathfrak{F}_\infty)$. Since $\{Y_n, \mathfrak{F}_n, n \geq 1\} \subset L^1(\mathfrak{F}_\infty)$, it suffices to show that for any $\varepsilon > 0$ there exists a convergent martingale $\{Z_n, \mathfrak{F}_n, n \geq 1\} \subset L^1(\mathfrak{F}_\infty)$ such that $P(\sup_n |Y_n - Z_n| > \varepsilon) < \varepsilon$. This implies on taking $\varepsilon = 1/2^k$ that

$$P\left[\limsup_n |Y_n - Z_n| = 0\right] \leq \sum_{k > k_0} P\left[\sup_n |Y_n - Z_n| > 2^{-k}\right] \leq 2^{-k_0} \to 0$$

as $k_0 \to \infty$, and hence $Y_n \to Y$ a.e.

As noted in the proof of Theorem 1.9, the uniform integrability of $\{Y_n, \mathfrak{F}_n, n \geq 1\}$ implies $\nu_n: A \mapsto \int_A Y_n \, dP$, $A \in \mathfrak{F}_n$ is a signed measure and defines a σ-additive set function $\nu: \mathfrak{F}_\infty \to \mathbb{R}$ such that $\nu | \mathfrak{F}_n = \nu_n$, $n \geq 1$. Also ν is P-continuous, so that if $Y = d\nu/dP$, we have $E^{\mathfrak{F}_n}(Y) = Y_n$ a.e. Let $Z \in \bigcup_n L^1(\mathfrak{F}_n)$ be chosen such that $E(|Y - Z|) < \varepsilon^2$, where $\varepsilon > 0$ is given above. This is possible since the space $\bigcup_n L^1(\mathfrak{F}_n)$ is dense in $L^1(\mathfrak{F}_\infty)$. Hence $Z \in L^1(\mathfrak{F}_n)$ for some n. If $Z_k = E^{\mathfrak{F}_k}(Z)$, then $\{Z_k, \mathfrak{F}_k, k \geq 1\}$ is a martingale that clearly converges a.e. since $Z_k = Z$ if $k \geq n$. On the other hand, $\{Y_n - Z_n, \mathfrak{F}_n, n \geq 1\}$ is a martingale. So by the maximal inequality

$$P\left[\sup_k |Y_k - Z_k| > \varepsilon\right] < \frac{1}{\varepsilon} \sup_k E(|Y_k - Z_k|) = \sup_k \frac{E(|E^{\mathfrak{F}_k}(Y - Z)|)}{\varepsilon}$$

$$\leq \frac{\sup_k E(E^{\mathfrak{F}_k}(|Y - Z|))}{\varepsilon} \leq \frac{E(|Y - Z|)}{\varepsilon} < \varepsilon, \tag{9}$$

where the conditional Jensen inequality and contractivity of $E^{\mathfrak{F}_k}$ are used. Since $\varepsilon > 0$ is arbitrary, this shows that $Y_n \to Y$ a.e. The uniform integrability and Vitali's theorem then imply that $Y_n \to Y$ in $L^1(P)$ also. This completes the proof of Step (ii) and hence of the theorem.

The argument of Step (ii) above contains more information than that needed for Theorem 4. A more general version of Step (ii) appears in the proof of Theorem 9 below. We give here a characterization of L^1-convergence of martingales.

5. Theorem *Let $\{X_n, \mathfrak{F}_n, n \geq 1\}$ be a martingale and $\mathfrak{F}_\infty = \sigma(\bigcup_n \mathfrak{F}_n)$. The following statements are equivalent*:

(i) $\{X_n, n \geq 1\} \subset L^1(P)$ *is (in a ball and) uniformly integrable*;

(ii) $\{X_n, n \geq 1\} \subset L^1(P)$ *is a Cauchy sequence*;

(iii) $\{X_n, \mathfrak{F}_n, 1 \leq n \leq \infty\}$ *is a martingale*;

(iv) $\sup_n E(|X_n|) < \infty$ *and* (*so* $X_n \to X_\infty$ *a.e. by Theorem* 4) $E(|X_\infty|) = \lim_n E(|X_n|)$.

(v) *There exists a convex function* $\varphi: \mathbb{R}^+ \to \mathbb{R}^+$, $\varphi(x)/x \uparrow \infty$ *as* $x \uparrow \infty$, *and* $\varphi(0) = 0$, *such that* $\sup_n E(\varphi(|X_n|)) < \infty$.

Proof (i) $\Rightarrow$ (ii) This was proved in Step (ii) of the preceding (second) proof.

(ii) $\Rightarrow$ (i) Since each Cauchy sequence in $L^1(P)$ is uniformly integrable, (i) holds, and this fact has nothing to do with martingale theory.

(i) $\Rightarrow$ (iii) In Step (ii) of the above proof it was shown that $X_n \to X$ a.e. and in $L^1(P)$. Also $E^{\mathfrak{F}_n}(X) = X_n$ a.e., and this implies $\{X_n, \mathfrak{F}_n, 1 \leq n \leq \infty\}$ is a martingale, which is (iii).

(iii) ⇒ (i) Since $X_n = E^{\mathfrak{F}_n}(X_\infty)$ and $\{E^{\mathscr{B}}(|X_\infty|) : \mathscr{B} \subset \Sigma, \mathscr{B}$ a σ-algebra$\}$ is uniformly integrable, (i) holds. That the latter collection has the stated integrability property follows from the definition (if $Y_{\mathscr{B}} = E^{\mathscr{B}}(|X_\infty|)$):

$$\int_{[Y_{\mathscr{B}} > \lambda]} Y_{\mathscr{B}}\, dP = \int_{[Y_{\mathscr{B}} > \lambda]} E^{\mathscr{B}}(|X_\infty|)\, dP = \int_{[Y_{\mathscr{B}} > \lambda]} |X_\infty|\, dP \to 0, \qquad \text{as} \quad \lambda \to \infty, \tag{10}$$

uniformly in $\mathscr{B}$ since $P[Y_{\mathscr{B}} > \lambda] \leq (1/\lambda)E(E^{\mathscr{B}}(|X_\infty|)) = (1/\lambda)E(|X_\infty|) \to 0$, as $\lambda \to \infty$, uniformly in $\mathscr{B}$. This proves (i).

(i) ⇒ (iv) $X_n \to X_\infty$ a.e., so $|X_n| \to |X_\infty|$ a.e. But $\{|X_n|, n \geq 1\}$ is also uniformly integrable since $\{X_n, n \geq 1\}$ is. Hence by Vitali's theorem $|X_n| \to |X_\infty|$ in $L^1(P)$. This statement yields (iv).

(iv) ⇒ (i) $|X_n| \to |X_\infty|$ a.e. as before. Then $E(|X_n|) \to E(|X_\infty|)$ implies uniform integrability of $\{X_n, n \geq 1\}$, and this again does not depend on martingale theory. We have proved this (in a more general form) in Theorem I.4.12.

(i) ⇔ (v) This is also established in Theorem I.4.4.

This completes the equivalence of all implications and the proof of the theorem is complete.

Let us now turn to the convergence theory of submartingales. This will be deduced from the preceding two theorems complementing the result of Theorem 3.

6. Theorem *If* $\{X_n, \mathfrak{F}_n, n \geq 1\}$ *is an* $L^1(P)$*-bounded submartingale, then* $X_n \to X_\infty$ *a.e. and* $E(|X_\infty|) \leq \liminf_n E(|X_n|)$.

Proof By Proposition 2, we have $X_n = X_n' + A_n$, $n \geq 1$, where $A_n \geq 0$ is increasing. Then using the fact that $\{X_n', \mathfrak{F}_n, n \geq 1\}$ is a martingale, one has

$$0 \leq E(A_n) \leq E(|X_n|) - E(X_n') \leq \sup_n E(|X_n|) - E(X_1') < \infty.$$

Hence $\lim_n E(A_n) < \infty$ so that

$$\sup_n E(|X_n'|) \leq \sup_n E(|X_n|) + \lim_n E(A_n) < \infty. \tag{11}$$

By Theorem 4, $X_n' \to X_\infty'$ a.e., since clearly $A_n \uparrow A_\infty$ a.e., and $E(A_\infty) = \lim_n E(A_n) < \infty$, we get $X_n = X_n' + A_n \to X_\infty' + A_\infty = X_\infty$ a.e. By the Fatou inequality it then follows that $E(|X_\infty|) \leq \underline{\lim}_n E(|X_n|)$. This completes the proof.

Remark Since $E(|X_n|) = -E(X_n) + 2E(X_n^+) \leq -E(X_1) + 2E(X_n^+)$ for a submartingale, $\sup_n E(|X_n|) < \infty$ is equivalent to $\sup_n E(X_n^+) < \infty$, which thus could be substituted in the above theorem. For martingales, the condition becomes $E(|X_n|) = E(X_n) + 2E(X_n^-) = E(X_1) + 2E(X_n^-)$, and hence $\sup_n E(|X_n|) < \infty$ is equivalent to $\sup_n E(X_n^-) < \infty$.

It is now clear that one can present an analogous characterization as in Theorem 5, for submartingales. This is stated as follows.

7. Theorem *Let $\{X_n, \mathfrak{F}_n, n \geq 1\}$ be a submartingale and $\mathfrak{F}_\infty = \sigma(\bigcup_n \mathfrak{F}_n)$. The following statements are equivalent:*

(i) *the process is uniformly integrable;*
(ii) *the process is Cauchy in $L^1(P)$;*
(iii) *$K = \sup_n E(|X_n|) < \infty$ so that $X_n \to X_\infty$ a.e., and $(X_n, \mathfrak{F}_n, 1 \leq n \leq \infty\}$ is a submartingale such that $K = E(|X_\infty|)$.*
(iv) *There exists a convex function $\varphi\colon \mathbb{R}^+ \to \mathbb{R}^+$, $\varphi(x)/x \uparrow \infty$, $\varphi(0) = 0$, and* $\sup_n E(\varphi(|X_n|)) < \infty$.

The proof is the same as that of Theorem 5 and will be omitted. One should note that the statement $\{X_n, \mathfrak{F}_n, 1 \leq n \leq \infty\}$ is a submartingale does *not* imply uniform integrability (as simple counterexamples show) and so is not equivalent to the other conditions, as distinct from the martingale case.

8. Remark If (Ω, Σ, P) is a probability space, $\nu\colon \Sigma \to \mathbb{R}^+$ is a measure, and $\{\mathfrak{F}_n, n \geq 1\}$ is an increasing sequence of σ-algebras in Σ, let $\nu_n = \nu | \mathfrak{F}_n$ and $X_n = d\nu_n^c/dP$, where ν_n^c is the P-continuous part of ν_n. It is then not difficult to check that $\{X_n, \mathfrak{F}_n, n \geq 1\}$ is an L^1-bounded supermartingale so that by Theorem 6, $X_n \to X_\infty$ a.e. Also $X_\infty = d\nu_\infty^c/dP$ a.e. This was shown by Andersen and Jessen [2]. Even though the key convergence statement is a consequence of Theorem 6, which is due to Doob, it can be shown that both these approaches are equivalent. This was discussed by the author [8].

We now prove the following result, using Theorem III.5.1, and then indicate how the hypothesis can be relaxed. This is our first vector martingale convergence assertion.

9. Theorem *Let (Ω, Σ, P) be a topological probability space, where Ω is a Hausdorff space and Σ is a σ-algebra containing all the compact sets of Ω. Let $\mathfrak{F}_n \subset \mathfrak{F}_{n+1} \subset \Sigma$ be σ-algebras such that each $P_n = P|\mathfrak{F}_n$ is regular for the compact sets in $\mathfrak{F}_n$, i.e., $P_n(A) = \sup\{P_n(C) : C \subset A, C \in \mathfrak{F}_n, \text{compact}\}$, $A \in \mathfrak{F}_n$. Suppose $\{X_n, \mathfrak{F}_n, n \geq 1\}$ is a stochastic process such that (i) $X_n\colon \Omega \to \mathscr{X}$ where $\mathscr{X}$ is a separable reflexive Banach space and each X_n is strongly measurable,* *i.e.,* X_n^{-1}(open set of $\mathscr{X}$) $\in \mathfrak{F}_n$, *(ii)* $\|X_n\|_1 = \int_\Omega \|X_n\|\, dP \le K_0 < \infty$ *for all n, where* $\|\cdot\|$ *is the norm of* $\mathscr{X}$, *and (iii) the process is a martingale, i.e.,* $\int_A X_n\, dP_n = \int_A X_{n+1}\, dP_{n+1}$, $A \in \mathfrak{F}_n$, $n \ge 1$, *where the integrals are taken in the strong (or Bochner) sense as we discussed in Section I.4. [Thus X is P-integrable iff there is a sequence of simple functions* Y_k *such that* $\|Y_k - X\| \to 0$ *a.e. and* $\int_\Omega Y_n\, dP = \int_\Omega \sum_{i=1}^n a_i^n \chi_{A_i^n}\, dP = \sum_1^n a_i^n P(A_i^n)$, $a_i^n \in \mathscr{X}$, $\{\int_\Omega Y_n\, dP\}_1^\infty$ *is Cauchy in* $\mathscr{X}$, *so that* $\int_\Omega X\, dP = \lim_n \int_\Omega Y_n\, dP$ *uniquely.] Then* $X_n \to X_\infty$ *a.e.*

Proof Regarding the Bochner integral and its properties to be used here, the reader may refer to Dunford–Schwartz [1, Chapter III]. We shall discuss later on the existence of conditional expectations for any Bochner integrable random variable (with values in a Banach space) so that hypothesis (iii) above is nonvacuous.

If we define $v_n: \mathfrak{F}_n \to \mathscr{X}$ by the usual equation

$$v_n(A) = \int_A X_n\, dP_n, \qquad A \in \mathfrak{F}_n, \tag{12}$$

then v_n is a vector measure and by (iii), $v_n = v_{n+1} | \mathfrak{F}_n$. Thus $\{(\Omega, \mathfrak{F}_n, v_n, g_{mn})_{m \le n}, n \ge m \ge 1\}$ is a projective system with g_{mn} = identity. Since we have the variations $|v_n|(A) = \int_A \|X_n\|\, dP_n \le \sup_n \|X_n\|_1 < \infty$, $A \in \mathfrak{F}_\infty = \sigma(\bigcup_{n \ge 1} \mathfrak{F}_n)$, and since the projective system is indexed by the integers ($D = \mathbb{N}$), the hypothesis of Theorem III.5.1 is satisfied (because $\mathscr{X}$ is reflexive and separable). Here we take the class $\mathscr{C}_n \subset \mathfrak{F}_n$ to be the compact sets and note that $v_n(\cdot)$ is also regular on $\mathfrak{F}_n$ because P_n is. In fact, for each $A \in \mathfrak{F}_n$ and $\varepsilon > 0$ there is a (compact) $C \in \mathscr{C}_n$, $A \supset C$, and $P_n(A - C) < \varepsilon / \|X_n\|_1$. Hence

$$\|v_n(A - C)\| \le \int_{A-C} \|X_n\|\, dP_n \tag{13}$$

and the right side is arbitrarily small by the absolute continuity of the Lebesgue integral since $\|X_n\|$ is integrable and $P_n(A - C) < \varepsilon / \|X_n\|_1$. Thus v_n has the key approximation property required of the above-quoted theorem. Hence $v = \varprojlim v_n$ is a vector measure on $\mathfrak{F}_\infty$ to $\mathscr{X}$ and has an approximation property. In particular, $|v|(\Omega) \le \sup_n \|X_n\|_1 < \infty$. However v being σ-additive on $\mathfrak{F}_\infty$ need not be P-continuous even though each v_n is.

Let $\mu(\cdot) = |v|(\cdot)$, the variation of v. Then $\mu(\cdot)$ is σ-additive since v is. Define $\alpha = \mu + P_\infty$, $P_\infty = P | \mathfrak{F}_\infty$, a finite measure on $\mathfrak{F}_\infty$, and then clearly both v and P_∞ are α-continuous. Let $\alpha_n = \alpha | \mathfrak{F}_n$. By the Radon–Nikodým theorem then $p_n = dP_n / d\alpha_n$ a.e., and since $\mathscr{X}$ is reflexive, the vector measure v

has the Radon–Nikodým property, i.e., $\nu(A) = \int_A f\,d\alpha$ for a unique strongly integrable (relative to α) $f: \Omega \to \mathscr{X}$. [We shall discuss, more generally, such results on differentiation of vector measures and their close relation with the martingale property in Chapter V.] Similarly, let $f_n = d\nu_n/d\alpha_n$. Then by the chain rule for the Radon–Nikodym derivatives, which holds in this context (cf. Dunford–Schwartz [1, III.10.5]), we have

$$\int_A f\,d\alpha = \nu_n(A) = \int_A X_n\,dP_n = \int_A X_n p_n\,d\alpha_n = \int_A f_n\,d\alpha_n,\ A \in \mathfrak{F}_n. \tag{14}$$

The properties of the strong integral allow us to identify the integrands so that $f_n = X_n p_n$, a.e. $[\alpha_n]$. Moreover, by Theorem 4, $p_n \to p_\infty$ a.e. (where $p_\infty = dP_\infty/d\alpha$ a.e. $[\alpha]$) and $p_\infty > 0$ a.e. $[P]$ since if $B = \{\omega : p_\infty(\omega) = 0\}$, then $P_\infty(B) = \int_B p_\infty\,d\alpha = 0$. We shall now show that $f_n \to f$ a.e. $[\alpha]$ so that $X_n = f_n/p_n \to f/p_\infty = X_\infty$ a.e. $[P]$ follows. [If $\mathscr{X} = \mathbb{R}$, this is simply Theorem 4, and together with the following proposition yields an alternate proof of the convergence statement there.] Since this special case has independent interest, we present it separately for later reference. Let us express (14) suggestively as $E^{\mathfrak{F}_n}(f) = f_n$ a.e. $[\alpha]$ and show $f_n \to f$ a.e.

10. Proposition *Let $\{f_n = E^{\mathfrak{F}_n}(f), \mathfrak{F}_n, n \geq 1\}$ be an $\mathscr{X}$-valued martingale on $(\Omega, \mathfrak{F}_\infty, \alpha)$, where $\mathfrak{F}_\infty = \sigma(\bigcup_{n\leq 1} \mathfrak{F}_n)$ and $\alpha(\Omega) < \infty$. Then $f_n \to f$ a.e. and in $L^1(\Omega, \Sigma, \alpha, \mathscr{X})$, the B_1-space of Theorem I.4.11.*

Proof Since $f_n \in L^1(\Omega, \mathfrak{F}_n, \alpha, \mathscr{X}) \subset L^1(\Omega, \mathfrak{F}_\infty, \alpha, \mathscr{X})$, we show that $\|f_n - f_m\| \to 0$ a.e. and in L^1-norm as $n, m \to \infty$, and then identify the limit of f_n as f. One uses the easily verifiable facts that $\bigcup_{n\geq 1} L^1(\Omega, \mathfrak{F}_n, \alpha, \mathscr{X})$ is a dense linear subspace of $L^1(\Omega, \mathfrak{F}_\infty, \alpha, \mathscr{X})$, that for $g \in L^1(\Omega, \mathfrak{F}_m, \alpha, \mathscr{X})$, $E^{\mathfrak{F}_n}(g) = g$ holds a.e. for all $n \geq m$, and that $\|E^{\mathfrak{F}_n}(f)\| \leq \tilde{E}^{\mathfrak{F}_n}(\|f\|)$ a.e. Here the second $\tilde{E}^{\mathfrak{F}_n}$ acts on the scalar function $\|f\|$ and the first one on the vector function f. This is the "conditional Jensen inequality" since the norm is a convex functional. (Or see Theorem V.1.2.)

For $\varepsilon > 0$ there exists a $g_\varepsilon \in \bigcup_{n\geq 1} L^1(\Omega, \mathfrak{F}_n, \alpha, \mathscr{X})$ such that $\|f - g_\varepsilon\|_1 < \varepsilon$. So $g_\varepsilon \in L^1(\Omega, \mathfrak{F}_{n_0}, \alpha, \mathscr{X})$ for some n_0 $(= n_0(\varepsilon))$. Thus $E^{\mathfrak{F}_n}(g_\varepsilon) = g_\varepsilon$ a.e. for $n \geq n_0$. If $n, m \geq n_0$, then we have on using $f_n = E^{\mathfrak{F}_n}(f)$

$$\begin{aligned}\|f_n - f_m\| &\leq \|E^{\mathfrak{F}_n}(f) - g_\varepsilon\| + \|E^{\mathfrak{F}_m}(f) - g_\varepsilon\| \quad \text{a.e.} \\ &\leq \|E^{\mathfrak{F}_n}(f - g_\varepsilon)\| + \|E^{\mathfrak{F}_m}(f - g_\varepsilon)\| \\ &\qquad \text{since} \quad g_\varepsilon = E^{\mathfrak{F}_{n_0}}(g_\varepsilon) \quad \text{a.e.} \\ &\leq 2 \sup_{k\geq 1} \tilde{E}^{\mathfrak{F}_k}(\|f - g_\varepsilon\|) \quad \text{a.e.}\end{aligned} \tag{15}$$

But $\{\tilde{E}^{\mathfrak{F}_n}(\|f-g_\varepsilon\|), \mathfrak{F}_n, n \geq 1\}$ is obviously a positive martingale. Hence by Theorem 1.3 and (15) we get for each $\lambda > 0$

$$P\left[\limsup_{m,n\to\infty}\|f_n - f_m\| \geq \lambda\right] \leq P\left[\limsup_{n\to\infty} \tilde{E}^{\mathfrak{F}_n}(\|f-g_\varepsilon\|) \geq \tfrac{1}{2}\lambda\right]$$

$$= \lim_{n\to\infty} P\left[\sup_{1\leq k\leq n} \tilde{E}^{\mathfrak{F}_k}(\|f-g_\varepsilon\|) \geq \tfrac{1}{2}\lambda\right]$$

$$\leq \lim_{n\to\infty} (2/\lambda)E(\tilde{E}^{\mathfrak{F}_n}(\|f-g_\varepsilon\|))$$

$$= 2/\lambda\|f-g_\varepsilon\|_1 << 2\varepsilon/\lambda.$$

Letting $\varepsilon \downarrow 0$, we see that $\|f_n - f_m\| \to 0$ a.e. as $n, m \to \infty$. So from the completeness of $\mathscr{X}$ there exists a measurable (for $\mathfrak{F}_\infty$) $h = \lim_n f_n$ a.e. $[\alpha]$, and by Fatou's lemma, $E(\|h\|) \leq \underline{\lim}_{n\to\infty} E(\|f_n\|) < \infty$. It remains to show that $h = f$ a.e., and $\|f - f_n\|_1 \to 0$.

Both these assertions follow if we show that $\{E^{\mathfrak{F}_n}(f), n \geq 1\}$ is uniformly integrable. First note that if $\int_A f_n \, d\alpha = v_n(A)$, then the variation $|v_n|(A) = \int_A \|f_n\| \, d\alpha$ by the classical theory. Hence for any $\lambda > 0$

$$\int_{[\|f_n\| > \lambda]} \|f_n\| \, d\alpha = |v_n|[\|f_n\| > \lambda] \leq |v|[\|f_n\| > \lambda], \tag{16}$$

where $v(A) = \int_A f \, d\alpha$ so that $v_n = v \mid \mathfrak{F}_n$. But we have

$$\alpha[\|f_n\| > \lambda] \leq \frac{1}{\lambda}\int_\Omega \|f_n\| \, d\alpha = \frac{1}{\lambda}|v_n|(\Omega) \leq \frac{1}{\lambda}|v|(\Omega) \to 0 \tag{17}$$

as $\lambda \to \infty$ uniformly in n. Also $|v|$ is α-continuous. Thus (17) and the right side of (16) imply that $\int_{[\|f_n\| > \lambda]} \|f_n\| \, d\alpha \to 0$ as $\lambda \to \infty$ uniformly in n. Since the measure α is finite, this implies $\{E^{\mathfrak{F}_n}(f), n \geq 1\}$ is uniformly integrable. Therefore for any $A \in \bigcup_{n\geq 1} \mathfrak{F}_n$ we have by the Vitali theorem (which holds here)

$$\int_A h \, d\alpha = \lim_{n\to\infty} \int_A f_n \, d\alpha = \lim_{n\to\infty} \int_A E^{\mathfrak{F}_n}(f) \, d\alpha = \int_A f \, d\alpha. \tag{18}$$

Hence $\int_A h \, d\alpha = \int_A f \, d\alpha$ also for $A \in \mathfrak{F}_\infty$ by the Hahn extension theorem so that $f = h$ a.e. It is now trivial that the uniform integrability implies $\int_\Omega \|f_n - f\| \, d\alpha \to 0$ as $n \to \infty$. This proves the proposition and completes the proof of Theorem 9.

The simple proof of the above proposition is based on the scalar version given by Billingsley [1, p. 116].

11. Remark We note that the reflexivity and separability of $\mathscr{X}$ in Theorem 9 were not used fully. The fact that $\mathscr{X}$ has the Radon–Nikodým property is clearly crucial in deducing (14). All reflexive spaces and all dual (or adjoint) separable spaces are known to have this property. In Theorem III.5.1 this was used in showing that $v = \overleftarrow{\lim}\, v_n$ is σ-additive in the norm topology of $\mathscr{X}$. But by the special nature of v_n, this can be shown to be true for any Banach space. Indeed, by hypothesis $v_n = v_{n+1} | \mathfrak{F}_n$, and $v: \bigcup_{n \geq 1} \mathfrak{F}_n \to \mathscr{X}$ is defined and additive as in Proposition III.1.2. But $|v_n|(A) = \int_A \|X_n\| \, d\alpha_n$ is a positive regular measure by the hypothesis on α_n, and $|v_{n+1}|(A) \geq |v_n|(A)$ so that $\mu(A) = \lim_{n \geq k, n \to \infty} |v_n|(A)$ gives μ to be an additive set function independent of k and is defined on $\bigcup_{n \geq 1} \mathfrak{F}_n$. Since $\mu_n = \mu | \mathfrak{F}_n \leq |v_n|$, it is σ-additive for each n and regular on $\mathfrak{F}_n$. Hence by Theorem III.1.5, $\mu = \overleftarrow{\lim}\, \mu_n$ exists (i.e., μ is σ-additive). But $\|v(A)\| \leq \mu(A)$ so that v is uniformly continuous on the algebra $\Sigma_0 = \bigcup_n \mathfrak{F}_n$ under the Fréchet metric $\rho(A, B) = \mu(A \Delta B)$, with Δ as the symmetric difference. This implies v is σ-additive on the algebra Σ_0 and then has a unique σ-additive extension to $\mathfrak{F}_\infty = \sigma(\Sigma_0)$ by Theorem I.2.2. Thus one has only to assume the Radon–Nikodým property for $\mathscr{X}$ for the validity of Theorem 9.

We now present a result on the decreasing indexed martingales, which have not been considered thus far. There are some differences between this and the increasing index case, particularly if the underlying measure space is σ-finite but nonfinite. These are required for the martingale formulation of ergodic theorems to be considered later.

Recall that if $\mathfrak{F}_n \supset \mathfrak{F}_{n+1}$ is a sequence in Σ, $\{X_n, \mathfrak{F}_n, n \geq 1\}$ is an adapted process, then it is a (decreasing) martingale if $E^{\mathfrak{F}_{n+1}}(X_n) = X_{n+1}$ a.e. Equivalently, if $\mathfrak{F}_{-n} \supset \mathfrak{F}_{-n+1}$ and $\{X_{-n}, \mathfrak{F}_{-n}, n \geq 1\}$ is adapted, then it is a martingale when $E^{\mathfrak{F}_{-n}}(X_{-(n+1)}) = X_{-n}$ a.e. We have the following result due to Doob [1] and Andersen–Jessen [1] independently.

12. Theorem *Let $\{X_n, \mathfrak{F}_n, n \leq 0\}$ be a martingale. If $\mathfrak{F}_{-\infty} = \bigcap_{n \leq 0} \mathfrak{F}_n$, then $X_n \to X_{-\infty}$ a.e. and in $L^1(P)$-norm, and $\{X_n, \mathfrak{F}_n, -\infty \leq n \leq 0\}$ is a martingale (hence uniformly integrable).*

Proof Let $X_* = \liminf_{n<0} X_n$, $X^* = \limsup_{n<0} X_n$, and for $a < b$ we have $[X_* < X^*] = \bigcup_{a,b,\text{rationals}} [X_* < a < b < X^*] = \bigcup C_{ab}$ (say). It is enough to prove that $P(C_{ab}) = 0$. The argument here follows that of Andersen–Jessen [2].

Let $H_a = [X_* < a]$, $H^b = [X^* > b]$. Then $\{H_a, H^b, C_{ab}\} \subset \mathfrak{F}_\infty$ and $C_{ab} = H_a \cap H^b$. Let $v_n: A \mapsto \int_A X_n \, dP$, $A \in \mathfrak{F}_n$. Then the martingale property implies $v_n | \mathfrak{F}_{n+1} = v_{n+1}$ and hence the function $v: \mathfrak{F}_\infty \to \mathbb{R}$ given by the equation $\nu(A) = \nu_n(A)$, $A \in \mathfrak{F}_\infty$ is well defined, does not depend on n, and is σ-additive. We prove the key inequalities

$$\nu(H_\alpha \cap A) \le \alpha P(H_\alpha \cap A), \qquad \nu(H^\beta \cap A) \ge \beta P(H^\beta \cap A), \qquad A \in \mathfrak{F}_\infty, \tag{19}$$

for any real α, β.

In fact, let $H_n = [\inf_{k \le -n} X_k < \alpha]$ so that $H_n \uparrow H_\alpha$ $(= \bigcup_{n=1}^{\infty} H_n)$. If now $H_{nn} = [X_n < \alpha]$, and $H_{kn} = [X_k < \alpha, X_j \ge \alpha, k+1 \le j \le n]$ for $k < n$, then $H_{kn} \in \mathfrak{F}_k$, $H_n = \bigcup_{k \le n} H_{kn}$ (disjoint union), and $H_{kn} \subset [X_k < \alpha]$. Hence for each $A \in \mathfrak{F}_\infty$ ($A \in \mathfrak{F}_k$, all $k < 0$), one has

$$\begin{aligned} \nu(H_n \cap A) &= \sum_{k \le n} \nu_k(H_{kn} \cap A) = \sum_{k \le n} \int_{H_{kn} \cap A} X_k \, dP \\ &\le \alpha \sum_{k \le n} P(H_{kn} \cap A) = \alpha P(H_n \cap A). \end{aligned} \tag{20}$$

This implies the first half of (19). Replacing X_n and α by $-X_n$ and $-\beta$ in this established result, the second half, hence (19), follows. (The latter can also be proved by a similar procedure.)

Let $a < b$ and $a = \alpha$, $b = \beta$ in (19). Take $A = C_{ab}$. Then the two inequalities of (19) together imply

$$aP(C_{ab}) \ge \nu(C_{ab}) \ge bP(C_{ab}). \tag{21}$$

Since $a < b$, this can hold only if $P(C_{ab}) = 0$, which implies $X_* = X^*$ as noted initially. Thus $X_n \to X_{-\infty}$ a.e. Since $|X_n| \le E^{\mathfrak{F}_n}(|X_0|)$, $n \le 0$, the sequence $\{X_n, n \le 0\}$ is uniformly integrable. Hence $X_n \to X_{-\infty}$ in $L^1(P)$ also, by Vitali's theorem and $X_{-\infty} = d\nu/dP$ a.e. It then is immediate that $\{X_n, \mathfrak{F}_n, -\infty \le n \le 0\}$ is a martingale. This completes the proof.

Remark The above method can be used to establish Theorem 4 in yet another way. There are still other proofs of this result. In terms of set martingales, Theorem 4 says that if $\{\nu_n, \mathfrak{F}_n, n \ge 1\}$ is a set martingale and P on $\mathfrak{F}_\infty$ $(= \sigma(\bigcup_n \mathfrak{F}_n))$ is a probability, and $\nu_n \ll P$, with $f_n = d\nu_n/dP$, then $f_n \to f_\infty$ a.e. Since $\nu\colon \mathfrak{F}_\infty \to \mathbb{R}$ defined by the ν_n (cf. Proposition III.1.2) is only additive, one cannot write $d\nu/dP$. However with the Yosida–Hewitt decomposition of $\nu = \nu^c + \nu^p$, one can show that $f_\infty = d\nu^c/dP$ a.e. Indeed, if $\nu_n = \nu_n^c + \nu_n^p$ are the restrictions of ν, ν^c, ν^p to $\mathfrak{F}_n$, and $g_n = d\nu_n^c/dP$, $h_n = d(\nu_n^p)^c/dP$, then $f_n = g_n + h_n$ a.e. $[P]$, and $\{g_n, \mathfrak{F}_n, n \ge 1\}$ is a uniformly integrable martingale since $\int_A g_n \, dP = \nu_n^c(A) = \nu^c(A) = \int_A g \, dP$ and g is $\mathfrak{F}_\infty$-measurable. So by Theorem 5, $g_n \to g$ a.e. Hence $h_n = f_n - g_n \to f_\infty - g_\infty = h_\infty$ a.e. But $\int_\Omega |h_n| \, dP \le |\nu_n^p|(\Omega) \le |\nu^p|(\Omega)$. So $\int_\Omega |h_\infty| \, dP \le |\nu|^p(\Omega)$ by

Fatou's lemma. Hence $\int_A |h_\infty|\,dP \le |\nu|^p(A)$, $A \in \bigcup_n \mathfrak{F}_n$. Since $|\nu^p|(\cdot)$ is purely finitely additive also, $h_\infty = 0$ a.e. and $f_\infty = g = d\nu^c/dP$ a.e. Applications of these results will be given for Gaussian processes in the following sections.

4.3 EXTENSIONS TO INFINITE MEASURES

The work on a unified account of ergodic and martingale theories to be treated in Chapter V and other applications point up the need to extend the results of the above sections to infinite measures. This, however, is not difficult, and an account will be useful. We include it here. For this it is necessary to state the conditions under which various Radon–Nikodým derivatives used above exist. This will also show the differences between the increasing and decreasing indexed martingales more clearly.

The increasing index case is substantially easier than the decreasing one, and so we dispose of the former. Recall that if (Ω, Σ, μ) is a measure space and $\mathfrak{F} \subset \Sigma$ is a σ-algebra, then $\mathfrak{F}$ *is μ-rich* if $\mu|\mathfrak{F}$ is σ-finite. [More generally, let ν be a semibounded measure on $\mathfrak{F}$ (i.e., $\nu: \mathfrak{F} \to [a, +\infty]$, $a > -\infty$ and σ-additive), which is μ_1 $(=\mu|\mathfrak{F})$-continuous. If $\nu(A) = \int_A f\,d\mu_1$ for a μ_1-unique $\mathfrak{F}$-measurable $f: \Omega \to \bar{\mathbb{R}}^+$, then one says that $\mathfrak{F}$ is *μ-rich*. So the Radon–Nikodým theorem is true for measures that are $\mu|\mathfrak{F}$-continuous.] Thus if $\mu(\Omega) < \infty$, every σ-subalgebra of Σ is μ-rich, and one sees the usefulness of this concept. It was introduced by Hunt [1]. Essentially $\mu|\mathfrak{F}$ is localizable.

With this terminology we have the following general result for the increasing indexed set martingales.

1. Theorem *Let (Ω, Σ, μ) be a complete measure space, $\mathfrak{F}_n \subset \mathfrak{F}_{n+1} \subset \Sigma$ be a sequence of σ-algebras such that $\mathfrak{F}_1$ is μ-rich. Let $\nu_n: \mathfrak{F}_n \to \bar{\mathbb{R}}$ be σ-additive and $\nu_n = \nu_{n+1}|\mathfrak{F}_n$, $n \ge 1$, $\nu_n \ll \mu$. (Hence ν_n must not take both $+\infty$ and $-\infty$.) If $\mu_n = \mu|\mathfrak{F}_n$, let $f_n = d\nu_n/d\mu_n$, which exists. If $\Sigma_0 = \{A \in \Sigma : \mu(A) < \infty\}$ and $\sup_n|\nu_n|(A) < \infty$ for each $A \in \Sigma_0 \cap \mathfrak{F}_1$, then $f_n \to f_\infty$ a.e. (μ), and f_∞ is determined in the following manner. If $\nu_A(B) = \lim_n \nu_n(B)$ for $B \in \mathfrak{F}_\infty(A)$, the trace of $\mathfrak{F}_\infty$ on A, which exists as noted in Proposition III.1.2 for each $A \in \Sigma_0$ as an additive bounded set function, then $f_{\infty,A} = d\nu_A^c/d\mu'$ a.e. $[\mu]$ on A where ν_A^c is the σ-additive part of ν_A in the Yosida–Hewitt decomposition (as remarked at the end of the last section) and $\mu' = \mu|\mathfrak{F}_\infty$. Moreover, $\{f_{\infty,A} : A \in \Sigma_0 \cap \mathfrak{F}_1\}$ determines an a.e. $[\mu]$ unique $\mathfrak{F}_\infty$-adapted function f_∞ which may not be integrable but which is finite a.e. $[\mu]$; or f_∞ is locally integrable in that we have $\int_A |f_\infty|\,d\mu \le \underline{\lim}_n \int_A |f_n|\,d\mu < \infty$, for each $A \in \Sigma_0 \cap \mathfrak{F}_1$. In particular, using the above $\{\mathfrak{F}_n\}_1^\infty$ sequence, if $\{X_n, \mathfrak{F}_n, n \ge 1\}$*

is a martingale with $\sup_n \int_A |X_n| \, d\mu < \infty$ *for each* $A \in \Sigma_0 \cap \mathfrak{F}_1$, *then* $X_n \to X_\infty$ *a.e.* (*finite*) *and* $\int_A |X_\infty| \, d\mu \le \underline{\lim}_n \int_A |X_n| \, d\mu < \infty$.

Proof We first observe that the richness of the σ-algebra $\mathfrak{F}_1$ implies the same for every $\mathfrak{F}_n \supset \mathfrak{F}_1$, $n \ge 1$. Hence f_n exists for each n. But each $\Sigma_0 \cap \mathfrak{F}_1 \subset \mathfrak{F}_n$, $n \ge 1$, so that $\{f_n \chi_A, \mathfrak{F}_n, n \ge 1\}$ is a sequence that satisfies the hypothesis of Theorem 2.4 because, by considering the traces $\mathfrak{F}_n(A) = \{A \cap B : B \in \mathfrak{F}_n\}$ which are σ-algebras, the problem reduces to the case of finite measures. Hence $f_n \chi_A \to f_{\infty,A}$ a.e. by that result, and the limit is identifiable as in the statement. [We thus have $\{f_{\infty,A} : A \in \Sigma_0 \cap \mathfrak{F}_1\}$ as a "cross section," or a "quasi-function."] However the richness of $\mathfrak{F}_1$ implies (in the σ-finite case, for example) that $\{A \in \mathfrak{F}_1 : \mu(A) < \infty\} \subset \Sigma_0$ and determines Ω, i.e., for σ-finite μ we can find a countable subset such that their union is Ω. In any case the uniqueness of the limit implies $f_{\infty,A} = f_{\infty,B} = f_{\infty,A \cap B}$ a.e. on $A \cap B$. Consequently, there is a unique $\mathfrak{F}_\infty$-measurable $f_\infty : \Omega \to \bar{\mathbb{R}}$ such that $f_{\infty,A} = f_\infty \chi_A$ a.e. by a standard result in measure theory (e.g., see Zaanen [1, Chapter 7, Theorem 5]). From this we may deduce that $f_{\infty,A}$ is finite a.e. and so f_∞ is finite a.e. The last inequality is immediate from Fatou's lemma.

In the martingale version, if $\nu_n(A) = \int_A X_n \, d\mu$, then $\nu_n : \mathfrak{F}_n \to \mathbb{R}$ satisfies the hypothesis of the above paragraph. Hence $X_n \to X_\infty$ a.e. and the other statements are likewise simple translations of the first part. This proves the result.

We leave the discussion of the submartingale case for the increasing index to the reader since the modifications needed are simple and the corresponding Doob decomposition is clearly valid. The decreasing index case presents some new features, and we consider that problem now in a more detailed manner than the above. Moreover, this version has immediate applications in the ergodic martingale applications to be treated later.

Suppose we have $\mathfrak{F}_n \supset \mathfrak{F}_{n+1}$, a sequence of σ-algebras in (Ω, Σ, μ) such that for *each* n, $\mathfrak{F}_n$ is μ-rich. Since $\mu | \mathfrak{F}_n$ is σ-finite does not imply the same of $\mu | \mathfrak{F}_{n+1}$, we have to make this stronger assumption to begin with. Note that this still does not give the μ-richness of $\mathfrak{F}_\infty = \bigcap_{n=1}^\infty \mathfrak{F}_n$, in contrast to the increasing index case. For instance, if $\Omega = \mathbb{R}$, $\mathfrak{F}_n = \sigma\{(0, n], (k, k+1], k > n\}$, and μ is the Lebesgue measure, then each $\mathfrak{F}_n$ is μ-rich but $\mathfrak{F}_\infty$ is not.

In view of the problems associated with $\mathfrak{F}_\infty$, one cannot meaningfully restrict μ if it is to be a nonfinite measure. To compensate for this, we assume that $\nu : \Sigma \to \bar{\mathbb{R}}$, the (dominated) signed measure, satisfies $|\nu_1|(\Omega) < \infty$ so that $|\nu_n|(\Omega) < \infty$ for all n, where $\nu_n = \nu | \mathfrak{F}_n$ as before. One can then obtain a result analogous to Theorem 1 in some respects.

Recall that the derivation of the crucial inequalities in Eq. (19) of Theorem 2.12 used the finiteness of measure P in the following way. The inequalities are

$$\nu(H_\alpha \cap A) \leq \alpha P(H_\alpha \cap A), \qquad \nu(H^\beta \cap A) \geq \beta P(H^\beta \cap A), \qquad A \in \mathfrak{F}_\infty, \quad (1)$$

where

$$H_\alpha = \left\{\omega : f_*(\omega) = \liminf_n f_n(\omega) < \alpha\right\},$$

$$H^\beta = \left\{\omega : f^*(\omega) = \limsup_n f_n(\omega) > \beta\right\},$$

and where $f_n = d\nu_n/dP_n$, with $P_n = P|\mathfrak{F}_n$. When P is replaced by the non-finite μ, then there may be no $A \in \mathfrak{F}_\infty$ $(= \bigcap_{n\geq 1} \mathfrak{F}_n)$ such that $0 < \mu(A) < \infty$; the inequalities (1) may be trivialities, i.e., $\nu(H_\alpha \cap A) = -\infty$ or 0, $\nu(H^\beta \cap A) = +\infty$ or 0. However, the proof of that step holds when $|\nu_1|(\Omega) < \infty$ (since $\lim_n \mu(H_n \cap A) = \mu(H_\alpha \cap A)$ for increasing sequences H_n which appear in the first inequality and the second one is then deduced from the first). Since $|\nu_n|(\Omega) \leq |\nu_1|(\Omega) < \infty$ for all $n \geq 1$, the work there holds if ν satisfies this hypothesis. Thus if $A = C_{ab} = \{\omega : f_*(\omega) < a < b < f^*(\omega)\}$ in (1), then $H_a \cap A = H^b \cap A = A$ and one has

$$a\mu(C_{ab}) \geq \nu(C_{ab}) \geq b\mu(C_{ab}), \qquad a < b. \quad (2)$$

Since $|\nu|(C_{ab}) < \infty$, $\mu(C_{ab}) = 0$ must be true again. However, $d\nu_\infty/d\mu_\infty$ need not exist since $\mathfrak{F}_\infty$ need not be μ-rich, and so no Radon–Nikodým derivative may exist. We can thus state the following slightly weaker form of Theorem 2.12:

2. Theorem *Let (Ω, Σ, μ) be a measure space, $\mathfrak{F}_n \downarrow \subset \Sigma$ be a sequence of μ-rich σ-subalgebras, and $\mathfrak{F}_\infty = \bigcap_{n\geq 1} \mathfrak{F}_n$. Let $\nu: \Sigma \to \overline{\mathbb{R}}$ be σ-additive and $|\nu_1|(\Omega) < \infty$ where $\nu_n = \nu|\mathfrak{F}_n$, $\nu_n \ll \mu_n$, $\mu_n = \mu|\mathfrak{F}_n$. If $f_n = d\nu_n/d\mu_n$, then $f_n \to f_\infty$ a.e. $[\mu]$ even though $d\nu_\infty/d\mu_\infty$ need not exist. If $\mathfrak{F}_\infty$ is μ-rich, then $f_\infty = d\nu_\infty/d\mu_\infty$ a.e. holds true. In particular, using only the point functions, let $\{f_n, \mathfrak{F}_n, n \geq 1\}$ be a martingale. Then for the μ-rich case, we have $f_\infty = \lim_n f_n = E^{\mathfrak{F}_n}(f_1)$ a.e., and then the martingale is uniformly integrable when $\mu(\Omega)$ is finite in addition.*

The fact that $\mathfrak{F}_\infty$ is not μ-rich implies a peculiarity for f_∞, and this can be analyzed as follows. Suppose there is no $A \in \mathfrak{F}_\infty$ such that $0 < \mu(A) < \infty$, i.e., $\mu(A) = 0$ or $\mu(A) = \infty$ are the only possibilities. Hence there is no subset of A in $\mathfrak{F}_\infty$ that has positive finite measure. Then under the hypothesis $|v_1|(\Omega) < \infty$, (1) becomes for $A \in \mathfrak{F}_\infty$ with $a = -\varepsilon$, $\varepsilon > 0$,

$$-\infty < v(H_{-\varepsilon} \cap A) \le -\varepsilon\mu(H_{-\varepsilon} \cap A) \le 0 \tag{3}$$

since $\mu \ge 0$. But $\mu(H_{-\varepsilon} \cap A) > 0$ is implossible unless $\mu(H_{-\varepsilon} \cap A) = +\infty$, which contradicts (3). Thus $\mu(H_{-\varepsilon} \cap A) = 0$. In particular, if $H_{-\varepsilon} = A$, then $\mu(H_{-\varepsilon}) = 0$ so that by the arbitrariness of $\varepsilon > 0$ we deduce that $f_* \ge 0$ a.e. By a similar argument (or by considering $-f_n$), one deduces that $f^* \le 0$ a.e. so that $f_\infty = 0$ a.e. One sees that this is essentially the most general case in the following sense. If we consider the measure algebra of (Ω, Σ, μ), then the richness of μ on Σ (σ-finiteness here) implies that every subfamily has a supremum, and it is the supremum of a countable subfamily. (This is a consequence of Theorem I.5.2, and Definition I.5.1.) Hence the class of sets of finite μ-measure in $\mathfrak{F}_\infty$ has a supremum that is σ-finite. Its complement then has no sets of finite positive measure. Thus the following result is true.

3. Corollary *If in the hypothesis of the above theorem $\mathfrak{F}_\infty$ has the additional property that the μ-measure of each of its members is either* 0 *or* $+\infty$, *then* $f_n \to f_\infty = 0$ *a.e.* $[\mu]$.

Proof We present an alternative proof of this corollary, which was proved in the above discussion. Recall that v is μ-continuous. It suffices to show that $f_* \ge 0 \ge f^*$ a.e. Let $H^\beta = \{\omega : f^*(\omega) > \beta\}$. Then $\mu(H^\beta) = 0$ or ∞ for all $\beta > 0$. We show that the second case is impossible. Since $-f_* = -(-f_n)^*$, the same result yields the first part of the inequality and hence the proof will be complete. The argument is due to Jerison [1].

Let $A_n = \{\omega : \max_{1 \le k \le n} f_k(\omega) \ge \beta\}$. Then $A_{n+1} \supset A_n$ and $\lim_n A_n \supset H^\beta$. Thus $\mu(A_n) \uparrow \infty$. So there is an n such that $\mu(A_n) > [|v_1^c|(\Omega)/\beta] = (1/\beta)\int_\Omega |f_1|\, d\mu = \alpha$ (say). But $A^n \in \mathfrak{F}_n$ (and μ_n is σ-finite). So there is a $B \in \mathfrak{F}_n$, $0 < \mu(B) < \infty$ and $\mu(A_n \cap B) > \alpha$. However, one has

$$\alpha < \mu(A_n \cap B) \le \mu(A_n) \le \frac{1}{\beta} \int_{A_n} |f_1|\, d\mu = \alpha. \tag{4}$$

It is a contradiction. Here in the last inequality we used the fact that $v_n \ll \mu_n$ (and $v \ll \mu$) for all n (i.e., f_n is a martingale sequence). Thus $\mu(A_n) = 0$ for all n so that $\mu(H^\beta) = 0$ for any $\beta > 0$. This is the desired result.

One may hope that by another method the hypothesis that $|v_1|(\Omega) < \infty$ may be omitted. However, in such a case the result need not be true as the following counterexample shows.

4. Counterexample Let $\Omega = [0, \infty)$, $\mu =$ Lebesgue measure, and Σ the Lebesgue σ-algebra of Ω. Let $\mathfrak{F}_n = \sigma\{[0, k), k > n\}$ be the σ-algebra shown. Consider the martingale $\{f_n, \mathfrak{F}_n, n \geq 1\}$ defined as follows:

$$f_1(x) = \begin{cases} -2, & 0 \leq x \leq 1 \\ 0, & 2^{2n} < x \leq 2^{2n+1}, \quad n = 0, 1, \ldots \\ -3, & 2^{2n+1} < x \leq 2^{2n+2}, \quad n = 0, 1, \ldots \end{cases}$$

and if $k \geq 2$,

$$f_k(x) = \begin{cases} -1 - \dfrac{[1 - (-1)^k]}{2}, & 0 < x \leq 2^{k-1}, \quad k \geq 2 \\ f_{k-1}(x), & 2^{k-1} < x < \infty, \quad k \geq 2. \end{cases}$$

Then, leaving the simple verification of the martingale property to the reader (it suffices to check this for the generators), we get $\lim_n \sup f_n = -1$, $\lim_n \inf f_n = -2$ a.e. Thus the martingale does not converge. However, $\int_\Omega |f_1| \, dP = +\infty$. This example is essentially due to Chow [1].

The above work shows that the behavior of the martingale $\{f_n, \mathfrak{F}_n, n \geq 1\}$ on (Ω, Σ, μ), ($\mathfrak{F}_n$ are μ-rich), depends on the finiteness of ν_1 on sets $A \in \mathfrak{F}_1$. If $\Sigma_0 = \{A \in \Sigma : \mu(A) < \infty\}$, and $|\nu_1|(A) < \infty$ for $A \in \Sigma_0 \cap \mathfrak{F}_\infty$, then $f_n \to f_\infty$ a.e. on each such A and may diverge on sets $B \notin \Sigma_0 \cap \mathfrak{F}_\infty$. This is the most general result that can be presented without further restrictions.

Now we deduce the results on the decreasing index submartingales from the above martingale theory since the case of increasing index presents no problems.

5. Theorem *Let (Ω, Σ, μ) be a measure space and $\mathfrak{F}_{n+1} \subset \mathfrak{F}_n \subset \Sigma$ be μ-rich σ-algebras. Suppose $\{f_n, \mathfrak{F}_n, n \geq 1\}$ is a submartingale such that $\sup_n \int_\Omega |f_n| \, d\mu < \infty$. Then $f_n \to f_\infty$ a.e. and $\int_\Omega |f_\infty| \, d\mu < \infty$. More generally, if $\int_\Omega f_1^+ \, d\mu < \infty$ and $A \in \mathfrak{F}_\infty = \bigcap_{n \geq 1} \mathfrak{F}_n$, $\mu(A) < \infty$, then $f_n \to f_\infty$ a.e. on A, and $-\infty \leq f_\infty \chi_A < \infty$ a.e.*

Proof By the submartingale hypothesis we have $E(f_n) = \int_\Omega f_n \, d\mu_n \geq \int_\Omega f_{n+1} \, d\mu_{n+1} = E(f_{n+1})$. Thus if $\sup_n E(|f_n|) < \infty$, then $\lim_n E(f_n)$ exists and is finite. Hence writing E for "expectation" even in the case of infinite measures, we have

$$0 \leq E\left[\sum_{n=1}^{\infty} (E^{\mathfrak{F}_{n+1}}(f_n) - f_{n+1})\right] = \lim_{n \to \infty} \sum_{k=1}^{n} [E(E^{\mathfrak{F}_{k+1}}(f_k)) - E(f_{k+1})]$$

$$= \lim_{n \to \infty} \sum_{k=1}^{n} [E(f_k) - E(f_{k+1})] = E(f_1) - \lim_{n \to \infty} E(f_n) < \infty. \tag{5}$$

Here we used the standard properties of conditional expectations since the $\mathfrak{F}_k$ are μ-rich and the results of Section II.1 are clearly valid in this case. Thus (5) implies

$$0 \leq \sum_{n=1}^{\infty} (E^{\mathfrak{F}_{n+1}}(f_n) - f_{n+1}) \qquad \text{converges a.e.} \tag{6}$$

Let us define the Doob decomposition of the submartingale as

$$f_n = \tilde{f}_n + \sum_{k=n}^{\infty} [E^{\mathfrak{F}_{k+1}}(f_k) - f_{k+1}]. \tag{7}$$

Then $\tilde{f}_n$ is $\mathfrak{F}_n$-adapted since $\mathfrak{F}_n \supset \mathfrak{F}_{n+1}$, and by (6) it is well defined.

To see that $\{\tilde{f}_n, \mathfrak{F}_n, n \geq 1\}$ is a decreasing martingale, consider

$$\begin{aligned} E^{\mathfrak{F}_{n+1}}(\tilde{f}_n) &= E^{\mathfrak{F}_{n+1}}(f_n) - \sum_{k=n}^{\infty} [E^{\mathfrak{F}_{k+1}}(f_k) - f_{k+1}] \\ &= f_{n+1} - \sum_{k=n+1}^{\infty} [E^{\mathfrak{F}_{k+1}}(f_k) - f_{k+1}] = \tilde{f}_{n+1} \quad \text{a.e.} \end{aligned} \tag{8}$$

Thus it is a martingale, and since $\sup_n E(f_n) \leq \sup_n E(|f_n|) < \infty$, one has

$$\begin{aligned} E(|\tilde{f}_n|) &\leq \sup_n E(|f_n|) + E\left(\sum_{k=1}^{\infty} [E^{\mathfrak{F}_{k+1}}(f_k) - f_{k+1}]\right) \\ &= \sup_n E(|f_n|) + E(f_1) - \lim_n E(f_n) < \infty \end{aligned} \tag{9}$$

by (5). Hence by Theorem 2 $\tilde{f}_n \to \tilde{f}_\infty$ a.e. Then by (7) and (6) $f_n \to f_\infty$ a.e. and $f_\infty = \tilde{f}_\infty$ a.e. It is then immediate by Fatou's lemma that $E(|f_\infty|) < \infty$. Note also that if $\mathfrak{F}_\infty$ is μ-rich, then $\{f_n, \mathfrak{F}_n, 1 \leq n \leq \infty\}$ is a submartingale which follows from the inequality

$$E^{\mathfrak{F}_\infty}(f_n) \geq E^{\mathfrak{F}_\infty}(\tilde{f}_n) = \tilde{f}_\infty = f_\infty \quad \text{a.e.}$$

To prove the last part, since

$$\infty > E(f_1^+) \geq E(f_1) \geq \cdots \geq \lim_n E(f_n) \geq -\infty, \tag{10}$$

we can still conclude that $f_n \to f_\infty$ a.e., but then $f_\infty = -\infty$ on a set of positive measure is possible. If $A \in \mathfrak{F}_\infty$, $\mu(A) < \infty$, then $\{f_n \chi_A, \mathfrak{F}_n(A), n \geq 1\}$ is a submartingale on the *finite* measure space $(A, \Sigma(A), \mu)$, using the notation of trace σ-algebras $\mathfrak{F}_n(A)$, $\Sigma(A)$. Hence by the first part $f_n \chi_A \to f \chi_A$ a.e. and by the maximal inequality (cf. Theorem 1.7)

$$\mu\left(\left\{\omega \in A : \sup_n f_n(\omega) \geq \lambda\right\}\right) \leq (1/\lambda) E(f_1^+ \chi_A) \to 0,$$

as $\lambda \to \infty$. Hence $\sup_n f_n \chi_A < \infty$ a.e., and $-\infty \leq f_\infty \chi_A < \infty$ a.e. The point here is that we can carry out the preceding computations on A, but the subtractions involved in (6)–(9) may be meaningless off A. This completes the proof.

The method of proof of the result is essentially that of Doob [1], and the infinite measure extension was given by Chow [1]. We remark that if the hypothesis $E(f_1^+) < \infty$ is omitted, then not only the proof but the result itself fails. An example can be given by modifying the preceding counterexample.

Our main interest in the infinite measure case is, as noted before, to consider the abstract ergodic and the (Radon–Nikodým) differentiation theory. However, before going to a detailed analysis of that aspect, it will be of interest to present some applications to the inference theory of stochastic processes. We give this in the next section to illustrate the key role of the martingale theory for some statistical problems of general interest.

4.4 APPLICATIONS TO LIKELIHOOD RATIOS

The concept of a likelihood ratio originates in the theory of statistical inference, and from an abstract viewpoint it is simply an aspect of the Radon–Nikodým density related to a pair of probability measures. The latter identification by itself is not of much use however. It is the construction of this density, by a limiting process from a suitable sequence of finite sets of random variables, which is useful. This is precisely where the martingale convergence and projective limit theories enter. We first motivate the concept, explain the relevance of likelihood ratios, and prove some results for its justification. As illustrations, we prove the dichotomy theorem (on equivalence or singularity) for Gaussian processes.

Suppose that (Ω, Σ) is a measurable space and $\{X_t, t \in T\}$ is a stochastic process, i.e., each $X_t: \Omega \to \mathbb{R}$ is measurable for Σ. From the physical situation we know that one of two probability measures P, Q on Σ governs the process. Thus either (Ω, Σ, P) or (Ω, Σ, Q) is the correct model describing the phenomenon. The inference problem consists of deciding or choosing one of the two models based on an "observation" (or evolution) of the process. By this we understand that a sample function ω (or $X_{(\cdot)}(\omega)$) is at our disposal and almost all such ωs obtained under natural experiments (i.e., observed "at random") should belong to sets A $(\in \Sigma)$ of high probability under P or Q whichever is the correct underlying probability governing the model. Thus when $P \neq Q$, there exist events A such that $P(A)$ is small and $Q(A)$ is large $(> P(A))$, which thus distinguish P and Q. If we have a collection $\{A_i, i \in I\}$ of such events distinguishing P and Q, then we should like to formulate a procedure or rule to designate a "good" A_0 from this collection that separates P and Q (i.e., $P(A_0)$ is very small and $Q(A_0)$ is very large) and then $\omega \in A_0$ implies that Q is the correct probability generating the process and P when $\omega \notin A_0$. We now turn this intuitively reasonable rule into a mathematically meaningful problem to solve.

In general it will not be possible to choose $P(A_0) = 0$ and $Q(A_0) = 1$ unless P and Q are mutually singular. For instance, if P and Q are two Gaussian measures having Ω for their support, then such an easy solution is not possible since they can be equivalent without being equal. Thus one wants to consider sets $\{A_i, i \in I\} \subset \Sigma$ such that $P(A_i) = \alpha$ for a prescribed $0 < \alpha < 1$ and then maximize $Q(A_i)$, $i \in I$. If there exists an $A_0 \in \{A_i, i \in I\} \subset \Sigma$ such that $P(A_0) = \alpha$ and $Q(A_0) \geq Q(A_i)$ for all $i \in I$, then we take A_0 as our "good critical region." This is a variational problem, and as one might suspect, it involves the Radon–Nikodým density dQ^c/dP. This function is the *likelihood ratio* referred to earlier. The number $\alpha > 0$, called the "size" of A_0, represents the probability of the "error of the first kind", namely, rejecting P when $\omega \in A_0$ if, in fact, P is the correct measure and $\beta = Q(A_0)$ indicates the "power", or the "goodness", of the rule. The number $1 - \beta$ represents the probability of the "error of the second kind", namely, rejecting Q or accepting P when $\omega \in \Omega - A_0$ if, in fact, Q is the correct measure. The point is that when one *repeats* the experiment and obtains ωs, one makes mistakes at most α and $(1 - \beta)$ proportions only. So we want to control the first and minimize the second error. In case $P \in \{P_i : i \in H\}$, $Q \in \{Q_j : j \in K\}$, where H and K are two disjoint index sets, then the above problem has a natural generalization, but now it is necessary to impose other conditions in order that the basic (simple) variational problem have a solution. It is even necessary to prove that the two families are "distinguishable." These questions belong to the inference theory of stochastic processes, and we shall not enter into too many details here. We only consider the above simple case and show that it admits an easy solution and then illustrate the nontriviality of the problem in the context of stochastic processes and calculate some associated likelihood ratios.

Let $f = dQ^c/dP$, the Radon–Nikodým derivative of the absolutely continuous part of Q for P, and suppose $E_0 \in \Sigma$ is a P-null set on which the P-singular part Q^s lives. Consider for $\alpha > 0$,

$$A_{0,k} = \{\omega : f(\omega) \geq k\} \cup E_0, \tag{1}$$

where $k \geq 0$ is chosen so that $P(A_{0,k}) = P(\{\omega : f(\omega) \geq k\}) \leq \alpha$. Since $P(A_{0,k}) \to 0$ as $k \to \infty$, this is possible. We now prove that the set $A_{0,k}$ is the desired critical region.

Let us note that $A_{0,k}$ of (1) can be given a different form. Thus if $\mu = P + Q$, and $Q = Q^c + Q^s$, the Lebesgue decomposition relative to P, then all these measures are defined on Σ and are μ-continuous. Let $f_0 = dP/d\mu$, $g_1 = dQ^c/d\mu$, $g_2 = dQ^s/d\mu$, and $g = g_1 + g_2 = dQ/d\mu$ a.e. $[\mu]$. Clearly, $\text{supp}(g_1) \subset \text{supp}(f_0)$, while $\text{supp}(g_2) \cap \text{supp}(f_0)$ is P-null. Hence

$\{\omega : (g_2/f_0)(\omega) > 0\} = \{\omega : (g_2/f_0)(\omega) = \infty\} = E_0$, which is a P-null set. On the other hand,

$$f = \frac{dQ^c}{dP} = \frac{dQ^c}{d\mu}\bigg/\frac{dP}{d\mu} = \frac{g_1}{f_0} \quad \text{a.e. } [\mu].$$

Hence for any $k > 0$

$$\begin{aligned}\{\omega : g(\omega) \geq kf_0(\omega)\} &= \left\{\omega : \frac{g_1 + g_2}{f_0}(\omega) \geq k\right\} \\ &= \left\{\omega : f(\omega) = \frac{g_1}{f_0}(\omega) \geq k\right\} \cup E_0 = A_{0,k}, \end{aligned} \tag{2}$$

and $A_{0,k}$ is defined by the *ratio* of the densities: g/f_0. Thus $A_{0,k}$ does not depend on the auxiliary measure μ, as seen from (1) and (2).

The maximal property of $A_{0,k}$ follows from the next result.

1. Lemma *The region $A_{0,k}$ of* (1) *is the best critical or distinguishing set in the sense that for any other $A \in \Sigma$ such that $P(A) \leq P(A_{0,k}) \leq \alpha$ we have $Q(A_{0,k}) \geq Q(A)$.*

Proof Consider A_k $(=A_{0,k})$ and A. Since $A_k - A_k \cap A$ is disjoint from $A - A_k \cap A$ $(\subset A_k^c)$, we have, using (2) and μ $(=P + Q)$,

$$\begin{aligned}Q(A_k) &= Q(A_k - A_k \cap A) + Q(A \cap A_k) \\ &= \int_{A_k - A \cap A_k} g\, d\mu + Q(A \cap A_k) \\ &\geq k \int_{A_k - A \cap A_k} f_0\, d\mu + Q(A \cap A_k) = kP(A_k - A \cap A_k) + Q(A \cap A_k) \\ &= k[P(A_k) - P(A \cap A_k)] + Q(A \cap A_k) \\ &\geq kP(A - A \cap A_k) + Q(A \cap A_k) \qquad \text{since} \quad P(A) \leq P(A_{0,k}) = P(A_k) \\ &= k \int_{A - A \cap A_k} f_0\, d\mu + Q(A \cap A_k) \geq \int_{A - A \cap A_k} g\, d\mu + Q(A \cap A_k) \quad \text{by} \quad (2) \\ &= Q(A - A \cap A_k) + Q(A \cap A_k) = Q(A). \end{aligned} \tag{3}$$

This completes the proof.

This result, due to Grenander [1], is an abstract rendering of a finite dimensional case which together with the principle of likelihood ratio has its origins in the ancient works of J. Neyman and E. Pearson. Looking at it as the solution of a variational problem, it has been generalized in various directions by Mann [1], and to certain vector measures by Wald and others. We shall present a sample extension in the Complements and Problems Section. The following technical remark, noted before, is detailed since it is used in proofs below.

Remark If (Ω, Σ) is a measurable space $\Omega = \mathbb{R}^T$ given by the Kolmogorov representation (cf. Theorem I.3.1), then $A \in \Sigma$ implies the existence of a countable set $J \subset T$ such that for every pair $\omega \in \Omega$, $\omega' \in A$ with $\omega(t) = \omega'(t)$, $t \in J$ we have $\omega \in A$, i.e., A is determined by a *countable* collection of indices (the collection depends on the set A). This was noted in the discussion for Theorem I.3.5. We include a short proof here. Thus let $\mathscr{B}$ be the family of subsets of Ω $(=\mathbb{R}^T)$ having the above property, i.e., a set B of Σ satisfies $B \in \mathscr{B}$ iff there is a countable $J_B \subset T$ such that $\omega \in \Omega$, $\omega' \in B$ with $\omega = \omega'$ on J_B implies $\omega \in B$. Then $\mathscr{B}$ is a σ-algebra. In fact, $\Omega \in \mathscr{B}$; $A_i \in \mathscr{B}$ implies $\bigcup_{i=1}^{\infty} A_i \in \mathscr{B}$ (because $\bigcup_{i \geq 1} J_{A_i} \subset T$ is countable). If $A \in \mathscr{B}$ and J_A is the corresponding countable set, then evidently A^c also has the same J_A and $\omega \in \Omega$, $\omega' \in A^c$, $\omega = \omega'$ on J_A implies (since $\omega' \notin A$) $\omega \notin A$, or $\omega \in A^c$. So $A^c \in \mathscr{B}$. If $I_n = \{t_1, \ldots, t_n\} \subset I$, $\mathscr{A} = \mathscr{A}_i$ is the Borel σ-algebra of $\mathbb{R}$, $A(I_n) \in \bigotimes_{i \in I_n} \mathscr{A}_i = \mathscr{A}_{I_n}$, and (the coordinate projection) $\pi_{I_n}: \Omega \to \mathbb{R}^{I_n}$, then $\pi_{I_n}^{-1}(A(I_n)) \in \mathscr{B}$ by definition. Hence $\pi_{I_n}^{-1}(\mathscr{A}_{I_n}) \subset \mathscr{B}$, and thus $\Sigma = \sigma(\bigcup_{I_n \subset I} \pi_{I_n}^{-1}(\mathscr{A}_n)) \subset \mathscr{B} \subset \Sigma$.

The above observation implies that we may be able to approximate the critical region A_k by families of the finite dimensional cylinder sets or by simple combinations of such sets. More explicitly: let $X_1, X_2, \ldots$ be a sequence of real random variables (coordinate functions in terms of Theorem I.3.1) on (Ω, Σ). Let $\mathfrak{F}_n = \sigma(X_1, \ldots, X_n) \uparrow$ be the σ-algebra generated by the X_is shown and $\mathfrak{F}_\infty = \sigma(\bigcup_{n \geq 1} \mathfrak{F}_n) \subset \Sigma$. If P, Q are measures on $\mathfrak{F}_\infty$ governing the process and $\mu = P + Q$, let P_n, Q_n, μ_n be the restricitions to $\mathfrak{F}_n$ and $f_n = dP_n/d\mu_n$, $g_n = dQ_n/d\mu_n$ a.e., the Radon–Nikodým densities relative to μ_n. Then $f_n = \Phi_n(X_1, \ldots, X_n)$, $g_n = \Psi_n(X_1, \ldots, X_n)$ for some positive Borel functions Φ_n, Ψ_n (by Theorem I.2.3) on $\mathbb{R}^n$. Since $\int_\Omega f_n \, d\mu_n = 1 = \int_\Omega g_n \, d\mu_n$, it follows by Theorem 2.4 that $f_n \to f_\infty$ a.e. $[\mu]$ and $g_n \to g_\infty$ a.e. Moreover, $g_n/f_n \to g_\infty/f_\infty$ a.e. $[P]$ and $f_n/g_n \to f_\infty/g_\infty$ a.e. $[Q]$. But by the same result (or translating to set martingale versions) we have $f = dQ^c/dP = g_\infty/f_\infty$ a.e. $[P]$ and $f_\infty/g_\infty = dP^c/dQ$ a.e. $[Q]$. Hence if one has the function space representation of the process so that $X_n(\omega) = \omega(n)$, then $g_n(\omega) = g_n(X_1(\omega), \ldots, X_n(\omega))$, etc. If one lets

$$\tilde{A}_{k,n} = \{g_n/f_n \geq k\}, \tag{4}$$

then $P(\tilde{A}_{k,n}) \to P(A_k)$ and similarly $Q(\tilde{A}_{k,n}) \to Q(A_k)$, as $n \to \infty$. Now we can express g_n, f_n and hence $\tilde{A}_{k,n}$ in terms of a set in $\mathbb{R}^n$ ($\tilde{A}_{k,n} = \pi_n^{-1}(B_{k,n})$, where $B_{k,n}$ is the base of $\tilde{A}_{k,n}$ in $\mathbb{R}^n$). Thus we conclude that with the discussion of the preceding paragraph, the likelihood ratios (and the associated critical regions) can be approximated by the finite dimensional densities. Since the sets involved are determined by countable collections of t points, the martingale theory of the preceding sections for real stochastic processes applies.

However, the approximation noted above needs a further and more precise description. For instance, the random variables involved may not be present in a progressively increasing sequence. In fact, if $\{X_t, a \leq t \leq b\}$ is a process, we may start with $a = t_0 < t_1 < \cdots < t_n = b$ and as $n \to \infty$, the subdivision of $[a, b]$ is refined. Then one only has a directed index set of the process derived from the X_t for the Radon–Nikodým derivatives. This is already clear in our discussion of A_k determining $[f \geq k]$. Hence it is necessary to consider a more general approximation. This will be done below. In case the process is indexed by integers, as above, then the simpler procedure suffices. As a motivation for the general case, we present an example to indicate that the likelihood ratio principle is "reasonable." This application is due to Doob [1].

2. Example Let $\{X_n, n \geq 1\}$ be a sequence of independent random variables each with the same distribution P or Q. This means $P[X_k < x] = F(x)$ and $Q[X_k < x] = G(x)$ for all $x \in \mathbb{R}$ and $k \geq 1$, and the X_k are mutually independent for either measure. Under these conditions we assert that either $P = Q$ or P and Q are singular, i.e., if $f = dQ^c/dP$, then $f = 0$ or 1 a.e. $[P]$ and thus $P(A_k) = 0$ and $Q(A_k) = 1$ in terms of the above discussion. (This will be a consequence of Theorem 9 below, but the present treatment has a motivational interest. Here we prove only that $f > 0$ a.e. $[P]$ implies $f = 1$ a.e. $[P]$.)

Let $\mu = P + Q$ and, using the notation introduced above, $g_n = dP_n/d\mu_n$, $f_n = dQ_n/d\mu_n$ a.e. We now consider the function space representation so that each $X_n(\omega)$ is the nth coordinate of $\omega \in \Omega = \mathbb{R}^{\mathbb{N}}$. Hence $g_n(\omega), f_n(\omega)$ can be identified with $g_n(x_1, \ldots, x_n), f_n(x_1, \ldots, x_n)$, where $X_n(\omega) = x_n$. (These are called Ψ_n and Φ_n above.) But the random variables are independent. Thus P_n, Q_n are product measures, and so $P_n = P|\pi_n^{-1}(\mathscr{B}_n)$, etc. The identically distributedness implies if f_0 and g_0 are the densities relative to μ of X_1 under P and Q, respectively, then $f_n(x_1, \ldots, x_n) = \prod_{i=1}^n f_0(x_i)$, $g_n(x_1, \ldots, x_n) = \prod_{i=1}^n g_0(x_i)$. By Theorem 2.6 (or the supermartingale convergence theorem) we conclude that $g_n \to g_\infty$ a.e., $f_n \to f_\infty$ a.e. $[\mu]$, and $f = g_\infty/f_\infty$ a.e. $[P_\infty]$, where f_n, g_n are *identified* with the densities on $(\Omega, \mathfrak{F}_n)$. We shall show that either $P = Q$ or P, Q are mutually singular. Thus either $f = 0$ or 1 a.e. $[P]$.

This calculation of limit depends on some careful computation since the martingale theory only tells the existence of the limit but not its functional form. Also the "identification" needs an explanation, and it is discussed later. (See the paragraph below with computations (18), (19), and the one following it.)

Clearly the calculation is needed only on sets of positive P-measure. So we may and do assume that the Borel function $f_1 > 0$. Then $h = g_1/f_1 \geq 0$ and is well defined. Let $Y_n = \log h(X_n)$. Since $0 < f_1 < \infty$ a.e. $[P]$, we may assume that $-\infty < Y_n < \infty$. The identity of distributions of X_n implies the same property of Y_n, and they are also independent. Then $Z = \log f = \sum_{n=1}^{\infty} Y_n$ is a random variable by the preceding paragraph (and the martingale theory). Let φ and ξ be the characteristic functions of Z and Y_1 (cf. Problem I.6.1 and the discussion on such functions φ, ξ). Then for any $t \in \mathbb{R}$ we have by definition and independence,

$$\xi(t) = E(e^{itZ}) = \lim_{n\to\infty} E\left(\exp\left(it \sum_{k=1}^{n} Y_k\right)\right) \qquad \text{by bounded convergence}$$

$$= \lim_{n\to\infty} \prod_{k=1}^{n} \varphi(t) = \lim_{n\to\infty} [\varphi(t)]^n \qquad \text{by identically distributedness.} \tag{5}$$

But the first and last terms of (5) show that $\xi^2(t) = \xi(t)$ is also true for any $t \in \mathbb{R}$. Or since $|\varphi(t)| \leq 1, \xi(t) = 1$ on $\{t: |\varphi(t)| = 1\}$, $=0$ on $\{t: |\varphi(t)| < 1\}$, i.e., ξ is two valued. Since ξ is continuous and $\xi(0) = 1$, we must have $\xi(t) = 1$ for all $t \in \mathbb{R}$. By the uniqueness theorem, $Z = 0$ a.e. $[P]$. Since $\xi(t) = 1$, it follows that $|\varphi(t)| = 1$ in (5). Thus $\varphi(t) = e^{i\theta(t)}$ and (5) becomes

$$1 = \lim_{n\to\infty} e^{in\theta(t)}, \qquad t \in \mathbb{R}. \tag{6}$$

But this is possible iff $\theta(t) = 0$ and hence $Y_k = 0$ a.e. $[P]$. Thus $f = 1$ a.e. $[P]$ and $g_1 = f_1$ a.e. $[\mu]$ or $P = Q$. (In the contrary case $f = 0$ a.e. $[P]$ follows from Theorem 9 below; so P, Q are singular.) This means, unless the distributions are identical, the likelihood ratio sequence $g_n(\omega)/f_n(\omega) \to 0$ a.e. $[P]$, as $n \to \infty$. Hence by Lemma 1, if we consider $A_k = [f(= g_\infty/f_\infty) \geq k]$ or, in the approximation procedure, if the observation $(X_1(\omega), \ldots, X_n(\omega))$ falls in $[g_n/f_n \geq k]$ for large enough n, we decide that P is the correct probability; in the opposite case we decide on Q.

An important point of the above case is to decide (and exclude) the singularity of a pair of probability measures P, Q on (Ω, Σ). Let us present a general criterion for this, interpreting the sequences $\left\{\left(\Omega, \mathfrak{F}_n, \begin{matrix} P_n \\ Q_n \end{matrix}\right), n \geq 1\right\}$ as two projective systems, by introducing a "distance" between (P_n, Q_n) and relating it to (P, Q). Since each of these is a set martingale, the work of Section III.2 finds a nice application here. We then use this to get the dichotomy theorem for Gaussian processes.

The distance between two measures referred to is the classical Hellinger distance, which was introduced into the stochastic theory by Kakutani [1] and which has been used in various modified forms ever since. Thus let P, Q be a pair of probability measures on (Ω, Σ) and $\mu = P + Q$. If $f = dP/d\mu$, $g = dQ/d\mu$ a.e., then one defines the "generalized" Hellinger integral as

$$H_\alpha(Q, P) = \int_\Omega f^\alpha g^{1-\alpha}\, d\mu, \qquad 0 \le \alpha \le 1. \tag{7}$$

If $\alpha = \frac{1}{2}$, this is the classical Hellinger integral—the most important case. It should be noted that $H_\alpha(\cdot,\cdot)$ depends only on P, Q but not on the dominating measure μ since if v is another (σ-finite) measure on Σ dominating both P, Q, then $f_0 = d\mu/dv$, $\tilde{f} = dP/dv$, $\tilde{g} = dQ/dv$ a.e. $[v]$ implies $H_\alpha(Q, P) = \int_\Omega \tilde{f}^\alpha \tilde{g}^{1-\alpha}\, dv$. Also by the Hölder inequality we note that $0 \le H_\alpha(Q, P) \le 1$ and $H_{1/2}(P, Q) = H_{1/2}(Q, P)$. Moreover, $H_\alpha(Q, P) = 0$ iff $f \cdot g = 0$ a.e. $[\mu]$, i.e., P, Q are mutually singular ($P \perp Q$) and $H_\alpha(Q, P) = 1$ iff there is equality in Hölder's inequality in (7), which implies that $(f^\alpha)^{(1-\alpha)/\alpha} = (g^{1-\alpha})$ a.e. $[\mu]$ or $f = g$ a.e., so that $P = Q$. We note that if $\alpha = \frac{1}{2}$, then $H_{1/2}(Q, P)$ is an inner product in $(L^2(\mu))^+$. It is seen that

$$H_{1/2}(Q, P) = (\sqrt{f}, \sqrt{g}), (\sqrt{f}, \sqrt{f})^{1/2} = \|\sqrt{f}\|_2 = 1 = \|\sqrt{g}\|_2, \tag{8}$$

$$\|\sqrt{f} - \sqrt{g}\|_2 = [(\sqrt{f} - \sqrt{g}, \sqrt{f} - \sqrt{g})]^{1/2} = [2(1 - H_{1/2}(Q, P))]^{1/2}. \tag{8'}$$

If $\mathfrak{F}_i \subset \Sigma$, $i \in I$ is an increasing generalized sequence (or net) of σ-algebras such that $\Sigma = \sigma(\bigcup_{i \in I} \mathfrak{F}_i)$, let $Q_i = Q|\mathfrak{F}_i$, $P_i = P|\mathfrak{F}_i$, $\mu_i = \mu|\mathfrak{F}_i$, and $f_i = dP_i/d\mu_i$, $g_i = dQ_i/d\mu_i$ a.e. Then Theorem 2.5 implies, in the case $I = \mathbb{N}$, that $f_n \to f$, $g_n \to g$ a.e., and these are uniformly integrable so that $H_\alpha(Q_n, P_n) \to H_\alpha(Q, P)$. In the following applications "I" will only be a directed set so that the above statement cannot be deduced (in fact, for directed index I the pointwise convergence is usually false). However, $\{f_i, \mathfrak{F}_i, i \in I\}$ is a martingale. It follows from a general theorem below on mean convergence for martingales that $f_i \to f$ in $L^1(\mu)$ and similarly for the g_i sequence. This will be sufficient for the above conclusion. But, as in Example 2, we should like to consider the g_i, f_i on finite dimensional spaces $(\mathbb{R}^i, \mathscr{B}_i, P_i)$ of the Kolmogorov representation. This means one has to first "identify" $(\mathbb{R}^i, \mathscr{B}_i)$ as $(\pi_i^{-1}(\mathbb{R}^i), \pi_i^{-1}(\mathscr{B}_i), P \circ \pi_i^{-1})$ before the martingale theory can be invoked. This identification was briefly noted in that example. One must recognize this conceptual distinction. So we prove the above result for projective systems and then emphasize the identification, which is usually slurred over in the literature.

The following formula for $H_\alpha(Q, P)$ is needed in computations:

3. Lemma *If Q, P are two probability measures on (Ω, Σ), then*

$$H_\alpha(Q, P) = \inf\left\{\sum_k (P(A_k))^\alpha (Q(A_k))^{1-\alpha} : \{A_k\}_{-\infty}^{\infty} \subset \Sigma, \quad \text{partition of} \quad \Omega\right\}. \tag{9}$$

Proof This is a slight modification of the classical upper Darboux sum evaluation of the Lebesgue integral (7). Thus if $\{A_k\}_{-\infty}^{\infty}$ is any partition of Ω, then using the Hölder inequality we have

$$\begin{aligned} H_\alpha(Q, P) &= \int_{\bigcup_k A_k} f^\alpha g^{1-\alpha}\, d\mu \\ &= \sum_k \int_{A_k} f^\alpha g^{1-\alpha}\, d\mu \leq \sum_k \left(\int_{A_k} f\, d\mu\right)^\alpha \left(\int_{A_k} g\, d\mu\right)^{1-\alpha} \\ &= \sum_k (P(A_k))^\alpha (Q(A_k))^{1-\alpha}. \end{aligned} \tag{10}$$

Hence the infimum on the right-hand side is not less than $H_\alpha(Q, P)$.

For the opposite inequality, let $1 < t < \infty$ and consider

$$A_{nm} = \{\omega : t^{n-1} \leq f^\alpha(\omega) < t^n,\ t^{m-1} \leq g^{1-\alpha}(\omega) < t^m\}, \qquad \text{all integers} \quad \text{n, m}.$$

Consequently,

$$\int_{A_{nm}} f^\alpha g^{1-\alpha}\, d\mu \geq t^{n+m-2}\mu(A_{nm}). \tag{10$'$}$$

Since $P(A_{nm}) \leq t^{n/\alpha}\mu(A_{nm})$, $Q(A_{nm}) \leq t^{m/(1-\alpha)}\mu(A_{nm})$, we have from the disjointness of $\{A_{nm}\}_{m,n}$:

$$\begin{aligned} H_\alpha(Q, P) &= \sum_{n,m} \int_{A_{nm}} f^\alpha g^{1-\alpha}\, d\mu + 0 \cdot \mu((\bigcup A_{nm})^c) \\ &\geq \frac{1}{t^2} \sum_{n,m} (P(A_{nm}))^\alpha (Q(A_{nm}))^{1-\alpha} \qquad \text{by} \quad (10'). \end{aligned} \tag{11}$$

Taking infima on both sides and then letting $t \to 1$, we get $H_\alpha(Q, P)$ to be at least as large as the right side of (9). But $\{A_{nm}\}_{m,n}$ is a partition of Ω if we add the sets $A_1 = [f = 0, g > 0]$, $A_2 = [f > 0, g = 0]$, and $A_3 = [f = g = 0]$. Clearly $P(A_1) = 0 = Q(A_2) = P(A_3) = Q(A_3)$ so that this addition to the last term of (11) leaves it unchanged. This together with (10) implies (9).

As a consequence, we have the following monotonicity property of H_α:

4. Corollary *If* $\left(\Omega_i, \Sigma_i, \begin{matrix} P_i \\ Q_i \end{matrix}\right)$, $i = 1, 2$, *are probability spaces and* $T\colon \Omega_1 \to \Omega_2$ *is a measurable mapping such that* $P_1 = P_2 \circ T^{-1}$, $Q_1 = Q_2 \circ T^{-1}$, *then* $H_\alpha(Q_1, P_1) \geq H_\alpha(Q_2, P_2)$. *In particular, if* $\Omega_1 = \Omega_2$, $\Sigma_2 \subset \Sigma_1$, $Q_2 = Q_1 | \Sigma_2$, $P_2 = P_1 | \Sigma_2$, *then* $H_\alpha(Q_1, P_1) \leq H_\alpha(Q_2, P_2)$.

It is sufficient to note that $T^{-1}(\Sigma_2) \subset \Sigma_1$ and $P_2 \circ T^{-1} = P_1 | (T^{-1}(\Sigma_2))$, $Q_2 \circ T^{-1} = Q_1 | (T^{-1}(\Sigma_2))$, and that there are fewer partitions in a subalgebra than in a bigger one.

We can now prove the analog of Theorem 2.4 for projective systems, which is a generalization of a result of Kakutani [1], who proved it for the Fubini–Jessen system with $I = \mathbb{N}$, but who obtained more precise information in that case (see Theorem 9 below).

5. Theorem *Let* $\{(\Omega_\alpha, \Sigma_\alpha, P_\alpha, g_{\alpha\beta})_{\alpha<\beta} : \alpha, \beta \text{ in } D\}$ *be a projective system of probability spaces admitting a projective limit* (Ω, Σ, P). *Suppose that* $Q_\alpha\colon \Sigma_\alpha \to [0, 1]$ *is a probability,* $\alpha \in D$, *and let* $\{(Q_\alpha, g_{\alpha\beta})_{\alpha<\beta} : \alpha, \beta \text{ in } D\}$ *be another projective system which admits the projective limit* (Ω, Σ, Q). *Then* $H_\beta(Q, P) = \lim_\alpha H_\beta(Q_\alpha, P_\alpha)$ *for each* $0 < \beta < 1$. *Hence* $P \perp Q$ *iff* $\lim_\alpha H_\beta(Q_\alpha, P_\alpha) = 0$, *and in particular* $P \perp Q$ *if for some* $\alpha \in D$ *we have* $P_\alpha \perp Q_\alpha$. [Note that unlike in (7)–(11), α is an index here and β is a number.]

Proof For any $i < j$ by the hypothesis of a projective system (cf. Definition III.1.1) $P_i = P_j \circ g_{ij}^{-1}$, $Q_i = Q_j \circ g_{ij}^{-1}$, where $g_{ij}\colon \Omega_j \to \Omega_i$, and hence by Corollary 4, $H_\beta(Q_i, P_i) \geq H_\beta(Q_j, P_j)$ so that $H_\beta(Q_i, P_i)$ is monotone decreasing and hence has a limit $\geq H_\beta(Q, P)$. We need to show that there is equality when $\Sigma = \sigma(\bigcup_{i \in D} g_i^{-1}(\Sigma_i))$, where $g_i\colon \Omega \to \Omega_i$ and $g_i = g_{ij} \circ g_j$ for $i \geq j$. This follows from a standard (but nontrivial) computation that we now present.

Let $\varepsilon > 0$ be given and choose a countable partition $\{A_k\}_{-\infty}^{\infty}$ of Ω, in accordance with Lemma 3, from Σ such that

$$\sum_{k=-\infty}^{\infty} (P(A_k))^\beta (Q(A_k))^{1-\beta} \leq H_\beta(Q, P) + \tfrac{1}{4}\varepsilon. \tag{12}$$

Since $\bigcup_k A_k = \Omega$ is a disjoint union and P, Q are probability measures, there exists n_0 $(=n_0(\varepsilon))$ such that

$$\sum_{|k|>n_0} P(A_k) < \tfrac{1}{4}\varepsilon, \qquad \sum_{|k|>n_0} Q(A_k) < \tfrac{1}{4}\varepsilon. \tag{13}$$

Let us estimate the measures of the *finite* collection $\{A_k, |k| \le n_0\}$. By continuity and finiteness, one can find for each $-n_0 \le k \le n_0$, η_k such that $0 < \eta_k < \varepsilon/3 \cdot 2^{|k|+2}$ and

$$(P(A_k) + \eta_k)^\beta (Q(A_k) + \eta_k)^{1-\beta} \le (P(A_k))^\beta (Q(A_k))^{1-\beta} + \varepsilon/3 \cdot 2^{|k|+2}. \tag{14}$$

But $\Sigma = \sigma(\bigcup_{\alpha \in D} g_\alpha^{-1}(\Sigma_\alpha))$ and $\{A_k\}_{-\infty}^{\infty} \subset \Sigma$. Hence for each $|k| \le n_0$ there exists a cylinder set $B_k \in g_{\alpha_k}^{-1}(\Sigma_{\alpha_k})$ such that

$$P(A_k \triangle B_k) < \eta_k, \qquad Q(A_k \triangle B_k) < \eta_k. \tag{15}$$

As usual $A \triangle B$ is the symmetric difference of A, B. Since we have only a finite collection $\{\alpha_k, |k| \le n_0\}$ by directedness of D, there is a $\gamma \in D$ such that $\alpha_k < \gamma$ and $g_{\alpha_k}^{-1}(\Sigma_{\alpha_k}) \subset g_\gamma^{-1}(\Sigma_\gamma)$, $|k| \le n_0$. So let C_k $(\in \Sigma_\gamma)$ be such that $B_k = g_\gamma^{-1}(C_k)$. Now we produce a partition of Ω_γ in Σ_γ and show that $H_\beta(Q_\gamma, P_\gamma)$ differs from $H_\beta(Q, P)$ by at most ε. This will complete the argument.

Since $\{C_k\}_{-n_0}^{n_0} \subset \Sigma_\gamma$ need not be disjoint, let $\{E_k\}_{-n_0}^{n_0} \subset \Sigma_\gamma$ be a disjunctification, i.e., $E_{-n_0} = C_{-n_0}$ and $E_k = C_k - \bigcup_{m=-n_0}^{k-1} C_m$ so that $E_k \subset C_k$ for $|k| \le n_0$. Hence using the fact that $P_\gamma = P \circ g_\gamma^{-1}$, one has

$$\begin{aligned}
P_\gamma\Big(\bigcup_{|k|\le n_0} E_k\Big) &= P_\gamma\Big(\bigcup_{|k|\le n_0} C_k\Big) = P\Big(\bigcup_{|k|\le n_0} g_\gamma^{-1}(C_k)\Big) \ge P\Big(\bigcup_{|k|\le n_0} B_k \cap A_k\Big) \\
&= \sum_{|k|\le n_0} P(B_k \cap A_k\Big) \qquad \text{since} \quad A_k \quad \text{are disjoint} \\
&\ge \sum_{|k|\le n_0} (P(A_k) - P(A_k \triangle B_k)) > \sum_{|k|\le n_0} P(A_k) - \sum_k \eta_k \\
&\qquad \text{by} \quad (13) \quad \text{and} \quad (15) \\
&\ge (1 - \tfrac{1}{4}\varepsilon) - \sum_k [\varepsilon/(3 \cdot 2^{|k|+2})] = 1 - \tfrac{1}{2}\varepsilon.
\end{aligned} \tag{16}$$

Replacing P by Q in this computation, one can deduce:

$$Q_\gamma\Big(\bigcup_{|k|\le n_0} E_k\Big) \ge 1 - \tfrac{1}{2}\varepsilon. \tag{17}$$

Consider the partition $\{E_k, -n_0 \le k \le n_0+1\}$ of Ω_γ in Σ_γ, where $E_{n_0+1} = (\bigcup_{|k|\le n_0} E_k)^c$. Then $P_\gamma(E_{n_0+1}) < \frac{1}{2}\varepsilon$, $Q_\gamma(E_{n_0+1}) < \frac{1}{2}\varepsilon$ by (16) and (17). Hence using Lemma 3, one has

$$
\begin{aligned}
H_\beta(Q_\gamma, P_\gamma) &\le \sum_{k=-n_0}^{n_0+1} (P_\gamma(E_k))^\beta (Q_\gamma(E_k))^{1-\beta} \\
&\le \sum_{|k|\le n_0} (P_\gamma(C_k))^\beta (Q_\gamma(C_k))^{1-\beta} + \tfrac{1}{2}\varepsilon \\
&= \sum_{|k|\le n_0} (P(B_k))^\beta (Q(B_k))^{1-\beta} + \tfrac{1}{2}\varepsilon \\
&\le \sum_{|k|\le n_0} (P(A_k)+\eta_k)^\beta (Q(A_k)+\eta_k)^{1-\beta} + \tfrac{1}{2}\varepsilon \qquad \text{by} \quad (15) \\
&\qquad \text{since} \quad |P(A_k) - P(B_k)| < \eta_k, \text{ and similarly for } Q(A_k), \\
&\le \sum_{|k|\le n_0} [(P(A_k))^\beta (Q(A_k))^{1-\beta} + [\varepsilon/(3\cdot 2^{|k|+2})]] + \tfrac{1}{2}\varepsilon \qquad \text{by} \quad (14), \\
&\le \sum_k (P(A_k))^\beta (Q(A_k))^{1-\beta} + \tfrac{1}{4}\varepsilon + \tfrac{1}{2}\varepsilon \\
&\le H_\beta(Q, P) + \varepsilon \qquad \text{by} \quad (12).
\end{aligned}
$$

Since $\varepsilon > 0$ is arbitrary, this implies $\lim_\alpha H_\beta(Q_\alpha, P_\alpha) = H_\beta(Q, P)$. The last statement on the singularity is now obvious since $Q \perp P$ iff $H_\beta(Q, P) = 0$. This completes the proof.

This useful extension of Kakutani's theorem is due to Brody [1]. We remarked at the beginning that this can be obtained from the mean convergence of a martingale with a directed index set. However, one can deduce the latter from the above theorem also. Before giving this, let us note how these results can be *identified since there is a system of spaces for the projective case and there is just one probability space for a martingale*, with a "filtration" or a stochastic base on it. The relations are given in detail as follows.

Let $\{f_i, \mathfrak{F}_i, i \in I\}$ be a directed indexed martingale on a probability space $(Z, \mathscr{E}, \mu)$. Thus for $i < i'$, $\mathfrak{F}_i \subset \mathfrak{F}_{i'}$ and $E^{\mathfrak{F}_i}(f_{i'}) = f_i$ a.e. (μ). Consider the spaces $(\mathbb{R}^\alpha, \mathscr{B}_\alpha)$ where α is a finite subset of I, and $\mathbb{R}^\alpha = \times_{i\in\alpha} \mathbb{R}_i$, $\mathbb{R}_i = \mathbb{R}$, and $\mathscr{B}_\alpha = \bigotimes_{i\in\alpha} \mathscr{B}_i$, $\mathscr{B}_i = \mathscr{B}$ the Borel σ-algebra of $\mathbb{R}$. If $(\Omega, \Sigma) = \varprojlim(\mathbb{R}^\alpha, \mathscr{B}_\alpha)$ as in the Kolmogorov–Bochner theorem ($g_{\alpha\beta} = \pi_{\alpha\beta}: \mathbb{R}^\beta \to \mathbb{R}^\alpha$, $g_\alpha = \pi_\alpha: \Omega \to \mathbb{R}^\alpha$ are coordinate projections), consider the mapping $h: z \mapsto (f_i(z), i \in I) \in \Omega$. Since each f_i is $(\mathscr{E}, \mathscr{B})$-measurable, we conclude that h is $(\mathscr{E}, \Sigma)$-measurable, i.e., for each $A \in \Sigma$, $\{z : h(z) = (f_i(z), i \in I) \in A\} \in \mathscr{E}$ (this is clear if A is a cylinder and the general case is then immediate). If $P = \mu \circ h^{-1}$, then (Ω, Σ, P) is a probability space and $\pi_i \circ h = f_i \colon \Omega \to \mathbb{R}$. Moreover, one may identify the f_i-process with the π_i-process because π_i is $\pi_i^{-1}(\mathscr{B})$ $(\subset \Sigma)$-measurable and

$$\int_\Omega |\pi_i(\omega)|\, dP(\omega) = \int_{h^{-1}(\Omega)} |\pi_i \circ h|(z)\, d\mu(z) = \int_Z |f_i|(z)\, d\mu(z) \tag{18}$$

and for each $A \in \pi_i^{-1}(\mathscr{B}_i)$ and $i < i'$ one has $h^{-1}(A) \in \mathfrak{F}_i$,

$$\begin{aligned}\int_A \pi_j(\omega)\, dP(\omega) &= \int_{h^{-1}(A)} (\pi_j \circ h)(z)\, d\mu(z) = \int_{h^{-1}(A)} f_j(z)\, d\mu(z) \\ &= \int_{h^{-1}(A)} f_i(z)\, d\mu(z), \quad i < j, \\ &= \int_{h^{-1}(A)} (\pi_i \circ h)(z)\, d\mu(z) = \int_A \pi_i\, d(\mu \circ h^{-1}).\end{aligned} \tag{19}$$

Thus $\{\pi_i, \pi_i^{-1}(\mathscr{B}_i), i \in I\}$ is a martingale on (Ω, Σ, P) which is equal to the given one on $(Z, \mathscr{E}, \mu)$, and one converges iff the other does.

To use this reduction in the above theorem, we note that if $f_i = dP_i/d\mu_i$, $u_i = dQ_i/d\mu_i$, where $Q_i = Q \circ g_i^{-1}$, $P_i = P \circ g_i^{-1}$, then the compatibility relations on the mappings $\{g_{ij}, g_i \colon i < j,\, i, j \in I\}$ imply that (on the space Ω) $\{f_i \circ g_i, g_i^{-1}(\Sigma_i), i \in I\}$ and $\{u_i \circ g_i, g_i^{-1}(\Sigma_i), i \in I\}$ are martingales, respectively, on (Ω, Σ, P) and (Ω, Σ, Q). Letting $\tilde{f}_i = \mathrm{f}_i \circ g_i$, $\tilde{u}_i = u_i \circ g_i$, one gets the martingales on (Ω, Σ, P) and then one may use the procedure of (18) and (19) to replace these spaces and the mappings g_i, g_{ij} by $\{\pi_i, \pi_i^{-1}(\mathscr{B}_i), i \in I\}$ on $(\mathbb{R}^I, \mathscr{B}_I, \tilde{P})$ and similarly the second one on the space $(\mathbb{R}^I, \mathscr{B}_I, \tilde{Q})$. *It is important to note this distinction, which shows how the function space representation decisively enters into the problems.* In applications (cf. Example 2 above) one "identifies" the f_i (and u_i)-process with the coordinate variable or π_i-process and proceeds as though one is working on the same function-space-represented process all the time. Following custom, we shall also use the same change of measure spaces whenever convenient without further explanation. In particular, the construction of likelihood ratios for processes depends *essentially* on this representation..

From Theorem 5 one deduces the following result.

6. Theorem *Let $\{f_i, \mathfrak{F}_i, i \in I\}$ be a directed index martingale on (Ω, Σ, P) $(\mathfrak{F}_i \uparrow \subset \Sigma)$ $\|f_i\|_1 \le K < \infty$, which is terminally uniformly integrable. Then there exists $f \in L^1(\Omega, \Sigma_0, P)$ such that $\|f_i - f\|_1 \to 0$ as "$i \to \infty$," where $\Sigma_0 = \sigma(\bigcup_{i \in I} \mathfrak{F}_i)$, and then $f_i = E^{\mathfrak{F}_i}(f)$ a.e.* [Terminal uniform integrability is recalled below.]

Proof By the decomposition of Proposition 2.1, we may assume that $f_i \ge 0$ a.e. since $\sup_i \int_\Omega |f_i|\, dP < \infty$ by hypothesis. (See particularly Proposition

III.5.3.) Let $\mu_i(\cdot) = \int_{(\cdot)} f_i \, dP\colon \mathfrak{F}_i \to \mathbb{R}^+$. Then $\mu_i = \mu_{i'} | \mathfrak{F}_i$ for $i < i'$ by the martingale property and $|\mu_i|(\Omega) \le k_0 < \infty$ all $i \in I$. Moreover, by Proposition III.1.2 there is an additive function $\mu\colon \bigcup_{i \in I} \mathfrak{F}_i \to \mathbb{R}^+$. But by definition of terminal uniform integrability (the set $\{f_i\}_{i \in I}$ is bounded in $L^1(\Omega, \Sigma, P)$; let $I' \subset I$ be a terminal set)

$$\lim_{P(A) \to 0} \int_A f_i \, dP = 0 \qquad \text{uniformly in} \qquad i \in I', \quad A \in \Sigma.$$

This can be written as $\mu(A) \to 0$ as $P(A) \to 0$ for each $A \in \bigcup_{i \in I'} \mathfrak{F}_i$ and hence μ is σ-additive there. Thus it has a unique σ-additive extension to Σ_0 and is P-continuous. By the Radon–Nikodým theorem there is a P-unique $f \in L^1(\Omega, \Sigma_0, P)$, $f = d\mu/dP$ a.e. We show that $f_i \to f$ in L^1 to complete the proof. It is enough to prove this for a *sequence* (why?).

Note that $\bigcup_{i \in I} L^2(\Omega, \mathfrak{F}_i, P)$ is dense in $L^2(\Omega, \Sigma_0, P)$. This simple proof is left to the reader. Let $0 \le g_0 \in L^1(\Omega, \mathfrak{F}_{i_0}, P) \subset L^1(\Omega, \Sigma, P)$ for some $i_0 \in I$. If $\nu(\cdot) = \int_{(\cdot)} g_0 \, dP$, then ν is a positive finite measure on Σ, and $\nu | \mathfrak{F}_i = \nu_i$ $(=\nu$, for $i \ge i_0)$. It is clear that

$$H_{1/2}(\nu, \mu) = \int_\Omega \sqrt{f g_0} \, dP, \qquad H_{1/2}(\nu_i, \mu_i) = \int_\Omega \sqrt{f_i g_0} \, dP, \qquad i \ge i_0. \tag{20}$$

Using (8) and Theorem 5, it follows that $(\sqrt{f_i}, \sqrt{g_0}) \to (\sqrt{f}, \sqrt{g_0})$ as $i \to \infty$. From this one may deduce that $(\sqrt{f_i}, h) \to (\sqrt{f}, h)$ for each $h \in \bigcup_{i \in I} [L^2(\Omega, \mathfrak{F}_i, P)]^+$, and since L^2 is a lattice, this implies that $\sqrt{f_i} \to \sqrt{f}$ weakley, i.e., $(\sqrt{f_i}, h) \to (\sqrt{f}, h)$ for all $h \in L^2(\Omega, \Sigma_0, P)$. Also $\|\sqrt{f_i}\|_2^2 = \int_\Omega f_i \, dP = \mu_i(\Omega) = \mu(\Omega) = \int_\Omega f \, dP = \|\sqrt{f}\|_2^2$ by the projective limit property (or a martingale has a constant expected value). Hence $\sqrt{f_i} \to \sqrt{f}$ in L^2-norm. This follows from a classical result for uniformly rotund (or uniformly convex) Banach spaces: Weak convergence plus convergence of norms implies strong convergence.

In the present case the proof is simple and is as follows. Since clearly $(\sqrt{f_i} + \sqrt{f}) \to 2\sqrt{f}$ weakly, one has

$$\begin{aligned} |(2\sqrt{f}, h)| &= \lim_i |(\sqrt{f_i} + \sqrt{f}, h)| = \limsup_i |(\sqrt{f_i} + \sqrt{f}, h)| \\ &\le \limsup_i \|h\|_2 \|\sqrt{f_i} + \sqrt{f}\|_2 \\ &\le 2\|h_2\|(\mu(\Omega))^{1/2} \end{aligned} \tag{21}$$

since $\|\sqrt{f_i}\|_2^2 = \|f\|_2^2 = \mu(\Omega)$. Hence taking "sup" on $\|h\|_2 \le 1$, one has

$$2(\mu(\Omega))^{1/2} = \sup\{|2(\sqrt{f}, h)| : \|h\|_2 \le 1\} \le 2(\mu(\Omega))^{1/2} \qquad \text{by} \quad (21).$$

Thus there is equality in (21), and it follows that $\|\sqrt{f_i}+\sqrt{f}\|_2 \to 2(\mu(\Omega))^{1/2}$. Using the parallelogram identity, one has

$$\|\sqrt{f_i}-\sqrt{f}\|_2^2+\|\sqrt{f_i}+\sqrt{f}\|_2^2 = 2(\|\sqrt{f_i}\|_2^2+\|\sqrt{f}\|_2^2) = 4[\mu(\Omega)]. \quad (22)$$

This clearly implies $\|\sqrt{f_i}-\sqrt{f}\|_2 \to 0 \quad \text{as} \quad i \to \infty$.

From the preceding result one has

$$\int_\Omega |f_i - f|\,dP = \int_\Omega (|\sqrt{f_i}-\sqrt{f}|)(\sqrt{f_i}+\sqrt{f})\,dP \leq \|\sqrt{f_i}-\sqrt{f}\|_2 \cdot 2(\mu(\Omega))^{1/2}$$

(by Hölder's inequality), and hence the right side tends to zero as $i \to \infty$. Thus $f_i \to f$ in $L^1(\Omega, \Sigma_0, P)$. The projective property implies $\mu_i = \mu \mid \mathfrak{F}_i$ and hence $f_i = E^{\mathfrak{F}_i}(f)$ a.e. This completes the proof.

This is an "extension" of Theorem 2.5 in that the index set is merely directed, but the conclusion is weaker—only norm convergence. This theorem will again be deduced as a consequence of general results in Chapter V (cf. Corollary 2.3 there). However, the pointwise convergence statement is generally false, even if I is countable (but not linearly ordered) and $0 \leq f_i \leq k_0 < \infty$, as shown by an example due to Dieudonné (1950). Note that the above result is sufficient for the proof of Theorem III.2.8 (see Eq. (27) there). The deduction of Theorem 6 from 5 seems unnoticed in the literature.

It is now possible to prove the dichotomy theorem for Gaussian processes. We already noted that $H_\beta(Q, P) = 0$ iff $Q \perp P$. Since $H_\beta = H_\beta(Q, P) = \int_\Omega f^\beta g^{1-\beta}\,d\mu$, this is so iff $f \cdot g = 0$ a.e. $[\mu]$. But equivalence of P, Q means that $[f > 0]$ and $[g > 0]$ must be equal sets a.e. To deduce this therefore, we need to consider g/f (and f/g), which will distinguish them if any one vanishes on a set of positive measure while the other does not. The appropriate function is obtained from information theory (cf. p. 164 on entropy functions), and the following functional seems to be due to Gel'fand and Yaglom [1] which will serve the purpose:

$$I(Q, P) = \int_\Omega (g - f)\log(g/f)\,d\mu, \quad (23)$$

where $\mu = P + Q$, $g = dQ/d\mu$, $f = dP/d\mu$ a.e. as before. If $0 \leq g/f < 1$, then considering Ω as the union of this set and its complement, we see that $0 \leq I = I(Q, P) \leq \infty$. The importance of this function is that $Q \equiv P$ (i.e., Q is *equivalent* to P) iff $f \cdot g > 0$ a.e. $[\mu]$ and hence if $I(Q, P) < \infty$, since the latter holds only if $0 \leq (g - f)\log(g/f) < \infty$ a.e. $[\mu]$, then $[f > 0] \triangle [g > 0]$ is μ-null. Thus $H_\beta = 0$ iff $Q \perp P$ (but $H_\beta > 0$ does not necessarily imply $Q \equiv P$), and $I < \infty$ only if $Q \equiv P$ (but $I = \infty$ does not necessarily imply $Q \perp P$). Hereafter we write H for H_β when $\beta = \frac{1}{2}$.

In the Gaussian case one has the following basic result, which was first obtained by Feldman [1] and Hájek [1] independently. This result leads to many others of interest both in theory and applications of statistical inference. Thus we have with the above notation:

7. Theorem *Let $\{X_t, t \in T\}$ be a (real) stochastic process on (Ω, Σ). If P, Q are two probability measures on Σ and if the process is Gaussian relative to both P and Q, then either $Q \perp P$ or $Q \equiv P$. They satisfy $Q \perp P$ iff $H = 0$ or $I = +\infty$ and $Q \equiv P$ iff $H > 0$ or $I < \infty$. Moreover, if (m, r) and (n, s) are the (mean, covariance) functions of the process under P and Q, then $P \equiv Q$ iff both the measures satisfy the conditions: $P(0, r) \equiv Q(0, s)$ and $P(m, r) \equiv Q(n, r) \equiv P(n, r)$ (i.e., the two pairs with zero means and with the same covariances are equivalent). Here T is just an index set.*

Proof We first recall that (for the last part) a Gaussian process (i.e., the family of its finite dimensional distributions) is uniquely determined by its mean and covariance functions (cf. I.6.1). Consequently $P(m, r)$ is a Gaussian measure on (Ω, Σ). This is meaningful and similarly for $Q(n, s)$.

Next we replace the given process by its function space representation so that (Ω, Σ) is now $(\mathbb{R}^T, \mathscr{B}^T)$, where $\mathscr{B}$ is the Borel σ-algebra of $\mathbb{R}$ and $\mathscr{B}^T = \bigotimes_{i \in T} \mathscr{B}_i$, $\mathscr{B}_i = \mathscr{B}$, $\mathbb{R}^T$ being the space of real functions on T to $\mathbb{R}$. Then for each finite subset $\alpha = (t_1, \ldots, t_n) \subset T$, one has the finite dimensional Gaussian probabilities P_α, Q_α on $(\mathbb{R}^\alpha, \mathscr{B}^\alpha)$, $\mathscr{B}^\alpha = \bigotimes_{i \in \alpha} \mathscr{B}_i$ defined as

$$P_\alpha\left\{\bigcap_{i=1}^{n} \{\omega : X_{t_i}(\omega) < x_i\}\right\} = P_\alpha\left\{\omega \in \mathbb{R}^T : \bigcap_{i=1}^{n} [\omega(t_i) < x_i]\right\}$$
$$= \int_{-\infty}^{x_1} \cdots \int_{-\infty}^{x_n} f_\alpha(u_1, \ldots, u_n)\, du_1, \ldots, du_n, \tag{24}$$

where

$$f_\alpha(x) = (2\pi)^{-n/2}[\det(r_\alpha)]^{-1/2} \exp[-\tfrac{1}{2}(r_\alpha^{-1}(x - m_\alpha), (x - m_\alpha)], \tag{25}$$

with $r_\alpha = (r(t_i, t_j),\ i, j = 1, \ldots, n)$ as the covariance matrix and $m_\alpha = (m(t_1), \ldots, m(t_n))$ the mean vector, $(x = (x_1, \ldots, x_n))$. Similarly, Q_α on $\mathscr{B}^\alpha$ is defined by replacing f_α by g_α and (m_α, r_α) by (n_α, s_α) in (24) and (25). Then by Theorem I.3.2 if $\pi_{\alpha\beta}: \mathbb{R}^\beta \to \mathbb{R}^\alpha$, $\pi_\alpha: \mathbb{R}^T \to \mathbb{R}^\alpha$ ($\alpha < \beta$ being finite subsets of T) are the coordinate projections, then $\bar{P} = \lim_{\leftarrow}(P_\alpha, \pi_\alpha)$ and $\bar{Q} = \lim_{\leftarrow}(Q_\alpha, \pi_\alpha)$, i.e., the projective limits exist and $(\bar{P}, \bar{Q})$ and (P, Q) define the *same* families of finite dimensional distribution functions. Hence by Theorem 5

$$H(P, Q) = \lim_{\alpha} H(P_\alpha, Q_\alpha) = H(\bar{P}, \bar{Q}). \tag{26}$$

We thus transfer the problem to the function space $(\mathbb{R}^T, \mathscr{B}^T)$ and in what follows use P, Q for $(\bar{P}, \bar{Q})$ on this space. (The function space representation is also sometimes called the *canonical representation* of a process.) By the last part of Theorem 5, $H = 0$ implies $Q \perp P$ for any probability measures and hence for Gaussian measures. Thus we need to prove only that $H > 0$ implies $Q \equiv P$. We present this proof in steps for clarity and also because it is long. [For the rest of this section, $\mathscr{B}^\alpha$ is written for $\mathscr{B}_\alpha$ for notational ease.]

I. Let D be the directed set of all finite subsets of T (ordered by inclusion) and let $\mathfrak{F}_\alpha = \pi_\alpha^{-1}(\mathscr{B}^\alpha) \subset \mathscr{B}^T$. If $P_\alpha^* = P|\mathfrak{F}_\alpha$, $Q_\alpha^* = Q|\mathfrak{F}_\alpha$, and $\mu_\alpha^* = \mu|\mathfrak{F}_\alpha$ where μ is a dominating measure for P, Q, which (for computational convenience below) is taken to be the Gaussian measure whose mean and covariance functions are $m - n$ and $(r + s)/2$, respectively, then $f = dP/d\mu$, $g = dQ/d\mu$ exist a.e. $[\mu]$ for $\mathscr{B}^T$. Also if $f_\alpha^* = dP_\alpha^*/d\mu_\alpha^*$, $g_\alpha^* = dQ_\alpha^*/d\mu_\alpha^*$, then $\{f_\alpha^*, \mathfrak{F}_\alpha, \alpha \in D\}$, $\{g_\alpha^*, \mathfrak{F}_\alpha, \alpha \in D\}$ are martingales and $f_\alpha^* \to f$, $g_\alpha^* \to g$ in $L^1(\mu)$ by Theorem 6. Let

$$H_\alpha = H(P_\alpha^*, Q_\alpha^*) = \int_{\mathbb{R}^T} (f_\alpha^* g_\alpha^*)^{1/2}\, d\mu_\alpha^*$$

and

$$I_\alpha = I(Q_\alpha^*, P_\alpha^*) = \int_{\mathbb{R}^T} (g_\alpha^* - f_\alpha^*) \log(g_\alpha^*/f_\alpha^*)\, d\mu_\alpha^*.$$

It is clear that $P_\alpha = P_\alpha^* \circ \pi_\alpha^{-1}$, $Q_\alpha = Q_\alpha^* \circ \pi_\alpha^{-1}$ (as in Theorem I.3.2). Hence $H(P_\alpha, Q_\alpha) \le H(P_\alpha^*, Q_\alpha^*) = H_\alpha$ by Corollary 4. If μ_α is the Gaussian measure on $\mathscr{B}^\alpha$ whose mean is $m_\alpha - n_\alpha$ and covariance $(r_\alpha + s_\alpha)/2$, let $g_\alpha = dQ_\alpha/d\mu_\alpha$, $f_\alpha = dP_\alpha/d\mu_\alpha$ and define

$$J_\alpha = J(Q_\alpha, P_\alpha) = \int_{\mathbb{R}^\alpha} (g_\alpha - f_\alpha) \log(g_\alpha/\mathrm{f}_\alpha)\, \mathrm{d}\mu_\alpha. \tag{27}$$

II. We assert that $I_\alpha = J_\alpha$ and I_α is monotone nondecreasing, $I = \lim_\alpha I_\alpha$ for any P, Q on $(\mathbb{R}^T, \mathscr{B}^T)$, Gaussian or not; and I is as in (23).

For let $\alpha \le \beta$ so that $\mathfrak{F}_\alpha \subset \mathfrak{F}_\beta$ and let us first show that $I_\alpha \le I_\beta$. If $I_\beta = \infty$, this is true, and so let $I_\beta < \infty$. Then by our discussion following (23) $Q_\beta^* \equiv P_\beta^*$ and (since $Q_\alpha^* = Q_\beta^*|\mathfrak{F}_\alpha$, etc.) also $Q_\alpha^* \equiv P_\alpha^*$. But then $dQ_\beta^*/dP_\beta^* = g_\beta^*/g_\alpha^* = u_\beta$ a.e. $[P_\beta^*]$, and $\{u_\alpha, u_\beta\}$ is a martingale relative to $(\mathfrak{F}_\alpha, \mathfrak{F}_\beta)$. Hence I_β can be written on using the conditional Jensen inequality applied to $\varphi_1(x) = x \log x$ and $\varphi_2(x) = -\log x$, $x > 0$, as

$$\begin{aligned} I_\beta = \int_{\mathbb{R}^T} (u_\beta - 1) \log u_\beta\, dP_\beta^* &= E(E^{\mathfrak{F}_\alpha}[(u_\beta - 1)\log u_\beta]) \\ &\ge E([(E^{\mathfrak{F}_\alpha}(u_\beta) - 1) \log E^{\mathfrak{F}_\alpha}(u_\beta)]) \\ &= E((u_\alpha - 1)\log u_\alpha) = I_\alpha \\ &\qquad \text{by the martingale property.} \end{aligned} \tag{28}$$

Hence I_α is monotone so that $\lim_\alpha I_\alpha = \bar{I} \le I$.

To see that there is equality between $\bar{I}$ and I, one may assume that $\bar{I} < \infty$. Then $Q \equiv P$. In fact, if this is false, there exists a set $A \in \mathscr{B}^T$ such that $P(A) = 0$ and $Q(A) > 0$. But by the remark following Lemma 1 above, A is determined by a *countable* set J' $(=J'(A)) \subset T$, i.e., $A \in \pi_{J'}^{-1}(\mathscr{B}^{J'}) = \mathfrak{F}_{J'} \subset \mathscr{B}^T$. Let $P_{J'}^* = P|\mathfrak{F}_{J'}$, $Q_{J'}^* = Q|\mathfrak{F}_{J'}$, and $f_{J'}^*, g_{J'}^*$ be the corresponding densities. Then $f_{J'}^* = 0$ a.e. on A, $g_{J'}^* > 0$ a.e. on A for $[\mu]$. Hence by (23) $I_{J'} = +\infty$. Let $\alpha_1 < \alpha_2 < \cdots$ be a monotone sequence of finite subsets of J' such that $\lim_n \alpha_n = J'$. Then $f_{\alpha_n}^* = E^{\mathfrak{F}_{\alpha_n}}(f_{J'}^*)$, $g_{\alpha_n}^* = E^{\mathfrak{F}_{\alpha_n}}(g_{J'}^*)$ a.e. Hence $f_{\alpha_n}^* \to f_{J'}^*$ and $g_{\alpha_n}^* \to g_{J'}^*$ a.e. and in $L^1(\mu)$. If $P_{\alpha_n}^* \equiv Q_{\alpha_n}^*$ is false for some n, then $I_{\alpha_n} = \infty$ and $I_{\alpha_n} = I_{J'}$. If $P_{\alpha_n}^* \equiv Q_{\alpha_n}^*$ for all n, then $f_{\alpha_n}^* > 0$, $g_{\alpha_n}^* > 0$ a.e. $[\mu]$, and

$$(g_{\alpha_n}^* - f_{\alpha_n}^*) \log(g_{\alpha_n}^*/f_{\alpha_n}^*) \to (g_{J'}^* - f_{J'}^*) \log(g_{J'}^*/f_{J'}^*)$$

a.e. $[\mu]$. On the other hand, our assumption on $Q_{\alpha_n}^*$, $P_{\alpha_n}^*$ also implies that $\{g_{\alpha_n}^*/f_{\alpha_n}^*, \mathfrak{F}_{\alpha_n}, n \geq 1\}$ is a (positive) martingale on $(\mathfrak{F}_{J'}, P_{J'}^*)$. Hence $\{\varphi_i(g_{\alpha_n}^*/f_{\alpha_n}^*), \mathfrak{F}_{\alpha_n}, n \geq 1\}$ are submartingales, where $\varphi_1(x) = x \log x$, $\varphi_2(x) = -\log x$, $i = 1, 2$. Consequently, $E(\varphi_i(u_{\alpha_n})) \uparrow K_i$, where $u_{\alpha_n} = g_{\alpha_n}^*/f_{\alpha_n}^*$. If at least one $K_i = \infty$, then $\lim_n I_{\alpha_n} = I_{J'}$. If both $K_i < \infty$, $i = 1, 2$, then using an easy computation one can deduce that with $K_0 = K_1 + K_2$,

$$\begin{aligned}
\sup_n E(|\varphi_1^+(u_{\alpha_n})|) &\leq K_0 + \sup_n E(\varphi_1^-(u_{\alpha_n})) \\
&\leq K_0 + |K_3| + |K_4| \sup_n E(u_{\alpha_n}) \\
&= K_0 + |K_3| + |K_4| \sup_n \int_{\mathbb{R}^T} g_{\alpha_n}\, d\mu \\
&\leq K_0 + |K_3| + |K_4| < \infty,
\end{aligned}$$

where $\varphi_1(t) \geq K_3 + K_4 t$ (by the support line property of the convex φ_1) is used here; similarly for φ_2. Then by Theorem 2.4 $\varphi_i(u_{\alpha_n}) \to Y_\infty^{(i)}$ a.e. and $E(Y_\infty^{(i)}) \leq \lim_n E(\varphi_i(u_{\alpha_n})) < \infty$. But $u_{\alpha_n} \to u_{J'} = g_{J'}^*/f_{J'}^*$ a.e., and φ_i is continuous. So $\varphi_i(u_{J'}) = Y_\infty^{(i)}$ a.e. Thus

$$I_{J'} = E(\varphi_1(u_{J'})) + E(\varphi_2(u_{J'})) = E(Y_\infty^{(1)}) + E(Y_\infty^{(2)}) < \infty.$$

But this is a contradiction since $I_{J'} = +\infty$. Hence $\lim_n I_{\alpha_n} = I_{J'} = \infty$. Also I_α is increasing and

$$\bar{I} = \lim_\alpha I_\alpha = \sup_\alpha I_\alpha \geq \sup_n I_{\alpha_n} = \lim_n I_{\alpha_n} = I_{J'} = +\infty. \tag{29}$$

This contradicts the assumption that $\bar{I} < \infty$. So $Q \equiv P$. Now dQ/dP is measurable relative to $\mathfrak{F}_{J_0'}$ where $J_0' \subset T$ is countable and $I = I_{J_0'}$. But then the preceding analysis implies $I = I_{J_0'} = \lim_n I_{\beta_n} \leq \underline{\lim}_n I_{\beta_n} = \bar{I}$, where $\beta_n \uparrow J_0'$ are finite subsets. The last inequality follows from Fatou's lemma. Thus $I = \bar{I}$.

It remains to show that $J_\alpha = I_\alpha$. For this, recall that $P_\alpha^* = P_\alpha \circ \pi_\alpha^{-1}$ by definition and that π_α is $(\mathfrak{F}_\alpha, \mathscr{B}^\alpha)$-measurable. Now if $Q_\alpha^* \not\equiv P_\alpha^*$, then $I_\alpha = \infty$ and there is a set $A \in \mathfrak{F}_\alpha$ such that $P_\alpha^*(A) = 0$ and $Q_\alpha^*(A) > 0$. Hence there is a $B \in \mathscr{B}^\alpha$, $A = \pi_\alpha^{-1}(B)$ so that $P_\alpha(B) = P_\alpha^*(A) = 0$, $Q_\alpha(B) = Q_\alpha^*(A) > 0$, and $J_\alpha = \infty$ in this case. If $Q_\alpha^* \equiv P_\alpha^*$ and $u_\alpha = dQ_\alpha^*/dP_\alpha^* > 0$, one has, on using $P_\alpha^* \circ \pi_\alpha^{-1} = P_\alpha$ for any $F = \pi_\alpha^{-1}(G)$, $G \in \mathscr{B}^\alpha$,

$$\int_F (u_\alpha - 1) \log u_\alpha \, dP_\alpha^* = \int_G (u_\alpha \circ \pi_\alpha - 1) \log(u_\alpha \circ \pi_\alpha) \, dP_\alpha. \tag{30}$$

It is clear that $u_\alpha \circ \pi_\alpha = dQ_\alpha/dP_\alpha$ and hence replacing G by $\mathbb{R}^\alpha$, (30) becomes $J_\alpha = I_\alpha$. Hence all the assertions of this step are proved. This also shows that $H(P_\alpha, Q_\alpha) = H(P_\alpha^*, Q_\alpha^*)$. *Note that the result of this step is an analog of Theorem 5 for the I (or $I_{(\cdot)}$)-functional. Thus the H- and I-functionals can be used separately for the dichotomy problems.*

As yet we have not used the fact that the measures P, Q are Gaussian. Only that the system is a projective family admitting a limit sufficed. Let us use the additional information for P, Q on $\mathscr{B}^T$, and for a shorter computational procedure we employ *both* H_α and I_α. One has to prove that $H > 0$ implies $I < \infty$ when P, Q are Gaussian measures so that $Q \equiv P$.

III. Let P_α, Q_α be Gaussian probability measures on $\mathscr{B}^\alpha$, $\alpha \in D$, with (m_α, r_α) and (n_α, s_α) as the respective parameters. If r_α and s_α are strictly positive definite for each $\alpha \in D$, i.e., both matrices r_α, s_α are invertible, then $H > 0$ implies $I = \sup_\alpha I_\alpha < \infty$.

For by definition μ_α is a Gaussian measure determined by $(m_\alpha - n_\alpha, \frac{1}{2}(r_\alpha + s_\alpha)) \equiv (d_\alpha, v_\alpha)$, say, which dominates both P_α, Q_α, and hence

$$H_\alpha = H(Q_\alpha, P_\alpha) = \int_{\mathbb{R}^\alpha} (f_\alpha g_\alpha)^{1/2} \, d\mu_\alpha, \tag{31}$$

$$I_\alpha = I(Q_\alpha, P_\alpha) = \int_{\mathbb{R}^\alpha} (g_\alpha - f_\alpha) \log \frac{g_\alpha}{f_\alpha} \, d\mu_\alpha. \tag{32}$$

Since $g_\alpha = dQ_\alpha/d\mu_\alpha = (dQ_\alpha/d\lambda)(d\lambda/d\mu_\alpha)$, where λ is the Lebesgue measure in $\mathbb{R}^\alpha$, so that g_α is the likelihood ratio of the Gaussian densities of Q_α and μ_α, one may explicitly express it using Eq. (25); similarly for f_α. Substituting these expressions in (31), (32), and remembering the formula for the bilateral Laplace transform (and an evaluation of the first four moments of the Gaussian distribution), one gets the following after an elementary but standard computation which is left to the reader:

$$H_\alpha = \left(\frac{\det r_\alpha \det s_\alpha}{\det v_\alpha}\right)^{1/2} \exp(-\tfrac{1}{4}(v_\alpha^{-1} d_\alpha, d_\alpha)), \tag{33}$$

$$I_\alpha = \tfrac{1}{2} \operatorname{tr}\{(r_\alpha - s_\alpha)(s_\alpha^{-1} - r_\alpha^{-1})\} + \tfrac{1}{2}((r_\alpha^{-1} + s_\alpha^{-1}) d_\alpha, d_\alpha), \tag{34}$$

where tr is the trace and the scalar product notation $(\cdot,\cdot)$ is used. To simplify the algebra, let

$$\delta_\alpha = \frac{(\det v_\alpha)^2}{\det r_\alpha \det s_\alpha}, \qquad \eta_\alpha = (v_\alpha^{-1} d_\alpha, d_\alpha). \tag{35}$$

Then $H_\alpha^{-4} = \delta_\alpha \exp(\eta_\alpha)$. Since δ_α is independent of m and $H_\alpha \le 1$, we deduce that (cf. (33) since for all d_α, $\delta_\alpha = H_\alpha^{-4}\exp(-\eta_\alpha)$, including $d_\alpha = 0$), $\delta_\alpha \ge 1$ for all $\alpha \in D$. But $\eta_\alpha \ge 0$ so that $\exp(\eta_\alpha) \ge 1$. Thus $\delta_\alpha \le H_\alpha^{-4} \le H^{-4} < \infty$ (since $H > 0$) and $\exp(\eta_\alpha) = (H_\alpha^{-4}/\delta_\alpha) \le H^{-4} < \infty$ both uniformly in α. Hence there is a constant $K_0 \ge H^{-4}$ such that $\eta_\alpha \le K_0 < \infty$, and so

$$1 \le \delta_\alpha \le K_0 < \infty, \qquad 0 \le \eta_\alpha \le K_0 < \infty. \tag{36}$$

With this let us obtain a bound on $I_\alpha = \frac{1}{2}(M_\alpha + N_\alpha)$, where M_α is the trace term and N_α is the last term in (34) for I_α. To get a bound for M_α, we choose a (nonsingular) coordinate transformation such that r_α is reduced to the identity and s_α to the diagonal form with eigenvalues $(1/\lambda_i) > 0$, $i = 1, \ldots, n$; i.e., $\lambda_i > 0$ are the eigenvalues of $r_\alpha s_\alpha^{-1}$. Then

$$\begin{aligned} M_\alpha &= \operatorname{tr}((r_\alpha s_\alpha^{-1}) + (r_\alpha s_\alpha^{-1})^{-1} - 2I) \\ &= \sum_{i=1}^{n} (\lambda_i + \lambda_i^{-1} - 2) = \sum_{i=1}^{n} (\lambda_i - 1)^2/\lambda_i, \end{aligned} \tag{37}$$

and hence

$$\begin{aligned} \delta_\alpha &= \left(\frac{\det v_\alpha}{\det r_\alpha}\right)\left(\frac{\det v_\alpha}{\det s_\alpha}\right) = \det(v_\alpha r_\alpha^{-1}) \det(v_\alpha s_\alpha^{-1}) \\ &= \det(\tfrac{1}{2}[I + (r_\alpha s_\alpha^{-1})^{-1}]) \det[\tfrac{1}{2}(I + r_\alpha s_\alpha^{-1})] \\ &= \prod_{i=1}^{n} \left(\frac{1+\lambda_i^{-1}}{2}\right)\left(\frac{1+\lambda_i}{2}\right) = \prod_{i=1}^{n} \left[\frac{(1+\lambda_i)^2}{4\lambda_i}\right] = \prod_{i=1}^{n} \left[1 + \frac{(\lambda_i - 1)^2}{4\lambda_i}\right]. \end{aligned} \tag{38}$$

From (36)–(38) one has

$$M_\alpha = 4\sum_{i=1}^{n} \frac{(\lambda_i - 1)^2}{4\lambda_i} \le 4\prod_{i=1}^{n}\left(1 + \frac{(\lambda_i - 1)^2}{4\lambda_i}\right) \le 4K_0 < \infty, \qquad \alpha \in D. \tag{39}$$

To find a similar uniform bound on N_α, consider η_α. Both η_α, N_α are (positive) quadratic forms involving v_α^{-1} and $(r_\alpha^{-1} + s_\alpha^{-1})$. As before, one diagonalizes these matrices so that v_α^{-1} goes to the identity and the second matrix has $v_i > 0$, $i = 1, \ldots, n$, for its diagonal elements. Hence writing $\bar{d}_\alpha$ for d_α in the new coordinate system, one has

$$\eta_\alpha = \sum_{i=1}^{n} \bar{d}_{\alpha i}^2, \qquad N_\alpha = \sum_{i=1}^{n} v_i \bar{d}_{\alpha i}^2. \tag{40}$$

Here $v_i^{-1} > 0$ are the eigenvalues of $v_\alpha^{-1}(r_\alpha^{-1} + s_\alpha^{-1})^{-1}$ or v_i of $v_\alpha(r_\alpha^{-1} + s_\alpha^{-1}) = I + \frac{1}{2}(r_\alpha s_\alpha^{-1} + (r_\alpha s_\alpha^{-1})^{-1})$. Hence $v_i = 1 + \frac{1}{2}(\lambda_i + \lambda_i^{-1})$, $i = 1, \ldots, n$, in the earlier notation. So

$$v_i = \frac{(\lambda_i + 1)^2}{2\lambda_i} = 2\left[\frac{(\lambda_i - 1)^2}{4\lambda_i} + 1\right] \leq 2\delta_\alpha \qquad \text{by} \quad (38).$$

Now using (40), one has

$$N_\alpha = \sum_{i=1}^{n} v_i \bar{d}_{\alpha i}^2 \leq 2\delta_\alpha \sum_{i=1}^{n} \bar{d}_{\alpha i}^2 \leq 2K_0^2 < \infty, \qquad \alpha \in D. \tag{41}$$

Finally, (39) and (41) yield $I_\alpha \leq 2K_0 + K_0^2 < \infty$ uniformly in α. Hence $I = \sup_\alpha I_\alpha \leq 2K_0 + K_0^2 < \infty$ as asserted.

IV. If r_α or s_α is not strictly positive definite on T for some $\alpha \in D$, then either r_α or s_α is degenerate. Then $Q_\alpha \perp P_\alpha$ by Step II (or Theorem 5). If both are invertible on a set $T_0 \subset T$, which is maximal for this property, so that r, s are strictly positive definite on T_0, the work of Step III applies to T_0. On $T - T_0$ both are degenerate since $H > 0$ implies only one of them cannot be nondegenerate. Our assumption thus yields that on $T - T_0$ both P, Q have their finite dimensional distributions concentrating on the same set of points and hence are equivalent. Thus on $T - T_0$ one always has $P \equiv Q$ and $I = 0$ or $H = 1$. This means $Q \equiv P$ on T_0 iff it is so on T itself—similarly for mutual singularity. This proves that $Q \equiv P$ or $Q \perp P$ according as $H > 0$ or $H = 0$, or equivalently, $I < \infty$ or $I = \infty$.

V. To prove the last part, consider $H_\alpha = \delta_\alpha^{-1/4} \exp(-\frac{1}{4}\eta_\alpha)$. Then $H_\alpha \geq H$ and $H > 0$ iff both $\inf_\alpha \delta_\alpha^{-1/4} > 0$ *and* $\inf_\alpha \exp(-\frac{1}{4}\eta_\alpha) > 0$. Let $m = n = 0$ so that $\eta_\alpha = 0$, $\alpha \in D$. We have $\inf_\alpha \delta_\alpha^{-1/4} > 0$ iff $\inf_\alpha H_\alpha > 0$ by the first part. On the other hand, when $\inf_\alpha \delta_\alpha^{-1/4} > 0$, one can have $\inf_\alpha H_\alpha > 0$ iff $\inf_\alpha \exp(-\frac{1}{4}\eta_\alpha) > 0$. Thus $H > 0$ iff $P(0, r) \equiv Q(0, s)$ *and* $P(m, v) \equiv Q(n, v)$. But then one also has $P(0, r) \equiv Q(0, v)$ and $P(n, r) \equiv Q(n, v)$. Thus by transitivity of the equivalence $\equiv$, one concludes that $P(m, r) \equiv P(n, r)$. Hence $H > 0$ iff $P(0, r) \equiv Q(0, s)$ *and* $P(m, r) \equiv P(n, r) \equiv Q(n, r)$, which is precisely the final part of the assertion. This completes the proof of the theorem.

The preceding proof is essentially due to Shepp [1]. The last part of the result has been first given by C. R. Rao and Varadarajan [1]. What this theorem says is to strengthen the result of Lemma 1 considerably in that we can distinguish perfectly (i.e., probability of error is 0) between the two measures P, Q based on observing a sample path for sufficiently many points of T, when the distributions are Gaussian. (See a related result in the Complements and Problems section about a Poisson case.) It is now of interest to simplify and give easily verifiable conditions on the means and covariance functions if the latter belong to some familiar classes. A detailed study of several problems on this topic has first been given by Shepp [2] and recently by Rozanov [1]. To indicate the scope, let us state the following result from Shepp [2].

8. Theorem *Let P be a Gaussian measure on $(\mathbb{R}^T, \mathscr{B}^T)$ with mean function m and covariance function r, where $T = [0, a]$, $a > 0$. Let Q be the (Wiener or) Brownian motion measure, i.e., a Gaussian measure with mean zero and covariance r_0 given by $r_0(u, v) = \min(u, v)$. Then $Q \equiv P$ iff there exists a function $k \in L^2(T \times T, dx\, dy)$, real Lebesgue space, $k(u, v) = k(v, u)$, for which*

$$r(s, t) = \min(s, t) - \int_0^s \int_0^t k(u, v)\, du\, dv, \qquad s \in t, \quad t \in T,$$

$$\int_0^a k(t, u)\varphi(u)\, du \neq \varphi(t) \text{ a,a, } (t), \qquad \textit{for any} \quad \varphi \in L^2(T, dx),$$

and $m'(t) = (dm/dt)(t)$ exists a.e. and $m' \in L^2(T, dx)$. Moreover, k is unique and is given by $k(s, t) = -(\partial^2/\partial s\, \partial t)r(s, t)$ for almost all (s, t) in $T \times T$.

We omit the details referring the reader to the original paper. For a different proof, see the author's paper [9, Section 6].

All the computations were needed in Theorem 7 since one has to use special information from H to I. Can one say anything in the nonsingular case for general (not necessarily Gaussian) measures? Indeed, there is an analog if each P_α, Q_α is a *product* measure. This was the original theorem due to Kakutani [1]. We present it here for comparison as well as for applications.

9. Theorem *Let $\left\{\left(\Omega_n, \Sigma_n, \begin{matrix} P_n \\ Q_n \end{matrix}, \pi_{mn}\right), n \geq m \geq 1\right\}$ be two projective systems of probability measures such that $\Omega_n = \times_{i=1}^n \Omega^i$, $\Sigma_n = \bigotimes_{i=1}^n \Sigma^i$, $P_n = \bigotimes_{i=1}^n P^i$, and $Q_n = \bigotimes_{i=1}^n Q^i$, $\pi_{mn}: \Omega_n \to \Omega_m$, the coordinate projection. Here $\left(\Omega^i, \Sigma^i, \begin{matrix} P^i \\ Q^i \end{matrix}\right)$ is a probability space for each i and Q_n, P_n are product measures. Then (the Fubini–Jessen theorem) if $\Omega = \times_{i=1}^\infty \Omega^i$, $\Sigma = \bigotimes_{i=1}^\infty \Sigma^i$, $P = \bigotimes_{i=1}^\infty P^i$, $Q = \bigotimes_{i=1}^\infty Q^i$, we have $\left(\Omega, \Sigma, \begin{matrix} P \\ Q \end{matrix}\right) = \varprojlim \left(\Omega_n, \Sigma_n, \begin{matrix} P_n \\ Q_n \end{matrix}, \pi_{mn}\right)$. Furthermore, $Q \equiv P$ iff $H = \lim_n H_n = \lim_n H(Q_n, P_n) > 0$ and $Q \perp P$ iff $H = 0$.*

Proof That $Q \perp P$ iff $H = 0$ is true more generally has been noted. We only need to prove the equivalence when $H > 0$. By our earlier discussion, for this one may as well assume that $Q_n \equiv P_n$ for each $n \geq 1$ since otherwise one gets $H = 0$ and singularity results. Hence $Q^i \equiv P^i$, $i \geq 1$. It is now immediate from the definition of the product measure that

$$u_n(\omega_n) = \frac{dQ_n}{dP_n}(\omega_n) = \prod_{i=1}^{n} \frac{dQ^i}{dP^i}(\omega^i) \quad \text{a.e.} \qquad (\omega_n = (\omega^1, \ldots, \omega^n) \in \Omega_n). \quad (42)$$

Thus (20) becomes

$$H(Q_n, P_n) = \int_{\Omega_n} u_n^{1/2}(\omega_n)\, dP_n = \prod_{i=1}^{n} \int_{\Omega^i} \left(\frac{dQ^i}{dP^i}\right)^{1/2} dP^i = \prod_{i=1}^{n} H(Q^i, P^i). \quad (43)$$

Since $H(Q_n, P_n) \downarrow H > 0$, one has $\lim_{n\to\infty} \prod_{i=1}^{n} H(Q^i, P^i) = \prod_{i=1}^{\infty} H(Q^i, P^i) = H > 0$. But $H = H(Q, P) = \int_\Omega (dQ^c/dP)^{1/2}\, dP$, and one has to show that the positivity of the above infinite product implies $Q^c = Q$ relative to P (and then that $P^c = P$, relative to Q). [Note that $H > 0 \Rightarrow \prod_{i=m+1}^{n} H(Q^i, P^i) \to 1$, as $m, n \to \infty$.]

Let $g_n^2 = dQ^n/d\mu^n$, $f_n^2 = dP^n/d\mu^n$, where $\mu^n = P^n + Q^n$. Then our earlier discussion (cf. (19)) implies that $dQ^n/dP^n = (g_n/f_n)^2$ and $u_n(\omega_n) = \prod_{k=1}^{n} (g_k/f_k)^2(\omega_n)$ a.e., where $(g_k/f_k)(\omega_n) = g_k(\omega^k)/f_k(\omega^k)$, i.e., evaluated at the kth coordinate of $\omega_n \in \Omega_n$. By (8) one has $H(Q_n, P_n)$ to be an inner product in $(L^2(\Omega_n, \Sigma_n, \mu_n))^+$, defined as $H(Q_n, P_n) = (\prod_{i=1}^{n} g_i, \prod_{i=1}^{n} f_i)$. Hence, with the notation and work between (26)–(27), if $u_n^* = dQ_n^*/dP_n^*$ on $\mathfrak{F}_n = \pi_n^{-1}(\Sigma_n)$ so that $u_n^* = u_n \circ \pi_n$ a.e., one has by (8′) and (43) for $n > m \geq 1$:

$$\|\sqrt{u_n^*} - \sqrt{u_m^*}\|_2^2 = 2\left(1 - \prod_{k=m+1}^{n} \int_{\Omega^k} \left(\frac{g_k}{f_k}\right) dP^k\right) = 2\left(1 - \prod_{k=m+1}^{n} H(Q^i, P^i)\right). \quad (44)$$

Using the convergence of the infinite product above, it follows from (44) that $\{\sqrt{u_n^*}\}_1^\infty \subset L^2(\Omega, \Sigma, P)$ is a Cauchy sequence and hence $\sqrt{u_n^*} \to u$ in L^2. Evidently this is also a consequence of Theorem 6 above. Let us show that $Q \equiv P$ and $dQ/dP = u^2$ a.e., to complete the proof. But $\{u_n^*, \mathfrak{F}_n, n \geq 1\}$ is a martingale where $\mathfrak{F}_n = \pi_n^{-1}(\Sigma_n)$ and $\Sigma = \sigma(\bigcup_{n\geq 1} \mathfrak{F}_n)$. By the above result (as in Theorem 6), $u_n^* = E^{\mathfrak{F}_n}(u^2)$ since by finiteness of the measures P, Q, $u_n^* \in L^1(\Omega, \mathfrak{F}_n, P_n^*)$ and $u^2 \in L^1(\Omega, \Sigma, P)$. That the mean convergence implies uniform integrability is used in deducing that $u_n^* = E^{\mathfrak{F}_n}(u^2)$. Thus

$$\int_A u^2\, dP = \int_A E^{\mathfrak{F}_n}(u^2)\, dP = \int_A u_n^*\, dP_n^* = Q_n^*(A) = Q(A), \qquad A \in \mathfrak{F}_n, \quad (45)$$

and hence for all $A \in \bigcup_{n=1}^{\infty} \mathfrak{F}_n$. Since Q is σ-additive, Eq. (45) also holds for all $A \in \sigma(\bigcup_{n=1}^{\infty} \mathfrak{F}_n)$. Thus $dQ/dP = u^2$ a.e. $[P]$, so $Q^c = Q$ for P. By symmetry (interchanging Ps and Qs above) one has $P^c = P$ for Q and $Q \equiv P$, as asserted.

Remark As we noted before, without further hypothesis one cannot say anything about equivalence from nonsingularity of a pair of probability measures P, Q under the only hypothesis that $H(Q, P) > 0$. For instance, let P, Q on $\mathbb{R}$ be defined by $P(A) = \int_A f_1(x)\,dx$, $Q(B) = \int_B f_2(x)\,dx$, A, B Borel sets, and f_1 a Gaussian density and f_2 a gamma density, e.g., f_1 is given by (25) with $n = 1$ and

$$f_2(x) = \begin{cases} e^{-x} & \text{for} \quad x \geq 0, \\ 0 & \text{for} \quad x < 0. \end{cases}$$

Then $H(Q, P) > 0$, but $Q \equiv P$ is false. Also Theorem 6 (or even Theorem 2.5) says that the densities $u_n^* \to u^2$ a.e. and in L^1 so that Q is P-continuous but not when Q, P are interchanged. Thus independence was needed for this "symmetry" argument or a special class (such as Gaussian) is needed for the stronger conclusion that $Q \equiv P$. (See Problem 9 for a class of Poisson processes.)

Thus far we have considered the question of equivalence or singularity of Gaussian (and other) measures. But to use Lemma 1, we need to compute the likelihood ratios of these processes also in the case of equivalence. Let us present a result on this problem. It indicates the interest and need for further developments before one is able to treat more results of this kind in inference theory.

The following theorem adds information to that contained in Theorem 7.

10. Theorem *Let P, Q be two nondegenerate Gaussian measures on $(\mathbb{R}^T, \mathscr{B}^T)$ with means m, n and the same covariance functions r. Suppose that $Q \equiv P$ or equivalently (by Theorem 7) $P(0, r) \equiv Q(n - m, r)$. Then for each $a \in \mathbb{R}$, $P(0, r) \equiv Q(a(n - m), r)$. Moreover, there is a linear functional l: $\mathbb{R}^T \to \mathbb{R}$ and a constant $C \geq 0$ such that the likelihood ratio is given by*

$$\frac{dQ(a(n-m), r)}{dP(0, r)}(\omega) = \exp\{al(\omega) - \tfrac{1}{2}a^2 C\}. \tag{46}$$

Proof Let $f = n - m$ and set $Q(a(n - m), r) = P_{af}$. By hypothesis $dP_f/dP = g$ exists a.e. (and $g > 0$ a.e.) $[P]$. So by the remark following Lemma 1 (cf. also Step II of the proof of Theorem 7) there exists a countable set $J \subset T$ such that g is $\mathfrak{F}_J = \pi_J^{-1}(\mathscr{B}^J)$ $(\subset \mathscr{B}^T)$-measurable. Let $\alpha_1 < \alpha_2 < \cdots$ be an increasing sequence of finite subsets of J and let $\mathfrak{F}_n = \sigma\{\{\omega : \bigcap_{t \in \alpha_n}[\omega(t) < b]\}, b \in \mathbb{R}\} = \pi_{\alpha_n}^{-1}(\mathscr{B}^{\alpha_n})$ so that $\mathfrak{F}_n \subset \mathfrak{F}_{n+1}$ and $\sigma(\bigcup_{n \geq 1} \mathfrak{F}_n) = \mathfrak{F}_J$. We again denote by $r_n = (r(t_i, t_j) : t_i \in \alpha_n, t_j \in \alpha_n)$ the nth order covariance matrix which then satisfies $d_n = \det(r_n) > 0$ for each n because of the nondegeneracy of P.

Let $h: \mathbb{R}^{\alpha_n} \to \mathbb{R}$ be a bounded Baire function so that $G = h \circ \pi_{\alpha_n}$ is a bounded $\mathfrak{F}_n$-measurable function on $\mathbb{R}^T = \Omega$ since π_{α_n} is $(\mathfrak{F}_n, \mathscr{B}^{\alpha_n})$-measurable and in fact $P_n^* \ (= P | \mathfrak{F}_n) = P_n \circ \pi_{\alpha_n}^{-1}$ (i.e., measure preserving), where P_n is the n-dimensional Gaussian measure on $\mathbb{R}^{\alpha_n}$ with mean zero and covariance r_n. Hence one has with $x = (x_1, \ldots, x_n) \in \mathbb{R}^{\alpha_n} = \mathbb{R}^n$

$$\int_\Omega G(\omega)\, dP = \int_{\mathbb{R}^n} h(x_1, \ldots, x_n)\, dP_n$$

$$= \frac{1}{(2\pi)^{n/2} d_n^{1/2}} \int_{\mathbb{R}^n} h(x_1, \ldots, x_n) \exp\left(-\frac{1}{2d_n} (r_n^{-1} x, x) \right) dx_1, \ldots, dx_n. \tag{47}$$

Define l_n and C_n (if $r_n^{-1} = (s_{ij}^n)$) as

$$l_n(\omega) = \frac{1}{d_n} \sum_{i,j=1}^{n} s_{ij}^n \omega(t_i) f(t_j),$$

$$C_n = \frac{1}{d_n} \sum_{i,j=1}^{n} s_{ij}^n f(t_i) f(t_j). \tag{48}$$

Then $l_n: \Omega \to \mathbb{R}$, and $C_n \geq 0$ are well defined for each n. Hence (47) becomes

$$\int_\Omega G(\omega) \exp[a l_n(\omega) - \tfrac{1}{2} a^2 C_n]\, dP$$

$$= \frac{1}{(2\pi)^{n/2} d_n^{1/2}} \int_{\mathbb{R}^n} h(x_1, \ldots, x_n) \exp\left[-\frac{1}{2d_n} (r_n^{-1}(x - af_n), (x - af_n)) \right] dP_n,$$

where $\quad f_n = (f(t_1), \ldots, f(t_n)) \in \mathbb{R}^n$,

$$= \frac{1}{(2\pi)^{n/2} d_n^{1/2}} \int_{\mathbb{R}^n} h(x + af_n) \exp\left[-\frac{1}{2d_n} (r_n^{-1} x, x) \right] dP_n$$

by a change of variable

$$= \int_\Omega G(\omega + af)\, dP = \int_\Omega G(\omega)\, dP_{af}. \tag{49}$$

Here we used the elementary result that $\exp[a l_n(\omega) - \frac{1}{2} a^2 C_n]$ is integrable for the Gaussian measure P so that the second part of the first line is finite, whence the left side is also and equality holds. If $a = 1$ (as (49) shows), since $G(\cdot)$ is an arbitrary bounded function belonging to a dense set in $L^1(\Omega, \mathfrak{F}_n, P)$, one concludes that (take $G = \chi_A$, $A \in \mathfrak{F}_n$) $d(P_f)_n^* / dP_n^* = \exp[a l_n(\cdot) - \frac{1}{2} a^2 C_n]$ a.e. Hence by Theorem 2.5 (or Theorem 6 or Theorem 2.4) this converges a.e. and in L^1-mean to $d(P_f)_J/dP_J$. In this a.e. statement if $\omega \in \Omega$ is a point of convergence, then so is $-\omega$ (since P has mean zero). Consequently, $\exp(al_n(-\omega) - \frac{1}{2}a^2C_n)$ also converges to a finite number; but $a \in \mathbb{R}$ and then $-(al_n(-\omega) - \frac{1}{2}a^2C_n) - (al_n(\omega) - \frac{1}{2}a^2C_n) = a^2C_n$. So C_n converges to a constant $C \geq 0$. It now follows that $l_n(\omega) \to l(\omega)$ for almost all ω and $l(\cdot)$ is a linear functional on $\Omega = \mathbb{R}^T$ since each l_n is. Hence one has $d(P_f)_J/dP_J$ to be equal a.e. $[P]$ to $\exp[l(\cdot) - \frac{1}{2}C]$. By (48) $l(\cdot)$ is the limit of sums and thus it can only be expressed as an integral on the function space Ω—a "stochastic integral." (We must therefore omit its characterization since these integrals have not been defined here.)

We use this result to show that $P_{af} \equiv P_f$ and then deduce their density. However, taking $G = \chi_A$, $A \in \mathfrak{F}_n$, one can express (49) as

$$P_{af}(A) = \int_A dP_{af} = \int_A \exp[al_n(\omega) - \tfrac{1}{2}a^2C_n]\, dP_n^*, \qquad a \in \mathbb{R}. \tag{50}$$

Hence $\{X_n, \mathfrak{F}_n, n \geq 1\}$ is a positive martingale where $X_n(\omega) = \exp(al_n(\omega) - \frac{1}{2}a^2C_n)$ so that $X_n \to X_\infty$ a.e. by Theorem 2.4 and $X_\infty = d(P_{af})_J^c/dP_J$ a.e. If it is shown that the X_n sequence is uniformly integrable, then one may interchange limits in (50) for all $A \in \mathfrak{F}_J$ and deduce that $(P_{af})_J = (P_{af})^c$. Note that $X_\infty = \exp[al(\cdot) - \frac{1}{2}a^2C]$ a.e., so that when this is proved (since $X_\infty > 0$ a.e. $[P]$), one can conclude that $P_{af} \equiv P$ also.

By Hölder's inequality (or by Theorem 1.14) it suffices to show for uniform integrability that $\{X_n\}_1^\infty$ is bounded in $L^p(\Omega, \mathscr{B}, P)$ for some $p > 1$. We establish this for $p = 2$ here. Thus let $G \equiv 1$ and $a = 2$ in (49). Then one has, since $P_{2f}(\Omega) = 1$,

$$\begin{aligned}\int_\Omega X_n^2(\omega)\, dP &= \int_\Omega \exp[2al_n(\omega) - 2a^2C_n]\exp[a^2C_n]\, dP \\ &= \exp(a^2C_n)P_{2f}(\Omega) = \exp(a^2C_n).\end{aligned}$$

Since $\exp(a^2C_n) \to \exp a^2C$, this sequence of numbers is bounded so that $\|X_n^2\|_2 \leq K_0 < \infty$ for all n. This completes the proof.

Let us indicate how one can combine this result with Lemma 1 to get a set $A_K \subset \Omega$ which is a critical region. Given $\varepsilon > 0$, find $K_\varepsilon > 0$ such that

$$A_K = \{\omega : \exp al(\omega) - \tfrac{1}{2}a^2C \geq K_\varepsilon\}, \qquad P(A_k) = \varepsilon,$$

where one can find an exact value for the probability since P is nonatomic. Thus we distinguish P and P_{af} for $a \neq 0$, by deciding on P_{af} if $\omega \in A_{K_1}$, where

$$A_{K_1} = \{\omega : al(\omega) \geq K_1\}, \qquad K_1 = K_\varepsilon + \tfrac{1}{2}a^2C,$$

and deciding on P if $\omega \in A_{K_1}^c$. In any inference problem, it is therefore important that one is able to calculate the likelihood ratios as here. The above theorem is due to Pitcher [1] where other interesting results may be found. The first systematic study of inference on processes is due to Grenander [1].

It is clearly possible to consider, in the Gaussian family, various processes when the covariance and mean functions are known to satisfy other conditions. However, there is a need for new tools to analyze such problems. For example, some problems are on "separability and measurability" questions of a process, i.e., the function theoretical aspects, and then the "stochastic integration." A further study of this subject will not be undertaken here. (Such a study has been included by the author in [10].) Instead, Section 4.5 is devoted to a generalized class of martingale processes.

4.5 ASYMPTOTIC MARTINGALES

As noted in Remark 2.11, a (sub) martingale $\{X_n, \mathfrak{F}_n, n \geq 1\}$ may be considered as a sequence such that the expected values of their increments satisfy $E^{\mathfrak{F}_n}(X_{n+1} - X_n) = 0\ (\geq 0)$ for each $n \geq 1$ and further, by Theorem 2.7, for each pair of simple stopping times τ_1, τ_2 of $\{\mathfrak{F}_n, n \geq 1\}$, $\tau_1 \leq \tau_2$, $E(X_{\tau_2} - X_{\tau_1}) = 0\ (\geq 0)$. The latter property may be taken as a definition, and consider processes for which these relations are valid thereby enlarging such classes for analysis. This new class is called *asymptotic martingales* and is stated as follows.

1. Definition Let $\{X_n, \mathfrak{F}_n, n \geq 1\}$ be an adapted sequence in $L^1(P)$ and $\mathscr{S}$ be the directed set of all simple stopping times of $\{\mathfrak{F}_n, n \geq 1\}$. Then the process $\{X_n, \mathfrak{F}_n, n \geq 1\}$ is called an *asymptotic martingale* ("*amart*" for short) if the net of real numbers $\{a_\tau, \tau \in \mathscr{S}\}$ converges where $a_\tau = E(X_\tau)$.

Recall that the net $\{a_\tau, \tau \in \mathscr{S}\}$ converges iff the Cauchy criterion is valid. So for $\varepsilon > 0$ there is an $\alpha \in \mathbb{R}$ and a $\tau_\varepsilon \in \mathscr{S}$ such that for $\tau > \tau_\varepsilon$ one has $|a_\tau - \alpha| < \varepsilon$. A similar definition holds for decreasing sequences also. A detailed treatment of amarts has been undertaken by Edgar and Sucheston [1]. The following account leans on their work. (Cf. also Baxter [1].) It is thus clear that the very generalization involves stopping times. One can consider the continuous time parameter (t in $\mathbb{R}^+$) processes by replacing $\mathscr{S}$ by the bounded set of all stopping times of $\{\mathfrak{F}_t, t \in \mathbb{R}^+\}$. Here only the discrete case is briefly discussed. As it will become clear, many of the basic ideas of proofs of martingale theory apply in the present extensions. The notations of Section 1 will be used.

One of the immediate advantages of this extension is that the space of amarts is a vector lattice. More precisely, one has:

2. Proposition *Let $\{\mathfrak{F}_n, n \geq 1\}$ be a fixed stochastic base in (Ω, Σ, P). Then the set of all $L^1(P)$-bounded amarts on this base is a (real) vector space. In particular, every truncation (i.e., $X_n^c = X_n$ if $|X_n| \leq c$, $X_n^c = +c$ if $X_n > c$, $X_n^c = -c$ if $X_n < -c$) of an amart is an amart.*

Proof Let $\{X_n^{(i)}, \mathfrak{F}_n, n \geq 1\}$, $i = 1, 2$, be a pair of amarts and $Y_n = X_n^{(1)} \vee X_n^{(2)}$. It is to be shown that $\{Y_n, \mathfrak{F}_n, n \geq 1\}$ is an amart. By hypothesis, given $\varepsilon > 0$, there exists a $\tau_0 \in \mathscr{S}$ such that for all τ, τ' in $\mathscr{S}$, $\tau, \tau' \geq \tau_0$, one has

$$\left| \int_\Omega (X_\tau^{(i)} - X_{\tau'}^{(i)})\, dP \right| < \tfrac{1}{2}\varepsilon, \qquad i = 1, 2. \tag{1}$$

Hence for all $\tau \geq \tau_0$, $\tau \in \mathscr{S}$, $i = 1, 2$,

$$\begin{aligned} \left| \int_\Omega X_\tau^{(i)}\, dP \right| &\leq \left| \int_\Omega (X_\tau^{(i)} - X_{\tau_0}^{(i)})\, dP \right| + \left| \int_\Omega X_{\tau_0}^{(i)}\, dP \right| \\ &\leq \tfrac{1}{2}\varepsilon + \int_\Omega |X_{\tau_0}^{(i)}|\, dP = K_i < \infty \end{aligned} \tag{2}$$

so that $\{\int_\Omega X_\tau^{(i)}\, dP, \tau \geq \tau_0\}$ is a bounded set of reals, $i = 1, 2$. On the other hand, since $(a_1 \vee a_2)^+ \leq a_1^+ + a_2^+$ for any real numbers a_i, $i = 1, 2$, consider ($\tau \geq \tau_0$),

$$\int_\Omega Y_\tau\, dP \leq \int_\Omega Y_\tau^+\, dP \leq \int_\Omega (X_\tau^{(1)})^+\, dP + \int_\Omega (X_\tau^{(2)})^+\, dP. \tag{3}$$

If $\tau \geq \tau_0$ is arbitrary, let N be a large integer ($\geq \tau$) and define τ_1, τ_2 by

$$\tau_1 = \tau \chi_{[X_\tau^{(1)} \geq 0]} + N \chi_{[X_\tau^{(1)} < 0]}, \qquad \tau_2 = \tau \chi_{[X_\tau^{(2)} \geq 0]} + N \chi_{[X_\tau^{(2)} < 0]}. \tag{4}$$

Then $\tau_i \in \mathscr{S}$, and (3) becomes

$$\begin{aligned} \int_\Omega Y_\tau\, dP &\leq \int_\Omega X_{\tau_1}^{(1)} - \int_{[X_\tau^{(1)} < 0]} X_N^{(1)}\, dP + \int_\Omega X_{\tau_2}^{(2)}\, dP - \int_{[X_\tau^{(2)} < 0]} X_N^{(2)}\, dP \\ &\leq K_1 + K_2 + \sup_n \int_\Omega |X_n^{(1)}|\, dP + \sup_n \int_\Omega |X_n^{(2)}|\, dP < \infty \end{aligned} \tag{5}$$

by (2) and the fact that $\{X_n^{(i)}, n \geq 1\}$ are in a ball of $L^1(P)$. Hence $\{\int_\Omega Y_\tau, \tau \geq \tau_0, \tau \in \mathscr{S}\}$ is bounded. So there exists a $\tau_0' \in \mathscr{S}$, $\tau_0' \geq \tau_0$ such that

$$\int_\Omega Y_\tau\, dP < \int_\Omega Y_{\tau_0'}\, dP + \varepsilon, \qquad \tau \geq \tau_0', \quad \tau \in \mathscr{S}. \tag{6}$$

To show that $\{\int_\Omega Y_\tau\,dP, \tau \in \mathcal{S}\}$ is Cauchy, we derive a lower inequality corresponding to (6) which then yields the desired conclusion. Let $A = [X_{\tau_0'}^{(1)} < X_{\tau_0'}^{(2)}]$, and for any $\tau^* \geq \tau_0'$ let $\sigma = \tau_0'\chi_A + \tau^*\chi_{A^c}$. Since $A \in \mathcal{G}(\tau_0')$ and the traces of $\mathcal{G}(\tau)$ and $\mathfrak{F}_n$ agree on $[\tau = n]$, it follows that $\sigma \in \mathcal{S}$. Hence noting that $Y_{\tau_0'} = X_{\tau_0'}^{(2)}$ on A, and $Y_{\tau_0'} = X_{\tau_0'}^{(1)}$ on A^c, one has [$\mathcal{G}(\tau)$ as in Eq. (1.5)]

$$\int_\Omega X_{\tau_0'}^{(1)}\,dP = \int_{A^c} Y_{\tau_0'}\,dP + \int_A X_{\tau_0'}^{(1)}\,dP,$$

$$\int_\Omega X_\sigma^{(1)}\,dP = \int_{A^c} X_{\tau^*}\,dP + \int_A X_{\tau_0'}^{(1)}\,dP.$$

Eliminating the common integral over A from these two, we have

$$\int_{A^c} Y_{\tau_0'}\,dP = \int_\Omega X_{\tau_0'}^{(1)}\,dP + \int_{A^c} X_{\tau^*}^{(1)}\,dP - \int_\Omega X_\sigma^{(1)}\,dP. \tag{7}$$

Since $\sigma \geq \tau_0' \geq \tau_0$, (1) and (7) imply

$$\int_{A^c} Y_{\tau_0'}\,dP \leq \int_{A^c} X_{\tau^*}^{(1)}\,dP + \tfrac{1}{2}\varepsilon = \int_{A^c} Y_{\tau^*}\,dP + \tfrac{1}{2}\varepsilon. \tag{8}$$

By a similar computation with $X_{\tau_0'}^{(2)}$ on A, one obtains

$$\int_A Y_{\tau_0'}\,dP \leq \int_A Y_{\tau^*}\,dP + \tfrac{1}{2}\varepsilon. \tag{8$'$}$$

Adding these two yields $\int_\Omega Y_{\tau_0'}\,dP \leq \int_\Omega Y_{\tau^*}\,dP + \varepsilon$. This and (6), with $\tau = \tau^*$ there, imply for $\tau^* \geq \tau_0'$ in $\mathcal{S}$

$$\left|\int_\Omega (Y_{\tau_0'} - Y_{\tau^*})\,dP\right| < \varepsilon. \tag{9}$$

It follows from (9) that $\{Y_n, \mathfrak{F}_n, n \geq 1\}$ is an amart. A similar procedure for $\{Z_n, \mathfrak{F}_n, n \geq 1\}$, $Z_n = X_n^{(1)} \wedge X_n^{(2)}$, shows that it is also an amart. Since the last statement is now clear from the preceding, this completes the proof.

On the other hand, the following result indicates that it is nontrivial to check that a given adapted process is an amart.

3. Proposition *Let $\{X_n, \mathfrak{F}_n, n \geq 1\}$ be an adapted process and suppose that $|X_n| \leq Y$ a.e., $n \geq 1$ and $Y \in L^1(P)$. Then the process is an amart iff $X_n \to X_\infty$ a.e.*

Proof Suppose $\tau_n \uparrow \infty$, $\tau_n \in \mathcal{S}$, is any sequence. If $X_n \to X_\infty$ a.e., then $X(\tau_n) \to X_\infty$ a.e. by a classical result in summability. The latter says that if $M = (a_{ij}, i \geq 1, j \geq 1)$ is an infinite matrix such that $a_{ij} \geq 0$, $\sum_{i=1}^\infty a_{ni} = 1$, $\lim_{n\to\infty} a_{nj} = 0$ for each j, then $\beta_n = \sum_{i=1}^\infty a_{ni}\alpha_i \to \alpha_\infty$ as $n \to \infty$ whenever the numbers $\alpha_n \to \alpha_\infty$. (Cf., e.g., Zygmund [1, p. 74].) Here we let

$a_{nj} = \chi_{[\tau_n = j]}(\omega)$, $\alpha_n = X_n(\omega)$, and $\alpha_\infty = X_\infty(\omega)$ so that $\beta_n = X(\tau_n)(\omega) = \sum_{k=1}^{N_n} X_k \chi_{[\tau_n = k]}(\omega)$. Since $|X_n| \le Y$ so that $|X(\tau_n)| \le Y_n$, it follows by the dominated convergence criterion that $\int_\Omega X(\tau_n)\, dP \to \int_\Omega X_\infty\, dP$. But $\{\tau_n \uparrow \infty\} \subset \mathscr{S}$ is arbitrary. So the net $\{\int_\Omega X_\tau\, dP, \tau \in \mathscr{S}\}$ must converge, and hence $\{X_n, \mathfrak{F}_n, n \ge 1\}$ is an amart.

Conversely, let the process be an amart and X_*, X^* be the "lim inf" and "lim sup" of the given sequence. Then by Theorem 1.9

$$0 \le \int_\Omega (X^* - X_*)\, dP \le \limsup_{\tau, \tau' \in \mathscr{S}} \int_\Omega (X_\tau - X_{\tau'})\, dP.$$

But as shown in the proof there (cf. Eq. (21)), this can be written also as

$$\int_\Omega (X^* - X_*)\, dP = \lim_{n \to \infty} E(X_{\tau_n}) - \lim_{\tau_n' \to \infty} E(X_{\tau_n'}) \tag{10}$$

for some $\tau_n \uparrow \infty$, $\tau_n' \uparrow \infty$, $\{\tau_n, \tau_n'\}_1^\infty \subset \mathscr{S}$. By the amart hypothesis the right-side limits of (10) are both finite (since $|X_{\tau_n}| \le Y \in L^1(P)$) and equal. (The whole net $\{\int_\Omega X_\tau\, dP, \tau \in \mathscr{S}\}$ converges so that each convergent subsequence shall have the same limit in the metric topology of $\mathbb{R}$.) Hence $X^* = X_*$ a.e., and $X_n \to X_\infty$ a.e. This completes the proof.

The above ideas and methods show that we are extending the (sub) martingale theory (L^1-bounded case). Hence the following result is the expected analog of Theorems 1.4 and 1.6, which thus unifies both the statements.

4. Theorem *Let $\{X_n, \mathfrak{F}_n, n \ge 1\}$ be an $L^1(P)$-bounded amart. Then $X_n \to X_\infty$ a.e. and $E(|X_\infty|) \le \lim_n \inf E(|X_n|)$.*

Proof Let X_n^c be the truncation of the X_n-process at $c > 0$. Then by Proposition 2 $\{X_n^c, \mathfrak{F}_n, n \ge 1\}$ is also an amart. Since $|X_n^c| \le c$ a.e. and $c \in L^1(P)$, it follows by Proposition 3 that $X_n^c \to \tilde{X}_\infty$ a.e. for each $0 < c < \infty$. Now one uses the argument of the classical Egorov theorem to complete the proof by showing that given $\varepsilon > 0$, there is a $\lambda_\varepsilon > 0$ (large) such that $P[\sup_n |X_n| > \lambda_\varepsilon] < \varepsilon$. Let us fill in the details here.

Since $\{X_n, \mathfrak{F}_n, n \ge 1\}$ and hence $\{-X_n, \mathfrak{F}_n, n \ge 1\}$ are amarts, and both are $L^1(P)$-bounded, it follows by Proposition 2 that $Y_n = X_n \vee (-X_n) = |X_n|$ defines an amart for $\{\mathfrak{F}_n, \text{n} \ge 1\}$. Let $A_\lambda^N = [\sup_{1 \le n \le N} |X_n| > \lambda]$. Define

$$\tau^N = \begin{cases} \inf\{n \le N : |X_n| > \lambda, \sup\limits_{n \le N} |X_n| > \lambda\} \\ N \quad \text{if} \quad \{\ \} = \varnothing. \end{cases} \tag{11}$$

Then $\tau^N \in \mathscr{S}$, and one has

$$\lambda P(A_\lambda^N) \leq \int_{A_\lambda^N} |X_{\tau^N}| \, dP \leq \int_\Omega |X_{\tau^N}| \, dP \leq K_0, \qquad N \geq 1, \tag{12}$$

where $K_0 < \infty$. Indeed, by the amart hypothesis (cf. Eq. (2)), there exists $\tau_0 \in \mathscr{S}$ such that (for the positive amart Y_n here) $\int_\Omega Y_\tau \, dP \leq \tilde{K} < \infty$, $\tau \geq \tau_0$. Hence for any $\tau' \in \mathscr{S}$ one has, if $n_0 \geq \tau_0$,

$$\begin{aligned}\int_\Omega Y_{\tau'} \, dP &\leq \int_\Omega Y_{\tau' \wedge \tau_0} \, dP + \left| \int_\Omega Y_{\tau' \vee \tau_0} \, dP - \int_\Omega Y_{\tau_0} \, dP \right| \\ &\leq \tfrac{1}{2}\varepsilon + \int_\Omega \max_{1 \leq n \leq n_0} |Y_n| \, dP = K_0 < \infty \qquad \text{for each} \quad \tau' \in \mathscr{S}. \end{aligned} \tag{13}$$

Consequently,

$$\lim_{N \to \infty} P(A_\lambda^N) = P\left[\sup_{n \geq 1} |X_n| > \lambda\right] \leq K_0/\lambda, \tag{14}$$

which can be made arbitrarily small if λ is large enough. The last part follows by Fatou's lemma. This completes the proof.

5. Remarks The arguments for (2) and (13) actually show that the net $\{\int_\Omega X_\tau \, dP : \tau \in \mathscr{S}\}$ is a *bounded set* of numbers for any amart $\{X_n, \mathfrak{F}_n, n \geq 1\}$. Hence while every martingale is an amart, not all sub- (or super-) martingales are amarts. The L^1-bounded ones are. Another important class of processes is called *semimartingales*. A process $\{X_n, \mathfrak{F}_n, n \geq 1\}$ is a semimartingale if it can be expressed as $X_n = Y_n + Z_n$, $n \geq 1$, where $\{Y_n, \mathfrak{F}_n, n \geq 1\}$ is a martingale and $\{Z_n, \mathfrak{F}_n, n \geq 1\}$ is a process such that for a.a. (ω), $\{Z_n(\omega), n \geq 1\}$ is of bounded variation. It is clear that (due to the Doob decomposition, cf. Proposition 2.2) every submartingale is a semimartingale. A subclass of semimartingales called *$(*)$-processes* (or *F-processes* by Orey [1]) consists of those defined for each $\{X_n, \mathfrak{F}_n, n \geq 1\}$ satisfying the *$(*)$-condition*

$$E\left[\sum_{i=1}^{\infty} |E^{\mathfrak{F}_n}(X_n - X_{n+1})|\right] < \infty. \tag{15}$$

It can be shown that every $(*)$-process is a semimartingale (i.e., one that admits the above decomposition). An account of semimartingales explaining this distinction is given in the author's monograph [10, Chapter V]. But a $(*)$-process is also an amart though a general semimartingale is not. Let us prove the former assertion here.

By (15), for each $\varepsilon > 0$ there is an n_0 $(=n_0(\varepsilon))$ such that $\sum_{n \geq n_0} E(|X_n - E^{\mathfrak{F}_n}(X_{n+1})|) < \frac{1}{2}\varepsilon$. Let $\tau \in \mathscr{S}$, $\tau \geq n_0$. Since τ is simple, there is an $n_1 > \tau$, and then

$$\begin{aligned}
\left|\int_\Omega (X_\tau - X_{n_1})\,dP\right| &= \left|\sum_{k=n_0}^{n_1} \int_{[\tau=k]} (X_k - X_{n_1})\,dP\right| \\
&= \left|\sum_{k=n_0}^{n_1} \sum_{n=k}^{n_1-1} \int_{[\tau=k]} (X_n - X_{n+1})\,dP\right| \\
&= \left|\sum_{n=n_0}^{n_1-1} \sum_{k=n_0}^{n} \int_{[\tau=k]} (X_n - E^{\mathfrak{F}_n}(X_{n+1}))\,dP\right| \\
&\qquad \text{since} \quad [\tau = k] \in \mathfrak{F}_n \\
&\leq \sum_{n=n_0}^{n_1-1} \sum_{k=n_0}^{n} \int_{[\tau=k]} |X_n - E^{\mathfrak{F}_n}(X_{n+1})|\,dP \\
&\leq \sum_{n \geq n_0} \int_\Omega |X_n - E^{\mathfrak{F}_n}(X_{n+1})|\,dP < \tfrac{1}{2}\varepsilon.
\end{aligned}$$

Thus if $\tau, \tau' \geq n_0$, $\{\tau, \tau'\} \subset \mathscr{S}$, then one may take $n_1 \geq \tau \vee \tau'$ so that

$$\left|\int_\Omega (X_\tau - X_{\tau'})\,dP\right| \leq \left|\int_\Omega (X_\tau - X_{n_1})\,dP\right| + \left|\int_\Omega (X_{\tau'} - X_{n_1})\,dP\right| < \tfrac{1}{2}\varepsilon + \tfrac{1}{2}\varepsilon = \varepsilon$$

by (16). Hence $\{\int_\Omega X_\tau\,dP, \tau \in \mathscr{S}\}$ is convergent, and so $\{X_n, \mathfrak{F}_n, n \geq 1\}$ is an amart whenever it is a $(*)$-process.

In Theorem 4.6 we have proved that a terminally uniformly integrable martingale $\{f_i, \mathfrak{F}_i, i \in I\}$ for any directed index I converges in $L^1(P)$, which implies convergence in probability. The following is an analog of the latter (weaker) result for amarts.

6. Proposition *Let $\{X_i, \mathfrak{F}_i, i \in I\}$ be an $L^1(P)$-bounded amart with I as a directed set. Then $X_i \to X$ in probability. This convergence is pointwise if I is countable and linearly ordered. In general, if there is a $Y \in L^1(P)$ such that $|X_i| \leq Y$ a.e., for the amart, then $X_i \to X$ in $L^1(P)$ also.*

Proof Recall that $X = \{X_i, \mathfrak{F}_i, i \in I\}$ is an amart iff X_i is integrable and for each $\tau \in \mathscr{S}$ (a directed set of simple stopping times of $\{\mathfrak{F}_i, i \in I\}$ as before) the net $\{\int_\Omega X_\tau\,dP : \tau \in \mathscr{S}\}$ converges in $\mathbb{R}$. The argument is essentially an adaption of that of Dunford and Schwartz [1, III.3.7]. In fact, by this last result, if $|X_i| \leq Y$, $Y \in L^1(P)$, then a generalized sequence $\{X_i, i \in I\}$ converges to X in $L^1(P)$ iff that sequence converges in probability to X. Thus if the sequence is an amart, we only need to establish the last property. But this can be done with a small modification of the proof of these authors.

If the generalized sequence is replaced by an ordinary sequence (i.e., X is an amart *sequence* $\{X_n, \mathfrak{F}_n, n \geq 1\}$), then $X_n \to X$ a.e. (hence in probability) by Theorem 4 since it is $L^1(P)$-bounded. It will be shown that the generalized sequence case is deduced from the special case. Suppose the X_i do not converge to X in probability. Then there is an $\varepsilon > 0$, and for each $i \in I$ there is a $j_i \in I$, $j_i \geq i$ such that

$$P[|X_{j_i} - X_i| > \varepsilon] > \varepsilon, \qquad \text{or equivalently}$$

$$d(X_{j_i}, X_i) = E\left(\frac{|X_{j_i} - X_i|}{1 + |X_{j_i} - X_i|}\right) \geq \varepsilon > 0,$$

where E is the expectation. Set $\alpha = j_i$ and consider the sequence $\{X_\alpha, \mathfrak{F}_\alpha, \alpha \in J\}$, where $J \subset I$ is a subdirected set. For each n, choose an $\alpha_n \in J$ such that for all $\tau \in \mathscr{S}$, $\tau \geq \alpha_n$, $|E(X_\tau - X_{\alpha_n})| < (1/n)$. This is possible since the original process is an amart and hence $\{E(X_\tau), \tau \in \mathscr{S}\}$ is a convergent net. Let a be the limit of this net in $\mathbb{R}$. For each n, the above-noted α_n are chosen such that $|E(X_{\alpha_n}) - a| < (1/n)$ and $\alpha_n < \alpha_{n+1}$. Then $\{X_{\alpha_n}, \mathfrak{F}_{\alpha_n}, n \geq 1\}$ is a (linearly ordered) sequence such that for all simple stopping times τ, τ', $|E(X_\tau - X_{\tau'})| \leq |E(X_\tau - X_{\alpha_n})| + |E(X_{\tau'} - X_{\alpha_n})| < (2/n)$ whenever $\tau, \tau' \geq \alpha_n$. Thus this sequence is an amart. Hence $X_{\alpha_n} \to$ a limit a.e. by Theorem 4, implying that $d(X_{\alpha_n}, X_{\alpha_{n+1}}) \to 0$. But this contradicts the fact that $d(X_{\alpha_n}, X_{\alpha_m}) \geq \varepsilon > 0$ for all $n, m \geq 1$ in this subnet. Thus the original supposition is false and $X_i \to X$ in probability holds.

The totally ordered case is left to the reader. This completes the proof.

The next result deals with the *Riesz decomposition* of an amart which is based on a corresponding result for supermartingales, the latter in turn being an extension of a classical result of Riesz for superharmonic functions. It is remarkable that all the extensions are obtained by mere modifications of the classical result and of its proof. Regarding variations and related results of the following theorem, one may refer to Astbury [1]. His definition of "stopping time" in the directed index case is different from the one used here. However, they agree when I is linearly ordered, which is the case considered in the next result.

7. Theorem *Let $\{X_i, \mathfrak{F}_i, i \in I\}$ be an amart where I is linearly ordered. Then X_i admits a unique decomposition as*

$$X_i = Y_i + Z_i, \qquad i \in I, \tag{17}$$

where $\{Y_i, \mathfrak{F}_i, i \in I\}$ is a martingale and $\{Z_i, \mathfrak{F}_i, i \in I\}$ is an amart satisfying (a) $\int_A Z_i \, dP \to 0$ as $i \uparrow$ for each $A \in \bigcup_i \mathfrak{F}_i$ and (b) the net $\{Z_\tau : \tau \in \mathscr{S}\}$ converges to zero in $L^1(P)$.

Proof The construction of Y_i is classical, and its modification to the present case is as follows. Now by definition (cf. (1)) for each $\varepsilon > 0$ there is a $\tau_0 \in \mathscr{S}$ such that for $\{\tau, \tau'\} \subset \mathscr{S}$, $\tau, \tau' > \tau_0$, implies

$$|E(X_\tau - X_{\tau'})| < \varepsilon. \tag{18}$$

For any $\tau > \tau_1 > \tau_0$ and $A \in \mathscr{G}(\tau_1)$, define τ_2 as $\tau_2 = \tau_1 \chi_A + \tau \chi_{A^c}$. [$\mathscr{G}(\tau)$ denotes again the σ-algebra of events prior to τ, cf. p. 174.] It is seen that $\tau_2 \in \mathscr{S}$ so that

$$\begin{aligned}\int_\Omega (X_{\tau_2} - X_\tau)\, dP &= \int_A X_{\tau_1}\, dP + \int_{A^c} X_\tau\, dP - \int_{A^c} X_\tau\, dP - \int_A X_\tau\, dP \\ &= \int_A (X_{\tau_1} - E^{\mathscr{G}(\tau_1)}(X_\tau))\, dP.\end{aligned} \tag{19}$$

One can deduce from this that $\{E^{\mathscr{G}(\tau_1)}(X_\tau), \tau > \tilde{\tau}_1, \tau \in \mathscr{S}\}$ is a Cauchy net in $L^1(P)$. In fact, let $\{\sigma_1, \sigma_2\} \subset \mathscr{S}$, $\sigma_1 > \sigma_2 > (\tilde{\tau}_1 \vee \tau_0)$ where $\tilde{\tau}_1 \in \mathscr{S}$ is arbitrarily fixed. Then

$$\begin{aligned}\|E^{\mathscr{G}(\tilde{\tau}_1)}(X_{\sigma_1} - X_{\sigma_2})\|_1 &= \|E^{\mathscr{G}(\tilde{\tau}_1)}(X_{\sigma_1} - E^{\mathscr{G}(\sigma_1)}(X_{\sigma_2}))\|_1 \\ &\le \|X_{\sigma_1} - E^{\mathscr{G}(\sigma_1)}(X_{\sigma_2})\|_1 \\ &= \sup_{A \in \mathscr{G}(\sigma_1)} \left| \int_\Omega \chi_A (X_{\sigma_1} - E^{\mathscr{G}(\sigma_1)}(X_{\sigma_2}))\, dP \right| \quad \text{by} \\ &\qquad \text{Proposition I.4.2 since such } \chi_A \\ &\qquad \text{are dense in } L^\infty(\mathscr{G}(\sigma_1)), \\ &\le \varepsilon \quad \text{by (18) and (19).}\end{aligned} \tag{20}$$

Next, for any $\sigma \in \mathscr{S}$ define $Y_\sigma = L^1 - \lim_{\tau \in \mathscr{S}} E^{\mathscr{G}(\sigma)}(X_\tau)$. This exists by the above analysis. Set $Z_\sigma = X_\sigma - Y_\sigma$. It is asserted that the equation $X_\sigma = Y_\sigma + Z_\sigma$ gives the desired decomposition if $\sigma(\omega) = i \in I$, $\omega \in \Omega$, is the trivial time from $\mathscr{S}$. Let us check that $\{Y_\sigma, \mathscr{G}(\sigma), \sigma \in \mathscr{S}\}$ is a martingale. Since the traces of $\mathfrak{F}_i$ and $\mathscr{G}(\sigma)$ on $[\sigma = i]$ agree, this will prove that $\{Y_i, \mathfrak{F}_i, i \in I\}$ is also a martingale. Thus for $\sigma' \ge \sigma$ from $\mathscr{S}$, one has, on recalling that the conditional expectation is a continuous operator on $L^1(P)$ and using its commutativity properties,

$$\begin{aligned}\|E^{\mathscr{G}(\sigma)}(Y_{\sigma'} - Y_\sigma)\|_1 &= \lim_{\tau, \tau' \in \mathscr{S}} \|E^{\mathscr{G}(\sigma)}(E^{\mathscr{G}(\sigma')}(X_\tau) - E^{\mathscr{G}(\sigma)}(X_{\tau'}))\|_1 \\ &= \lim_{\tau, \tau' \in \mathscr{S}} \|E^{\mathscr{G}(\sigma)}(X_\tau - X_{\tau'})\|_1 = 0 \qquad \text{by (20).}\end{aligned}$$

Hence $E^{\mathscr{G}(\sigma)}(Y_{\sigma'}) = Y_\sigma$ a.e. Since I is linearly ordered, it is clear that $\{X_i - Y_i, \mathfrak{F}_i, i \in I\}$ is also an amart. Moreover, for $\sigma > \tau$

$$E^{\mathscr{G}(\tau)}(Z_\sigma) = E^{\mathscr{G}(\tau)}(X_\sigma - Y_\sigma) = E^{\mathscr{G}(\tau)}(X_\sigma) - Y_\tau. \tag{21}$$

As $\sigma \uparrow$, $E^{\mathscr{G}(\tau)}(X_\sigma) \to Y_\tau$ in L^1 by definition of Y_τ, so that the right side of (21) tends to zero in $L^1(P)$. Further, (21) yields

$$\|Z_\tau\|_1 \leq \|E^{\mathscr{G}(\tau)}(Z_\tau - Z_\sigma)\|_1 + \|E^{\mathscr{G}(\tau)}(Z_\sigma)\|_1 \to 0 \qquad \sigma\uparrow. \tag{22}$$

Now (22) implies (by the duality argument used in (20)) both (a) and (b).

Regarding uniqueness, let $X_i = \tilde{Y}_i + \tilde{Z}_i$ be another decomposition with the properties of (a) and (b) of the statement. Then for $A \in \mathfrak{F}_i$,

$$\begin{aligned}\int_A Y_i\,dP &= \lim_{j\in I} \int_A Y_j\,dP = \lim_{j\in I} \int_A X_j\,dP \\ &= \lim_{j\in I} \int_A \tilde{Y}_j\,dP + \lim_{j\in I} \int_A \tilde{Z}_j\,dP \\ &= \lim_{j\in I} \int_A \tilde{Y}_j\,dP = \int_A \tilde{Y}_i\,dP \qquad \text{by the martingale property.}\end{aligned}$$

Since Y_i and $\tilde{Y}_i$ are $\mathfrak{F}_i$-measurable and $A \in \mathfrak{F}_i$, this implies $Y_i = \tilde{Y}_i$ a.e., and hence $Z_i = \tilde{Z}_i$ a.e. This completes the proof.

If the $\{X_n, \mathfrak{F}_n, n \geq 1\}$ is an L^1-bounded supermartingale, so that it is an amart, then $\{Z_n, \mathfrak{F}_n, n \geq 1\}$ is also a supermartingale since

$$Y_n + E^{\mathfrak{F}_n}(Z_{n+1}) = E^{\mathfrak{F}_n}(X_{n+1}) \leq X_n = Y_n + Z_n \quad \text{a.e.}$$

But $Z_n \geq 0$, in addition to $\|Z_n\|_1 \to 0$, because

$$Z_n = X_n - Y_n = X_n - \lim_{k\to\infty} E^{\mathfrak{F}_n}(X_k) \geq X_n - X_n = 0,$$

by the supermartingale inequality. A positive supermartingale $\{Z_n, \mathfrak{F}_n, n \geq 1\}$ is called a *potential* if $Z_n \to 0$ in $L^1(P)$, as $n \to \infty$. Thus the preceding theorem implies the following result:

8. Corollary (A form of the Riesz decomposition) *Let $\{X_n, \mathfrak{F}_n, n \geq 1\}$ be an $L^1(P)$-bounded supermartingale. Then it admits an a.e. unique decomposition $X_n = Y_n + Z_n$, $n \geq 1$, where $\{Y_n, \mathfrak{F}_n, n \geq 1\}$ is a martingale and $\{Z_n, \mathfrak{F}_n, n \geq 1\}$ is a potential.*

With this result we essentially established the basic useful decompositions of the (discrete) martingale theory; namely, the Doob decomposition and a Riesz decomposition. These may be used as starting points for numerous applications and extensions of the theory. Some of these will be indicated in the Complements and Problems Section below.

Complements and Problems

1. Let $\{X_n, \mathfrak{F}_n, n \geq 1\}$ be a (sub-) martingale and $\{\alpha_{n+1}, \mathfrak{F}_n, n \geq 1\}$ be an adapted sequence of random variables. If $(\alpha \circ X)_n = \sum_{k=1}^{n} \alpha_k(X_k - X_{k-1})$, where one takes $X_0 = 0$ a.e., then the α_n-process is said to be *predictable* for the X-process and $\{(\alpha \circ X)_n, \mathfrak{F}_n, n \geq 1\}$ is a *predictable transform* of the X-process. Show that every stopping time T of $\{\mathfrak{F}_n, n \geq 1\}$ determines a two-valued decreasing predictable process and conversely. [*Hint*: Let $\alpha_k = \chi_{[T \geq k]}$. If α_k is a two-valued decreasing predictable process, then define $T = \inf\{n > 0 : \alpha_{n+1} = 0\}$ with $\inf\{\varnothing\} = +\infty$.] Show that for α_k taking 0 and 1, the transformed process is a (sub)martingale and that $E((\alpha \circ X)_n) \leq E(X_n)$ with equality in the martingale case. [*Hint*: Use induction for the last part.]

2. Let $\{X_n, n \geq 1\}$ be an arbitrary sequence of random variables on (Ω, Σ, P) and set $\mathfrak{F}_n = \sigma(X_k, 1 \leq k \leq n)$, $\mathscr{B}_n = \sigma(X_k, k \geq n)$, the generated σ-algebras. Let $\mathfrak{F}_\infty = \sigma(\bigcup_n \mathfrak{F}_n)$, $\mathscr{B}_\infty = \bigcap_n \mathscr{B}_n$. If Y is a function of $\{X_n, n \geq 1\}$, and $Y \in L^1(P)$, show that (a) $\lim_{n\to\infty} E^{\mathfrak{F}_n}(Y) = Y$ a.e. and (b) $\lim_{n\to\infty} E^{\mathscr{B}_n}(Y) = \tilde{Y}$ a.e. exist. If further the X_ns are mutually independent, show that $\tilde{Y} =$ a constant a.e., and calculate the constant. [*Hint*: For the last part, verify that every set in $\mathscr{B}_\infty$ has probability zero-or-one.]

3. If $X = \{X_n, \mathfrak{F}_n, n \geq 1\}$ is an $L^1(P)$-bounded martingale and $\varphi_n = X_n - X_{n-1}$ ($X_0 = 0$ a.e.) the increment, then $s(X) = [\sum_{n=1}^{\infty} \varphi_n^2]^{1/2}$ is called the "martingale square function." Show that $s(X) < \infty$ a.e. even though $s(X)$ may not be integrable. [The square function $s(X)$ plays an important role in harmonic analysis, and it corresponds to the "Lusin s-function" in the classical theory on the torus. *Hints for proof*: Consider $\sum_{k=1}^{n} \varphi_k^2 = X_n^2 - 2\sum_{k=1}^{n-1} \varphi_{k+1}X_k$. Since by Theorem 2.4, $X_n^2 \to X_\infty^2$ a.e., the result follows if $(\sum_{k=1}^{n-1} \varphi_{k+1}X_k)(\omega)$ is bounded for a.a. (ω). If $A_n^\lambda = \{\omega : |X_k|(\omega) < \lambda, 1 \leq k \leq n\}$, then using the fact that $P[\sup_n |X_n| \geq \lambda] \to 0$ as $\lambda \to \infty$, verify that $\sum_{n=1}^{\infty} E^{\mathfrak{F}_n}(|\tilde{\varphi}_{n+1}|^2) < \infty$ a.e., where $\hat{\varphi}_k = \hat{X}_k - \hat{X}_{k-1}$ with $\hat{X}_k = X_k \chi_{A_k^\lambda}$. Deduce that $\hat{Y}_n = \sum_{k=1}^{n} \hat{\varphi}_k \hat{X}_{k-1} \to \hat{Y}_\infty$ a.e. $(\leq \lambda)$, and hence $(X_n^2 - \sum_{k=1}^{n} \varphi_k^2)\chi_{A^\lambda}$ tends to zero a.e. where $A^\lambda = \lim_n A_n^\lambda$. Since $\lambda > 0$ is arbitrary, the result follows. This useful observation is due to Austin [1].]

4. Let $\xi_1, \xi_2, \ldots$ be independent identically distributed random variables with mean α. Then they obey the *strong law of large numbers*, i.e., $(1/n)\sum_{k=1}^{n} \xi_k \to \alpha$ as $n \to \infty$, with probability one. [*Hints*: Let $S_n = \sum_{k=1}^{n} \xi_k$, $\mathscr{B}_n = \sigma(S_k, k \geq n)$. Then $E^{\mathscr{B}_n}(\xi_1) \to \tilde{X}$ a.e. by Theorem 2.12. Show that using the definition of S_n, $E^{\sigma(S_n)}(\xi_1) = E^{\mathscr{B}_n}(\xi_1)$ a.e., where $\tilde{\mathscr{B}}_n = \sigma(S_n, \xi_k, k \geq n+1)$. Hence $E^{\sigma(S_n)}(\xi_1) \to \tilde{X}$ a.e. By symmetry, $E^{\sigma(S_n)}(\xi_1) = E^{\sigma(S_n)}(\xi_j)$, $j \leq n$, and so it is equal to S_n/n a.e. Now use the last part of Problem 2 above.]

5. Let $\tilde{\Omega} = \{\omega : \omega = (x_0, x_1, \ldots), x_i \in \mathbb{R}$, and $\|\omega\| = \sum_{i=0}^{\infty}|x_i| \leq 1\}$, the closed unit ball of l_1, the space of summable real sequences. Let $\Omega_n = \{\omega \in \tilde{\Omega} : \omega = (x_i, 0 \leq i \leq n, x_k = 0 \text{ for } k > n)\}$. Let $\pi_{mn} : \Omega_n \to \Omega_m$ be coordinate projections $m \leq n$, so that $\{(\Omega_n, \pi_{mn}) : n \geq m \geq 1\}$ is a projective system of spaces and $\Omega_\infty = \lim_{\leftarrow} \Omega_n$ their limit. Let $\pi_n : \Omega_\infty \to \Omega_n$ be the coordinate projection (Ω_∞ is compact). Considering l_1 as the adjoint space $(c_0)^*$, where c_0 is the Banach space of null convergent real sequences (with uniform norm), $\tilde{\Omega}$ becomes a compact Hausdorff space in its weak* topology. If we set $\rho(\omega, \omega') = \sum_{n=0}^{\infty}(1/2^n)|x_n - x'_n|[1 + |x_n - x'_n|]^{-1}$ for ω, ω' in $\tilde{\Omega}$, then it becomes a (compact) metric space and this topology is the same as the weak* topology. Let $\mathscr{B}_n \uparrow$ be the respective Borel σ-algebras of $\Omega_n \uparrow$. Let P_n: $\mathscr{B}_n \to [0, 1]$ be defined by

$$P_n(A_n) = \frac{1}{2^{n+1}} \int \cdots \int_{A_n \cap [|x_0| + \cdots + |x_n| \leq 1]} \{[(1 - |x_0|)] \cdots [(1 - (|x_0|) + \cdots + |x_{n-1}|)]\}^{-1} \, dx_0 \cdots dx_n$$

for $n \geq 1$ and $A_n \in \mathscr{B}_n$. Then P_n is a probability measure on $\mathscr{B}_n$.

(a) Show that $\{(\Omega_n, \mathscr{B}_n, P_n, \pi_{mn}) : m \geq n \geq 0\}$ is a projective family admitting a (Baire) projective limit $(\Omega, \mathscr{B}, P)$. [*Hint*: Check that P_ns are regular and the Kolmogorov–Bochner theorem applies. Actually, $\mathscr{B}_n \subset \mathscr{B}_{n+1}$ and the result is simpler since Ω_ns are cylinders. $\Omega = \Omega_\infty$ is isomorphic to $\tilde{\Omega}$, cf. III.2.5.]

(b) If $C(\Omega)$ is the space of all real continuous functions, the functional $E(\cdot)$ defined by $E(f) = \int_\Omega f\, dP$ is continuous. [Note that $f_n = f \circ \pi_n$ are dense in $C(\Omega)$. See also Problem 6 in Chapter I.]

(c) If $f \in C(\Omega)$, let $v(A) = \int_A f\, dP$, $A \in \mathscr{B}$, and let $v_n = v | \mathscr{B}_n$. If $g_n = dv_n/dP_n$, then $g_n = E^{\mathscr{B}_n}(f)$ and $g_n \to g_\infty = dv/dP$ $(=f)$ a.e.

(d) Let $\mathfrak{F}_n$ be the σ-algebra generated by the Borel sets of $\tilde{\Omega}_n = \{\omega \in \Omega : \omega = (x_i, x_i = 0 \text{ for } 1 \leq i \leq n)\}$. Thus $\mathfrak{F}_n \downarrow \subset \mathscr{B}$ and if $\tilde{v}_n = v | \mathfrak{F}_n$, $Q_n = P | \mathfrak{F}_n$, $h_n = d\tilde{v}_n/dQ_n$, then $h_n = E^{\mathfrak{F}_n}(f)$ and so $h_n \to h_\infty$ a.e. $[P]$. Moreover, $h_\infty = E(f)$. [Note that $\bigcap_n \mathfrak{F}_n$ is the trivial σ-algebra and apply Theorem 2.12, so that $E = E^{\mathfrak{F}_\infty}$.]

(e) If $f_0^t(\cdot) = \sum_{n=0}^{\infty} X_n(\cdot) t^n \in C(\Omega)$, considering $X_n(\cdot)$ as a coordinate function in $C(\Omega_n)$, show that (recalling the way the P_n were defined) $\int_\Omega X_n(\omega)\, dP = \int_{\Omega_n} X_n\, dP_n = 0$ and similarly $\int_\Omega \varphi(X_0 t, \ldots, X_n t^n) X_{n+1} t^{n+1}\, dP = 0$ for any bounded Baire function φ. By remark 1.5, the sequence $S_n(t) = \sum_{k=0}^{n} X_m t^k$ is a martingale relative to the $\mathscr{B}_n$. Show, moreover, that $E(|S_n(t)|) \leq \sum_{i=0}^{n} |t|^i E(|X_i|) \leq [2/(2 - |t|)]$ for $|t| < 2$ and deduce that $S_n(t) \to f_0^t(\cdot)$ a.e. for t in that range. [*Hint*: Verify by induction that $\int_{\Omega_n} |X_n|^k\, dP_n = (1 + k)^{-n-1}$ for $k > -1$. One may and does write the number $X_n(\omega)$ as x_n below.]

(f) For the f_0 of (e) with $|t| \leq 1$, the classical power series says that it converges when $R(\omega) = \lim_n \sup|x_n|^{1/n} \leq 1$. In the present setup, show that for almost all $\omega \in \Omega$, $R(\omega) = \underset{\sim}{e}^{-1}$ so $R(\omega) \leq \frac{1}{2}$, where $\underset{\sim}{e}$ is the base of the natural logarithm. [*Hints*: with (e) deduce that $\int_{\Omega_n}[|X_n|^{1/n} - \underset{\sim}{e}^{-1}]^2 \, dP_n \to 0$ implying $\lim_j |X_{n_j}|^{1/n_j} = \underset{\sim}{e}^{-1}$ a.e. for some subsequence $\{n_j\}_1^\infty$. So $\lim_n \sup|X_n|^{1/n} \geq \underset{\sim}{e}^{-1}$ a.e. For the opposite inequality, let $0 < r < \underset{\sim}{e}$, so $r < [1 + (1/n)]^{1/n} \uparrow \underset{\sim}{e}$ for $n \geq n_0$. If $\tilde{S}_n(r) = [\sum_{k=0}^n |X_k| r^k]^{1/n_0}$ show that

$$\int_{\Omega_n} \tilde{S}_n(r) \, dP_n \leq \frac{n_0}{n_0 + 1} \sum_{m=0}^{\infty} \left(\frac{n_0 r^{1/n_0}}{n_0 + 1} \right)^{1/m} < \infty,$$

so that $\lim_n \tilde{S}_n(r)$ exists a.e. $[P]$. Hence $R(\omega) \cdot r = \lim_n \sup|X_n|^{1/r} \cdot r \leq 1$ a.e. and $R(\omega) \leq \underset{\sim}{e}^{-1}$ if $r \uparrow \underset{\sim}{e}$. This result is essentially due to Eberlein [1].]

6. We indicate a proof of the *Kolmogorov–Sinaĭ theorem* on the calculation of entropy of an endomorphism considered in Problems 5–6 of Chapter III. The notation and results of these two problems will be used. Let T be an endomorphism of (Ω, Σ, P), i.e., T measurable and $P = P \circ T^{-1}$. If $\mathscr{A} \subset \Sigma$ is a *finite* algebra, then the "long time average information" in T relative to $\mathscr{A}$ is $h(T, \mathscr{A}) = \lim_n (1/n) H(\bigvee_{k=0}^{n-1} T^{-k}(\mathscr{A}))$, where $H(\mathscr{A}) = -\sum_{i=0}^n P(A_i) \log P(A_i)$, $\{A_i\}_1^n$ being a partition generating $\mathscr{A}$. The entropy of T is $h(T) = \sup\{h(T, \mathscr{A}) : \mathscr{A} \subset \Sigma, \mathscr{A} \text{ finite}\}$.

Theorem *If T is invertible and $\bigvee_{n=-\infty}^{\infty} T^n(\mathscr{A}) = \Sigma$, then $h(T) = h(\mathscr{A}, T)$ where $\mathscr{A}$ is a finite algebra. If T is only an endomorphism such that $\bigvee_{n=0}^{\infty} T^{-n}(\mathscr{A}) = \Sigma$, then again the same formula holds, i.e., $h(T) = h(T, \mathscr{A})$.*

Proof In view of the monotonicity properties (cf. the definition), it suffices to show that $h(T, \mathscr{B}) \leq h(T, \mathscr{A})$ for any finite algebra $\mathscr{B} \subset \Sigma$. If $\mathscr{A}_n = \bigvee_{k=-n}^{n} T^k(\mathscr{A})$, then $\mathscr{A}_n \uparrow \Sigma$. Now using the relations between $h(T, \mathscr{A})$ and $H(\mathscr{A})$, show that $h(T, \mathscr{A}) = h(T, \mathscr{A}_n)$. With a property of the conditional entropy (cf. p. 165), deduce that $h(T, \mathscr{B}) \leq h(T, \mathscr{A}_n) + H(\mathscr{B} | \mathscr{A}_n)$. But by the martingale convergence theorem (for the L^∞ functions!), $P^{\mathscr{A}_n}(A) \to P^{\Sigma}(A)$ a.e. for each $A \in \Sigma$. Considering a fixed partition $\{A_i\}_1^m$ generating $\mathscr{B}$, we get from this and the definition of $H(\mathscr{B} | \mathscr{A}_n)$ $(= E(\sum_{i=1}^m \varphi(P^{\mathscr{A}_n}(A_i)))$ by p. 165, $\varphi(t) = t \log t$), that $H(\mathscr{B} | \mathscr{A}_n) \to H(\mathscr{B}, \Sigma)$ because of the bounded convergence theorem. But $\mathscr{B} \subset \Sigma$ and $H(\mathscr{B} | \Sigma) \leq H(\mathscr{B} | \mathscr{B}) = 0$. Hence $h(T, \mathscr{B}) \leq h(T, \mathscr{A}_n)$, and this proves the result. Replacing $\mathscr{A}_n$ by $\bar{\mathscr{A}}_n = \bigvee_{k=0}^{n-1} T^{-k} \mathscr{A}$, the last part follows. This result enables calculations of entropy in a number of concrete cases. [See Billingsley [1] for further information on this question and related matters.]

7. Analogous to the concept of information in an endomorphism, we consider the corresponding concept of information of one probability measure relative to another.

Let (Ω, Σ) be a measurable space and $P_i: \Sigma \to [0,1]$, $i = 1, 2$, be two probability measures. If $\mathscr{A} \subset \Sigma$ is a finite algebra generated by a partition $\{A_i\}_1^n$, define the information in P_1 relative to P_2 on $\mathscr{A}$ as $I_{P_2}(\mathscr{A}, P_1) = \sum_{i=1}^n P_1(A_i) \log[P_1(A_i)/P_2(A_i)]$ where $0/0$ and $0 \cdot (\pm \infty)$ are taken as 0. If $\mathfrak{F} \subset \Sigma$ is any σ-algebra, then we define $I_{P_2}(\mathfrak{F}, P_1) = \sup\{I_{P_2}(\mathscr{A}, P_1) : \mathscr{A} \subset \mathfrak{F}$, $\mathscr{A}$ finitely generated$\}$. Clearly $I_{P_2}(\cdot, P_1)$ is increasing, and $I_{P_2}(\mathscr{A}, P_1) < \infty$ implies P_1 is P_2-continuous on $\mathscr{A}$. [We have discussed a similar information function in Section 4 in connection with the dichotomy problem.]

(a) Let $\mathfrak{F} \subset \Sigma$ be a countably generated σ-algebra. Then there exists a sequence $\mathfrak{F}_n \uparrow \mathfrak{F}$ of finitely generated algebras such that $I_{P_2}(\mathfrak{F}_n, P_1) \uparrow I_{P_2}(\mathfrak{F}, P_1)$. ⌊*Hints*: If $d(A, B) = P_2(A \triangle B)$ is the Fréchet distance on the symmetric differences, then $(\mathfrak{F}, d)$ is a separable metric space. Also $I_{P_2}(\cdot, P_1)$ is a lower semicontinuous function on $\mathfrak{F}$ relative to the order $(\mathfrak{F}_n, \subset)$.]

(b) If P_1 is P_2-continuous on $\mathfrak{F}$ where $\mathfrak{F}$ is as in (a) and $f_\infty = dP_1/dP_2$, then $I_{P_2}(\mathfrak{F}, P_1) < \infty$ iff $\int_\Omega f_\infty |\log f_\infty| \, dP_2 < \infty$. [*Hints*: If $P_{in} = P_i | \mathfrak{F}_n$ and $f_n = dP_{1n}/dP_{2n}$, then $f_n \to f_\infty$ a.e. and in L^1-mean by Theorems 2.4 and 4.6. Now use (a). Conversely, let $I_{P_2}(\mathfrak{F}, P_1) < \infty$. Then $I_{P_2}(\mathfrak{F}_n, P_1) = \int_\Omega f_n \log f_n \, dP_{2n}$ and $f_n \to f_\infty$ a.e. Since $\varphi(t) = t \log t$ is convex, $\{\varphi(f_n), \mathfrak{F}_n, n \geq 1\}$ is an integrable submartingale, and the result follows easily from Theorem 2.6.]

(c) If $\mathfrak{F} \subset \Sigma$ is a σ-algebra, P_1 is P_2-continuous on $\mathfrak{F}$, and $f = dP_1/dP_2$, let $\mathfrak{F}_f \subset \mathfrak{F}$ be the smallest σ-algebra relative to which f is measurable. Then $I_{P_2}(\mathfrak{F}_f, P_1) = I_{P_2}(\mathfrak{F}, P_1) = \int_\Omega f \log f \, dP_2$. [*Hints*: It suffices to show $I_{P_2}(\mathfrak{F}, P_1) \leq I_{P_2}(\mathfrak{F}_f, P_1)$ when the last term is finite. By definition there is a finite algebra $\mathscr{A}_\varepsilon \subset \mathfrak{F}_f$ such that $I_{P_2}(\mathscr{A}_\varepsilon, P_1)$ is within ε of $I_{P_2}(\mathfrak{F}_f, P_1)$ and hence $\mathfrak{F}_f$ may be assumed countably generated. Then by (b), $I_{P_2}(\mathfrak{F}_f, P_1) = \int_\Omega f \log f \, dP_2$. If $A \in \mathfrak{F}$, $P_1(A) > 0$, let $P_{iA}(\cdot) = [1/P_i(A)] P_i(\cdot)$, $i = 1, 2$, and $\mathfrak{F}(A)$ be the trace σ-algebra of $\mathfrak{F}$ on A. Then $f_A = dP_{1A}/dP_{2A} = [P_2(A)/P_1(A)] f$ a.e. and $\int_A f \log f \, dP_2 = P_1(A) \int_A \log f \, dP_{1A} \geq P_1(A) \log(\int_A f \, dP_{1A})$ (by Jensen's inequality for concave functions), $\geq P_1(A) \log(P_1(A)/P_2(A))$ since $\int_A f^2 \, dP_2 \geq P_1^2(A)/P_2(A)$ (Cauchy–Schwarz–Buniyakovsky inequality). So if $\mathscr{A} \subset \mathfrak{F}$ is generated by a partition, then $I_{P_2}(\mathscr{A}, P_1) \leq I_{P_2}(\mathfrak{F}_f, P_1)$ and, taking suprema on all $\mathscr{A}$s, one gets the desired inequality.]

(d) Let $(S_i, \mathscr{S}_i)$, $i = 1, \ldots, k$, be measurable spaces and $(S, \mathscr{S}) = \times_{i=1}^k (S_i, \mathscr{S}_i)$ be their product. If (Ω, Σ, P) is a probability space, $F_i: \Omega \to S_i$ is a measurable mapping, let $Q = P \circ F^{-1}: \mathscr{S} \to [0,1]$ be the induced probability where $F = (F_1, \ldots, F_k): \Omega \to S$. Consider the product measure $\mu = \bigotimes_{i=1}^{k} P \circ F_i^{-1}$ on $\mathscr{S}$, where $P \circ F_i^{-1}: \mathscr{S}_i \to [0,1]$. Define the information $I(F_1, \ldots, F_k) = I_Q(F^{-1}(\mathscr{S}), \mu)$ contained in Q relative to μ on $\mathscr{S}$. Then (a)–(c) imply: $I(F_1, \ldots, F_k) < \infty$ iff (i) Q is μ-continuous and (ii) $\int_S h(s)|\log h(s)|\, d\mu < \infty$, where $h = dQ/d\mu$. Finally, show that one always has $I(F_1, \ldots, F_k) = \int_S h(s) \log h(s)\, d\mu$ in the sense that both sides are finite or infinite simultaneously. [The concept of information between two measures and the result of (d) were originally formulated by Gel'fand, Kolmogorov, and Yaglom [1]. We essentially follow Kallianpur [1] for the present somewhat more general treatment.]

8. Here is a generalization of Lemma 4.1 with some novel applications. If μ_1, μ_2 are two measures on $(\mathbb{R}^n, \mathscr{B})$, where $\mathscr{B}$ is the Borel σ-algebra of $\mathbb{R}^n$, let us say that a set $A \in \mathscr{B}$ has *size* $\mu_1(A)$ and "power" $\mu_2(A)$. If $\mathscr{L} \subset \mathscr{B}$ is a ring, one says that a set $A \in \mathscr{L}$ is "most powerful" in $\mathscr{L}$ if $\mu_2(A) \geq \mu_2(B)$ for all $B \in \mathscr{L}$ for which $\mu_1(A) = \mu_1(B)$, i.e., for all $B \in \mathscr{L}$ of the same size as A. In what follows let μ_1 on $\mathscr{L}$ have the *Darboux property*, i.e., for any $A \in \mathscr{L}$, $0 \leq \alpha \leq \mu_1(A)$ implies the existence of $B \in \mathscr{L}$, $B \subset A$ with $\mu_1(B) = \alpha$. The measures need not be probabilities in this problem.

(a) Let $A \in \mathscr{L}$ be a set of size α and let $A_i \in \mathscr{L}$ be of sizes α_i, $i = 1, \ldots, k$ such that $\alpha = \sum_{i=1}^{k} p_i \alpha_i$ where $0 \leq p_i$, $\sum_{i=1}^{k} p_i = 1$. Then $\sum_{i=1}^{k} p_i \mu_2(A_i) \leq \mu_2(A)$. [If $k = 1$, this is essentially Lemma 4.1, and the same argument extends by the Darboux property of μ_1 and then by induction.]

(b) Let I be an indexing of $\mathscr{L}$, i.e., $\mathscr{L} = \{A_i : i \in I\}$ is the ring. Let $f_j(i) = \mu_j(A_i)$, $j = 1, 2$, and $\mathscr{E}$ be a σ-algebra of I containing all one point sets. Suppose that $f_j: I \to \mathbb{R}^+$ is $\mathscr{E}$-measurable. If $P: \mathscr{E} \to \mathbb{R}^+$ is a probability and $\alpha \geq 0$ is given such that $\int_I f_1(i)\, dP(i) = \alpha$, let us assume that $\int_I f_2(i)\, dP(i) < \infty$. If $A \in \mathscr{L}$ is a most powerful set of size α, then one has $\int_I f_2(i)\, dP(i) \leq \mu_2(A)$. [Note that if P concentrates on a single point, then one has Lemma 4.1 again. Observe that (a) is true if $k = \infty$ also, and use that result here for each countable partition of I. The general case is then obtained by approximation since the integral is finite. In general, if μ_1 has atoms, there may be no $A \in \mathscr{L}$ of a given size. This part says that one can achieve the size by enlarging the space $\Omega = \mathbb{R}^n$ with an adjunction of I. Then the $f_j(i) = \mu_j(A_i)$ become "conditional measures." In statistical contexts this is called a "randomization procedure," and (b) says that the "power" is not increased by randomization when the μ_j are probabilities.]

(c) Let $\Omega = [a, \infty)$ and F_X be the distribution function of a random variable $X: S \to \Omega$ on (S, Σ, P) and $\mathscr{B}$ be the Borel σ-algebra of Ω. So $F_X(x) = P \circ X^{-1}(-\infty, x)$. Then for any $\varphi: \Omega \to \mathbb{R}$ with its derivative $\varphi' \geq 0$ and nonincreasing (i.e., φ is concave) $\int_S \varphi(X)\, dP \leq \varphi(\int_S X\, dP)$. If φ, ψ are two monotone functions on $\Omega(=\mathbb{R}^+) \to \mathbb{R}^+$ ($a = 0$ here) such that $\psi \circ \varphi^{-1}$ has a nonincreasing nonnegative derivative, then one has $\psi^{-1}(\int_S \psi(X)\, dP) \leq$

$\varphi^{-1}(\int_S \varphi(X)\,dP)$. [*Hints*: One can consider μ_1, μ_2 to be defined as $\mu_1[(a,x)] = x - a$, $\mu_2[(a,x)] = \varphi(x) - \varphi(a)$, and $\mathscr{L} = \mathscr{B}$. Then the most powerful set of size $x - a$ is $(a,x) \in \mathscr{L}$, and the result follows from (b) on identifying the P there with the one given here, $I = S$, etc. The second part follows from this since one can apply the first part to $\psi \circ \varphi^{-1}$ and replace X by $\varphi(X)$. In connection with this problem, see Mann [1]. (c) is Jensen's inequality, and the interesting point is that it is a consequence of a form of Lemma 4.1.]

9. We now indicate that certain classes of Poisson processes have the dichotomy property as in Theorem 4.7. If $a(\lambda, k) = \lambda^k e^{-\lambda}/k!$ for $0 < \lambda, k < \infty$, and $=1$ for $\lambda = k = \infty$ or $\lambda = k = 0$ (and $=0$ otherwise), then $a(\lambda, \cdot)$ is a Poisson "density" of a random variable with parameter λ. If $(r_1, \ldots, r_n)$ are nonnegative integers, then $P_n(r_1, \ldots, r_n) = \prod_{i=1}^n a(\lambda_i, r_i)$ is an n-dimensional density on the positive orthant of $\mathbb{R}^n$ (cf. Eq. (4) of Section I.1). The family $\{p_n\}_1^\infty$ of densities is compatible as noted in Section I.1, and hence by the Kolmogorov existence theorem there is (Ω, Σ, P) supporting a stochastic process $\{X_t, t \in \mathbb{R}^+)$ whose n-dimensional distributions are the given n-dimensional Poisson distributions. So $X_t(\omega)$ is an integer for each $t \in \mathbb{R}$ and $\omega \in \Omega$. Let $L^1(\Omega, \Sigma, P)$ be the associated space and consider a measurable space $(\mathbb{E}, \mathscr{G})$ and a σ-additive set function $\eta: \mathscr{G} \to (L^1(\Omega, \Sigma, P))^+$ so that $\eta(A)$ is a Poisson random variable with parameter λ_A for each $A \in \mathscr{E}$. Its total variation $|\eta|(A) = \lambda_A$ as an easy check shows. But η is a vector measure so that $|\eta|(\cdot)$ is σ-additive and hence $\lambda(\cdot)$ is also a measure. The process $\{\eta(A), A \in \mathscr{E})$ is thus a set of Poisson random variables which are independent for disjoint A and whose parameters $\{\lambda(A), A \in \mathscr{E}\}$ define a measure. We express this as

$$P_\lambda[\eta(A_i) = k_i, i = 1, \ldots, r] = \prod_{i=1}^r a(\lambda(A_i), k_i) \qquad (*)$$

for disjoint A_i in $\mathscr{E}$, and assume that $\lambda: \mathscr{E} \to \bar{\mathbb{R}}^+$ is σ-finite.

(a) Suppose P_λ and P_μ are two Poisson measures determined as above where λ, μ are both finite on $\mathscr{E}$. Show that if λ is μ-continuous and $f = d\lambda/d\mu$ a.e., then P_λ is P_μ-continuous and the Radon–Nikodým density is given by

$$\frac{dP_\lambda}{dP_\mu} = \exp[-\lambda(\mathbb{E}) - \mu(\mathbb{E})] \prod_{i=1}^{\eta(\mathbb{E})} f(e_i),$$

where $\{e_i\}_1^\infty$ are the atoms for which $\eta(e_i)$ has positive P_λ-measure (since $\eta(A)(\omega)$ takes integer values, so that each $A \in \mathscr{E}$ can be a union of at most a countable set of atoms).

(b) Let P_λ, P_μ be as in (a) and λ, μ be σ-finite on $\mathscr{E}$. Then P_λ is P_μ-continuous iff (i) λ is μ-continuous, (ii) the set $B_c = \{x : |f(x) - 1| > c\} \in \mathscr{E}$

satisfies $\int_{B_c}|f-1|\,d\mu < \infty$ for all $c > 0$, and (iii) $\int_{\mathbb{E}-B_c}(f-1)^2\,d\mu < \infty$ for some $c > 0$.

(c) Suppose P_λ, P_μ, λ, μ are as in (b) and let $\lambda \equiv \mu$. Then either $P_\lambda \equiv P_\mu$ or $P_\lambda \perp P_\mu$. Moreover, $P_\lambda \equiv P_\mu$ iff there is a $c > 0$ such that $\int_{B_c}|f-1|\,d\mu + \int_{\mathbb{E}-B_c}(f-1)^2\,d\mu < \infty$, and $P_\lambda \perp P_\mu$ iff for all $c > 0$, $\int_{B_c}|f-1|\,d\mu + \int_{\mathbb{E}-B_c}(f-1)^2\,d\mu = \infty$. Thus when $\lambda \equiv \mu$, we have the dichotomy as in the Gaussian case. [The proof needs several computations. The result is obtained from Gikhman–Skorokhod [1, Theorem 7.3], and the present formulation is due to M. Brown [1]. The first paper has many results on this subject. See also Briggs [1] on related results.]

10. (a) Let $X = \{X_t, t \in T\}$ be a process with $(\mathbb{R}^T, \mathscr{B}^T, P)$ as its canonical representation. If $f\colon T \to \mathbb{R}$ is given and $Y_t = X_t + f(t)$, then $\{Y_t, t \in T\}$ has the same $(\mathbb{R}^T, \mathscr{B}^T)$ for the measurable space, and let P_f be its induced measure. Then f is an *admissible mean* for X if $P_f \ll P$. The class $M(X)$ of admissible means is analyzed here if X is a Gaussian process with mean zero, continuous covariance $r(\cdot,\cdot)$, and T is a compact interval. Set $\mathbb{R}^T = \Omega$. If $F(\cdot)$ is a function of bounded variation on T and $f(s) = \int_T r(s,t)\,dF(t)$, show that $f \in M(X)$ (so P_f is P-continuous) and that the likelihood ratio is given by

$$\frac{dP_f}{dP} = \exp\left\{\int_T X(t)\,dF(t) - \frac{1}{2}\int_T f(t)\,dF(t)\right\},$$

where $X(\cdot,\cdot)$ is jointly measurable by assumption (actually this is a consequence of the continuity of r as is easily checked) and $\int_T X(t)\,dF(t)$ is a random variable (by Fubini's theorem). Show conversely that if $f \in M(X)$ and $dP_f/dP = \exp\{\int_T X(t)\,dF(t) - \frac{1}{2}\alpha\}$, then α is the second term given above. [The proof is entirely similar to that of Theorem 4.10.]

(b) If $f \in M(X)$ and $r(\cdot,\cdot)$ are given, there need not exist an F as in (a). A positive solution is available for the following class of triangular covariances (cf. I.6.2): $r(s,t) = u(\min(s,t))\cdot v(\max(s,t))$ for s, t in $T = [a,b]$ with $u(a) \geq 0$, $v > 0$ on T, $(uv^{-1})(t)$ strictly increasing and u, v have continuous second derivatives on T. Show that the integral equation $f(s) = \int_T r(s,t)\,dF(t)$ has a solution F which is of bounded variation on T iff (i) $df/dt = f'$ exists and is of bounded variation on T and (ii) $f(a) = 0$ if $u(a) = 0$. When this holds, one can express F as $F(t) = -\int_a^t[d\lambda(t)/v(t)]$, where $\lambda(a) = f(a)u^{-1}(a)$ or $=0$ according as $u(a) > 0$ or $u(a) = 0$, and $\lambda(s) = [v(s)f'(s) - f(s)v'(s)][v(s)u'(s) - u(s)v'(s)]^{-1}$ for $a < s < b$, $\lambda(b) = 0$. [This follows by differentiating the integral equation and by direct verification thereafter.]

(c) For the Gaussian processes with triangular covariances as in (b), we can consider more general structures than (a) by allowing different covariances. This is illustrated by an application of Theorem 4.7 with explicit likelihood ratios. Thus let (Ω, Σ) be $(\mathbb{R}^T, \mathscr{B}^T)$ and let $P(f, r)$, $Q(g, \rho)$ be two Gaussian probabilities on $\mathscr{B}^T$ with means f, g and covariances $r(s, t) = u(\min(s, t))\cdot v(\max(s, t))$ and $\rho(s, t) = \theta(\min(s, t))\cdot \varphi(\max(s, t))$ as in (b). Let f', g' exist and be of bounded variation on T. Then prove that $P(f, r) \equiv Q(g, \rho)$ iff (i) $v(t)u'(t) - u(t)v'(t) = \varphi(t)\theta'(t) - \theta(t)\varphi'(t)$, $t \in T$, and (ii) $\theta(a), u(a)$ are both zero, in which case $f(a) = g(a) = 0$, or both >0 simultaneously. When these conditions hold, one has

$$\frac{dQ(g, \rho)}{dP(f, r)}(\omega) = \frac{dQ(0, \rho)}{dP(0, r)}(\omega - g)\,\frac{dP(g - f, r)}{dP(0, r)}(\omega - f).$$

The last one is obtained from (a), and the first factor is of the following form:

$$\frac{dQ(0, \rho)}{dP(0, r)}(\omega) = C_1 \exp\left\{C_2 X_a^2(\omega) + C_3 X_b^2(\omega) - \int_a^b h(t) X_t^2(\omega)\, dt\right\}, \qquad (*)$$

where $C_1 > 0$, C_2, C_3 are some constants (determined by $u(a)$, $v(a)$, $\theta(a)$, $\varphi(a)u(b)$, etc.) and

$$h(t) = \frac{1}{\varphi(t)v(t)}\cdot\frac{d}{dt}\left[\frac{v(t)\varphi'(t) - \varphi(t)v'(t)}{v(t)u'(t) - u(t)v'(t)}\right].$$

(This is a specialization of Theorem 4.7 but gives explicit formulas.)

(d) As an example to evaluate (a), let $T = [0, 1]$, $r_{b_0} = r(\cdot, \cdot)$, $r_{b_1} = \rho(\cdot, \cdot)$, where

$$r_b(s, t) = \cosh b(\min(s, t)) \cosh b(\max(1 - s, 1 - t))\cdot(b \sinh b)^{-1},\ 0 < b_0 < b_1.$$

Then show that the eigenvalues and eigenfunctions of $r_b(s, t)$ are $\lambda_n = n^2\pi^2 + b^2$, $n = 0, 1, \ldots$, and $g_0 = 1$, $g_n(t) = \sqrt{2}\cos n\pi t$, $n = 1, 2, \ldots$. Next the Fourier representation of $X(t, \cdot) = \sum_{n=0}^{\infty} X_n(\cdot)(g_n(t)/\sqrt{\lambda_n})$ is well defined and converges a.e. Deduce that $(*)$ becomes

$$\frac{dP_1}{dP_0}(\omega) = \frac{dP(0, r_{b_1})}{dP(0, r_{b_0})}(\omega) = C_1 \exp\left\{-\frac{1}{2}\sum_{n=0}^{\infty} X_n^2(\omega)(b_1^2 - b_0^2)\right\},$$

where

$$0 < C_1 = \prod_{i=1}^{\infty}\left[\left(1 + \frac{b_1^2}{n^2\pi^2}\right)^{-1/2}\right] < \infty$$

and

$$\int_\Omega X_n^2\, dP_i = \frac{1}{n^2\pi^2 + b_i^2}, \qquad i = 0, 1,$$

so that $\sum_{n=1}^{\infty} X_n^2$ converges a.e. [For the special but detailed results in (b) and (c), see Varberg [1]; (a) is due to Pitcher [1]. This problem shows what can be obtained by specializing the work of Section 4. An extensive account for the Gaussian case is to be found in Rozanov [1] and a comprehensive account in Gikhman and Skorokhod [1]. A general form of (b) is given by the author [4.V].]

11. In this problem we show how martingales are related to nonlinear prediction sequences. Let $\mathfrak{F}_n \uparrow \subset \Sigma$ and $f \in L^p(\Sigma)$ on (Ω, Σ, P), $1 < p < \infty$. For each n, let $f_n \in L^p(\mathfrak{F}_n)$ the unique element such that $\|f - f_n\|_p = \inf\{\|f - g\|_p : g \in L^p(\mathfrak{F}_n)\}$. The existence of f_n is a consequence of the uniform convexity of $L^p(\mathfrak{F}_n)$. The mapping $Q^{\mathfrak{F}_n}: f \to f_n$ is idempotent and $I - Q^{\mathfrak{F}_n}$ is contractive. Show that $Q^{\mathfrak{F}_n} = E^{\mathfrak{F}_n}$ if $p = 2$, and f_n is given for $1 < p < \infty$ as a unique solution of the following:

$$\int_\Omega |f_0 - f_n|^{p-1}\left(\left[\frac{d}{dt}(|f_0 - f_n + th|)\right]_{t=0}\right) dP = 0, \qquad h \in L^p(\mathfrak{F}_n). \qquad (**)$$

Show, however, that $f_n \to f_\infty = Q^{\mathfrak{F}_\infty}(f) \in L^p(\mathfrak{F}_\infty)$, where $\mathfrak{F}_\infty = \sigma(\bigcup_{n \geq 1} \mathfrak{F}_n)$, both in norm and a.e. [P]. [*Hints*: If $p = 2$, the result is a simple consequence of Theorem 2.4. The integral equation is deduced by a variational argument to minimize $G(g) = \int_\Omega |f_0 - g|^p \, dP$ for g over $L^p(\mathfrak{F}_n)$. The norm convergence is dependent on the weak compactness of the unit ball of $L^p(\mathfrak{F}_n)$, $1 < p < \infty$, and then the pointwise convergence is obtained with a method used in the proof of Theorem 1.7 and an argument as in the last half of the proof of Theorem 2.3. In connection with this problem, see Andô and Amemiya [1] and the author [3, 4.III]. $Q^{\mathfrak{F}}$ is called a *prediction operator*. Since

$$\frac{d}{dt}(|a + tb|)\bigg|_{t=0} = \frac{ab}{|a|} \qquad \text{for} \quad a \neq 0,$$

$(**)$ simplifies to

$$\int_{[f_0 > f_n]} (f_0 - f_n)^{p-1} \, dP = \int_{[f_n > f_0]} (f_n - f)^{p-1} \, dP,$$

which is useful for computing f_n. This form admits an extension if $\varphi(x) = |x|^p/p$ is replaced by a more general convex function leading to the use of Orlicz spaces $L^\varphi(\Sigma)$. For a discussion of its relation with estimation, see DeGroot and the author [1], and an abstract version in this sense is discussed in the author's paper [11]. But the linear prediction problem is not related to the above set of ideas. An account of the latter may be found in Rozanov [2].]

CHAPTER

V

Abstract Martingales and Applications

This final chapter is devoted to a treatment of vector martingale theory, including differentiation, contrasting it with the previous scalar case. Various applications of the results in analysis are given. These include the martingale formulations of ergodic theory, a unified treatment of ergodic and martingale theories by Reynolds operators, and certain martingales in harmonic analysis. A few other extensions including additive set martingales are given in the Complements and Problems Section.

5.1 INTRODUCTION

The role played by various hypotheses in the scalar theory (of Chapter IV) will be clarified and appreciated when we abstract their essential features. For instance, it is possible to develop the (scalar) convergence theory of Section IV.2 without the Radon–Nikodým theorem because of the properties of the real line. This is almost clear from Section IV.5. But such an attempt is of no value if the range space of martingales is a Banach space. Indeed, there is a close relationship between the geometry of the range (vector) space of these processes, the possibility of carrying out the differentiation of σ-additive set functions (relative to their variations), and the convergence (mean or pointwise) of martingales into these spaces. Also there are important applications that require such general results. We have seen in Theorem IV.4.6 that for the mean convergence, directed index sets appear in the theory of martingales that need not be functions defined on a fixed point set. Accordingly, we consider such general martingales and then present certain interesting specializations.

Let us introduce the abstract concept. To motivate this, recall that a martingale $\{f_n, \mathfrak{F}_n, n \geq 1\}$ on a probability space (Ω, Σ, P) is a process such that $E^{\mathfrak{F}_n}(f_{n+1}) = f_n$ a.e., $E^{\mathfrak{F}_n}E^{\mathfrak{F}_{n+k}} = E^{\mathfrak{F}_{n+k}}E^{\mathfrak{F}_n} = E^{\mathfrak{F}_n}$, $k \geq 1$, and the conditional expectations are contractive projections on $L^p(P)$, $p \geq 1$.

1. Definition Let $\mathscr{X}$ be a (real) Banach space and $\{E_i, i \in I\}$ be an indexed family of uniformly bounded projections on $\mathscr{X}$ (i.e., $E_i^2 = E_i \in B(\mathscr{X})$, where $(I, <)$ is a directed set. Suppose for each i, j in I with $i < j$ (or $j < i$) we have $E_iE_j = E_jE_i = E_i$. Then a family $\{f_i, E_i, i \in I\}$ with $f_i \in \mathscr{X}$ is an *increasing* (or respectively *decreasing*) *abstract martingale* if $E_i(f_j) = f_i$.

To define a submartingale, it is necessary that $\mathscr{X}$ be also a lattice. The problem is thus to find conditions on the martingale (and the space $\mathscr{X}$) in order that $f_i \to f$ in norm or weakly in $\mathscr{X}$. The latter term means that $x^*(f_i) \to x^*(f)$ for each $x^* \in \mathscr{X}^*$, the adjoint space. Note that the family $\{E_i, i \in I\}$ is not necessarily mutually commuting when I is not linearly ordered. We shall show below that such martingales exist in general-enough spaces for applications. These include the Banach spaces B_p of Section I.4. In the latter case, it is necessary to establish the existence of conditional expectations so that the operators E_i of the above definition are available. The following result provides this fact for the $B_p = L^p(\Sigma, \mathscr{X})$ for *any* Banach space $\mathscr{X}$. It is due to Scalora [1] and independently to Chatterji [1].

2. Theorem *Let (Ω, Σ, P) be a probability space and $\mathscr{X}$ a Banach space. If $f \in L^1(\Sigma, \mathscr{X})$ and $\mathscr{B} \subset \Sigma$ is a σ-algebra, then there exists a P-unique $\tilde{f} \in L^1(\mathscr{B}, \mathscr{X})$ such that the mapping $E^{\mathscr{B}}: f \mapsto \tilde{f}$ is a contractive linear projection on $L^1(\Sigma, \mathscr{X})$ with range $L^1(\mathscr{B}, \mathscr{X})$, and it satisfies*

$$\int_A f\,dP = \int_A E^{\mathscr{B}}(f)\,dP_{\mathscr{B}}, \qquad A \in \mathscr{B}. \tag{1}$$

Moreover, if $\tilde{E}^{\mathscr{B}}$ is the scalar conditional expectation on $L^1(\Sigma, \mathbb{R})$, then $\|E^{\mathscr{B}}(f)\| \leq \tilde{E}^{\mathscr{B}}(\|f\|)$ a.e., where $\|\cdot\|$ is the norm of $\mathscr{X}$, and if $\mathscr{B}_1 \subset \mathscr{B}_2 \subset \Sigma$ are two σ-algebras, then $E^{\mathscr{B}_1} = E^{\mathscr{B}_1}E^{\mathscr{B}_2} = E^{\mathscr{B}_2}E^{\mathscr{B}_1}$.

Proof Let $f \in L^p(\Sigma, \mathscr{X})$ be a step function, i.e., $f = \sum_{i=1}^n a_i\chi_{A_i}$, $A_i \in \Sigma$, disjoint, and $a_i \in \mathscr{X}$. Then we have for all $B \in \mathscr{B}$

$$\begin{aligned}\int_B f\,dP &= \sum_{i=1}^n a_i P(A_i \cap B) \qquad \text{by definition}\\ &= \sum_{i=1}^n a_i \int_B P^{\mathscr{B}}(A_i)\,dP_{\mathscr{B}} = \int_B \left(\sum_{i=1}^n a_i \tilde{E}^{\mathscr{B}}(\chi_{A_i})\right) dP_{\mathscr{B}},\end{aligned} \tag{2}$$

where $\tilde{E}^{\mathscr{B}}$ is the (scalar) conditional expectation as given in Chapter II. Let $\tilde{f} = \sum_{i=1}^n a_i \tilde{E}^{\mathscr{B}}(\chi_{A_i})$, A_i disjoint. Then $\tilde{f}$ is an $\mathscr{X}$-valued $\mathscr{B}$-measurable function uniquely defined such that (2) is true. Consider the operator $E^{\mathscr{B}}: f \mapsto \tilde{f}$. It is clear that $E^{\mathscr{B}}$ does not depend on the representation of f and is well defined. Thus for each step function one has

$$\int_B f\,dP = \int_B E^{\mathscr{B}}(f)\,dP_{\mathscr{B}}, \qquad B \in \mathscr{B}, \tag{3}$$

and $E^{\mathscr{B}}(f)$ is $\mathscr{B}$-measurable. Since A_i are disjoint, $\|f\| = \sum_{i=1}^{n} \|a_i\| \chi_{A_i}$ is a real measurable function. Hence

$$\|E^{\mathscr{B}}(f)\| = \left\| \sum_{i=1}^{n} a_i \tilde{E}^{\mathscr{B}}(\chi_{A_i}) \right\| \le \sum_{i=1}^{n} \|a_i\| \tilde{E}^{\mathscr{B}}(\chi_{A_i}) = \tilde{E}^{\mathscr{B}}(\|f\|) \quad \text{a.e.} \tag{4}$$

since $\tilde{E}^{\mathscr{B}}$ is linear. This implies $\int_\Omega \|E^{\mathscr{B}}(f)\|\,dP_{\mathscr{B}} \le \int_\Omega \|f\|\,dP = \|f\|_1$, and we conclude that $E^{\mathscr{B}}$ is a contraction. It is also clear that $E^{\mathscr{B}}(f) = f$ for any $\mathscr{B}$-measurable step function, so that $(E^{\mathscr{B}})^2 = E^{\mathscr{B}}$. Thus $E^{\mathscr{B}}$ is a contractive projection. It is linear since $\tilde{E}^{\mathscr{B}}$ is linear. This means $E^{\mathscr{B}}$ is defined on the linear subset of all step functions in $L^1(\Sigma, \mathscr{X})$ with range into $L^1(\mathscr{B}, \mathscr{X})$. Since such a set is norm dense in $L^1(\Sigma, \mathscr{X})$ and the range is a complete metric space, $E^{\mathscr{B}}$ has a unique norm preserving extension to all of $L^1(\Sigma, \mathscr{X})$ by the uniform continuity of the bounded linear operator. But $E^{\mathscr{B}}$ is also an identity on the dense subset of step functions of $L^1(\mathscr{B}, \mathscr{X})$ and hence by the same argument has a unique extension to be the identity on $L^1(\mathscr{B}, \mathscr{X})$. Thus $E^{\mathscr{B}}$ is a contractive linear projection on $L^1(\Sigma, \mathscr{X})$ with range $L^1(\mathscr{B}, \mathscr{X})$.

We already noted that (3) is (1) for step functions. If $f \in L^1(\Sigma, \mathscr{X})$ is arbitrary, then for each $\varepsilon > 0$ there is a step function f_ε with $\|f - f_\varepsilon\|_1 < \frac{1}{2}\varepsilon$ and hence for any $B \in \mathscr{B}$

$$\begin{aligned}
\left\| \int_B f\,dP - \int_B E^{\mathscr{B}}(f)\,dP_{\mathscr{B}} \right\| &\le \left\| \int_B (f - f_\varepsilon)\,dP \right\| + \left\| \int_B f_\varepsilon\,dP - \int_B E^{\mathscr{B}}(f_\varepsilon)\,dP_{\mathscr{B}} \right\| \\
&\quad + \left\| \int_B (E^{\mathscr{B}}(f_\varepsilon) - E^{\mathscr{B}}(f))\,dP_{\mathscr{B}} \right\| \\
&\le \int_B \|f - f_\varepsilon\|\,dP + 0 + \int_B \|E^{\mathscr{B}}(f - f_\varepsilon)\|\,dP_{\mathscr{B}} \quad \text{by (3)} \\
&\le \|f - f_\varepsilon\|_1 + \|E^{\mathscr{B}}(f - f_\varepsilon)\|_1 \le 2\|f - f_\varepsilon\|_1 < \varepsilon
\end{aligned}$$

because $E^{\mathscr{B}}$ is a contraction. Since $\varepsilon > 0$ is arbitrary, this proves (1).

The proof of the last part is analogous to the scalar case. If $B \in \mathscr{B}$, taking $\varepsilon = 1/n$ in the above, one has $\|f - f_n\|_1 \to 0$ as $n \to \infty$, where f_n is a step function. But

$$\begin{aligned}
\left| \int_B \|E^{\mathscr{B}}(f_n)\|\,dP - \int_B \|E^{\mathscr{B}}(f)\|\,dP \right| &\le \int_B \|E^{\mathscr{B}}(f - f_n)\|\,dP \le \|E^{\mathscr{B}}(f - f_n)\|_1 \\
&\le \|f - f_\varepsilon\|_1 \to 0 \qquad \text{as} \quad n \to \infty.
\end{aligned} \tag{5}$$

Similarly,

$$\left|\int_B \tilde{E}^{\mathscr{B}}(\|f_n\|)\,dP - \int_B \tilde{E}^{\mathscr{B}}(\|f\|)\,dP\right| \le \int_B \tilde{E}^{\mathscr{B}}(\|f-f_n\|)\,dP$$

$$= \int_B \|f-f_n\|\,dP$$

$$\le \|f-f_n\|_1 \to 0 \quad \text{as} \quad n \to \infty. \tag{6}$$

From (5) and (6), since $\|f_n - f\|_1 \to 0$, it follows that for each $B \in \mathscr{B}$,

$$\int_B \|E^{\mathscr{B}}(f)\|\,dP = \lim_n \int_B \|E^{\mathscr{B}}(f_n)\|\,dP \le \lim_n \int_B \tilde{E}^{\mathscr{B}}(\|f_n\|)\,dP$$

$$= \int_B \tilde{E}^{\mathscr{B}}(\|f\|)\,dP.$$

Since the integrands are $\mathscr{B}$-measurable and B is arbitrary in $\mathscr{B}$, one has $\|E^{\mathscr{B}}(f)\| \le \tilde{E}^{\mathscr{B}}(\|f\|)$ a.e.

Finally, since the range of $E^{\mathscr{B}_1}$ is contained in that of $E^{\mathscr{B}_2}$ and these are projections, it is obvious that $E^{\mathscr{B}_1} = E^{\mathscr{B}_2}E^{\mathscr{B}_1}$, a projection being the identity on its range. On the other hand, if $B \in \mathscr{B}_1 \subset \mathscr{B}_2$, one has by (1), $f \in L^1(\Sigma, \mathscr{X})$,

$$\int_B E^{\mathscr{B}_1}(E^{\mathscr{B}_2}(f))\,dP_{\mathscr{B}_1} = \int_B E^{\mathscr{B}_2}(f)\,dP_{\mathscr{B}_2} = \int_B f\,dP = \int_B E^{\mathscr{B}_1}(f)\,dP_{\mathscr{B}_1}.$$

The extreme integrands can be identified a.e., so that $E^{\mathscr{B}_1} = E^{\mathscr{B}_1}E^{\mathscr{B}_2}$. This completes the proof.

If one employs other "standard" properties of the strongly integrable functions, one can deduce the useful fact that $E^{\mathscr{B}}$ commutes with any bounded operator on $\mathscr{X}$. Indeed, the following more precise result can be stated. We recall that if $\mathscr{X}, \mathscr{Y}$ are Banach spaces and $T\colon D \to \mathscr{Y}$ is a linear operator with domain on a linear set $D \subset \mathscr{X}$ such that T is closed (i.e., the graph $\{(x, Tx) : x \in D\} \subset \mathscr{X} \times \mathscr{Y}$ is a closed set), then $f \in L^1(\Sigma, \mathscr{X})$ and $\{f(\omega), \omega \in \Omega\} \subset D$ implies that $\tilde{T}f$ defined by $(\tilde{T}f)(\omega) = T(f(\omega))$, $\omega \in \Omega$ is a measurable function; if $\tilde{T}f \in L^1(\Sigma, \mathscr{Y})$, and moreover for each $A \in \Sigma$, $\int_A f\,dP \in D$ implies that $T(\int_A f\,dP) = \int_A(\tilde{T}f)\,dP$. This important property can be used in our applications. (See Dunford–Schwartz [1, III.6.20], where the result is given with $A = \Omega$, but the proof works here. This is a classical result of E. Hille.) With this one has the following.

3. Theorem *Let (Ω, Σ, P) be a probability space, $\mathscr{B} \subset \Sigma$ a σ-algebra, and $\mathscr{X}, \mathscr{Y}$ a pair of Banach spaces. Let T be a closed linear transformation with domain $D \subset \mathscr{X}$ and range $\mathscr{Y}$. For any $f \in L^1(\Sigma, \mathscr{X})$ such that $\{f(\omega), \omega \in \Omega\} \subset D$*

if $\tilde{T}$ *is the associated mapping of* T *acting on functions,* $\tilde{T}f \in L^1(\Sigma, \mathscr{Y})$, $A \in \mathscr{B} \Rightarrow$ *the element* $\int_A f\,dP$ *is in* D *and*

$$T\left(\int_A E^{\mathscr{B}}(f)\,dP_{\mathscr{B}}\right) = \int_A E^{\mathscr{B}}(\tilde{T}f)\,dP_{\mathscr{B}}, \tag{7}$$

and if $E^{\mathscr{B}}(f)$ *takes its values in* D *also, then the left side is* $\int_A \tilde{T}E^{\mathscr{B}}(f)\,dP_{\mathscr{B}}$ *so that* $(\tilde{T}E^{\mathscr{B}})(f) = (E^{\mathscr{B}}\tilde{T})(f)$ *a.e. In particular, if* T *is bounded, then it follows that* $\tilde{T}E^{\mathscr{B}} = E^{\mathscr{B}}\tilde{T}$. *Thus* $E^{\mathscr{B}}$ *and* $\tilde{T}$ *commute if* $\mathscr{X} = \mathscr{Y}$. *[Here* $E^{\mathscr{B}}$ *is written for both operators on* $L^1(\Sigma, \mathscr{X})$ *and* $L^1(\Sigma, \mathscr{Y})$ *for simplicity. If* $\mathscr{Y} = \mathbb{R}$, *then these are written as* $E^{\mathscr{B}}$ *and* $\tilde{E}^{\mathscr{B}}$.*]*

Proof Using the preceding property of T, the proof follows at once. Indeed, by (1)

$$\int_A E^{\mathscr{B}}(\tilde{T}f)\,dP_{\mathscr{B}} = \int_A \tilde{T}f\,dP = T\int_A f\,dP = T\int_A E^{\mathscr{B}}(f)\,dP_{\mathscr{B}}, \qquad A \in \mathscr{B}.$$

If T is bounded on $\mathscr{X}$, then it is clearly closed and $(D = \mathscr{X})$ $E^{\mathscr{B}}(f)$ takes its values in D, so that the last integral above is $\int_A \tilde{T}E^{\mathscr{B}}(f)\,dP_{\mathscr{B}}$. One can then identify the integrands that are in $L^1(\mathscr{B}, \mathscr{Y})$. Since now f is arbitrary, $\tilde{T}E^{\mathscr{B}} = E^{\mathscr{B}}\tilde{T}$ as stated. [T and $\tilde{T}$ are usually identified, and we do so from now on.]

The most interesting application is when $T = x^*$, $\mathscr{Y} = \mathbb{R}$, and $E^{\mathscr{B}}$ and $\tilde{E}^{\mathscr{B}}$ are the operators as before on $L^1(\Sigma, \mathscr{X})$ and $L^1(\Sigma, \mathbb{R})$. Then by (7), $x^*(E^{\mathscr{B}}(f)) = \tilde{E}^{\mathscr{B}}(x^*(f))$ a.e. This connects the scalar and vector martingales. In the next section we treat the abstract case and combine it with the above result for convergence theorems.

5.2 ABSTRACT MARTINGALES AND CONVERGENCE

In the preceding section, the concept of an abstract martingale (of operators) is introduced and its vector version for the Bochner integrable functions is given. These will be of interest only when their convergence theory is established. We present this theory here, starting with the mean (strong) convergence.

The first result on mean convergence, due essentially to Uhl [1], is given by the following theorem.

1. Theorem *Let* $\{f_i, E_i, i \in I\}$ *be a decreasing or increasing martingale in a Banach space* $\mathscr{X}$. *Then it converges in norm to an element* $f \in \mathscr{X}$ *iff there exists a relatively weakly compact set* $K \subset \mathscr{X}$ *such that for each* $\varepsilon > 0$

there is an i_0 $(=i_0(\varepsilon))$ *with the property that* $i > i_0$ *implies* $f_i \in K + \varepsilon B$ $(=\{k + \varepsilon b : k \in K, b \in B\})$, *where* $B = \{b \in \mathscr{X} : \|b\| < 1\}$ *is the open unit ball of* $\mathscr{X}$ *and where* $\|\cdot\|$ *is the norm functional on* $\mathscr{X}$.

Proof For necessity, since the martingale converges to f, let $K = \{f\}$, which is trivially weakly (and strongly) compact, and for any $\varepsilon > 0$ there is i_0 $(=i_0(\varepsilon))$ such that $i > i_0$ implies $f_i \in K + \varepsilon B$ by the definition of convergence. Thus only the sufficiency is nontrivial. In fact, this depends on the classical Banach–Saks–Mazur theorem, which states that the weak and strong closures of a convex set in a Banach space are identical (cf. Dunford–Schwartz [1, V.3.13]). We present the proof of sufficiency using this result in steps.

I. Under the hypothesis, there is a subsequence of the martingale converging weakly to an element $f \in \mathscr{X}$, i.e., the martingale has a weak cluster point.

For since the $\{f_i, i \in I\}$ is not necessarily in K, we define a "nearby" sequence $\{g_i, i \in J\}$ in K with a cluster point f and then deduce $f_i \to f$ weakly. Thus let $i_1 \in I$ be chosen such that $i > i_1$ implies $f_i \in K + B$. If $i_{n-1} \in I$ is chosen (inductively), let $i_n > i_{n-1}$ in I such that $i > i_n$ implies $f_i \in K + (1/n)B$, using the hypothesis. Since $\{f_i, i > i_n, n \geq 1\}$ may not be in K, one defines a "nearby" sequence $\{g_j, j \in J\} \subset K$ as follows where $J \subset I$.

(a) If $i > i_n$ for all n, then $f_i \in K + (1/n)B$ for all n, so $f_i \in K$, and take $g_i = f_i$ in this case.

(b) If $i > i_{n_0}$ but $i \not> i_{n_0+1}$ for some n_0, then $f_i \in K + (1/n_0)B$, and so let $g_i \in K$ be such that $\|g_i - f_i\| < 1/n_0$.

(c) If (a) and (b) do not hold so that there is no $i > i_n$ for any n, then let $g_i \in K$ be any element.

Now consider the generalized sequence $\{g_i, i \in J\} \subset K$, which need not be a martingale. Since K is relatively weakly compact, the net $\{g_i, i \in J\}$ has a (weak) cluster point in $\mathscr{X}$, say f. Thus there is a subnet $\{g_j, j \in J_1\}$, $J_1 \subset J \subset I$ such that $g_j \to f$ weakly. If now we identify J_1 as a subset of I, so that f_τ, g_τ have the same index τ in J_1, then $\{f_\tau, \tau \in J_1\}$ is a subsequence of $\{f_i, i \in I\}$. But $g_\tau \to f$ weakly. Hence for any $x^* \in \mathscr{X}^*$ one has

$$\limsup_\tau |x^*(f_\tau - f)| \leq \limsup_\tau [|x^*(f_\tau - g_\tau)| + |x^*(g_\tau - f)|]$$

$$\leq \limsup_\tau \|x^*\| \, \|f_\tau - g_\tau\| + 0 \leq \|x^*\|(1/n) \to 0 \quad (1)$$

since n is arbitrary, (a)–(c) apply, and $g_\tau \to f$ weakly. (The arbitrary gs of (c) which did not correspond to any fs do not belong to this convergent subsequence.) Consequently $f_\tau \to f$ weakly. Thus the given martingale $\{f_i, i \in I\}$ has a weak cluster point.

Thus far the martingale property did not enter the argument. We use this additional property in deducing the full result.

II. The weak cluster point f above has the properties that (α) $f_i = E_i(f)$ for all $i \in I$ if the martingale is increasing and (β) $E_i(f) = f$ for all $i \in I$ if it is a decreasing martingale.

For in case (α) for each $i \in I$ there is a $\tau > i$, $\tau \in J_1 \subset I$ with $f_i = E_i(f_\tau)$ by the martingale property and the directedness of I. Hence for any $x^* \in \mathscr{X}^*$, if $y^* = x^* \circ E_i$ $(= E_i^* \circ x^* \in \mathscr{X}^*)$, we have

$$x^*(f_i - E_i(f)) = x^*(E_i(f_\tau) - E_i(f)) = y^*(f_\tau - f) \to 0 \tag{2}$$

as $\tau \to \infty$ by (1). Since x^* is arbitrary, it follows that $f_i = E_i(f)$.

In case (β) the argument is essentially the same. In fact, for any $i_0 \in I$ there is $i > i_0$, $E_i(f_{i_0}) = f_i$, and if $y^* = x^* \circ E_i$ as above, then for any $\varepsilon > 0$ there is a $\tau \in J_1$, $\tau > i$ such that $E_i E_\tau = E_\tau E_i = E_\tau$ and

$$\begin{aligned} &|x^*(f - E_\tau(f_{i_0}))| < \tfrac{1}{2}\varepsilon, \\ &|y^*(f - E_\tau(f_{i_0}))| = |x^*(E_i(f) - E_i E_\tau(f_{i_0}))| < \tfrac{1}{2}\varepsilon \end{aligned} \tag{3}$$

since $f_\tau \to f$ weakly. The above two inequalities yield

$$|x^*(f - E_i(f))| < \varepsilon, \qquad x^* \in \mathscr{X}^*. \tag{4}$$

Hence $E_i(f) = f$, $i \in I$.

III. To connect the weak and strong convergences, the Banach–Saks–Mazur theorem is needed here. First consider the decreasing case and deduce the increasing case from it later. Thus let

$$G = \left\{ \sum_{k=1}^{n} c_k E_{i_k} : i_k \in I,\, c_k \geq 0,\, i_k < i_{k+1},\, \sum_{k=1}^{n} c_k = 1 \right\}$$

be the convex hull of $\{E_i, i \in I\}$ and

$$H = \left\{ h : h = \sum_{k=1}^{n} c_k E_{i_k}(f_{i_0}) = \sum_{k=1}^{n} c_k f_{i_k} : i_k \in I,\, i_k < i_{k+1},\, c_k \text{ as above} \right\},$$

where n varies from element to element. By the directedness of I, $\{H, G\}$ are nonempty and are the convex hulls of the martingale and the projections. Then the preceding steps (I, II) imply that f is in the weak closure of H, and hence by the above-quoted theorem it is in the norm closure of H. Thus if $M = \sup_i \|E_i\|$, then $M < \infty$ by hypothesis, and for any $\varepsilon > 0$ there exists an $h \in H$ such that

$$\|f - h\| < \frac{\varepsilon}{M+1}, \qquad h = \sum_{k=1}^{n} c_k E_{i_k}(f_{i_0}) = \sum_{k=1}^{n} c_k f_{i_k}. \tag{5}$$

By directedness of I, there exists $i' \in I$, $i' > i_k$, $k = 1, \ldots, n$ so that $E_{i'} = E_{i'}E_{i_k} = E_{i_k}E_{i'}$. Hence by (5), $E_{i'}(h) = \sum_{k=1}^{n} c_k E_{i'}(f_{i_k}) = \sum_{k=1}^{n} c_k f_{i'} = f_{i'}$ since the martingale is decreasing. Thus

$$\|f - f_{i'}\| = \|E_{i'}(f) - E_{i'}(h)\| \le \|E_{i'}\| \, \|f - h\| \le M\varepsilon/(M+1) < \varepsilon. \qquad (6)$$

This shows that $f_i \to f$ strongly in the decreasing case.

Let $\{E_i\}_{i \in I}$ be increasing and define $Q_i = I - E_i$. Then $\{Q_i, i \in I\}$ is a uniformly bounded decreasing net of projections on $\mathscr{X}$. Also

$$Q_\tau f = f - E_\tau(f) = f - f_\tau \to 0, \qquad \tau \to \infty, \qquad (7)$$

weakly by Step II. Hence the above work applies, and by (6) one has

$$\|f - f_i\| = \|f - E_i(f)\| = \|Q_i(f)\| \to 0. \qquad (8)$$

Thus $f_i \to f$ in norm in this case also. This completes the proof.

2. Remark If $E_\infty \colon \mathscr{X} \to \mathscr{X}$ is defined by setting $E_\infty(f) = \lim_i E_i(f)$, $f \in \mathscr{X}$, then, by the fact that $\{E_i(f), i \in I\}$ is a martingale and that limits are unique in a Hausdorff space, one sees that E_∞ is a linear operator and is bounded. Indeed, for the last property note that (linearity being evident) by the Fatou inequality,

$$\|E_\infty(f)\| \le \varliminf_i \|E_i(f)\| \le M\|f\|, \qquad f \in \mathscr{X}.$$

Moreover, one has $E_\infty E_i = E_i E_\infty$, and hence $E_\infty^2 = E_\infty$. We leave the simple proof, which follows from the continuity of E_∞ and the given properties of E_i, to the reader. Thus $E_i \to E_\infty$ in the strong operator topology.

From this theorem one can get an alternative proof of Theorem IV.4.6.

3. Corollary *Let (Ω, Σ, P) be a probability space and $\mathfrak{F}_i \subset \mathfrak{F}_{i'} \subset \Sigma$ if $i < i'$, $i \in I$ (directed set) be σ-algebras. Let $\{f_i, \mathfrak{F}_i, i \in I\} \subset L^p(\Omega, \Sigma, P)$ be a martingale. Then $f_i \to f$ in L^p iff: (a) $\|f_i\|_p \le K_0 < \infty$, $i \in I$ for $1 < p < \infty$, (b) for $p = 1$, $\{f_i, i \in I\}$ is (bounded and) terminally uniformly integrable. For the decreasing index, every bounded martingale $\{f_i, \mathfrak{F}_i, i \in I\} \subset L^p(\Sigma)$, $1 \le p < \infty$ converges strongly ($\mathfrak{F}_i \supset \mathfrak{F}_{i'}$ for $i < i'$). [Again terminal uniformity means: given $\varepsilon > 0$, there is i_0 $(= i_0(\varepsilon))$ and a δ $(= \delta_\varepsilon)$ such that $i > i_0$ implies $\int_A |f_i| \, dP < \varepsilon$ for any $P(A) < \delta$ (uniformly in $i > i_0$).]*

Proof Since $L^p(\Sigma)$, $1 < p < \infty$ is reflexive, and then each bounded set is relatively weakly compact, the result follows from the above theorem. If $p = 1$, it is known (by Theorem I.4.7) that this integrability yields the relative weak compactness of $\{f_i, i > i_0\}$, and thus we again deduce the result. For the decreasing case, the uniform integrability is a consequence of the martingale property. So the result follows in all cases.

The proof of the theorem contains the following result.

4. Corollary *Let $\{f_i, E_i, i \in I\}$ be an abstract (increasing or decreasing) martingale. Consider the statements:*

(i) *The set $\{f_i, i \in I\} \subset \mathscr{X}$ has a weak cluster point;*
(ii) *the martingale is weakly convergent;*
(iii) *the martingale is strongly convergent;*
(iv) *the family $\{E_i, i \in I\}$ of uniformly bounded projections (such that $E_i E_{i'} = E_{i'} E_i = E_i$ for $i < i'$, etc.) converges to E_∞ in the strong operator topology, i.e., $\|E_i(x) - E_\infty(x)\| \to 0$ for each $x \in \mathscr{X}$.*

Then one has the implications: (i) $\Leftrightarrow$ (ii) $\Leftrightarrow$ (iii) $\Rightarrow$ (iv), *and if either the martingale is decreasing or there is an $f \in \mathscr{X}$ such that $f_i = E_i(f)$, $i \in I$, then* (iv) $\Rightarrow$ (iii) *also.*

As an interesting consequence of these propositions and ideas, a Radon–Nikodým theorem for vector measures relative to a finite (and then σ-finite) measure can be obtained. This is an important result, and we include it here. It will be useful to recall the p-variation of a vector measure. The p-semivariation was given in Definition II.3.6, and the 1-variation in Section I.4.

If $v: \Sigma \to \mathscr{X}$ is a σ-additive function (weak and strong additivity being the same as noted in Section I.4) and P on Σ is a probability measure, then v is of *p-bounded variation* on A, relative to P iff $|v|_p(A) < \infty$, $A \in \Sigma$, where

$$|v|_p^p(A) = \sup\left\{\sum_{i=1}^{n}\left(\frac{\|v(A_i)\|}{P(A_i)}\right)^p P(A_i) : \{A_i\}_1^n \subset \Sigma, \text{ disjoint}, A = \bigcup_{i=1}^{n} A_i\right\}. \tag{9}$$

Here and below we take 0/0 as 0. If $p = 1$, $|v|_1(\cdot)$ is denoted as $|v|(\cdot)$. It was given in Section I.4 (cf. Eq. (29)). Also $|v|(\cdot)$ is additive or σ-additive according as v is, and $v(\cdot)$ is P-continuous whenever $|v|(\cdot)$ is. We can now state the desired theorem, which is due to Phillips [1], with a slightly different proof.

5. Theorem *Let (Ω, Σ, P) be a complete probability space and $v: \Sigma \to \mathscr{X}$ be a σ-additive P-continuous vector measure. Then there exists uniquely an $f \in L^p(\Sigma, \mathscr{X})$, $1 \le p < \infty$, such that*

$$v(A) = \int_A f\, dP, \qquad A \in \Sigma \tag{10}$$

if (i) $|v|_p(\Omega) < \infty$ and (ii) for each $n \geq 1$, $\{v(A)/P(A) : \|v(A)/P(A)\| \leq n, A \in \Sigma\}$ is relatively weakly compact in $\mathscr{X}$.

Proof We present the proof in steps, as it is somewhat long.

I. Let $\pi = \{A_i\}_1^n \subset \Sigma$ be a partition of Ω, i.e., A_i are disjoint and $\Omega = \bigcup_{i=1}^n A_i$. Let $\mathscr{D}$ be the directed set of all such partitions of Ω ordered by inclusion in that if $\pi_i \in \mathscr{D}$, $i = 1, 2$, then $\pi_1 \leq \pi_2$ whenever for each $A \in \pi_2$ there is a $B \in \pi_1$, $A \subset B$ a.e., $(P(A - B) = 0)$. For each $\pi = \{A_i\}_1^n \in \mathscr{D}$ define a step function:

$$v_\pi = \sum_{i=1}^n \frac{v(A_i)}{P(A_i)} \chi_{A_i}, \qquad \left(\frac{0}{0} \text{ is taken } 0\right). \tag{11}$$

Clearly $v_\pi \in L^p(\Sigma, \mathscr{X})$. Let E_π: $L^p(\Sigma, \mathscr{X}) \to L^p(\Sigma, \mathscr{X})$ be the operator given by

$$E_\pi(f) = \sum_{i=1}^n \frac{1}{P(A_i)} \left(\int_{A_i} f \, dP\right) \chi_{A_i} = \tilde{f}_\pi. \tag{12}$$

Now E_π is evidently linear, $E_\pi(\tilde{f}_\pi) = \tilde{f}_\pi$, and then $E_\pi^2 = E_\pi$. It is also contractive. To see this, we apply the Jensen inequality, after discarding the terms in (12) for which $P(A_i) = 0$, since $\varphi(x) = |x|^p$, $p \geq 1$, is convex. Thus if $\sum'$ is the sum after dropping such zero terms in (12), one has

$$\|E_\pi(f)\|_p^p = \|\tilde{f}_\pi\|_p^p = \sum_{i=1}^{n}{}' \left| \int_{A_i} f \frac{dP}{P(A_i)} \right|^p \int_\Omega \chi_{A_i} \, dP$$

since the A_i are disjoint

$$\leq \sum_{i=1}^{n}{}' \varphi\left(\int_{A_i} \|f\| \, d\tilde{P}_i\right) P(A_i),$$

$$\tilde{P}_i = \frac{1}{P(A_i)} P \quad \text{is a probability}$$

$$\leq \sum_{i=1}^{n}{}' \left(\int_{A_i} \varphi(\|f\|) \, d\tilde{P}_i\right) P(A_i) \qquad \text{by Jensen's inequality}$$

$$= \sum_{i=1}^n \int_{A_i} \|f\|^p \, dP_i = \|f\|_p^p. \tag{13}$$

Also if $\pi_1 \leq \pi_2$, it is readily seen that $E_{\pi_1} E_{\pi_2} = E_{\pi_2} E_{\pi_1} = E_{\pi_1}$.

We note that $\{v_\pi, E_\pi, \pi \in \mathscr{D}\}$ is an increasing martingale, i.e., for each $\pi_1 \le \pi_2$, $E_{\pi_1}(v_{\pi_2}) = v_{\pi_1}$. In fact, let $\pi_1 = \{A_i\}_1^n$, $\pi_2 = \{B_i\}_1^m$ $(m \ge n)$, and $B_i \subset A_j$ a.e., for some j. Then

$$E_{\pi_1}(v_{\pi_2}) = \sum_{i=1}^{n} \frac{1}{P(A_i)}\left(\int_{A_i} v_{\pi_2}\, dP\right)\chi_{A_i} = \sum_{i=1}^{n}\sum_{j=1}^{m} \frac{v(B_j)}{P(B_j)} \cdot \frac{P(A_i \cap B_j)}{P(A_i)}\chi_{A_i}$$

$$= \sum_{i=1}^{n}\sum_{j=1}^{m} \frac{v(B_j \cap A_i)}{P(A_i)} \cdot \frac{P(A_i \cap B_j)}{P(B_j)}\chi_{A_i} = \sum_{i=1}^{n} \frac{v(A_i)}{P(A_i)}\chi_{A_i} = v_{\pi_1} \quad \text{a.e.}$$

II. We shall prove (10) in this step under a stronger condition, namely, $\|v(A)\| \le n_0 P(A)$ for all $A \in \Sigma$ and for some $n_0 \ge 1$. The general case will follow from this *key* reduction. Let $K_0 = \{v(A)/P(A) : \|v(A)\| \le n_0 P(A), A \in \Sigma\}$. By hypothesis this is a relatively weakly compact set in $\mathscr{X}$. Let K_1 be the closed convex hull determined by K_0. Then by a classical result of Kreĭn and Šmulian, K_1 is weakly compact. It is clear that $\{v_\pi(\omega), \omega \in \Omega\} \subset K_1$ for each $\pi \in \mathscr{D}$ (since $\sum_{i=1}^n \chi_{A_i} = 1$).

Let $\tilde{E}_\pi$ be the same operator as in (12) when $\mathscr{X} = \mathbb{R}$. Then the last computation of Step I shows that for each $x^* \in \mathscr{X}^*$, $\{x^*(v_\pi), \tilde{E}_\pi, \pi \in \mathscr{D}\}$ is a (scalar) martingale in $L^p(\Sigma, \mathbb{R})$. Moreover, if $\pi = \{A_i\}_1^n$,

$$\int_A (x^* v_\pi)\, dP = \sum_{i=1}^{n} \frac{(x^* v)(A_i)}{P(A_i)} P(A_i \cap A) = x^* v(A), \qquad A \in \pi. \tag{14}$$

Now let $\{\pi_i\}_1^\infty \subset \mathscr{D}$ be an increasing sequence of (increasingly a.e. finer) partitions, and consider the set $K_2 = \bigcup_{n=1}^\infty \{v_{\pi_n}(\omega) : \omega \in \Omega\}$. Since each set on the right is in K_1, it follows that $K_2 \subset K_1$. For each n, v_{π_n} takes only finitely many values so that K_2 is countable. As a subset of K_1, K_2 is also relatively weakly compact. If $\mathscr{X}_0$ is the closed linear space generated by K_2, then it is a separable subspace of $\mathscr{X}$. Hence it has a denumerable determining (or total) set $\Gamma = \{x_i^*\}_1^\infty \subset \mathscr{X}^* \cap S^*$, so for each $x \in \mathscr{X}_0$, $x_i^*(x) = 0$, all i imply $x = 0$ or equivalently $\sup\{|x_i^*(x)| : i \ge 1\} = \|x\|$. This follows from the fact that there is a dense denumerable set in $\mathscr{X}_0$, and then one obtains the desired determining set with an application of the Hahn–Banach theorem (cf., e.g., Dunford–Schwartz [1, V.7.35]). Here S^* is the unit sphere of $\mathscr{X}^*$.

Since $\{x^* v_{\pi_n}, E_{\pi_n}, n \ge 1\}$ is a martingale by (14), one may apply Theorem IV.2.4 and conclude that $x^* v_{\pi_n} \to x^* f$ a.e., and in L^p (see also Theorem IV.2.5). This means there exists a null set $N_{x_i^*}$ such that if $N = \bigcup_{i=1}^\infty N_{x_i^*}$, then $P(N) = 0$ and for all $\omega \in \Omega - N$, $(x^* f)(\omega) = \lim_n (x^* v_{\pi_n})(\omega)$, $x^* \in \Gamma$. But $\{v_{\pi_n}(\omega), n \ge 1\} \subset K_2$, and by the relative weak compactness of this set, there is a subsequence $\{\pi_{n'}\}_1^\infty$ such that $y^*v_{\pi_{n'}}(\omega) \to y^*f(\omega)$ for all $y^* \in \mathscr{X}^*$. Hence one can conclude that y^*f is measurable, or f is weakly measurable. To show that it is strongly measurable, we only need to show that $f(\Omega - N)$ is separable. Note that in the above line, the (weak) limit of $v_{\pi'_n}(\omega)$ is an element of K_3, the convex closure of K_2 in $\mathscr{X}_0$, so that $f(\omega) \in K_3$. If there exists an $\omega_0 \in \Omega - N$ such that $f(\omega_0) \notin \mathscr{X}_0$, then there is a subsequence $\{\pi_{n''}\}_1^\infty$ and $x_0 \in K_3 \subset \mathscr{X}_0$ such that $(x^*v_{\pi_{n''}})(\omega_0) \to x^*(x_0)$ for all $x^* \in \mathscr{X}^*$ and also to $x^*f(\omega_0)$ for $x^* \in \Gamma$. Thus $x^*(x_0 - f(\omega_0)) = 0$ for all $x^* \in \Gamma$. Since Γ is total, $x_0 - f(\omega_0) = 0$ contradicting the assumption that $f(\omega_0) \notin \mathscr{X}_0$. Thus $f(\Omega - N) \subset K_3 \subset \mathscr{X}_0$, and by the earlier remark on measurability, f is measurable. We now show that $\|f\|^p$ is integrable. Indeed, for $\omega \in \Omega - N$

$$\|f\|^p(\omega) = \sup\{|x^*f|^p(\omega) : x^* \in \Gamma\} = \sup\left\{\left(\lim_n(|x^*v_{\pi_n}|)^p(\omega)\right) : x^* \in \Gamma\right\}$$

$$\leq \limsup_n\{\sup|(x^*v_{\pi_n})(\omega)|^p : x^* \in \Gamma\} = \limsup_n \|v_{\pi_n}\|^p(\omega). \tag{15}$$

But

$$\|v_{\pi_n}\|^p \leq \sum_{k=1}^{m_n} \left\|\frac{v(A_k^n)}{P(A_k^n)}\right\|^p \chi_{A_k^n} \leq n_0^p, \tag{16}$$

where $\pi_n = \{A_k^n\}_1^{m_n} \in \mathscr{D}$ and where we used the special assumption of bounds stated at the beginning of this step. Hence by Fatou's lemma, one gets from (15) and (16)

$$\int_\Omega \|f\|^p \, dP \leq \int_\Omega \limsup_n \|v_{\pi_n}\|^p \, dP \leq n_0^p P(\Omega) < \infty.$$

Thus $f \in L^p(\Sigma, \mathscr{X})$, and $\{v_{\pi_n}, E_{\pi_n}, n \geq 1\}$ has f as a weak limit.

Let $\{\pi'_n\}_{n\geq 1} \subset \mathscr{D}$ be another increasing sequence. By the above argument it is seen that $\{v_{\pi'_n}, E_{\pi'_n}, n \geq 1\}$ is a martingale in $L^p(\Sigma, \mathscr{X})$ which has f' as a weak limit. Let $\{\tilde{\pi}_n\}_{n\geq 1}$ be a combined (refining) increasing sequence. Then by the preceding argument $\{v_{\tilde{\pi}_n}, E_{\tilde{\pi}_n}, n \geq 1\}$ has $\tilde{f} \in L^p(\Sigma, \mathscr{X})$ as a weak limit. Since one can also consider the convergence of the scalar martingales $\{x^*v_{\pi_n}, n \geq 1\}$, $\{x^*v_{\pi'_n}, n \geq 1\}$, and $\{x^*v_{\tilde{\pi}_n}, n \geq 1\}$, it follows easily that $\tilde{f} = f = f'$ a.e., must be true. Thus the martingale $\{v_\pi, E_\pi, \pi \in \mathscr{D}\}$ converges weakly. So by Corollary 4(ii), $v_\pi \to f$ strongly and f is the unique limit of the v_π. [We can deduce this conclusion from Corollary 4(i) soon after proving the first statement that $v_{\pi_n} \to f$ weakly, so that f is a weak cluster point of the above martingale.] To conclude that (10) is true, for any $A \in \Sigma$ we have, since $\nu(A) = \int_A \nu_\pi(\omega)\, dP$,

$$\left\| \int_A f\, dP - \nu(A) \right\| \leq \int_\Omega \|f - \nu_\pi\| \cdot \chi_A\, dP \leq \|f - \nu_\pi\|_p \cdot [P(A)]^{1/p'}$$

by Hölder's inequality

$$\leq \|f - \nu_\pi\|_p \to 0 \qquad \text{as} \quad \pi \uparrow \infty \tag{17}$$

since $\nu_\pi \to f$ in $L^p(\Sigma, \mathscr{X})$ by the above and $P(A) \leq 1$. This proves (10) in this case.

III. Suppose ν satisfies only (i) and (ii) of the theorem. For each n, and $x^* \in \mathscr{X}^*$, consider the signed measure $x^*\nu - n\|x^*\|P\colon \Sigma \to \mathbb{R}$. Then by the Hahn decomposition, $\Omega = \Omega^1_{x^*} \cup \Omega^2_{x^*}$ ($\Omega^i_{x^*} \in \Sigma$), a disjoint union where on $\Omega^1_{x^*}$ the above measure is positive and on $\Omega^2_{x^*}$ it is negative. Thus for all $A \subset \Omega^1_{x^*}$, $A \in \Sigma$, $x^*\nu(A) \geq n\|x^*\|P(A)$ is true, and the opposite inequality holds on $\Omega^2_{x^*}$. Let $\mathscr{A}_n = \{\Omega^1_{y^*}: \text{a positive set for } y^*\nu - n\|y^*\|P,\ y^* \in \mathscr{X}^*\} \subset \Sigma$. Since the measure space is finite (so localizable), $\mathscr{A}_n$ has a supremum A^n in Σ (by Theorem I.5.2). Then $A \subset A_0$, $A \in \Sigma$, implies $x^*\nu(A) \geq n\|x^*\|P(A)$ for some x^*. Hence $\|\nu(A)\| \geq nP(A)$ for all measurable $A \subset A^n$, and $\|\nu(A)\| \leq nP(A)$ for $A \subset \Omega - A^n$, $A \in \Sigma$. Next consider a similar decomposition of A^n such that $x^*\nu - (n+1)\|x^*\|P \geq 0$, on A^{n+1}, and $A^{n+1} \subset A^n$ so that $\|\nu(A)\| \geq (n+1)P(A)$ for all $A \subset A^{n+1}$, $A \in \Sigma$, and $\|\nu(B)\| < (n+1)P(B)$ for $B \subset \Omega - A^{n+1}$, $B \in \Sigma$. But a vector measure is bounded (because every signed measure $x^*\nu\colon \Sigma \to \mathbb{R}$ is bounded and then the mapping $x^* \mapsto x^*\nu$ is continuous so that by the uniform boundedness theorem, $\|\nu(A)\|$ is uniformly bounded for all $A \in \Sigma$ since $\mathscr{X}^*$ is complete). So $\|\nu(B)\| \geq nP(B)$ for all $B \subset A^n$ cannot hold for large enough n. Hence $A^n \downarrow \varnothing$, and for $A \subset \Omega - A^n$, $A \in \Sigma$, we have $\|\nu(A)\| \leq nP(A)$.

Now consider the trace σ-algebras $\Sigma(\Omega - A^n)$ and apply the result of Step II. Then there is a unique $f_n \in L^p(\Sigma(\Omega - A^n), \mathscr{X})$ such that (10) is true. If we set $f = f_n$ on $\Omega - A^n$, $n \geq 1$, then f is well defined. Since $\bigcap_{n=1}^\infty A^n = \varnothing$, one has $f_n \to f$ a.e. Moreover,

$$\int_\Omega \|f\|^p\, dP \leq \varliminf_n \int_\Omega \|f_n\|^p\, dP = \varliminf_n |\nu|_p^p(\Omega - A^n) \leq |\nu|_p^p(\Omega) < \infty. \tag{18}$$

But $\|f - f_n\|^p \leq \|f\|^p$ so that by the dominated convergence and the fact that $\|f - f_n\| \to 0$ a.e., it follows that $\|f - f_n\|_p \to 0$. So the argument of (17) applies for each $A \in \Sigma$. Hence (10) is true in the general case. This completes the proof.

The result is basic in abstract differentiation of vector measures, and since the Banach space $\mathscr{X}$ is not restricted, the details cannot be materially simplified. In fact, the conditions are close to being necessary. This is obtained by a slight modification of the above proof, and the precise statement is given in the Complements and Problems section (see Problem 2). Hereafter the simple modifications needed if (Σ, P) is not complete will be omitted.

The above result also shows that the martingales and differentiation are closely tied together (both in the scalar and vector cases). It is clear that the following set of conditions, related to those of Theorem 1, can be obtained immediately, as has been noted by Uhl [1].

6. Corollary *Let (Ω, Σ, P) be a probability space and $v: \Sigma \to \mathscr{X}$ be a vector measure that is P-continuous. Let v_π be defined by* (11) *for each partition $\pi \in \mathscr{D}$. Then there exists a unique $f \in L^p(\Sigma, \mathscr{X})$, $1 \leq p < \infty$, such that $v_\pi \to f$ in norm if there is a compact set $K \subset L^p(\Sigma, \mathscr{X})$ and, for each $\varepsilon > 0$, an element $\pi_0 \in \mathscr{D}$ such that for each $\pi \geq \pi_0$, $v_\pi \in K + \varepsilon B$, where B is the* (*open*) *unit ball of $L^p(\Sigma, \mathscr{X})$.*

It is known that $L^p(\Sigma, \mathscr{X})$, $1 < p < \infty$, is reflexive iff $\mathscr{X}$ is and that in a reflexive space a set A is weakly compact iff it is closed and bounded. Hence we can simplify the conditions of the above corollary (as well as Theorem 1) in this important case. The statement follows.

7. Corollary *Let $\mathscr{X}$ be a reflexive Banach space and $L^p(\Sigma, \mathscr{X})$, $1 < p < \infty$, be as above. Then a vector measure $v: \Sigma \to \mathscr{X}$ admits a unique representation*

$$v(A) = \int_A f\,dP, \qquad f \in L^p(\Sigma, \mathscr{X}), \quad A \in \Sigma, \tag{19}$$

if $|v|_p(\Omega) < \infty$, i.e., v has a finite p-variation relative to P.

In fact, we only have to note that the sets $\{v(A)/P(A) : \|v(A)\|/P(A) \leq n, A \in \Sigma\}$ are bounded in $\mathscr{X}$ for each n if $|v|_p(\Omega) < \infty$. This also implies that v is P-continuous and $f \in L^p(\Sigma, \mathscr{X})$.

Let us now consider briefly operator valued martingales. (See Section I.4 on measurability concepts.) The following two results indicate the possibilities. As usual, $B(\mathscr{X}, \mathscr{Y})$ is the set of all continuous linear operators on $\mathscr{X}$ into $\mathscr{Y}$.

8. Theorem *Let (Ω, Σ, P) be a probability space and let $\mathscr{X}, \mathscr{Y}$ be Banach spaces of which $\mathscr{X}$ is separable. If $f: \Omega \to B(\mathscr{X}, \mathscr{Y})$ is a strongly measurable and strongly integrable* (*i.e., $\int_\Omega \|fx\|\,dP \leq k < \infty$ for all $x \in \mathscr{X}$ and $\int_A fx\,dP \in \mathscr{Y}$, $A \in \Sigma$*) *operator function, let $v_f: \Sigma \to B(\mathscr{X}, \mathscr{Y})$ be the associated vector measure. Suppose that v_f has finite total variation. Let μ denote the variation measure* *of v_f so that μ is P-continuous. If $\mathscr{B} \subset \Sigma$ is a σ-algebra $v'_f = v_f | \mathscr{B}$, then there exists uniquely an operator $V_f: \Omega \to B(\mathscr{X}, \mathscr{Y})$ such that $\|V_f\|$ is P-integrable and*

$$\int_\Omega \|V_f\| h \, dP = \int_\Omega h \, dP, \qquad h \in L^1(\mathscr{B}, \mathbb{R}); \tag{20}$$

also for each $g \in L^1(\mathscr{B}, \mathscr{X})$, $y^* \in \mathscr{Y}^*$, $\langle V_f g, y^* \rangle$ is P-integrable, and one has

$$\left\langle \int_\Omega g \, dv'_f, y^* \right\rangle = \int_\Omega \langle V_f g, y^* \rangle \, dP, \tag{21}$$

where the left integral is defined in the Dunford–Schwartz sense, i.e., for step functions and then by Cauchy sequence procedure. If $E^{\mathscr{B}}: f \mapsto V_f$ is the mapping, then $E^{\mathscr{B}}$ is a linear operator and satisfies

$$\int_\Omega \langle E^{\mathscr{B}}(f) g, y^* \rangle \, dP = \int_\Omega \langle fg, y^* \rangle \, dP, \qquad g \in L^1(\mathscr{B}, \mathscr{X}), \quad y^* \in \mathscr{Y}^* \tag{22}$$

or, taking $g = \chi_A x$,

$$\int_A \langle E^{\mathscr{B}}(f) x, y^* \rangle \, dP = \int_A \langle fx, y^* \rangle \, dP, \qquad A \in \mathscr{B}, \quad x \in \mathscr{X}, \quad y^* \in \mathscr{Y}^*. \tag{23}$$

For each $A \in \mathscr{B}$, $f\chi_A$ is a strongly measurable operator function so that (23) can be reduced to the familiar form:

$$\int_A E^{\mathscr{B}}(f) x \, dP = \int_A fx \, dP, \qquad A \in \mathscr{B}, \quad x \in \mathscr{X}. \tag{24}$$

Moreover, if $\mathscr{B}_1 \subset \mathscr{B}_2 \subset \Sigma$ are σ-algebras and $E^{\mathscr{B}_i}$ are the corresponding (linear projection) operators given by (24), then

$$E^{\mathscr{B}_1}(E^{\mathscr{B}_2}(f))x = E^{\mathscr{B}_2}(E^{\mathscr{B}_1}(f))x = E^{\mathscr{B}_1}(f)x \qquad \text{a.e. } [P] \tag{25}$$

and

$$\|E^{\mathscr{B}_1}(f)x\| \le \tilde{E}^{\mathscr{B}_1}(\|fx\|) \quad \text{a.e.}$$

In particular, if $\mathscr{X} = \mathbb{R}$, the result reduces to Theorem 1.2.

The proof is completed by showing that for each $x \in \mathscr{X}$, $\{v_f(A)x/P(A) : A \in \mathscr{B}, P(A) > 0\}$ is relatively weakly compact and that there is a countable determining set in $\mathscr{Y}^*$ for $\mathscr{Y}_1$, the linear span of $\{f(\omega)\mathscr{X} : \omega \in \Omega\}$. One can then use the arguments of Dinculeanu [1, pp. 269–272 and Proposition 16 on p. 102]. We shall not reproduce the details, which are long but not difficult.

With this, one can define a *strong operator martingale* $\{f_n, \mathscr{B}_n, n \ge 1\}$ as follows: If each $f_n: \Omega \to B(\mathscr{X}, \mathscr{Y})$ ($\mathscr{X}$ separable) is strongly measurable and *strongly integrable* (i.e., v_{f_n} has *finite* total variation), then $E^{\mathscr{B}_n}(f_{n+1})x = f_n x$ a.e. $n \geq 1$. One can then prove the following result.

9. Theorem *Let (Ω, Σ, P) be a probability space, $\mathscr{X}$ a separable, and $\mathscr{Y}$ a general Banach space. Suppose $\mathscr{Y}$ has the Radon–Nikodým property (i.e., every vector measure into $\mathscr{Y}$, of finite variation has a strong density relative to its variation measure). Let $\{f_n, \mathscr{B}_n, n \geq 1\}$ be a strong operator martingale on (Ω, Σ, P), $\mathscr{B}_n \subset \mathscr{B}_{n+1} \subset \Sigma$. If $\sup_n \int_\Omega \|f_n x\| \, dP = K < \infty$, then there exists a strongly measurable $f_\infty : \Omega \to B(\mathscr{X}, \mathscr{Y})$ such that $\|f_n - f_\infty\|$ is measurable and $\|f_n - f_\infty\| \to 0$ a.e., even though f_∞, f_n are not necessarily uniformly measurable. Moreover, $\{f_n, \mathscr{B}_n, 1 \leq n \leq \infty\}$ is a strong operator martingale where $\mathscr{B}_\infty = \sigma(\bigcup_{n=1}^{\infty} \mathscr{B}_n)$ iff the set $\{\|f_n\|, n \geq 1\} \subset L^1(\Sigma, \mathbb{R})$ is uniformly integrable.*

This shows what kinds of conditions are needed for the general study. We omit the proof for this one also (it is sketched in the author's paper [8]). The point is that the operator valued martingales have not been much investigated, compared to the vector valued case. Note that to include the weakly measurable and integrable case, further work is needed, and there are some difficulties even here in an extension of the vector valued case treated earlier. On the other hand, with the uniformly measurable martingales, one should consider refinements of the case of vector martingales specifically to the operator valued case. Typically, the hypothesis should bring in some spectral properties of f_n as in the study of uniform ergodic theory. Virtually nothing is known in this area. It will be of interest to make a systematic study of these problems.

5.3 MARTINGALES AND ERGODIC THEORY

It was a classical remark due to Wiener that ergodic theory is really a part of integration. Since conditional expectations are also integrals relative to appropriate function space valued (i.e., conditional probability) measures, as seen from Chapter II, martingale theory is thus essentially a part of integration. Both are related to the theory of integration in infinitely many variables. It was felt for a long time that both of these theories should be obtainable from a single structure. In fact, there are many similarities in form as well as in proofs of the main convergence theorems in both cases for this observation. In this section, following Jerison [1], we formulate a key ergodic sequence as a martingale. This will not solve the unification problem, but it shows a need for further developments and refinements of certain martingales on (necessarily) infinite measure spaces. In Section 4 we shall show how a "superstructure" unifying both theories through Reynolds operators can be built and a solution to the above question can be given. However, it is of a somewhat different kind of result.

Let us briefly recall the ergodic problem. If (Ω, Σ, P) is a probability space, then a transformation $T: \Omega \to \Omega$ is measurable and measure preserving iff $T^{-1}(\Sigma) \subset \Sigma$ and $P \circ T^{-1} = P$. Writing $T^n = T(T^{n-1})$ and $T^0 =$ identity, a classical (1890) result due to H. Poincaré says that for almost every $\omega \in E \in \Sigma$ there are infinitely many n such that $T^n\omega \in E$. Thus almost every point of E is visited by the transformation infinitely often. This result has an intimate connection with certain phenomena in statistical mechanics. An important problem, there, is to find the mean sojourn time (if any) for the points of E visited by T. This can be restated as follows: Does the following limit exist in some sense?

$$\lim_{n\to\infty} \frac{1}{n} \sum_{i=0}^{n-1} \chi_E(T^i\omega), \qquad E \in \Sigma. \tag{1}$$

Mathematically, this is a special case of finding the class of measurable $f: \Omega \to \mathbb{R}$ for which $\lim_n(\sum_{i=0}^{n} f(T^i\omega)/n)$ exists in some sense. But then the transformation $U: f \mapsto f \circ T$ is a positive linear mapping on the (linear) class of such functions, and the above problem is subsumed under the one described as follows: Find conditions on $U: \mathscr{C} \to \mathscr{C}$ where $\mathscr{C} \subset L^p(\Omega, \Sigma, P)$, say, such that for each $f \in \mathscr{C}$

$$\lim_{n} \frac{1}{n} \sum_{i=0}^{n-1} U^i g \quad \text{exists in some sense.} \tag{2}$$

Thus ergodic theory concerns the convergence behavior of such sequences of averages as (2), involving continuous (in fact, contractive) linear operators U, the type of convergence being pointwise a.e., norm, and such others. It is interesting (but not obvious) that such sequences are identifiable with certain martingales on a (generally infinite) measure space, making the work of Section 2 (and particularly of Section IV.3) relevant. We shall now show how this can be done. It may be noted that the pointwise convergence of (1) is a form of the strong law of large numbers (see Problem 4 in the Complements and Problems section of Chapter IV) and can be regarded as a generalization of the latter.

Let $\mathbb{N}$ be the set of all positive integers, $\mathscr{N}$ the power set of $\mathbb{N}$, and $\zeta(\cdot): \mathscr{N} \to \bar{R}^+$ the counting measure. If (Ω, Σ, P) is a probability space, let $(S, \mathscr{A}, \mu)$ be the Cartesian product $(\mathbb{N}, \mathscr{N}, \zeta) \times (\Omega, \Sigma, P)$ so that $\mu: \mathscr{N} \otimes \Sigma \to \bar{R}^+$ is only σ-finite and nonfinite since $P(\Omega) > 0$. Let $\mathfrak{F}_n = \sigma[\{0, 1, \ldots, n-1\}, \{k\} : k \geq n]$ be the σ-algebra generated by the sets shown so that $\{k\}$, $k \geq n$, and $\{0, 1, \ldots, n-1\}$ are atoms of $\mathfrak{F}_n(\supset \mathfrak{F}_{n+1})$. Let $\mathscr{B}_n = \mathfrak{F}_n \otimes \Sigma \subset \mathscr{A}$. Then for each n, $\mu_n = \mu | \mathscr{B}_n$ is σ-finite and $\mathscr{B}_n \supset \mathscr{B}_{n+1}$. We now formulate an ergodic average on (Ω, Σ, P) as a martingale on $(S, \mathscr{A}, \mu)$. Thus let $T: L^1(\Sigma) \to L^1(\Sigma)$ $(= L^1(\Omega, \Sigma, P))$ be a continuous linear operator and $f \in L^1(\Sigma)$. For each $s = (k, \omega) \in S$ define $h(k, \omega) = h(s) = (T^k f)(\omega)$, and let $h_n: S \to \mathbb{R}$ be given by

$$h_n(k, \omega) = \begin{cases} \dfrac{1}{n} \displaystyle\sum_{i=0}^{n-1} (T^i f)(\omega) & \text{for} \quad (k, \omega) \in \{0, 1, \ldots, n-1\} \times \Omega \\ h(k, \omega) & \text{for} \quad (k, \omega) \in \{k : k \geq n\} \times \Omega. \end{cases} \tag{3}$$

As usual $(Tf)(\cdot)$ stands for $(T\tilde{f})(\cdot)$, where $\tilde{f}$ is a member of the equivalence class of f. Since h_n is constant on the generators of $\mathscr{B}_n$, it is measurable relative to this σ-algebra as it is measurable for $\mathscr{A}$. We now establish the stated result:

1. Theorem *With the above notation and assumptions of* (3), *h is locally integrable on S, and $\{h_n, \mathscr{B}_n, n \geq 1\}$ is a decreasing (locally integrable) martingale, i.e., $E^{\mathscr{B}_n}(h) = h_n$ a.e. $[\mu]$, and convergence a.e. of the martingale is equivalent to convergence a.e. of the sequence* (2) *with U replaced by T there.*

Proof Regarding the local integrability statement on h, we let $v_h(\cdot) = \int_{(\cdot)} h \, d\mu$ and show that $|v_h|(A) < \infty$ for each set of finite μ-measure. This follows at once when it is shown that $|v_h|(\cdot)$ is σ-finite. It was already noted that h and h_n are measurable for $\mathscr{A}$ and $\mathscr{B}_n$. Let $S_k = \{k\} \times \Omega \in \mathscr{A}$. So $S = \bigcup_{k=1}^{\infty} S_k$, a disjoint union, and $\mu(S_k) < \infty$. But $|v_k|(S_k) = \int_{S_k} |h| \, d\mu = \int_\Omega |T^k f| \, dP \leq \|T\|^k \int_\Omega |f| \, dP < \infty$ for each k. Hence $|v_h|(\cdot)$ is σ-finite. Thus we also deduce that $E^{\mathscr{B}_n}$ exists. However, h need not be μ-integrable, as simple examples indicate.

To show that $\{h_n, \mathscr{B}_n, n \geq 1\}$ is a decreasing martingale, we prove that $E^{\mathscr{B}_n}(h) = h_n$ a.e. $[\mu]$, and this implies the martingale property because

$$\begin{aligned} E^{\mathscr{B}_{n+1}}(h_n) &= E^{\mathscr{B}_{n+1}}(E^{\mathscr{B}_n}(h)) = E^{\mathscr{B}_n}(E^{\mathscr{B}_{n+1}}(h)) = E^{\mathscr{B}_n}(h_{n+1}) \\ &= h_{n+1} \quad \text{a.e. } [\mu]. \end{aligned} \tag{4}$$

Now to prove $E^{\mathscr{B}_n}(h) = h_n$, it suffices to establish $(\mu_n = \mu | \mathscr{B}_n)$

$$\int_A h_n \, d\mu_n = \int_A E^{\mathscr{B}_n}(h) \, d\mu_n = \int_A h \, d\mu, \qquad A \in \mathscr{B}_n, \tag{5}$$

or only for the generators A of $\mathscr{B}_n$. Since $S = \bigcup_{k=1}^{\infty} S_k$, it is clear that $S_k \notin \mathscr{B}_n$ for $0 \leq k \leq n-1$, but $\bigcup_{k=0}^{n-1} S_k \in \mathscr{B}_n$. Let $A \in \mathscr{B}_n$, and $A_1 = A \cap \bigcup_{k=0}^{n-1} S_k$, $A_2 = A - A_1$. Then $A_i \in \mathscr{B}_n$, and we show that (5) is true if A is replaced by A_i, $i = 1, 2$. But on A_2, $h_n = h$ by definition. So (5) is trivial (and true) in this case. On the other hand, $A_1 \subset \bigcup_{k=0}^{n-1} S_k$, so that $A_1 = \mathbb{N}_n \times E$ where $\mathbb{N}_n = \{0, 1, \ldots, n-1\} \in \mathfrak{F}_n$ and $E \in \Sigma$ since $\mathbb{N}_n$ is an atom of $\mathfrak{F}_n$. Hence

$$\begin{aligned}\int_{A_1} h_n \, d\mu_n &= \int_{\mathbb{N}_n} \int_E h_n(k, \omega) \, dP \, d\zeta = n \int_E h_n(k, \omega) \, dP, \qquad k \in \mathbb{N}_n, \\ &= \int_E \sum_{k=0}^{n-1} (T^k f)(\omega) \, dP \qquad \text{by} \quad (3). \end{aligned} \tag{6}$$

One also has

$$\int_{A_1} h(k, \omega) \, d\mu = \int_{\mathbb{N}_n} \int_E (T^k f)(\omega) \, dP \, d\zeta = \int_E \sum_{k=0}^{n-1} (T^k f)(\omega) \, dP. \tag{7}$$

Thus (6) and (7) show that (5) is true when $A = A_1$.

If $h_n \to \tilde{h}$ a.e. $[\mu]$, then $\tilde{h}$ is $\mathscr{B}(= \bigcap_{n=1}^{\infty} \mathscr{B}_n)$-measurable. But every set of $\mathscr{B}$ is of the form $\mathbb{N} \times B$, $B \in \Sigma$, so that $\tilde{h}$ is independent of the first variable. This means $(1/n) \sum_{i=0}^{n-1} (T^i f)(\omega) \to \tilde{h}(\omega)$ a.e. $[P]$. Note that if h_1 is μ-integrable, then by Corollary IV.3.3 $h = 0$ a.e. $[P]$. Since clearly this need not be true (e.g., T is the identity), we cannot assume the μ-integrability of h_1 in this problem. It follows that the convergence of ergodic averages takes place a.e. iff the locally integrable martingale $\{h_n, \mathscr{B}_n, n \geq 1\}$ converges a.e. This proves the theorem.

The convergence of the above martingale cannot be obtained from the martingale convergence theorems that we have had thus far. If T arises from a measure preserving transformation, then the validity of the convergence a.e. is the classical Birkhoff ergodic theorem. Again, more generally, if T is a contraction on both the $L^1(\Sigma)$ and $L^\infty(\Sigma)$, then the convergence a.e. takes place and this is the Hopf–Dunford–Schwartz ergodic theorem. This means under these conditions, there exists another class of martingale convergence theorems. Uncovering such results will fill an obvious gap in the martingale theory on infinite measure spaces, particularly for the decreasing index case.

5.4 A UNIFIED FORMULATION OF SOME ERGODIC AND MARTINGALE THEOREMS

The result of Section 3 shows that a mere formulation of an ergodic sequence as a martingale is not likely to lead to the desired convergence theorems. However, it is possible to present a result accompanied by its limit theorem which combines both the martingale and ergodic statements.

This is achieved through an application of Reynolds operators of Section II.6. Since the proof of Theorem II.6.2 uses the mean ergodic theorem as a key ingredient, the result can be regarded as a unified formulation of the two though not an independent one by itself (at least until a proof of Theorem II.6.2 without reference to ergodic-martingale theories is devised). Nevertheless, this approach has independent interest and is different from that of Section 3, with perhaps some applications elsewhere. The following presentation is based on the author's paper [12] and on Rota's announcement [3].

The idea behind the proposed "unified formulation" of these theories is to find a family of operators that have the built-in features of ergodic and martingale (i.e., conditional expectation) operations structurally. The work of Section II.6 shows that Reynolds operators have these properties. However, before presenting the definition of a "generalized martingale," which includes this family, we shall attempt to motivate the result from classical considerations, as otherwise the concept may appear completely ad hoc.

Recall that for a collection of numbers $\{a_n\}_1^\infty$, $\sum_{n=0}^\infty a_n$ converges in the sense of Cesàro, or $(C, 1)$-mean, if the sequence of averages $s_n = (1/n)\sum_{i=0}^{n-1} a_i$ converges; and it converges in the sense of Abel mean, or A-converges, if $\lim_{x\to 1-}(1-x)\sum_{n=0}^\infty a_n x^n$ exists. While $(C, 1)$-convergence implies A-convergence, the converse holds provided the $\{s_n, n \geq 1\}$ sequence is bounded in addition to A-convergence. This is classical. If one considers formally $a_n = (T^n f)(\omega)$, then the $(C, 1)$-means are the ergodic averages $s_n = (1/n)\sum_{i=0}^{n-1}(T^i f)(\omega)$, and the A-convergence part becomes (on writing $(y - T)$ for $(yI - T)$ for ease):

$$\lim_{x\to 1-}(1-x)\sum_{n=0}^\infty x^n(T^n f)(\omega) = \lim_{x\to 1-}(1-x)[(1-xT)^{-1}f](\omega)$$
$$= \lim_{y\to 1+}(y-1)[(y-T)^{-1}f](\omega). \tag{1}$$

Now we recognize that $(y - T)^{-1} = R_y(T)$ is the resolvent of the linear operator T. But then the pointwise a.e. limit of the ergodic averages exists whenever these are bounded and $\lim_{y\to 1+}(y-1)(R_y f)(\omega)$ exists a.e. The last limit can be expressed in the following more suggestive form at least formally:

$$\lim_{x\to 1-}(1-x)R_x = \lim_{x\to 1-}\sum_{n=0}^\infty (1-x)x^n T^n$$
$$= \lim_{\lambda\to 0+}\sum_{n=0}^\infty [1-e^{-\lambda}]e^{-n\lambda}T^n. \tag{2}$$

If we set $U(t) = T^n$ for $n \le t < n+1$, $n \ge 0$, then (2) can be written as

$$\lim_{\lambda \to 0+} \lambda R_\lambda f = \lim_{\lambda \to 0+} \lambda \int_0^\infty e^{-t\lambda} U(t) f \, dt, \qquad f \in L^p(\Sigma), \tag{3}$$

where the integral can be taken, for instance, in the abstract Riemann sense. Also $\{U(t), t \ge 0\}$ is a family of bounded commuting operators and in fact forms a strongly continuous semigroup on $L^p(\Sigma)$. If T were an operator arising from a measure preserving transformation, then it would be verified that $U(t)(fg) = (U(t)f)(U(t)g)$, for f, g in $L^1 \cap L^\infty(\Sigma)$, so that $U(t)$ is also a *homomorphism*. All the above formal computations can be justified by use of the operational calculus. A comparison of (3) and the structure of the Reynolds operators given by Theorem II.6.2 shows the relevance of the latter, which have been under investigation since the 1920s by J. Kampé de Fériet and others. Moreover, as the reader may verify by a straightforward but somewhat long computation, the operator $V_\lambda = \lambda R_\lambda$ satisfies the Reynolds identity for each $\lambda > 0$. But R_λ is a resolvent operator and as such satisfies the algebraic identity: $R_\lambda - R_\mu = (\lambda - \mu) R_\lambda R_\mu$ for $\lambda, \mu > 0$ by the standard results of linear analysis. With this background, let us now introduce the desired concept.

1. Definition A family $\{V_\lambda, \lambda > 0\}$ of mutually commuting Reynolds operators on $L^p(\Sigma)$ is called a *generalized martingale* if the following algebraic identity is satisfied:

$$(\mu V_\lambda - \lambda V_\mu) V_\mu f = (\mu - \lambda) V_\lambda V_\mu^2 f, \qquad 0 < \lambda < \mu, \quad f \in L^1 \cap L^\infty(\Sigma). \tag{4}$$

Evidently, the family $V_\lambda = E^{\mathscr{B}_\lambda}$, $\mathscr{B}_\mu \subset \mathscr{B}_\lambda \subset \Sigma$ for $0 < \lambda < \mu$, of conditional expectations satisfies (4) and $\{V_\lambda f, \lambda > 0\}$ is a (continuous parameter) martingale. Also if $V_\lambda f = \lambda R_\lambda f = \lambda \int_0^\infty e^{-\lambda t} U(t) f \, dt$, then $\{V_\lambda f, \lambda > 0\}$ represents an ergodic average (or flow) so that the convergence of $V_\lambda f$ as $\lambda \to 0+$ yields both the convergence theorems. This is the sought for result in the present context, and it is stated in the following form. Here it is necessary to use several results from linear analysis; precise references will be given.

2. Theorem *Let $\{V_\lambda, \lambda > 0\}$ be a generalized martingale on $L^p(\Omega, \Sigma, P)$, $1 \le p < \infty$, where P is a Carathéodory regular probability. Suppose that for each $\lambda > 0$ the Reynolds operator V_λ satisfies the hypothesis of Theorem II.6.2, i.e., (i) $V_\lambda f_0 = f_0$ for a weak unit (we may take $f_0 = 1$ here), (ii) V_λ is contractive on $L^p(\Sigma)$, and (iii) if $p = 1$, also let V_λ be a weakly compact operator. Then for each $f \in L^p(\Sigma) \cap L^{\varphi_0}(\Sigma)$, $\lim_{\lambda \to 0+} V_\lambda f$ exists a.e. and in norm, where $\varphi_0(x) = |x| \log^+ |x|$.*

Proof We first prove the norm convergence. By Theorem II.6.2, the closure of the range of each V_λ in $L^p(\Sigma)$ is precisely $L^p(\mathscr{B}_\lambda)$ where $\mathscr{B}_\lambda \subset \Sigma$ is a σ-algebra relative to which the weak unit f_0 is measurable. Moreover, on $L^p(\mathscr{B}_\lambda)$, V_λ is injective, as shown at the end of Step IV in the proof of Theorem II.6.2. Hence V_λ^2 is also reduced by the conditional expectation $E^{\mathscr{B}_\lambda}$ and has its range dense in $L^p(\mathscr{B}_\lambda)$. Since $\{V_\lambda, \lambda > 0\}$ is a generalized martingale, the identity (4) must hold for any $0 < \lambda < \mu$. We assert that these two facts imply $L^p(\mathscr{B}_\lambda) \supset L^p(\mathscr{B}_\mu)$. In fact, if $f \in L^p(\mathscr{B}_\mu)$, then $\{V_\mu f, V_\mu^2 f\} \subset L^p(\mathscr{B}_\mu)$, and by (4) one has

$$\lambda V_\mu^2 f = \mu V_\lambda(V_\mu f) - (\mu - \lambda) V_\lambda(V_\mu^2 f) \in L^p(\mathscr{B}_\lambda) \tag{5}$$

since $L^p(\mathscr{B}_\lambda) \subset L^p(\Sigma)$. Now from (5) it follows that the closure of the range of V_μ^2 which is $L^p(\mathscr{B}_\mu)$ must be contained in the complete space $L^p(\mathscr{B}_\lambda)$. Thus for $0 < \lambda < \mu$ *we can (and do) assume* $\mathscr{B}_\lambda \supset \mathscr{B}_\mu$. Let $\mathscr{B}_0 = \bigcup_{\lambda > 0} \mathscr{B}_\lambda$ and $\mathscr{B} = \sigma(\mathscr{B}_0)$. Then the weak unit f_0 is $\mathscr{B}$-measurable, and $\bigcup_{\lambda > 0} L^p(\mathscr{B}_\lambda)$ is a dense subspace of $L^p(\mathscr{B})$.

It was already noted that if $V_\lambda = \lambda R_\lambda$, then R_λ is a resolvent operator $R(\lambda, D)$, where $D = I - R^{-1}$ is a closed densely defined operator on $L^p(\mathscr{B}_\lambda)$ for each $\lambda > 0$, so that it is so on $L^p(\mathscr{B})$ (and in fact is the infinitesimal generator of the semigroup $\{V(t), t \geq 0\}$ appearing in the representation theorem—cf. Steps V and IX of the proof of Theorem II.6.2). We now use hypotheses (ii) and (iii) (and note that the weak compactness condition is automatic for $1 < p < \infty$ since L^p is then reflexive) and deduce that $\{V_\lambda f, \lambda > 0\}$ is a bounded set in $L^p(\mathscr{B})$, $1 \leq p < \infty$ and is relatively weakly compact there. Hence there is a weakly convergent subsequence $\{V_{\lambda_n} f, n \geq 1\}$ where $\lambda_n \to 0$ as $n \to \infty, f \in L^p(\mathscr{B})$. This implies by a well-known abstract ergodic theorem of E. Hille (cf. Yosida [1, p. 217]), $V_\lambda f \to \tilde{f}$ in norm as $\lambda \to 0+$. It remains to show that the same is true for all $f \in L^p(\Sigma)$.

Let $f \in L^p(\Sigma)$ and set $f = g + h$ where $g = E^{\mathscr{B}}(f) \in L^p(\mathscr{B})$ and $h = f - g$. Since $E^{\mathscr{B}} E^{\mathscr{B}_\lambda} = E^{\mathscr{B}_\lambda} E^{\mathscr{B}} = E^{\mathscr{B}_\lambda}$ for all $\lambda > 0$ by Proposition II.1.2 ($\mathscr{B}_\lambda \subset \mathscr{B}$), and V_λ commutes with $E^{\mathscr{B}_\lambda}$, we deduce that V_λ commutes with $E^{\mathscr{B}}$ for all λ, and $E^{\mathscr{B}}(h) = g - g = 0$ a.e., so that $E^{\mathscr{B}_\lambda}(h) = E^{\mathscr{B}_\lambda}(E^{\mathscr{B}}(h)) = 0$ a.e. Hence h is in the null space of $E^{\mathscr{B}_\lambda}$ for each $\lambda > 0$, so that it is also in the null space of V_λ (since V_λ is reduced by the projection $E^{\mathscr{B}_\lambda}$), i.e., $V_\lambda h = 0$ for each $\lambda > 0$. Hence

$$V_\lambda f = V_\lambda g + 0 \to \tilde{g} \qquad \text{as} \quad \lambda \to 0+, \tag{6}$$

in norm by the preceding paragraph. Since $f \in L^p(\Sigma)$ is arbitrary, the norm convergence of the generalized martingale $\{V_\lambda f, \lambda > 0\}$ follows for $p \geq 1$ without further restrictions.

Let us turn to the pointwise convergence. Consider the resolvent R_λ $(=R(\lambda, D))$ of D given by

$$R_\lambda f = \int_0^\infty e^{-\lambda t} U(t) f\,dt, \qquad \lambda > 0, \quad f \in L^p(\mathscr{B}_\lambda), \tag{7}$$

so that $V_\lambda = \lambda R_\lambda$ and the integral is taken in the strong, or the Bochner, sense. We now use the fact that the underlying probability space (Ω, Σ, P) is (Carathéodory regular and hence) strictly localizable. So by Theorem I.5.4. there exists a (set) lifting ρ (i.e., put $\tilde{\rho}(\chi_A) = \rho(A)$) on Σ. With this, it is possible to introduce a topology in Ω (cf., e.g., the books by Ionescu Tulcea [1] or the author [10]). Since one can assume that each $\mathscr{B}_\lambda$ is complete, we obtain for each $\lambda > 0$, $\rho(\mathscr{B}_\lambda) \subset \mathscr{B}_\lambda$. Then it is known and not hard to show that each $\mathscr{B}_\lambda$-measurable (bounded) function is equivalent to a (bounded) continuous function in this topology. Thus if $\bar{f}$ denotes such a continuous representation of f $(f = \bar{f}$ a.e.), then $\bar{f} \in L^p(\mathscr{B}_\lambda)$. By the strong continuity of the semigroup $\{U(t), t \geq 0\}$, we deduce from (7) that $R_\lambda f = R_\lambda \bar{f}$ a.e., and $R_\lambda \bar{f}$ is continuous in the lifting topology for each $\lambda > 0$. (Thus the integral can now be interpreted in the "abstract Riemann" sense.) Hence by a result from Dunford–Schwartz [1, VIII.9.21], we conclude that $\lim_{\lambda \to 0+}(\lambda R_\lambda \bar{f})(\omega) = \lim_{\lambda \to 0+}(\lambda R_\lambda f)(\omega)$ exists a.e. If $f \in M = \bigcup_{\lambda > 0} L^p(\mathscr{B}_\lambda)$, then $f \in L^p(\mathscr{B}_{\lambda_0})$ for some $\lambda_0 > 0$ and hence for $0 < \lambda < \lambda_0$, $f \in L^p(\mathscr{B}_\lambda)$ since $\mathscr{B}_\lambda \supset \mathscr{B}_{\lambda_0}$ (see remark following Eq. (5)). Consequently, $\lim_{\lambda \to 0+}(V_\lambda f)(\omega)$ exists a.e. for each $f \in M$, which is a dense subspace of $L^p(\mathscr{B})$. It is necessary now to extend the result to all of $L^p(\mathscr{B})$. This is nontrivial, and we establish the result where all the σ-algebras can be (and are) taken complete.

Let $f \in L^p(\mathscr{B})$. Then $\{E^{\mathscr{B}_\lambda}(|f|), \mathscr{B}_\lambda, \lambda > 0\}$ is a terminally uniformly integrable positive martingale. The measurability problems are not difficult (one may apply Theorem I.5.2, or may also assume that the martingale is "separable," cf., e.g., the books noted in the above paragraph on the latter) to conclude that $f^* = \sup_{\lambda > 0} E^{\mathscr{B}_\lambda}(|f|)$ is measurable. If $p > 1$, it follows immediately from Theorem IV.1.12 that $f^* \in L^p(\mathscr{B})$. In the case of $p = 1$, using the compactness hypothesis, the same representation of V_λ holds, and since $L^1(\Sigma) \supset L^{\varphi_0}(\Sigma)$ (the latter is an Orlicz space) due to the finiteness of measures, again Theorem IV.1.12 is applicable and one concludes that $f^* \in L^1(\mathscr{B})$. With this the proof can be completed as follows.

By the positivity of V_λ and of $U(t)$, we have (on using the above fact) that

$$|V_\lambda f| \leq V_\lambda(|f|) \leq \lambda \int_0^\infty e^{-\lambda t} U(t) E^{\mathscr{B}_\lambda}(|f|)\,dt \leq \int_0^\infty e^{-\lambda t} U(t) f^*\,dt \quad \text{a.e.} \tag{8}$$

If $\beta_\lambda(t) = \lambda e^{-\lambda t}$, then β_λ is a decreasing positive function on $\mathbb{R}^+$ for each $\lambda > 0$. Since $\{U(t), t \geq 0\}$ is a strongly continuous, positivity preserving, contractive semigroup on all $L^p(\mathscr{B})$, $1 \leq p \leq \infty$ (arising from a measure preserving transformation), we may apply another result from Dunford–Schwartz [1, VIII.9.3], according to which there exists a measurable and a.e. finite function g_f such that (with (8))

$$V_\lambda(|f|) \leq \int_0^\infty U(t)f^*\beta_\lambda(t)\,dt \leq g_f \int_0^\infty \beta_\lambda(t)\,dt = g_f \quad \text{a.e.} \tag{9}$$

Hence one has $\sup_{\lambda>0}|V_\lambda f| \leq g_f < \infty$ a.e., for each $f \in L^p(\mathscr{B})$, and $\lim_{\lambda \to 0+} V_\lambda f$ exists for each $f \in M$, a dense subspace of $L^p(\mathscr{B})$. Consequently, by Banach's theorem (cf. Dunford–Schwartz [1, p. 332]) we must have $\lim_{\lambda \to 0+} V_\lambda f$ existing a.e., for all $f \in L^{\varphi_0}(\mathscr{B})$. Finally, if $f \in L^{\varphi_0}(\Sigma)$, then one can express f as $f = f_1 + f_2$ where $f_1 = E^{\mathscr{B}}(f) \in L^{\varphi_0}(\mathscr{B})$ and $f_2 = f - f_1$ as in the case of norm convergence above. Then $V_\lambda f_2 = 0$, and hence $V_\lambda f = V_\lambda f_1 \to$ a limit a.e. This completes the proof of the theorem.

3. Remark It is of some interest to remark that the above argument admits an extension to certain Orlicz spaces of the type entering in Theorem I.4.4. This is however nontrivial, and we omit its detail here.

While the formulation of the result of Theorem 2 includes the L^{φ_0} (or $L \log L$) class martingales, it contains only the Abel limits in the ergodic theory. We thus need to show that the traditional $(C, 1)$, or Cesàro, averages can also be covered in this setup. The following result shows how the Birkhoff–Khintchine form of the individual ergodic theorem can be obtained for the continuous parameter case (i.e., flows).

4. Proposition *Let $\{U(t), t \geq 0\}$ be a strongly continuous semigroup of positive contractive operators arising from a measure preserving map of a probability space (Ω, Σ, P) as in Theorem 2. Then for each $f \in L^p(\Sigma) \cap L^{\varphi_0}(\Sigma)$, $\lim_{t\to\infty}(1/t)\int_0^t U(s)f\,ds$ exists a.e., where the integral is taken in the strong or the Bochner sense, and $\varphi_0(x) = |x|\log|x|$, $1 \leq p < \infty$.*

Proof As noted in the introduction of this section, we recall that the result on Abel averages implies the $(C, 1)$-convergence provided the following two conditions hold:

(i) $\displaystyle \sup_{t>0}\left|\frac{1}{t}\int_0^t (U(s)f)(\omega)\,ds\right| < \infty$ for almost all ω;

(ii) $\displaystyle \left\{\frac{1}{t}\int_0^t (U(s)f)(\omega)\,ds\right\}$ is feebly oscillating as $t \to \infty$ for almost all ω (definition of "feebly oscillating" is recalled below).

Now, to prove (i), note that $U(t): f \mapsto f \circ \tau^t$ is a positive contraction on $L^1(\Sigma)$ and $L^\infty(\Sigma)$ for all $t \geq 0$ where τ is the measurable and measure preserving transformation. As in the proof of the theorem, for such a family, for each $f \in L^p(\Sigma)$, there exists an f^* (measurable and finite a.e.) such that

$$\int_0^\infty |U(t)f| \alpha_s(t)\, dt \leq f^* \int_0^\infty \alpha_s(t)\, dt \quad \text{a.e.,} \tag{10}$$

where $\alpha_s = (1/s)\chi_{(0,s)}$ is a positive decreasing function on $\mathbb{R}^+$ for each fixed but arbitrary $s > 0$ (cf. Eq. (9)). Since the integral on the right of (10) is unity, (i) follows.

As regards (ii), recall that a function $g: \mathbb{R}^+ \to \mathbb{R}$ is *feebly oscillating* as $t \to \infty$ if $\lim_{t\to\infty} |g(t) - g(r)| = 0$ where $t/r \to 1$. In the present case let $g(t) = (1/t) \int_0^t (U(s)f)(\omega)\, ds$. Then for $r < t$,

$$\begin{aligned} |g(t) - g(r)| &\leq \left|\frac{1}{t} - \frac{1}{r}\right| \int_0^r |U(s)f|(\omega)\, ds + \frac{1}{t}\int_r^t |U(s)f|(\omega)\, ds \\ &\leq \frac{t-r}{t}\frac{1}{r}\int_0^r |U(s)f|(\omega)\, ds + \frac{1}{t}\int_0^{t-r} |U(s+r)f|(\omega)\, ds \\ &\leq \frac{t-r}{t} f^*(\omega) + \frac{t-r}{t} f^*(\omega) \qquad \text{by} \quad (10), \end{aligned} \tag{11}$$

where, if $W_r(t) = U(t + r)$, then $\{W_r(t), t \geq 0\}$ satisfies (10). Hence letting $t \to \infty$ such that $t/r \to 1$, (11) implies that $g(\cdot)$ is feebly oscillating for almost all ω. This completes the proof.

5.5 MARTINGALES IN HARMONIC ANALYSIS

In this section we indicate how the Walsh–Fourier series and its generalizations are interwoven with the martingale and projective limit theories. Our interest will again be the pointwise convergence of "partial Fourier series," but there are limitations for a direct application of martingales, and we discuss both aspects. This will further illuminate the subject.

We recall some facts from harmonic analysis to prepare the setting for the martingale applications and give precise references for their proofs. This part of the theory starts with the Walsh series. It is convenient to state a few concepts. Let $(I, \mathscr{B}, \lambda)$ be the Lebesgue unit interval $I = [0, 1)$, and $L^p(I) = L^p(I, \mathscr{B}, \lambda)$. Then the Rademacher family $\{\varphi_n\}_1^\infty \subset L^2(I)$ is an (incomplete) orthonormal set defined by $\varphi_n(x) = \operatorname{sgn}(\sin(2^{n+1}\pi x))$, $x \in I$, $n \geq 0$. If

$n = 2^{n_1} + \cdots + 2^{n_r}$, where $n_1 > n_2 > \cdots > n_r \geq 0$, is the (unique) dyadic expansion of n, then the Walsh functions $\{\psi_n\}_0^\infty \subset L^2(I)$ are given by $\psi_n(x) = \prod_{i=1}^r \varphi_{n_i}(x)$, $n \geq 1$. This is a complete orthonormal set, where $\psi_0 = 1$, in $L^2(I) \subset L^1(I)$. For any $f \in L^1(I)$, let $c_n = \int_I f\psi_n \, d\lambda$ and $S_n(f) = \sum_{k=0}^{n-1} c_k \psi_k$. Then J. L. Walsh showed in 1923 that $S_{2^n}(f) \to f$ a.e. and in L^1-norm, but that the full sequence $S_n(f) \not\to f$ without further restrictions. We analyze the special subsequence and show that it is a martingale, so that the result is a consequence of Theorem IV.2.4 and then prove that a similar phenomenon occurs in a much more general situation involving locally compact groups.

The special nature of $\{\varphi_n\}_1^\infty$ can be used to simplify $S_{2^n}(f)$ as

$$S_{2^n}(f)(t) = \sum_{k=0}^{2^n-1} c_k \psi_k(t) = \int_I f(x) K_n(x, t) \, d\lambda(x)$$

$$= 2^n \int_{\alpha_n(t)}^{\beta_n(t)} f(x) \, d\lambda(x), \tag{1}$$

where $K_n(x, t) = \sum_{i=0}^{2^n-1} \psi_i(x)\psi_i(t) = \prod_{i=0}^{n-1}(1 + \varphi_i(x)\varphi_i(t))$ and where $\alpha_n(t) = m/2^n$ for $m/2^n \leq t < (m+1)/2^n$, $=0$ otherwise, and $\beta_n(t) = (m+1)/2^n$ for $m/2^n \leq t < (m+1)/2^n$, $=0$ otherwise (so that $K_n(x, t) = 2^n$ on $[\alpha_n, \beta_n)$, $=0$ outside). Now, if one writes $I_{n-1}^t = [\alpha_n(t), \beta_n(t)]$, an interval containing t, and if $F(x) = \int_0^x f(t) \, d\lambda(t)$, then $\lambda(I_{n-1}^t) = 2^{-n}$, and (1) says that $S_{2^n}(f)(t) = F(I_{n-1}^t)/\lambda(I_{n-1}^t)$. Since $F'(x) = f(x)$ a.e. (in fact, $F'(x) = f(x)$ for each x that is not a dyadic rational and we are writing $F(I_{n-1}^t)$ for $F(\beta_n(t)) - F(\alpha_n(t)))$ by the classical Lebesgue theorem on differentiation [or the fundamental theorem of calculus], one deduces that $\lim_{n\to\infty} S_{2^n}(f)(t) = F'(t) = f(t)$ a.e. $[\lambda]$. Let us now show that actually $\{S_{2^n}(f), n \geq 1\}$ can be identified with a martingale, and for this the special nature of the sequence $\{S_{2^n}\}_1^\infty$ is needed. [The full sequence is not a martingale; it already shows the limitations of the latter for a straightforward application in harmonic analysis since it is now known that $S_n(f) \to f$ a.e., for all $f \in L^p(I)$, $1 < p < \infty$.]

Let $D_n = D = \{0, 1\}$ and $G = \times_{n=1}^\infty D_n$, the dyadic group of 0's and 1's with group operation as componentwise addition (mod 2). With discrete topology for D and the resulting product topology for G, the latter becomes a compact abelian (metric) group. Its unique (normalized) Haar measure ζ is the product measure induced by μ on D where $\mu(\{0\}) = \mu(\{1\}) = \frac{1}{2}$. Then ζ is related to the Lebesgue measure λ on $I = [0, 1)$ as follows. Let $t\colon G \to I \cup \{1\}$ be the mapping defined by

$$t(x) = \sum_{i=1}^\infty \frac{x_i}{2^i}, \qquad x = (x_1, x_2, \ldots) \in G, \quad x_i = 0 \quad \text{or} \quad 1. \tag{2}$$

Then except for dyadic rationals, t is one-to-one and onto. Taking finite expansions of such rationals, t becomes one-to-one, measure preserving, and $\zeta = \lambda \circ t$, where we also use t for the induced set mapping. It can be shown that (from a well-known result of J. von Neumann) each Walsh function ψ is given by $\psi_\chi = \chi \circ t^{-1}$ for a unique character χ of G, i.e., $\chi \in \hat{G}$ the dual group of all homomorphisms on G onto the unit circle in the complex plane. We may now express (1) in this group context. Let $H_n = \{x \in G : x = (x_1, x_2, \ldots), x_i = 0 \text{ for } 1 \le i \le n\}$ and let $\Gamma_n = G/H_n$ the quotient space. Since H_n is a closed normal subgroup of G, Γ_n is also a group and can be "identified" with a subgroup, namely $\{x \in G : x = (x_1, x_2, \ldots), x_i = 0 \text{ for } i > n\}$. Let μ_n be the normalized Haar measure of H_n which can be related to λ on $[\alpha_n, \beta_n)$ of (1) as follows. If $I_n^p = [p/2^n, (p+1)/2^n)$, $p = 0, 1, \ldots, 2^n - 1$, then $t(H_n) = I_n^p$ for some p. For each $x \in G$, the coset $x + H_n \in \Gamma_n$ is mapped onto I_n^p for a unique p. Hence $\lambda(t(H_n)) = \lambda(I_n^p) = 2^{-n}$. Since $\mu_n(H_n) = 1$, we get $\mu_n = 2^n(\lambda \circ t)$. Thus if $T_{H_n}(f \circ t)(\hat{x}) = \int_{H_n}(f \circ t)(x + y)\, d\mu_n(y)$ for $x \in \hat{x} = x + H_n$, then $T_{H_n}: (f \circ t) \mapsto T_{H_n}(f \circ t)$ is a mapping onto functions on Γ_n. Let $\tau_{H_n}(\hat{f}) = \hat{f} \circ \Pi_{H_n}$ where $\Pi_{H_n}: G \to \Gamma_n$ is the canonical (or quotient) mapping. Then $(\tau_{H_n} \circ T_{H_n})(f \circ t)$ is again a function on G. We now have

$$T_{H_n}(f \circ t)(\hat{x}) = \int_{H_n} (f \circ t)(x + y)\, d\mu_n(x)$$

$$= 2^n \int_{v = t(x) \in I_n^p} f(v)\, d\lambda(v), \qquad x \in \hat{x} = x + H_n. \tag{3}$$

Comparing the right side of (3) with the right side of (1), one obtains

$$S_{2^n}(f)(v) = ((\tau_{H_n} \circ T_{H_n})(f \circ t))(v), \qquad v \in I. \tag{4}$$

It is of interest to note that as $n \to \infty$, the compact subgroups Γ_n "fill" all of G. The last term means the following. If $\Pi_n = \Pi_{H_n}: G \to \Gamma_n$, and $\Pi_{mn}: \Gamma_n \to \Gamma_m$ where $m \le n$ $(H_m \supset H_n)$, then $\Pi_{mn} \circ \Pi_{nr} = \Pi_{mr}$ and $\Pi_{mn} \circ \Pi_n = \Pi_m$ (Π_{mm} is the identity) for $m \le n \le r$ so that $\{\Gamma_n, \Pi_{mn}, n \ge m \ge 1\}$ is a projective system of closed groups and $G = \lim_{\leftarrow}(\Gamma_n, \Pi_n)$ the projective limit (cf. Chapter III). It can be shown, moreover, that each character χ of Γ_n induces a character $\bar{\chi} = \chi \circ \Pi_n$ of G, and, conversely, every (irreducible) character of G can be so obtained from some Γ_n. ($\hat{G}$ may be identified with the direct limit of $\hat{\Gamma}_n$, i.e., $\hat{G} = \lim_{\rightarrow} \hat{\Gamma}_n$ here.) Thus for an application of martingale theory it suffices to prove that $\tau_{H_n} \circ T_{H_n}: L^1(G) \to L^1(G)$ is a conditional expectation for each n. The main result below is essentially a proof of this fact. However, this is true much more generally (with only a little more effort) for any locally compact group if H_n are suitable subgroups, including the above result, which now serves as a motivation. We shall therefore consider the general case as it is found useful for some analyses with the lifting operator. For the results on such groups that are needed, reference may be made to Reiter [1, Chapter 4 and particularly Chapter 8]. The following account is taken from the author's paper [13]. (See also Jerison and Rabson [1], Edwards and Hewitt [1], and Ionescu Tulcea [2]. The results of the latter paper need this generality!)

Let G be an arbitrary locally compact group and $H \subset G$ a closed subgroup. Since H is also locally compact, let dy be a left Haar measure on H. If $\Gamma = G/H$ is the quotient (or homogeneous) space, let $\Pi_H: G \to \Gamma$ be the canonical mapping (so it is an open map). Let $\mathscr{B}$ and $\mathscr{G}$ be the Borel σ-algebras of G and Γ. We define T_H as above, i.e., $(T_H f)(\hat{x}) = \int_H f(xy)\,dy$, $\hat{x} = xH \in \Gamma$, $f \in C_{00}(G)$, the space of real continuous functions on G with compact supports. Then $(T_H f) \in C_{00}(\Gamma)$ and is an onto mapping. Let $\mu: \mathscr{B} \to \bar{\mathbb{R}}^+$ be a Radon measure (so by hypothesis, μ is finite on each compact set in $\mathscr{B}$). Then there exists a Radon measure $\hat{\mu}: \mathscr{G} \to \bar{R}^+$ such that

$$\int_\Gamma \left(\int_H f(xy)\,dy \right) d\hat{\mu}(\hat{x}) = \int_\Gamma (T_H f)(\hat{x})\,d\hat{\mu}(\hat{x})$$
$$= \int_G f(x)\,d\mu(x), \qquad f \in C_{00}(G) \tag{5}$$

iff μ satisfies

$$\int_G f(xy^{-1})\,d\mu(x) = \delta_H(y) \int_G f(x)\,d\mu(x), \qquad y \in H, \quad f \in C_{00}(G), \tag{6}$$

where $\delta_H(\cdot)$ is the left (Haar) modular function of H. When this condition holds, $\hat{\mu}$ satisfying (5) is also unique. (See Reiter [1, p. 157] for a simple proof.) If H is a normal subgroup, then the existence of $\hat{\mu}$ satisfying the fundamental formula (5) is originally established by A. Weil in 1938, with μ as a left Haar measure. It then follows that $\delta_H = \delta_G | H$, and hence (6) is true. The existence of $\hat{\mu}$, when μ is a left Haar measure and H is merely a closed subgroup but G is second countable, is established by G.W. Mackey in 1952, and it is extended to general G by F. Bruhat in 1956. Each of these extensions is nontrivial. So we call any such formula giving (5) the *Weil–Mackey–Bruhat formula*, and the pair $(\hat{\mu}, \mu)$ the *WMB-pair*. If $d\mu = f\,dx$, where f is any Haar integrable function (dx being a left Haar measure), then $\hat{\mu}$ always exists. Thus the class of WMB pairs is nonvoid.

When (5) holds for $f \in C_{00}(G)$, it can be shown by a standard argument that the formula is also true for all $f \in L^1(G, \mathscr{B}, \mu)$. Then $T_H: L^1(G, \mathscr{B}, \mu) \to L^1(\Gamma, \mathscr{G}, \hat{\mu})$ is a positive contractive operator, whose range contains $C_{00}(\Gamma)$. Let $\tau_H: \hat{f} \to \hat{f} \circ \Pi_H$ be defined for each $\hat{f} \in L^1(\Gamma, \mathscr{G}, \hat{\mu})$ so that $\tau_H \hat{f}$ is a function on G, and it is constant on cosets. In fact, it is measurable for $\mathfrak{F} = \Pi_H^{-1}(\mathscr{G}) \subset \mathscr{B}$, as shown below. However $\tau_H \hat{f}$ is not necessarily μ-integrable even though τ_H preserves the boundedness properties of f. This will be clear if one takes $H = G$ and μ as a left Haar measure. Since then $\Gamma = \{e\}$ and each $\hat{f} \in L^1(\Gamma, \mathscr{G}, \hat{\mu})$ is a constant, $\tau_H \hat{f}$ is also a constant, $\notin L^1(G, \mathscr{B}, \mu)$ if the constant $\neq 0$. Thus to consider the operator in (4), we are led to assume that $\tau_H \hat{f} \in L^1(\mathscr{B})$. In case μ is a Haar measure, then the preceding assumption is essentially the same as requiring that H be compact. Since $\hat{f} \in C_{00}(\Gamma)$ implies in this case $\hat{f} \circ \Pi_H$ has compact support, $\hat{f} \circ \Pi_H \in C_{00}(G)$. We therefore assume hereafter that H is a compact subgroup with its normalized Haar measure and prove the following key result. It connects many seemingly different topics together.

1. Theorem *Let G be a locally compact group, $H \subset G$ a compact subgroup, and $\Gamma = G/H$. If $\mathscr{B}, \mathscr{G}$ are Borel σ-algebras of G and Γ, then suppose $(\hat{\mu}, \mu)$ is a WMB-pair on $(\mathscr{B}, \mathscr{G})$. If $L^1(\mathscr{B})$, $L^1(\mathscr{G})$ are the corresponding Lebesgue spaces and T_H, τ_H are the associated mappings, i.e., T_H: $L^1(\mathscr{B}) \to L^1(\mathscr{G})$ and τ_H: $L^1(\mathscr{G}) \to L^1(\mathscr{B})$, then*

(a) *τ_H, T_H are positive linear contractions and τ_H is an isometry onto a subspace $\mathscr{L}$ of $L^1(\mathscr{B})$;*

(b) *if $\mathfrak{F} = \Pi_H^{-1}(\mathscr{G}) \subset \mathscr{B}$, then $\tau_H \circ T_H = E^{\mathfrak{F}}$ the conditional expectation operator on $L^1(\mathscr{B})$ with range $L^1(\mathfrak{F}) = \mathscr{L} \subset L^1(\mathscr{B})$, and $T_H \circ \tau_H$ is the identity on $L^1(\mathscr{G})$;*

(c) *if μ is a left Haar measure on G, and $\hat{\mu}$ on Γ is the associated part of the WBM pair $(\hat{\mu}, \mu)$ which then becomes (left) invariant (as, e.g., if H is also normal), then $E^{\mathfrak{F}}$ commutes with the left translation, i.e.,*

$$(L_y f)(x) = f(y^{-1}x), \qquad (R^y f)(x) = f(xy^{-1})\delta_G(y^{-1}), \qquad f \in L^1(\mathscr{B}), \quad y \in G$$
$$\Rightarrow E^{\mathfrak{F}}(L_y f) = L_y E^{\mathfrak{F}}(f) \text{ a.e.}, \qquad E^{\mathfrak{F}}(R^y f) = \delta_G(y^{-1}) R^y E^{\mathfrak{F}}(f) \text{ a.e.} \tag{7}$$

Remark The importance of the WMB pair $(\hat{\mu}, \mu)$ is that if μ is a left Haar measure on G, then $\hat{\mu}$ on Γ need not be (left) invariant, but it is (left) *quasi-invariant*, i.e., if $\Lambda_y(\hat{x}) = y^{-1}\hat{x}$ is the action of G on Γ on the left, then one has

$$\int_\Gamma \hat{f}(\Lambda_y \hat{x})\, d\hat{\mu}(\hat{x}) = \int_\Gamma \hat{f}(\hat{x})\lambda_y(\hat{x})\, d\hat{\mu}(\hat{x}), \qquad \hat{f} \in L^1(\hat{\mu}), \quad y \in G, \tag{8}$$

where λ_y is a positive function on Γ. If $\lambda_y(\hat{x}) = a_y$, independent of $\hat{x}$, then $\hat{\mu}$ is called *relatively invariant*, and it is invariant if $\lambda_y(\hat{x}) = 1$, $\hat{x} \in \Gamma$, $y \in G$. These properties are very important for harmonic analysis on homogeneous spaces. For a discussion on this set of results, see Reiter [1].

Proof To prove the various parts, we first make a simple reduction which includes details of some of the less obvious facts stated just prior to the theorem. Recall that (by the Doob–Dynkin lemma (Theorem I.2.3)) if $h: \Omega \to \tilde{\Omega}$ is a mapping of Ω onto $\tilde{\Omega}$ and $\Sigma, \tilde{\Sigma}$ are their σ-algebras and $f: \Omega \to \mathbb{R}$ is a (Σ)-measurable function, then f is measurable for $h^{-1}(\tilde{\Sigma}) \subset \Sigma$ iff there exists a $(\tilde{\Sigma})$-measurable $g: \tilde{\Omega} \to \mathbb{R}$ such that $f = g \circ h$. We specialize this in the present case with $\Omega = G$, $\Sigma = \mathscr{B}$, $h = \Pi_H$, and $\tilde{\Sigma} = \mathscr{G}$, $\tilde{\Omega} = \Gamma$. If $\mathfrak{F} = \Pi_H^{-1}(\mathscr{G}) \subset \mathscr{B}$, then it is a σ-algebra generated by the cosets $\{\hat{x} = xH : x \in \mathscr{G}\}$, so that each coset is an atom of $\mathfrak{F}$. Hence by the Doob–Dynkin lemma stated above, $f\colon G \to \mathbb{R}$ is $\mathfrak{F}$-measurable iff there exists $\hat{f}\colon \Gamma \to \mathbb{R}$, which is $\mathscr{G}$-measurable, such that $f = \hat{f} \circ \Pi_H$. Since Π_H is an open onto mapping and $\mathscr{G}, \mathscr{B}$ are Borel σ-algebras, we have $f^{-1}(\text{open}) = \Pi_H^{-1}(\hat{f}^{-1}(\text{open}))$, or $\Pi_H(f^{-1}(\text{open})) = \hat{f}^{-1}(\text{open})$ since for any mapping $T\colon S \to S'$, $T(T^{-1}(B)) = B$ for each $B \subset S'$ iff T is onto. Thus f is continuous iff $\hat{f}$ is. Let us make a further simplification about the hypothesis on the subgroup H.

From the fact that Π_H is a point mapping on G to Γ, we can define the image measure $\nu = \mu \circ \Pi_H^{-1}$ of μ on $\mathscr{G}$. Then from standard results on image measures (see Section I.2), for any measurable $g\colon \Gamma \to \mathbb{R}$ ($g \circ \Pi_H$ is $\mathscr{B}$-measurable and) one has

$$\left(\int_A g\,d\nu =\right) \int_A g\,d(\mu \circ \Pi_H^{-1}) = \int_{\Pi_H^{-1}(A)} (g \circ \Pi_H)\,d\mu, \qquad A \in \mathscr{G}, \tag{9}$$

in the sense that if either side exists, so does the other and there is equality. But we also have by (5), (since $(\hat{\mu}, \mu)$ is a WMB pair) on replacing f by $f\chi_{\Pi_H^{-1}(A)}$ there, because that formula is valid on $L^1(\mathscr{B})$,

$$\int_A (T_H f)(\hat{x})\,d\hat{\mu}(\hat{x}) = \int_{\Pi_H^{-1}(A)} f(x)\,d\mu(x), \qquad A \in \mathscr{G}. \tag{10}$$

On comparing (9) and (10), one notes that in the former, μ and the (Ω, Σ) can be abstract while in (10) the given group structure is essential. Also $T_H\colon L^1(\mathscr{B}) \to L^1(\mathscr{G})$ is onto for the WMB pair $(\hat{\mu}, \mu)$. Thus if $g \in L^1(\mathscr{G})$, then there exists $f \in L^1(\mathscr{B})$ such that $g = T_H(f)$, but *f need not be in $L^1(\mathfrak{F})$*, whereas $g \circ \Pi_H$ is $\mathfrak{F}$-measurable and hence by (9) it is in $L^1(\mathfrak{F})$ for any μ. This shows $(\mu \circ \Pi_H^{-1}, \mu)$ generally do not form a WMB pair on (G, Γ). But if $f = g \circ \Pi_H \in L^1(\mathfrak{F})$, then unless μ is finite, or H is compact, every coset $\hat{x} = xH (\in \mathscr{B})$ has infinite measure and $g \circ \Pi_H = 0$, i.e., $g = 0$ a.e. when μ is a nonfinite Radon measure. In particular, if μ is a left Haar measure on G, this is the case. If H is a compact group, then $T_H(1) = 1$ a.e., and since $T_H(C_{00}(G)) = C_{00}(\Gamma)$ which is dense in $L^1(\mathscr{G})$ and if $g \in C_{00}(\Gamma)$, then $g \circ \Pi_H \in C_{00}(G)$ (since $g \circ \Pi_H$ has also compact support). This implies that $\tau_H(g) \in L^1(\mathfrak{F})$ for each $g \in L^1(\mathscr{G})$ whatever WMB pair $(\hat{\mu}, \mu)$ is. Thus for any $f \in C_{00}(G)$, which is constant on cosets, $f = g \circ \Pi_H$, $g \in C_{00}(\Gamma)$, and for any $h \in L^1(\mathscr{B})$ one has

$$\begin{aligned} T_H(fh)(\hat{x}) &= \int_H (fh)(xy)\,dy = g(\hat{x}) \int_H h(xy)\,dy \\ &= g(\hat{x})\,T_H(h)(\hat{x}), \qquad (= g(\hat{x}) \quad \text{if} \quad h = \chi_H) \end{aligned} \tag{11}$$

since the Haar measure on the compact group is normalized by hypothesis, and so $T_H(\chi_H) = 1$. Hence (9), (10), and (11) imply (since (11) is also true if $f \in L^1(\mathfrak{F}) \cap L^\infty$) for $(\hat{\mu}, \mu)$:

$$\int_A g d(\mu \circ \Pi_H^{-1}) = \int_{\Pi_H^{-1}(A)} (g \circ \Pi_H) \, d\mu = \int_A T_H(g \circ \Pi_H) \, d\hat{\mu} = \int_A g \, d\hat{\mu}. \tag{12}$$

But $g \in L^1(\mathscr{G})$ is arbitrary, and one concludes that $\mu \circ \Pi_H^{-1} = \hat{\mu}$. (Thus in the absence of compactness of H, one has to restrict τ_H and T_H further. It is therefore natural to let H be compact.)

(a) It is clear that T_H is a positive linear mapping on $L^1(G)$, and (5) implies it is a contraction, i.e., $\|T_H f\|_{1,\hat{\mu}} \leq \|f\|_{1,\mu}$. Similarly, the mapping τ_H is positive and linear. If $g \in L^1(\mathscr{G})$, then (12) implies (with $A = \Gamma$)

$$\|g\|_{1,\hat{\mu}} = \int_\Gamma |g| \, d\hat{\mu} = \int_G |g \circ \Pi_H| \, d\mu = \int_G |\tau_H g| \, d\mu = \|\tau_H g\|_{1,\mu}. \tag{13}$$

Hence τ_H is an isometry. By the preceding discussion, $\tau_H(L^1(\mathscr{G})) \subset L^1(\mathfrak{F})$ since every $\tau_H g$ is $\mathfrak{F}$-measurable. To see that equality must hold here, we note by the earlier discussion that every $f \in L^1(\mathfrak{F})$ is constant on cosets $\{xH : x \in G\}$ so that $f = \hat{f} \circ \Pi_H$ for a unique $\hat{f} \in L^1(\mathscr{G})$. Thus $f = \tau_H(\hat{f})$ and τ_H is onto, or $\tau_H(L^1(\mathscr{G})) = L^1(\mathfrak{F})$. Note that $\tau_H(\hat{f}\hat{g}) = \tau_H(\hat{f})\tau_H(\hat{g})$ if $\hat{f}$ or $\hat{g}$ is bounded so that it is also a homomorphism.

(b) By the preceding analysis (cf. (11)), for each $\hat{f} \in L^1(\mathscr{G})$, one has $T_H(\tau_H(\hat{f}) = T_H(f) = \hat{f}$ a.e. $(\hat{\mu})$. Hence $T_H \circ \tau_H$ is the identity. So $E^{\mathscr{L}} = \tau_H \circ T_H$ is a positive contractive projection on $L^1(\mathscr{B})$ with range $L^1(\mathfrak{F})$ since both T_H and τ_H are onto. It remains to show that $E^{\mathscr{L}} = E^{\mathfrak{F}}$, the conditional expectation. If μ is finite, then this is a simple consequence of Theorem II.2.6. In the general case, we note that since H is compact, for each $f \in L^1(\mathscr{B}) \cap L^\infty$,

$$|T_H(f)| \leq \int_H |f(xy)| \, dy \leq \|f\|_\infty, \qquad |\tau_H(\hat{f})| = |\hat{f} \circ \Pi_H|$$

so that $\|T_H f\|_\infty \leq \|f\|_\infty$ and $\|\tau_H(\hat{f})\|_\infty = \|\hat{f}\|_\infty$. Thus $\|E^{\mathscr{L}}(f)\|_\infty \leq \|f\|_\infty$ also holds. To deduce that $E^{\mathscr{L}} = E^{\mathfrak{F}}$ on L^1, note that by the Riesz interpolation theorem (cf. Zygmund [1, Part II, p. 95] or by a direct computation), both τ_H and T_H are defined and are contractions on all $L^p(\mathscr{B})$ and $L^p(\mathscr{G})$, $1 \leq p \leq \infty$. These properties are inherent for $E^{\mathfrak{F}}$.

Let $h \in L^1(\mathscr{B})$ and $A \in \mathfrak{F}$. Then $A = \Pi_H^{-1}(B)$ for a $B \in \mathscr{G}$, and one has (since $\chi_A = \tau_H(\chi_B)$)

$$\begin{aligned} E^{\mathscr{L}}(h\chi_A) &= \tau_H[T_H(h\tau_H(\chi_B))] = \tau_H[\chi_B T_H(h)] \quad \text{by} \quad (11) \\ &= \tau_H(\chi_B)(\tau_H \circ T_H(h)) \qquad \text{since} \quad \tau_H \text{ is a homomorphism} \\ &= \chi_A E^{\mathscr{L}}(h). \end{aligned}$$

Hence

$$\begin{aligned}\int_A E^{\mathscr{L}}(h)\,d\mu &= \int_{A=\Pi_H^{-1}(B)} E^{\mathscr{L}}(h\chi_A)\,d\mu = \int_B T_H(E^{\mathscr{L}}(h\chi_A))\,d\hat{\mu} \qquad \text{by} \quad (10)\\ &= \int_B T_H \circ \tau_H(T_H(h\chi_A))\,d\hat{\mu}\\ &= \int_B T_H(h\chi_A)\,d\hat{\mu} \qquad \text{since} \quad T_H \circ \tau_H \text{ is the identity by (a)}\\ &= \int_{\Pi_H^{-1}(B)} h\chi_A\,d\mu = \int_A h\,d\mu \qquad \text{by} \quad (10). \end{aligned} \tag{14}$$

If $\nu_h(\cdot) = \int_{(\cdot)} h\,d\mu$, then (14) says that $E^{\mathscr{L}}(h)$ coincides with the Radon–Nikodým derivative of ν_h relative to μ on $\mathfrak{F}(C)$ where $C \subset G$ is compact. But $h \in L^1(B)$ implies its support is σ-compact, so that on its support $d\nu_h/d\mu$ exists and equals $E^{\mathscr{L}}(h)$. By uniqueness, this correspondence $h \mapsto E^{\mathscr{L}}(h)$ is well defined and is the conditional expectation. It follows that $E^{\mathscr{L}} = E^{\mathfrak{F}}$, as asserted.

(c) The hypothesis on Γ implies that $\hat{\mu}$ is (left) invariant. Now for any $g \in L^\infty(\mathfrak{F})$ we use the averaging property of $E^{\mathfrak{F}}$, namely $E^{\mathfrak{F}}(fg) = gE^{\mathfrak{F}}(f)$ a.e., for $f \in L^1(\mathscr{B})$. But μ is a left Haar measure on G. Thus if $g_y(x) = g(yx)$, then $g_y \in L^\infty(\mathfrak{F})$ also so that (the invariance of $\hat{\mu}$ is easy, but is not needed here)

$$\begin{aligned}\int_G g(x)E^{\mathfrak{F}}(L_y f)(x)\,d\mu(x) &= \int_G E^{\mathfrak{F}}(gL_y f)\,d\mu\\ &= \int_G g(L_y f)\,d\mu \qquad \text{by} \quad (14)\\ &= \int_G L_y(g_y f)\,d\mu \qquad \text{since} \quad g_y(x) = g(yx)\\ &= \int_G (g_y f)\,d\mu \qquad \text{by left invariance of } \mu\\ &= \int_G E^{\mathfrak{F}}(g_y f)\,d\mu \qquad \text{by} \quad (14)\\ &= \int_G g_y E^{\mathfrak{F}}(f)\,d\mu = \int_G L_y(g_y E^{\mathfrak{F}}(f))\,d\mu\\ &= \int_G gL_y(E^{\mathfrak{F}}(f))\,d\mu, \qquad L_y \text{ being a homomorphism.}\end{aligned}$$

Since g is arbitrary in $L^\infty(\mathfrak{F})$, this implies $E^{\mathfrak{F}}(L_y f) = L_y E^{\mathfrak{F}}(f)$ a.e. The assertion regarding R^y is similar and is left to the reader.

It is clear that with $g \in L^\infty(\mathfrak{F})$, g_y is also bounded, and to see that it is $\mathfrak{F}$-measurable, note that $g = \hat{g} \circ \Pi_H$ with $\hat{g} \in L^\infty(\mathscr{G})$. Since τ_H is a homomorphism, we have $g_y(x) = g(yx) = \hat{g}(\Pi_H(yx)) = \hat{g}(\hat{y}\hat{x}) = \hat{g}_{\hat{y}}(\hat{x})$. So $g_y = \hat{g}_{\hat{y}} \circ \Pi_H$. But $\hat{g}$ and $\hat{g}_{\hat{y}}$ are $\mathscr{G}$-measurable. In fact, $\hat{g} \in C_{00}(\Gamma)$ implies $\hat{g}_{\hat{y}} \in C_{00}(\Gamma)$, and then the general case follows by approximation. From this one deduces that $g_y \in L^\infty(\mathfrak{F})$ because of the Doob–Dynkin lemma. This completes the proof of the theorem.

In order to prove a generalization of Walsh's result, we need to note the monotonicity of $\{E^{\mathfrak{F}_i}\}_1^\infty$. This may be stated as follows.

2. Proposition *Let $H_1 \subset H_2 \subset G$ be compact subgroups of the locally compact group G, and let $\Gamma_i = G/H_i$, Π_i: $G \to \Gamma_i$, $i = 1, 2$, and $(\hat{\mu}_i, \mu)$ be the WMB pairs on (Γ_i, G), $i = 1, 2$. If $\mathscr{G}_i$ is the Borel σ-algebra of Γ_i, $\mathfrak{F}_i = \Pi_i^{-1}(\mathscr{G}_i) \subset \mathscr{B}$, then $\mathfrak{F}_1 \supset \mathfrak{F}_2$ and hence $E^{\mathfrak{F}_1}E^{\mathfrak{F}_2} = E^{\mathfrak{F}_2}E^{\mathfrak{F}_1} = E^{\mathfrak{F}_2}$, in the notation of Theorem 1 above.*

Proof Since $H_1 \subset H_2$, the canonical mapping Π_2 is constant on each coset xH_1.

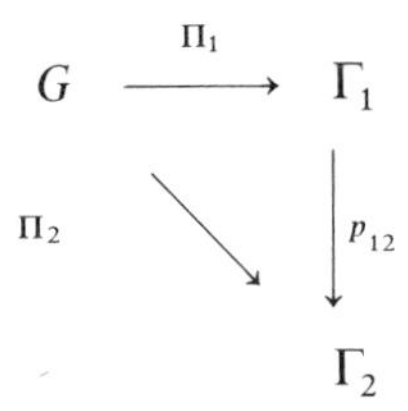

As partitions of G, Γ_1 refines Γ_2. So let p_{12}: $\Gamma_1 \to \Gamma_2$ be defined as the natural projection such that $\Pi_2 = p_{12} \circ \Pi_1$. This means p_{12} is an identity on each coset xH_1 into xH_2, and thus p_{12} is uniquely defined. To see that it is continuous (and open), since Π_i are onto, we have for any open set $O \subset \Gamma_2$,

$$\Pi_2(\Pi_2^{-1}(p_{12}^{-1}(O))) = p_{12}^{-1}(O), \qquad \Pi_1^{-1}(p_{12}^{-1}(O)) = \Pi_2^{-1}(O). \tag{15}$$

Since $\Pi_2^{-1}(O)$ is open and Π_2(open) is open, we conclude that $p_{12}^{-1}(O)$ is open. Hence p_{12} is continuous. Also that $\Pi_2(O) = p_{12}(\Pi_1(O))$ is open implies immediately that p_{12} is an open mapping.

To prove the inclusions, let $A \in \mathfrak{F}_2$. Since Γ_i are locally compact, $\mathscr{G}_i$ are generated by all closed subsets of Γ_1. So it suffices to show now for each closed $C \in \mathscr{G}_2$ that $\Pi_2^{-1}(C) \in \Pi_1^{-1}(\mathscr{G}_1) = \mathfrak{F}_1$ since the sets on the left (are closed and) generate $\mathfrak{F}_2$. Hence let $A = \Pi^{-1}(C)$ with C closed in $\mathscr{G}_2$. Then $p_{12}^{-1}(C)$ is a closed subset of Γ_1 so that $p_{12}^{-1}(C) \in \mathscr{G}_1$. Thus $\Pi_1^{-1}(p_{12}^{-1}(C)) = \Pi_2^{-1}(C) \in \Pi_1^{-1}(\mathscr{G}_1)$. Hence $\mathfrak{F}_2 \subset \mathfrak{F}_1$. Since $E^{\mathfrak{F}_i}$ are conditional expectations, the verification of the last statement on commutativity is the same as (and follows from) Proposition II.1.2. This completes the proof.

The above proposition and theorem enable us to obtain a considerable generalization of the convergence statement of the sequence (6). Note that we have now shown that $S_{2^n}(f) = E^{\mathfrak{F}_n}(f \circ t)$, where $\mathfrak{F}_n$ is determined by H_n there. The convergence a.e. and in L^1-norm follow from Theorem IV.2.4. This identification "justifies" calling the sequence $\{E^{\mathfrak{F}_i}(f)\}_1^\infty$ again a (generalized) "Fourier partial sum" of f on the group G. However, instead of a Haar measure on G, we are now able to admit any WMB pair in this statement. This is the generalization that applies to a whole class of WMB pairs.

The operators τ_H, T_H determining $E^{\mathfrak{F}}$, $\mathfrak{F} = \Pi_H^{-1}(\mathscr{G})$, have some special properties. We mention the following. If $p \geq 1$ and $p' = p/(p-1)$, and $h \in L^p(\mathscr{B})$, $g \in L^{p'}(\mathfrak{F})$, so that $g = \hat{g} \circ \Pi_H$ for a $\hat{g} \in L^{p'}(\mathscr{G})$, one has for any WMB pair $(\hat{\mu}, \mu)$ on (Γ, G) where H is a compact subgroup of G,

$$\begin{aligned}
\int_G hg\, d\mu &= \int_G h \cdot \tau_H \hat{g}\, d\mu = \int_G E^{\mathfrak{F}}(h(\tau_H \hat{g}))\, d\mu \qquad \text{by} \quad (14)\\
&= \int_G (\tau_H \hat{g}) E^{\mathfrak{F}}(h)\, d\mu \qquad \text{since} \quad g = \tau_H \hat{g} \text{ is } \mathfrak{F}\text{-measurable}\\
&= \int_G \tau_H(\hat{g}(T_H h))\, d\mu \qquad \text{since} \quad E^{\mathfrak{F}} = \tau_H \circ T_H \quad \text{and}\\
&\qquad\qquad \tau_H \text{ is a homomorphism}\\
&= \int_\Gamma (\hat{g} T_H h)\, d\hat{\mu} \qquad \text{since} \quad (\hat{\mu}, \mu) \text{ is a (WMB) pair.}
\end{aligned} \tag{16}$$

Thus writing $\langle \cdot, \cdot \rangle_G$ and $\langle \cdot, \cdot \rangle_\Gamma$ for the linear functional notations on $L^p(\mathscr{B})$ and $L^p(\mathscr{G})$, (16) is expressible as

$$\langle h, \tau_H \hat{g} \rangle_G = \langle T_H h, \hat{g} \rangle_\Gamma, \qquad h \in L^p(\mathscr{B}), \quad \hat{g} \in L^{p'}(\mathscr{G}), \quad 1 \leq p < \infty. \tag{17}$$

This shows that $\tau_H^* = T_H$ and $T_H^* = \tau_H$, i.e., each is the adjoint of the other, because $(L^p(\mathscr{B}))^*$ and $(L^p(\mathscr{G}))^*$ can be identified with $L^{p'}(\mathscr{B})$ and $L^{p'}(\mathscr{G})$, respectively.

Let $\{G, H_\alpha, \alpha \in I\}$ be a directed family of groups such that (i) G is locally compact, (ii) $H_\alpha \subset G$ is a compact subgroup of G for each α, and (iii) $\alpha < \beta$ in I, a directed index set, implies $H_\alpha \supset H_\beta$. If $H = \bigcap_{\alpha \in I} H_\alpha$, a compact subgroup of G, let $\Gamma_\alpha = G/H_\alpha$, $\Gamma = G/H$ be the corresponding factor spaces. If $\Pi_\alpha\colon G \to \Gamma_\alpha$ and $\Pi_{\alpha\beta}\colon \Gamma_\beta \to \Gamma_\alpha (\alpha < \beta)$ are the canonical mappings, then, with the quotient topologies, these become locally compact spaces and $\{\Pi_\alpha, \Pi_{\alpha\beta}\}$ are continuous (open) onto mappings by Proposition 2 above. The latter also shows that for $\alpha < \beta < \gamma$ in I, $\Pi_{\alpha\alpha}$ is the identity, $\Pi_\alpha = \Pi_{\alpha\beta} \circ \Pi_\beta$, and $\Pi_{\alpha\beta} \circ \Pi_{\beta\gamma} = \Pi_{\alpha\gamma}$. Hence $\{\Gamma_\alpha, \Pi_{\alpha\beta}, \alpha < \beta \text{ in } I\}$ is a projective system of spaces. Let us say that the family $\{G, H_\alpha, \alpha \in I\}$ has *the* $(*)$*-property* with respect to $\{H_\alpha, \alpha \in I\}$ iff

$$G/H = \Gamma = \varprojlim(\Gamma_\alpha, \Pi_{\alpha\beta}). \tag{18}$$

Thus G/H is the projective limit of the system $\{(\Gamma_\alpha, \Pi_{\alpha\beta})_{\alpha<\beta} : \alpha, \beta \text{ in } I\}$. Clearly the dyadic group in Walsh's problem has the $(*)$-property even with $I = \mathbb{N}$. Although a simple usable characterization of groups G with the property (18) is not known, there are large collections of groups that satisfy (18). We state the following classical result, due to A. Weil and H. Cartan, indicating the nature of the preceding remark.

If G is a compact group, then there exist closed (hence compact) subgroups $\{H_\alpha, \alpha \in I\}$, $H_\alpha \supset H_\beta$ for $\alpha \leq \beta$, $\bigcap_{\alpha \in I} H_\alpha = \{e\}$, e the identity of G, such that $\Gamma_\alpha = G/H_\alpha$ (is a compact Lie group and), $G = \varprojlim(\Gamma_\alpha, \Pi_\alpha)$. If, moreover, G is also first countable, then $I = \mathbb{N}$ may be taken in the above. Hence every compact G has the $(*)$-property relative to such a net $\{H_\alpha, \alpha \in I\}$. (If also G is totally disconnected and first countable, then each Γ_n is finite and thus the special problem on Walsh series with G dyadic and its generalizations are included here.)

We can now present the desired result on convergence:

3. Theorem *Let $\{G, H_\alpha, \alpha \in I\}$ be a system of locally compact groups and $H = \bigcap_{\alpha \in I} H_\alpha$. Suppose that G has the $(*)$-property relative to $\{H_\alpha, \alpha \in I\}$ (so that each H_α is compact). For each $\alpha \in I$, let $(\hat{\mu}_\alpha, \mu)$ be a WMB pair on (Γ_α, G) where $\Gamma_\alpha = G/H_\alpha$. If $f \in L^p(G, \mathscr{B}, \mu)$, $1 \leq p < \infty$, and $f_\alpha = (\tau_{H_\alpha} \circ T_{H_\alpha})(f) = E^{\mathfrak{F}_\alpha}(f)$, then $\{f_\alpha, \alpha \in I\}$ is a martingale that converges in norm to $\tilde{f} = (\tau_H \circ T_H)(f)$. If, moreover, $I = \mathbb{N}$ (or only countable but linearly ordered), then the convergence is also pointwise a.e. $[\mu]$. In either case $\tilde{f} = f$ a.e. iff $H = \{e\}$.*

Proof In what follows, let $E_\alpha = \tau_{H_\alpha} \circ T_{H_\alpha} = E^{\mathfrak{F}_\alpha}$, the conditional expectation. It is clear that $\{E_\alpha, \alpha \in I\}$ is a uniformly bounded (by 1) family of projections, which by Proposition 2 has the necessary commutativity and monotone properties required of Corollary 2.4. Since $f_\alpha = E_\alpha(f) \in L^p(G, \mathscr{B}, \mu) = \mathscr{X}_p$ and $\|f_\alpha\|_p \leq \|f\|_p$, $\alpha \in I$, the set $\{f_\alpha, \alpha \in I\}$ is bounded in $\mathscr{X}_p$. If $1 < p < \infty$, then the space is reflexive so that every bounded set is relatively weakly compact. So taking $K = \{f_\alpha, \alpha \in I\}$ in that statement, its hypothesis is trivially satisfied and the strong convergence statement follows immediately from that result (i.e., Corollary 2.4). It remains to consider the case that $p = 1$.

To prove this, we use a special argument, similar to that already used in the proof of Proposition IV.2.10. Thus if $\mathfrak{F} = \sigma(\bigcup_{\alpha \in I} \mathfrak{F}_\alpha)$, then $E^{\mathfrak{F}} = \tau_H \circ T_H$. Also $L^1(G, \mathfrak{F}, \mu)$ has $\bigcup_{\alpha \in I} L^1(G, \mathfrak{F}_\alpha, \mu)$ as a norm dense subspace. Hence from the fact that $\tilde{f} = E^{\mathfrak{F}}(f) \in L^1(G, \mathfrak{F}, \mu)$ (and density) for each $\varepsilon > 0$ there exists an $\alpha_0 \in I$ and $f_\varepsilon \in L^1(G, \mathfrak{F}_{\alpha_0}, \mu)$ such that $\|\tilde{f} - f_\varepsilon\|_1 < \varepsilon/2$. If $\alpha > \alpha_0$, then $E_{\alpha_0}(f_\varepsilon) = E_\alpha E_{\alpha_0}(f_\varepsilon) = E_{\alpha_0} E_\alpha(f_\varepsilon)$, and since $E_{\alpha_0} | L^1(G, \mathfrak{F}_{\alpha_0}, \mu)$ is the identity, $E_{\alpha_0}(f_\varepsilon) = f_\varepsilon$. However, by (14) $E_{\alpha_0}(f) = E_{\alpha_0}(\tilde{f}) \in L^1(G, \mathfrak{F}_{\alpha_0}, \mu)$. So we have for $\alpha > \alpha_0$

$$\begin{aligned} \|\tilde{f} - f_\alpha\|_1 &\leq \|\tilde{f} - f_\varepsilon\|_1 + \|f_\varepsilon - f_\alpha\| \leq \varepsilon/2 + \|E_\alpha(f_\varepsilon - \tilde{f})\|_1 \\ &\qquad\qquad \text{since} \quad E_\alpha(\tilde{f}) = f_\alpha \\ &\leq \varepsilon/2 + \|f_\varepsilon - \tilde{f}\|_1 \leq \varepsilon/2 + \varepsilon/2 = \varepsilon \end{aligned} \tag{19}$$

by contractivity of E_α. Hence $f_\alpha \to \tilde{f}$ in norm. (The above proof also holds for all $p \geq 1$, and there is no need to use the result of Corollary 2.4 here!)

Finally, if I is countable and linearly ordered, we may set $I = \mathbb{N}$. Then one sees that the support of f_α is σ-finite, being integrable, and $\mathfrak{F}_n \uparrow \mathfrak{F}$. Thus the hypothesis of Theorem IV.2.5 is satisfied, and we conclude that $f_n \to \tilde{f}$ a.e. That $f = \tilde{f}$ a.e. iff $f \in L^1(G, \mathfrak{F}, \mu)$ or iff $f = \tau_H \circ T_H(f)$ for each f implies $\tau_H \circ T_H$ is the identity, and hence ($T_H \circ \tau_H$ is the identity always) $H = \{e\}$ is now immediate. This completes the proof of the theorem.

Remark As Theorem IV.2.4 implies, the pointwise convergence holds here locally a.e., if f is only locally integrable, i.e., $f\chi_A$ is integrable for each compact $A \subset G$. Since Π_H is also closed for compact H, and A $(\subset \Pi_H^{-1}(\Pi_H(A)) = A_0)$ is compact, one may assume that $\tilde{f}$ is only locally integrable.

It is evident that the hypothesis of the theorem is needed only in deducing that $\{f_\alpha = \tau_{H_\alpha} \circ T_{H_\alpha}(f), \alpha \in I\}$ is a martingale in $L^p(\mathscr{B})$. The proof on convergence is valid for any martingale on an abstract measure space (Ω, Σ, μ). However, using the group structure further, one may strengthen Theorem 3 as follows.

4. Theorem *Let $f \in L^p(G, \mathscr{B}, \mu)$ and suppose f is continuous at a point $x_0 \in G$. Let $\mu(O) > 0$ for each nonempty open set $O \subset G$. (In particular μ may be a left Haar measure on G.) Then, under the hypothesis of Theorem 3, with $I = \mathbb{N}$, $f_n(x_0) \to \tilde{f}(x_0)$. If f is uniformly continuous on a set $A \in \mathscr{B}$, then $f_n \to \tilde{f}$ uniformly on that A.*

Proof Since H is compact, let V be a neighborhood of H. Now $H = \lim_n H_n = \bigcap_{n \geq 1} H_n$ implies, for large enough n, that each $H_n \subset V$. In fact, the collection $\{H_n \cap V^c\}_1^\infty$ is decreasing and each set is compact. But $\bigcap_n (H_n \cap V^c) = H \cap V^c = \varnothing$. By compactness there is an n_0 such that

$H_{n_0} \cap V^c = \varnothing$. Hence for $n \geq n_0$ we have $H \subset H_n \subset H_{n_0} \subset V$. On the other hand, $\tilde{f}$ is measurable for $\mathfrak{F} = \Pi_H^{-1}(\mathscr{G}) \supset \Pi_{H_n}^{-1}(\mathscr{G}) = \mathfrak{F}_n$. So $\tilde{f}$ is constant on cosets xH, $x \in G$. Since $H \subset H_n$ and f_n is constant on cosets of H_n, it is also constant on cosets of H in particular. But f is continuous at $x_0 \in G$ implies that it must be continuous on the coset $\hat{x}_0 = x_0 H$.

Let $\varepsilon > 0$ be given. Then there exists a neighborhood $V \supset H$, depending on x_0 (so $x_0 V \supset x_0 H$) such that

$$|\tilde{f}(x) - \tilde{f}(x_0)| < \varepsilon, \qquad x \in x_0 V,$$

or $\qquad$ (20)

$$|\tilde{f}(xy) - \tilde{f}(x_0)| < \varepsilon, \qquad x \in x_0 V, \quad y \in H.$$

However, by the first paragraph, there is an n_0 such that $n \geq n_0$ implies $H_n \subset V$. We deduce that (20) also holds for all $x \in x_0 H_n \subset x_0 V$ and hence for $x \in x_0 V$ and $y \in H_n$. Note that $T_{H_n}(\chi_{H_n}) = 1$ a.e. since H_n is compact and its Haar measure is normalized. Also $f_n = E^{\mathfrak{F}_n}(f) = E^{\mathfrak{F}_n}E^{\mathfrak{F}}(f) = E^{\mathfrak{F}_n}(\tilde{f})$ since $\mathfrak{F}_n \subset \mathfrak{F}$. Consider for $n \geq n_0$,

$$\begin{aligned} |E^{\mathfrak{F}_n}(\tilde{f})(x_0) - \tilde{f}(x_0)| &= |\tau_{H_n} \circ T_{H_n}\tilde{f} - \tilde{f}|(x_0) \\ &= |(T_{H_n}(\tilde{f}) - \tilde{f})(\Pi_{H_n}(x_0))| \\ &\leq \int_{H_n} |\tilde{f}(x_0 y) - \tilde{f}(x_0)|\, dy, \qquad x_0 \in \Pi_{H_n}(x_0), \\ &\leq \varepsilon \int_{H_n} dy = \varepsilon \quad \text{by (20) and } H_n \subset V. \end{aligned}$$

Since $\varepsilon > 0$ and μ gives positive measure to V by hypothesis, it follows that $f_n(x_0) \to \tilde{f}(x_0)$.

Now, if $\tilde{f}$ is uniformly continuous on a set A, we can choose V that works for all $\{xH, x \in A\}$ and then as before all $H_n \subset V$ for $n \geq n_0$. The above procedure then holds for all $x_0 \in A$ uniformly. Hence $f_n \to \tilde{f}$ uniformly on A. This completes the proof.

Finally, we present some complements and applications of the foregoing results in the Complements and Problems Section.

Complements and Problems

1. Let (Ω, Σ, P) be a probability space and $T_n: L^p(P) \to L^p(P)$ be a linear operator such that it is defined and is a contraction on both $L^p(P)$ and its conjugate $L^q(P)$ where $q = p/(p-1)$, $1 < p < 2$. Let $\{T_n, n \geq 1\}$ be an increasing sequence of such operators, $T_n^2 = T_n$. Show that each T_n is defined on all spaces $L^p(P)$, $1 \leq p \leq \infty$, and $\lim_n (T_n f)(\omega)$ exists for a.a. (ω). A similar statement holds for the decreasing sequences also. [*Hints*: If $\mathscr{M}_n \subset \mathscr{M}_{n+1}$, $\mathscr{M}_n = T_n(L^p(P))$, then there exists a maximal set $B_n \in \Sigma$ outside of which each element of $\mathscr{M}_n$ vanishes a.e. Let $\mathscr{B}_n = \Sigma(B_n)$ be the σ-algebra relative to which each f in $\mathscr{M}_n$ is measurable. Verify that $T_n f = \varphi_n \chi_{B_n} E^{\mathscr{B}_n}(\varphi_n f), f \in L^p(\Sigma)$ for some measurable φ_n which satisfies $|\varphi_n| = 1$ a.e. on B_n and set $\varphi_n = 1$ on $\Omega - B_n$. This representation of T_n is nontrivial and uses a result of Andô [1, p. 402]. Note that in this computation it can be assumed that $B_n \subset B_{n+1}$ and $\varphi_n = \varphi_{n+1} | B_n$ outside of a null set. Deduce the result from the martingale convergence theorem. The decreasing case is similar. For more on this, see the author's paper [12].]

2. In this problem an alternative form of Theorem 2.5 giving a necessary and sufficient condition will be presented. Suppose that $\mathscr{X}$ is a Banach space and $\nu: \Sigma \to \mathscr{X}$ is a vector measure where (Ω, Σ) is a measurable space. If $P: \Sigma \to \mathbb{R}^+$ is a probability and ν is P-continuous, then there exists a strongly measurable Bochner integrable P-unique $f: \Omega \to \mathscr{X}$ such that $\nu(A) = \int_A f\, dP$, $A \in \Sigma$ iff (i) ν is of finite total variation (i.e., $|\nu|(\Omega) < \infty$) and (ii) for each $\varepsilon > 0$ there exists a set $B_\varepsilon \in \Sigma$, $P(B_\varepsilon) \geq 1 - \varepsilon$ such that $\{\nu(A)/P(A): A \in \Sigma(B_\varepsilon), P(A) > 0\}$ is relatively weakly compact in $\mathscr{X}$. [*Hints*: The proof of sufficiency is essentially the same as that given in the text with only simple modifications. The necessity of (i) is a manipulation of the definitions; (ii) is proved by using the Egorov theorem, which gives the B_ε and a sequence of simple $f_n: \Omega \to \mathscr{X}$ such that $\|f_n - f\|(\omega) \to 0$ uniformly in $\omega \in B_\varepsilon$. Then the operator $T_n: L^1(P) \to \mathscr{X}$ defined by $T_n(g) = \int_\Omega g f_n\, dP$ is compact, and $T_n \to T$ in the uniform norm; so T is compact. Deduce (ii) from this.] If instead of (i) above, we have $|\nu|_p(\Omega) < \infty$ for $1 \leq p < \infty$, so that f is strongly measurable and $\|f\|(\cdot) \in L^p(P) \subset L^1(P)$, formulate the corresponding result (in which (ii) remains the same). [The proof is analogous or can be reduced to the case $p = 1$. Regarding this statement and generalizations, see Dinculeanu and Uhl [1].]

3. In this problem we consider finitely additive "probabilities" (i.e., those with unit mass). Such set functions and their importance in gambling theory have been emphasized in the work of Dubins and Savage [1]. Thus let Ω be a nonempty set, Σ be an algebra of Ω, $\mu: \Sigma \to \bar{\mathbb{R}}^+$ be additive, and $P: \Sigma \to [0, 1]$ be additive. Let Σ_μ be the ring of sets of finite μ-measure in Σ. If $\varphi: \mathbb{R} \to \mathbb{R}^+$ is a continuous Young function and $F: \Sigma_\mu \to \mathscr{X}$ (a Banach space) is additive, then F has φ-bounded variation (cf. Eq. (2.9)) on Σ relative to μ (or P) on A iff $I_\varphi(F, A) < \infty$, where (0/0 is taken as 0, and $\Sigma(A)$ is the traces of Σ on A)

$$I_\varphi(F, A) = \sup\left\{\sum_{k=1}^{n} \varphi\left(\frac{\|F(A_k)\|}{\mu(A_k)}\right)\mu(A_k) : \{A_k\}_1^n \subset \Sigma_\mu(A), \text{ disjoint}\right\}.$$

Let $I_\varphi(F) = I_\varphi(F, \Omega)$ and $N_\varphi(F) = \inf\{k > 0 : I_\varphi(F/k) \leq 1\}$. The space $V^\varphi(\Sigma, \mathscr{X})$, of all $\mathscr{X}$-valued additive set functions F with $N_\varphi(F) < \infty$, becomes a Banach space with $N_\varphi(\cdot)$ as norm. Let F_π denote a step function where $F_\pi(A) = \sum_{k=1}^n [F(A_k)/\mu(A_k)]\mu(A_k \cap A)$, $A \in \Sigma_\mu$, and $\pi = \{A_k\}_1^n \subset \Sigma_\mu$ is a "partition," i.e., a disjoint collection of sets. Let $S^\varphi(\Sigma, \mathscr{X})$ be the closed subspace determined by the "step set functions" of the form F_π, $F \in V^\varphi(\Sigma, \mathscr{X})$, π is any "partition." It can be checked that $S^\varphi = V^\varphi$ for $\mathscr{X}$ reflexive and $\varphi(2x) \leq C\varphi(x)$, $x \geq 0$. If $f \in L^\varphi(\Omega, \Sigma, \mu)$ and $\lambda : f \mapsto \int_{(\cdot)} f\, d\mu$, then $\lambda(f) \in V^\varphi(\Sigma, \mathscr{X})$ and $\lambda(L^\varphi(\Sigma, \mathscr{X})) \subset V^\varphi(\Sigma, \mathscr{X})$ is an isometric embedding which is not closed if μ is not σ-additive. If $\mathscr{B} \subset \Sigma$ is a subalgebra and $F \in S^\varphi(\Sigma, \mathscr{X})$, define $Q^{\mathscr{B}}: F \mapsto F|\mathscr{B}$, restriction to $\mathscr{B}$. Then $Q^{\mathscr{B}}$ is a positive linear contraction, indempotent and $N_\varphi(Q^{\mathscr{B}}(F) - \tilde{F}_\pi) \to 0$ as π is refined where $\tilde{F}_\pi$ is F_π when π is a "partition" contained in $\mathscr{B}$. If μ is replaced by P (a σ-additive probability), $f \in L^1(\Sigma, \mathscr{X})$, $\mathscr{B} \subset \Sigma$ a σ-algebra, and $E^{\mathscr{B}}$ is the conditional expectation, show that $Q^{\mathscr{B}}(\lambda f) = \lambda E^{\mathscr{B}}(f)$. If $\{\mathscr{B}_\tau, \tau \in I\}$ is an increasing net of algebras contained in Σ, $F_\tau \in S^\varphi(\mathscr{B}_\tau, \mathscr{X})$ for each $\tau \in I$, then $\{F_\tau, \mathscr{B}_\tau, \tau \in I\}$ is an (increasing) *S^φ-set martingale* (as in Definition III.1.6), i.e., $\tau < \tau'$ implies $Q^{\mathscr{B}_\tau}(F_{\tau'}) = F_\tau$ or $F_{\tau'}|\mathscr{B}_\tau = F_\tau$. Then verify the following convergence statements:

(a) Let $\{F_\tau, \mathscr{B}_\tau, \tau \in I\}$ be an S^φ-set martingale, ($\varphi(2x) \leq C\varphi(x)$), and $\mathscr{B} = \bigcup_{\tau \in I} \mathscr{B}_\tau$. If $F(A) = \lim_\tau F_\tau(A)$, $A \in \mathscr{B} \cap \Sigma_\mu$, then $F \in S^\varphi(\mathscr{B}, \mathscr{X})$ and $N_\varphi(F - F_\tau) \to 0$ "$\tau \to \infty$." [A standard but a detailed argument is needed.]

(b) For the martingale of (a), the following statements are equivalent:

(i) the martingale converges strongly in $S^\varphi(\Sigma, \mathscr{X})$;
(ii) the martingale converges weakly in $S^\varphi(\Sigma, \mathscr{X})$;
(iii) there exists $x_A \in \mathscr{X}$ for each $A \in \mathscr{B} \cap \Sigma_\mu$ such that $x^*(F_\tau(A)) \to x^*(x_A)$, and in this case $F(A) = x_A$ defines F as an element of $S^\varphi(\mathscr{B}, \mathscr{X})$.

If $\varphi'(x) \uparrow \infty$, $\mathscr{X}$ is reflexive, then each of (i)–(iii) is equivalent to the boundedness of the martingale in the N_φ-norm. [Note that this result includes Corollary 2.4, and both these results show how the limit procedures can be minimized. The proofs can be based on the arguments of that section. In connection with this problem and its relation to the point function case, see Uhl [2]. This general approach includes the main mean convergence results of Krickeberg and Pauc [1].]

4. We present an abstract version of the convergence Theorem 4.3 when the measure spaces have no group structure but have the projective limit property.

(a) Let $\{(\Omega_\alpha, \Sigma_\alpha, P_\alpha, g_{\alpha\beta})_{\alpha<\beta} : \alpha, \beta \text{ in } D\}$ be a projective system of probability spaces where $g_{\alpha\beta}: \Omega_\beta \to \Omega_\alpha$ is onto. (See Section III.2 for concepts and results.) If $L^p(\Omega_\alpha, \Sigma_\alpha, P_\alpha) = L^p(P_\alpha)$, $p \geq 1$ is a Lebesgue space, $f \in L^p(P_\alpha)$, let $v_\alpha^f: A \mapsto \int_A f\, dP_\alpha$, $A \in \Sigma_\alpha$, be a mapping. Show that there is a consistent family of positive mappings $\tau_{\alpha\beta}: L^p(P_\alpha) \to L^p(P_\beta)$ if we define $v_\alpha^f = v_\beta^{\tau_{\alpha\beta}(f)} \circ g_{\alpha\beta}^{-1}$ for any $\alpha < \beta$. [*Hint*: If $h = \sum_{i=1}^n a_i \chi_{A_i} \in L^p(P_\alpha)$, define $\tau_{\alpha\beta}: h \mapsto \tau_{\alpha\beta}(h) = \sum_{i=1}^n a_i \chi_{g_{\alpha\beta}^{-1}(A_i)}$ $(\in L^p(P_\beta))$ and extend it to the desired mapping.]

(b) Show that there is a bounded linear onto mapping $T_{\alpha\beta}: L^p(P_\beta) \to L^p(P_\alpha)$ such that $\tau_{\alpha\beta} \circ T_{\alpha\beta} = E^{\alpha\beta}: L^p(P_\beta) \to L^p(P_\beta)$ is a conditional expectation. Hence for any $f \in L^p(P_\beta)$, $\{E^{\alpha\beta}(f), \alpha < \beta \text{ in } D\}$ is a decreasing martingale in α for each $\alpha < \beta$ fixed, relative to the σ-algebras $\{g_{\alpha\beta}^{-1}(\Sigma_\alpha)\}_{\alpha<\beta}$ in Σ_β. [*Hints*: Since P_β is finite, the contractive projection on $L^p(P_\beta)$, with range $L^p(g_{\alpha\beta}^{-1}(\Sigma_\alpha), P_\beta)$, is a conditional expectation by Theorem II.2.8. But $\tau_{\alpha\beta}$ is one-one on the latter space. Hence there exists a mapping $T_{\alpha\beta}$ such that the following diagram commutes:

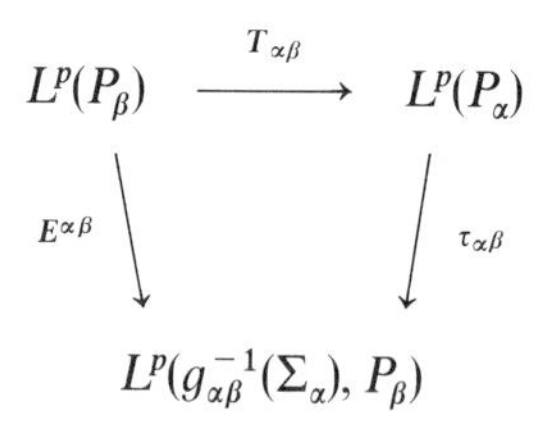

The martingale property is then a consequence. Finally, check that $E^{\alpha\beta} \circ E^{\beta\gamma} = E^{\alpha\gamma}$, $T_{\alpha\beta} \circ T_{\beta\gamma} = T_{\alpha\gamma}$, $\alpha < \beta < \gamma$, hold as a result of the coherency of $g_{\alpha\beta}$s and compatibility of P_αs.]

(c) Let $\bar{\mathscr{M}} = \{(E^{\alpha\beta}(f), \beta > \alpha \text{ in } D): f \in L^p(P_{\alpha_0}), \alpha > \alpha_0\}$ be the space of martingales. Define the seminorm $|||\cdot|||: f \mapsto \sup_{\beta > \beta_0} \|E^{\alpha_0\beta}(f)\|_p$. Show that $\{\bar{\mathscr{M}}, |||\cdot|||\}$ is a seminormed vector space (componentwise addition), which need not be complete. The elements of $\bar{\mathscr{M}}$ are called "vector fields" on D into the Banach spaces $L^p(P_\beta)$. The completion $\tilde{\mathscr{M}}$ contains elements which may be only additive set functions (and set martingales of the form considered in the preceding problem). [Standard computation yields the result.]

(d) Suppose the basic projective system admits the limit (Ω, Σ, P) and let $g_\alpha: \Omega \to \Omega_\alpha$ be the corresponding onto mapping. If P_α^* is defined on $g_\alpha^{-1}(\Sigma_\alpha)$ by $P_\alpha^* \circ g_\alpha^{-1} = P_\alpha$ (i.e., $P_\alpha^* = P \mid f_\alpha^{-1}(\Sigma_\alpha)$), let $E^\alpha: L^p(\Omega, \Sigma, P) \to L^p(\Omega, g_\alpha^{-1}(\Sigma), P_\alpha^*)$, $\tau_\alpha: L^p(P_\alpha) \to L^p(P_\alpha^*)$, and $T_\alpha: L^p(P) \to L^p(P_\alpha)$ be the corresponding mappings so that $\tau_\alpha \circ T_\alpha = E^\alpha$, $1 \leq p < \infty$. Show that $\tilde{\mathscr{M}}$ of (c) now consists of martingales of point functions and that $\{E^\alpha(f), g_\alpha^{-1}(\Sigma_\alpha), \alpha > \alpha_0\} \subset L^p(P) = L^p(\Omega, \Sigma, P)$ is a convergent martingale.

(e) Suppose $T_{\alpha\beta}: L^p(P_\beta) \to L^p(P_\alpha)$ is as in (b) so that $T_{\alpha\beta} \circ T_{\beta\gamma} = T_{\alpha\gamma}$, $\alpha < \beta < \gamma$. Let $\mathbf{f} = \{f_\alpha, \alpha \in D\} \in \bar{\mathscr{M}}$ be a vector field, so $f_\alpha \in L^p(\Sigma_\alpha)$ and $T_{\alpha\beta}(f_\beta) = f_\alpha$. Then $\{f_\alpha, T_{\alpha\beta}, \alpha < \beta \text{ in } D\}$ may be termed a *projective martingale family*. Let $|||\cdot|||: \mathbf{f} \mapsto |||\mathbf{f}||| = \sup_{\alpha \in D} \|f_\alpha\|_p$ be the (norm) functional;

then $(\bar{\mathscr{M}}, |||\cdot|||)$ becomes a normed linear space of all $\mathbf{f}$ with $|||\mathbf{f}||| < \infty$. Show that $\bar{\mathscr{M}}$ is complete under this norm for $1 \le p \le \infty$ whenever $\mathbf{f} \in \bar{\mathscr{M}}$ implies $(f_\alpha, \alpha \in D)$ is terminally uniformly integrable. [This follows from Section 4 and (d) above.]

(f) Let the projective system be as in (d) admitting a limit. With the same symbols defined there, show that $L^p(\Omega, \Sigma, P) = \lim_{\leftarrow}(L^p(P_\alpha), T_{\alpha\beta})$. Suppose $\mathscr{X}$ is a Banach space and there is a family $(Q_\alpha, \alpha \in D)$ of mappings $Q_\alpha: \mathscr{X} \to L^p(P_\alpha)$ such that $Q_\alpha = T_{\alpha\beta} \circ Q_\beta$ for $\alpha < \beta$. Show that $Q = \lim_{\leftarrow}(Q_\alpha, T_{\alpha\beta})$ exists and $Q \circ T_\alpha = Q_\alpha$. (See Theorem III.3.5 and discussion following its statement on the projective limit of mappings.) Also Q is bounded if $\|Q_\alpha\| \le K_0 < \infty$, $\alpha \in D$, and if $\mathscr{X}$ is moreover a lattice and Q_α is an isotone mapping for each $\alpha \in D$, then so is Q. [*Hints*: By (e), $f_\alpha = Q_\alpha x \in L^p(P_\alpha)$, $\alpha \in D$ for $x \in \mathscr{X}$. So $\{f_\alpha, T_{\alpha\beta}, \alpha \le \beta \text{ in } D\}$ is a projective martingale. Define $Q: \mathscr{X} \to \prod_{\alpha \in D} L^p(P_\alpha)$ by the equation $\pi_\alpha(Qx) = Q_\alpha x \in L^p(P_\alpha)$ where π_α is the coordinate projection. It suffices to show that $\pi_\alpha(Qx) = T_{\alpha\beta}(\pi_\beta Q(x)) = T_{\alpha\beta}(Q_\beta x)$, and this follows as in Theorem III.3.5. The rest uses (e). In connection with this result, see Scheffer [1]. If $\Omega_\alpha = \Omega$, $\Sigma_\alpha = \Sigma$, $\alpha \in D$, then $(\bar{\mathscr{M}}, |||\cdot|||)$ was defined and analyzed by Pitcher [3] and the topological properties were considered when the L^ps are replaced by Orlicz spaces and the "sup" is replaced by a convex functional by Rosenberg [1]. For some further generalizations on projective and direct limits, see Vasilach [1].]

5. Finally, we present an extension of the martingale convergence theorem to a pair of vector valued measures which generalize the conditional probabilities of Chapter II.

(a) Let (Ω, Σ, P) be a probability space, $\mathscr{A} \subset \Sigma$ a σ-algebra, and $P^{\mathscr{A}}: \Sigma \to \mathscr{X} = L^\infty(\Omega, \mathscr{A}, P)$ be the conditional probability measure in the sense of Chapter II. Let $\mathscr{Y}_p = L^p(\Omega, \Sigma, P^{\mathscr{A}})$ be the space of measurable pth power integrable real functions relative to $P^{\mathscr{A}}$ (with the Dunford–Schwartz integral as in Chapter II), $1 \le p \le \infty$, so that if $\pi: \mathscr{X} \to \mathscr{Y}_p$ is an algebra homomorphism, one has $\int_\Omega \pi(f) g \, dP^{\mathscr{A}} = f \int_\Omega g P^{\mathscr{A}}$, $f \in \mathscr{X}$ (by the averaging property of $E^{\mathscr{A}}$). If we define $\nu^h(A) = \int_A h \, dP^{\mathscr{A}}$, $A \in \Sigma$, $h \in \mathscr{Y}_1$, then ν^h is an $\mathscr{X}$-valued σ-additive (in terms of the order in $\mathscr{X}$, as in Section I.4) set function. If $\mathscr{B}_n \subset \mathscr{B}_{n+1} \subset \Sigma$ and $\mathscr{B}_\infty = \sigma(\bigcup_{n \ge 1} \mathscr{B}_n)$ are σ-algebras and $\nu_n^h = \nu^h | \mathscr{B}_n$, let $h_n = d\nu_n^h / dP_n^{\mathscr{A}}$, where $P_n^{\mathscr{A}} = P^{\mathscr{A}} | \mathscr{B}_n$. (That such h_n exists (and $\mathscr{B}_n$-adapted) follows easily in the present case, cf., e.g., the author's paper [7].) Show that $h_n \to h_\infty$ a.e. and $h_\infty = h$ a.e. $[P]$ iff $\mathscr{B}_\infty = \Sigma$. [Argument is similar to that of Theorem IV.2.12.]

(b) In the preceding part, both $\mathscr{X}$ and $\mathscr{Y}_\infty$ are Stone algebras satisfying the countable chain condition; namely, each bounded subset contains a countable subset such that both have the same upper bounds. This property admits an extension as follows: Let (Ω, Σ) be a measurable space, $\mathscr{X}$ a Stone algebra $C(S)$, of real continuous functions on a compact Stone space S. Let $\mu: \Sigma \to \mathscr{X}^+$ be a σ-additive set function in terms of the order in $\mathscr{X}$ as above. Assume that $\mathscr{X}$ satisfies the countable chain condition and that there is an algebra homomorphism $\pi: \mathscr{X} \to \mathscr{L}_\infty$ where $\mathscr{L}_p = L^p(\Omega, \Sigma, \mu)$, $1 \leq p \leq \infty$, the space of real pth power μ-integrable functions (as in the specialization of the Dunford–Schwartz integral with order taken into account) such that $\int_\Omega \pi(x) f \, d\mu = x \int_\Omega f \, d\mu$, $x \in \mathscr{X}$. If $\mathscr{B}_n \subset \mathscr{B}_{n+1} \subset \Sigma$ and $\mathscr{B}_\infty = \sigma(\bigcup_{n \geq 1} \mathscr{B}_n)$ is as before, let $\pi(\mathscr{X}) \subset L^\infty(\Omega, \mathscr{B}_1, \mu)$. If $f \in L^1(\Omega, \Sigma, \mu)$ and $v^f(A) = \int_A f \, d\mu$, $A \in \Sigma$, then v^f is an $\mathscr{X}$-valued σ-additive set function, and if $v_n^f = v^f | \mathscr{B}_n$, $\mu_n = \mu | \mathscr{B}_n$, then there exists a unique $f_n \in L^1(\Omega, \mathscr{B}_n, \mu_n)$ such that $v_n^f(A) = \int_A f_n \, d\mu_n$, $A \in \mathscr{B}_n$. (This is a consequence of a Radon–Nikodým theorem due to Wright [3].) The mapping $E^{\mathscr{B}_n}: f \mapsto f_n$ is then a "generalized conditional expectation." Show that this $E^{\mathscr{B}}$ has the usual properties given in Section II.1. For this sequence $\{f_n, \mathscr{B}_n, n \geq 1\}$ show that $f_n \to f_\infty$ a.e. $[\mu]$. [Again with the properties of this new integral (monotone convergence theorem being true), the proof of Theorem IV.2.12 extends and only that $v_\infty^f: \bigcup_{n=1}^\infty \mathscr{B}_n \to \mathscr{X}$ has a unique extension to $\mathscr{B}_\infty$ needs an additional argument. For simplicity let $f \geq 0$ and consider $\mathscr{C} = \{A \in \mathscr{B}_\infty : v^f(A) \geq 0\} \supset \bigcup_{n \geq 1} \mathscr{B}_n = \mathscr{D}_0$, an algebra. Then by Zorn's lemma, there is a maximal $\mathscr{D}$ such that $\mathscr{D}_0 \subset \mathscr{D} \subset \mathscr{C}$. Let $\tilde{\mathscr{D}} = \{A \in \mathscr{B}_\infty : \lim_n \chi_{A_n} = \chi_A, A_n \in \mathscr{D}\}$. Observe that $\tilde{\mathscr{D}} \subset \mathscr{C}$ is an algebra and then note that it is a σ-algebra containing $\mathscr{D}$. So $\mathscr{D} = \tilde{\mathscr{D}}$ and $\tilde{\mathscr{D}} = \sigma(\mathscr{D}_0) = \mathscr{B}_\infty$. We can also use the monotone class theorem instead of Zorn's lemma here.]

(It is possible to prove an analogous result if $\mathscr{L}_p$ is replaced by a Stone algebra and $\{E^{\mathscr{B}_n}\}_1^\infty$ is replaced by a monotone sequence of averaging operators. These and the necessary integration theory were discussed by Wright in [2–4]. Another interesting application of abstract methods for a class of stochastic problems is surveyed in the paper by Kingman [1].)

Bibliography

The following abbreviations are used:

AM	Annals of Mathematics
AMS	Annals of Mathematical Statistics
AP	The Annals of Probability
PAMS	Proceedings of the American Mathematical Society
TAMS	Transactions of the American Mathematical Society
JMA	Journal of Multivariate Analysis
PJM	Pacific Journal of Mathematics
TV	Teorija Verojatnosti i ee Primenenija
ZW	Zeitschrift für Wahrscheinlichkeitstheorie

Andersen, E. S., and Jessen, B.

[1] "Some limit theorems on integrals in an abstract set." *Danske Vid. Selsk. Mat.-Fys. Medd.* **22**, No. 14, 29 pp. (1946).

[2] "On the introduction of measures in infinite product sets," *ibid.* **25**, No. 4, 8 pp. (1948).

Andô, T.

[1] "Contractive projections in L^p-spaces." PJM **17**, 391–405 (1966).

Andô, T., and Amemiya, I.

[1] "Almost everywhere convergence of prediction sequences in L^p, $1 < p < \infty$." ZW **4**, 113–120 (1965).

Astbury, K. A.

[1] "Amarts indexed by directed sets." AP **6**, 267–278 (1978).

Austin, D. G.

[1] "A sample function property of martingales." AMS **37**, 1396–1397 (1966).

Bartle, R. G.

[1] "A general bilinear vector integral." *Studia Math.* **15**, 337–352 (1956).

Baxter, J. R.

[1] "Pointwise in terms of weak convergence." PAMS **46**, 395–398 (1974).

Billingsley, P.

[1] Ergodic Theory and Information. Wiley & Sons, New York, 1965.

Birkhoff, G.

[1] "Moyennes des fonctions bornées." Congrès d'Algèbre et Théorie des Nombres, Paris, 143–154 (1949).

Bochner, S.

[1] Harmonic Analysis and the Theory of Probability. Univ. Calif. Press, 1955.

Bourbaki, N.

[1] Theory of Sets. Addison-Wesley (and Hermann), Reading, 1967.

[2] Élements de mathématique VI. Chapitre IX (also Chs. 3–5). Hermann, Paris, 1969.

Briggs, V. D.

[1] "Densities for infinitely divisible processes." JMA **5**, 178–205 (1975).

Brody, E. J.

[1] "An elementary proof of the Gaussian dichotomy theorem." ZW **20**, 217–226 (1971).

Brown, J. R.

[1] "Inverse limits, entropy and weak isomorphism for discrete dynamical systems." TAMS **164**, 55–66 (1972).

Brown, M.

[1] "Discrimination of Poisson processes." AMS **42**, 773–776 (1971).

Chacón, R. V.

[1] "A 'stopped' proof of convergence." *Adv. in Math.* **14**, 365–368 (1974).

Chatterji, S. D.

[1] "Martingales of Banach-valued random variables." *Bull. Amer. Math. Soc.* **66**, 395–398 (1960).

Chi, G. Y. H., and Dinculeanu, N.

[1] "Projective limits of measure preserving transformations on probability spaces." JMA **2**, 404–417 (1972).

Choksi, J. R.

[1] "Inverse limits of measure spaces." *Proc. London Math. Soc.* **8** (Ser 3), 321–342 (1958).

Chow, Y. S.

[1] "Martingales in a σ-finite measure space indexed by directed sets." TAMS **97**, 254–285 (1960).

Császár, A.

[1] "Sur la structure des espaces de probabilité conditionnelle." *Acta Math. Hung.* **6**, 337–361 (1955).

DeGroot, M. H., and Rao, M. M.

[1] "Bayes estimation with convex loss." AMS **34**, 839–846 (1963).

Dieudonné, J.

[1] "Sur un théorème de Jessen," *Fund. Math.* **37**, 242–248 (1950).

Dinculeanu, N.

[1] Vector Measures. Pergamon Press, London, 1967.

[2] "Conditional expectations for general measure spaces." JMA **1**, 347–364 (1971).

Dinculeanu, N., and Foiaş, C.

[1] "Algebraic models for measures." *Illinois J. Math.* **12**, 340–351 (1968).

Dinculeanu, N., and Kluvánek, I.

[1] "On vector measures." *Proc. London Math. Soc.* **17** (Ser 3), 505–512 (1967).

Dinculeanu, N., and Rao, M. M.

[1] "Contractive projections and conditional expectations." JMA **2**, 362–381 (1972).

Dinculeanu, N., and Uhl, Jr., J. J.

[1] "A unifying Radon–Nikodým theorem for vector measures." JMA **3**, 184–203 (1973).

Doob, J. L.

[1] Stochastic Processes. Wiley and Sons, New York, 1953.

Douglas, R. G.

[1] "Contractive projections in an L^1-space." PJM **15**, 443–462 (1965).

Dubins, L. E.

[1] "Conditional probability distributions in the wide sense." PAMS **8**, 1088–1092 (1957).

Dubins, L. E., and Savage, L. J.

[1] How to Gamble if You Must. McGraw-Hill, New York, 1965.

Dunford, N., and Schwartz, J. T.

[1] Linear Operators, Part I: General Theory. Interscience, New York, 1958.

Dynkin, E. B.

[1] Foundations of the Theory of Markov Processes. Pergamon Press, London, 1960.

Eberlein, W. F.

[1] "An integral over function space." *Canad. J. Math.* **14**, 379–384 (1962).

Edgar, G. A., and Sucheston, L.

[1] "Amarts: A class of asymptotic martingales. A. Discrete parameter." JMA **6**, 193–221 (1976).

Edwards, R. E., and Hewitt, E.

[1] "Pointwise limits for sequences of convolution operators." *Acta Math.* **113**, 181–218 (1965).

Feldman, J.

[1] "Equivalence and perpendicularity of Gaussian processes." PJM **8**, 699–708 (1959).

Frolik, Z., and Pachl, J.

[1] "Pure measures." *Comment. Math. Universitatis Carolinae* **14**, 279–293 (1973).

Gel'fand, I. M., Kolmogorov, A. N., and Yaglom, A. M.

[1] "Amount of information and entropy for continuous distributions." (in Russian). *Trans. Third Math. Congress USSR* **3**, 300–320 (1958).

Gel'fand, I. M., and Vilenkin, N. Ya.

[1] Generalized Functions, Volume 4. Academic Press, New York, 1964.

Gel'fand, I. M., and Yaglom, A. M.

[1] "Calculation of the amount of information about a random function contained in another random function." (in Russian). *Uspekhi Mat. Nauk* **12**, No. 1 (73), 3–52 (1957).

Gikhman, I. I., and Skorokhod, A. V.

[1] "On the densities of probability measures in function spaces." *Russian Math. Surveys* **21**, No. 6, 83–156 (1966).

Grenander, U.

[1] "Stochastic processes and statistical inference." *Ark. Mat.* **1**, 195–277 (1950).

Haimo, F.

[1] "Some limits of Boolean algebras." PAMS **2**, 566–576 (1951).

Hájek, J.

[1] "On a property of normal distribution of any stochastic process." (in Russian). *Čech. Math. J.* **8** (Ser 2), 610–618 (1958).

Halmos, P. R.

[1] Measure Theory. Van Nostrand, Princeton, 1950.

Hardy, G. H., Littlewood, J. E., and Pólya, G.
[1] Inequalities. Cambridge Univ. Press, London, 1934.

Harpain, F., and Sion, M.
[1] "A representation theorem for measures on infinite dimensional spaces." PJM **30**, 47–58 (1969).

Hewitt, E.
[1] "A note on measures in Boolean algebras." *Duke Math. J.* **20**, 253–256 (1953).

Hewitt, E., and Stromberg, K.
[1] Real and Abstract Analysis. Springer-Verlag, New York, 1965.

Hunt, G. A.
[1] Martingales et Processus de Markov. Dunad, Paris, 1966.

Ionescu Tulcea, A., and Ionescu Tulcea, C.
[1] Topics in the Theory of Lifting. Springer-Verlag, Berlin, 1969.
[2] "On the existence of a lifting commuting with the left translations of an arbitrary locally compact group." *Proc. Fifth Berkeley Symp. Math. Statist. and Prob.* **2**, 63–97 (1967).

Ionescu Tulcea, C.
[1] "Measures dans les espaces produits." *Atti Acad. Naz. Lincei Rend. cl. Sci. Fis. Mat. Nat.* **7** (Ser 8), 208–211 (1949/1950).

Isaac, R.
[1] "A proof of the martingale convergence theorem." PAMS **16**, 842–844 (1965).

Jerison, M.
[1] "Martingale formulation of ergodic theorems." PAMS **10**, 531–539 (1959).

Jerison, M., and Rabson, G.
[1] "Convergence theorems obtained from induced homomorphisms of a group algebra." AM **63** (Ser 2), 176–190 (1956).

Jiřina, M.
[1] "On regular conditional probabilities." *Čech. Math. J.* **9** (Ser 2), 445–450 (1959).

Kaç, M., and Slepian, D.
[1] "Large excursions of Gaussian processes." AMS **30**, 1215–1228 (1959).

Kakutani, S.
[1] "On equivalence of infinite product measures." AM **49** (Ser 2), 214–224 (1948).

Kallianpur, G.
[1] "On the amount of information contained in a σ-field." *Contrib. Prob. and Statist.*, Stanford Univ. Press, Stanford, California, 265–273 (1960).

Kampé de Fériet, J.
[1] "Sur un problem d'algébre abstrait posé par la definition de la moynne dans la theorie de la turbulence." *An. Soc. Sci. Bruxelles* **63**, 156–172 (1949).
[2] "Problémes mathématiques posés par la mecanique statistique de la turbulence." *Proc. Inter. Cong. Math., Amsterdam,* **3**, 237–242 (1954).

Kappos, D. A.
[1] Probability Algebras and Stochastic Spaces. Academic Press, New York, 1969.

Kingman, J. F. C.
[1] "Subadditive ergodic theory." AP **1**, 883–909 (1973).

Kirk, R. B.
[1] "Kolmogorov type consistency theorems for products of locally compact, B-compact spaces." *Proc. Acad. Sci., Amsterdam,* Ser A **73**, 77–81 (1970).

Kolmogorov, A. N.

[1] Grundbegriffe der Wahrscheinlichkeitsrechnung. Springer-Verlag, Berlin, 1933.

Kopp, E., Strauss, D., and Yeadon, F. J.

[1] "Positive Reynolds operators on Lebesgue spaces." *J. Math. Anal. Appl.* **44**, 350–365 (1973).

Krasnoselskiĭ, M. A., and Rutickiĭ, Ya. B.

[1] Convex Functions and Orlicz Spaces. Noordhoff, Groningen, 1961.

Krickeberg, K.

[1] "Convergence of conditional expectation operators." TV **9**, 538–549 (1964).

[2] "Convergence of martingales with a directed index set." TAMS **83**, 313–337 (1959).

Krickeberg, K., and Pauc, C. Y.

[1] "Martingales et dérivation." *Bull. Soc. Math. France* **91**, 455–543 (1963).

Lamb, C. W.

[1] "A short proof of the martingale convergence theorem." PAMS **38**, 215–217 (1973).

Loève, M.

[1] Probability Theory. (3d edition). Van Nostrand, Princeton, New Jersey, 1963.

Loomis, L. H.

[1] Introduction to Abstract Harmonic Analysis. Van Nostrand, Princeton, New Jersey, 1953.

Mann, H. B.

[1] "An inequality suggested by the theory of statistical inference." *Illinois J. Math.* **6**, 131–136 (1962).

Marczewski, E.

[1] "On compact measures." *Fund. Math.* **40**, 113–124 (1953).

McShane, E. J.

[1] "Families of measures and representations of algebras of operators." TAMS **102**, 328–345 (1962).

[2] Order-Preserving Maps and Integration Processes. *Ann. Math. Studies* No 31, Princeton Univ. Press, Princeton, 1953.

Métivier, M.

[1] "Limites projectives de mesures. Martingales. Applications." *Ann. Mat. Pura Appl.* **63** (Ser 4), 225–352 (1963).

Meyer, P. A.

[1] Probability and Potentials. Blaisdell Co., Waltham, 1966.

[2] Martingales and Stochastic Integrals. Math. Lecture Notes, Springer, No. 284, 1972.

Millington, H., and Sion, M.

[1] "Inverse systems of group-valued measures." PJM **44**, 637–650 (1973).

Moy, S.-C.

[1] "Characterizations of conditional expectation as a transformation on function spaces." PJM **4**, 47–64 (1954).

Nelson, E.

[1] "Regular probability measures on function spaces." AM **69** (Ser 2), 630–643 (1959).

Olson, M. P.

[1] "A characterization of conditional probability." PJM **15**, 971–983 (1965).

Orey, S.

[1] "*F*-processes." *Proc. Fifth Berkeley Symp. Math. Statist. and Prob.* **2**, 301–313 (1967).

Ornstein, D. S.
[1] "An application of ergodic theory to probability theory." AP **1**, 43–65 (1973).

Parthasarathy, K. R.
[1] Probability Measures on Metric Spaces. Academic Press, New York, 1967.

Pettis, B. J.
[1] "On extension of measures." AM **54** (Ser 2), 186–197 (1951).

Phillips, R. S.
[1] "On weakly compact subsets of a Banach space." *Am. J. Math.* **65**, 108–136 (1943).

Pitcher, T. S.
[1] "Likelihood ratios of Gaussian processes." *Ark. Mat.* **4**, 35–44 (1959).
[2] "Parameter estimation for stochastic processes." *Acta Math.* **112**, 1–40 (1964).
[3] "A more general property than domination for sets of probability measures." PJM **15**, 597–611 (1965).

Prokhorov, Yu. V.
[1] "Convergence of random processes and limit theorems in probability theory." TV **1**, 157–214 (1956).

Rao, C. R., and Varadarajan, V. S.
[1] "Discrimination of Gaussian processes." *Sankhyā*, Ser A **25**, 303–330 (1963).

Rao, M. M.
[1] "Linear functionals on Orlicz spaces: General theory." PJM **25**, 553–585 (1968).
[2] "Conditional measures and operators." JMA **5**, 330–413 (1975).
[3] "Conditional expectations and closed projections." *Proc. Acad. Sci., Amsterdam*, Ser A **68**, 100–112 (1965).
[4] "Inference in stochastic processes—I–V." I, TV **8**, 282–298 (1963); II, ZW **5**, 317–335 (1966); III, ZW **8**, 49–72 (1967); IV, *Sankhyā*, Ser A **36**, 63–120 (1974); V, *Sankhyā*, Ser A **37**, 538–549 (1975).
[5] "Two characterizations of conditional probability." PAMS **59**, 75–80 (1976).
[6] "Abstract Lebesgue–Radon–Nikodým theorems." *Ann. Mat. Pura Appl.* **76** (Ser 4), 107–132 (1967).
[7] "Remarks on a Radon–Nikodým theorem for vector measures," *Proc. Symp. on Vector & Operator-valued Measures & Appl.*, Academic Press, New York, 303–317, 1973.
[8] "Abstract nonlinear prediction and operator martingales." JMA **1**, 129–157 (1971) and Erratum **9**, 614 (1979).
[9] "Covariance analysis of nonstationary time series." Developments in Statist. Volume I, Academic Press, New York, 171–225 (1978).
[10] Stochastic Processes and Integration. Sijthoff and Noordhoff, Alphen ann den Rijn, The Netherlands, 1979.
[11] "Prediction sequences in smooth Banach spaces." *Ann. Inst. H. Poincaré* **8**, 319–332 (1972).
[12] "Abstract martingales and ergodic theory." *Proc. Third Symp. Multivariate Anal.*, Academic Press, New York, 45–60 (1973).
[13] "Conjugate series, convergence, and martingales." *Rev. Roum. Math. Pures et Appl.* **22**, 219–254 (1977).
[14] "Stochastic processes and cylindrical probabilities," *Sankhyā*, Ser A (to appear).

Raoult, J.-P.
[1] "Limites projectives de mesures σ-finites et probabilités conditionnelles.." *C. R. Acad. Sci.*, Ser A (Paris) **260**, 4893–4896 (1965).

Reiter, H.

[1] Classical Harmonic Analysis and Locally Compact Groups. Oxford Univ. Press, London, 1968.

Rényi, A.

[1] "On a new axiomatic theory of probability." *Acta Math. Hung.* **6**, 285–333 (1955).

[2] Foundations of Probability. Holden-Day, San Francisco, California, 1970.

Riesz, F., and Sz.N-agy, B.

[1] "Über Kontractractionen des Hilbertschen Raumen." *Acta Sci. Math. Szeged* **10**, 202–205 (1943).

Rohlin, V. A.

[1] "Lectures on the entropy theory of measure preserving transformations." *Russian Math. Surveys* **22**, 1–52 (1967).

Rosenberg, R. L.

[1] "Orlicz spaces based on families of measures." *Studia Math.* **35**, 15–49 (1970). (Cf. also Ph.D. thesis, Carnegie-Mellon Univ., Pittsburgh, Pennsylvania, 1968.)

Rota, G.-C.

[1] "On the representation of averaging operators." *Rend. Sem. Mat. Univ. Padova* **30**, 52–64 (1960).

[2] "Reynolds operators." *Proc. Symp. Appl. Math. (Am. Math. Soc.)* **16**, 70–83 (1964).

[3] "Une théorie unifiée martingales et des moyennes ergodiques." *C. R. Acad. Sci.* (Paris), Ser A **252**, 2064–2066 (1961).

Royden, H. L.

[1] Real Analysis. MacMillan and Co., New York, 1969.

Rozanov, Yu. A.

[1] Infinite-dimensional Gaussian Distributions. *Proc. Steklov Inst. Math.* **108**, 1971 (English translation, Amer. Math. Soc.).

[2] Stationary Random Processes. Holden-Day, San Francisco, California, 1967 (English translation).

Ryll-Nardzewski, C.

[1] "On quasi-compact measures." *Fund. Math.* **40**, 125–130 (1953).

Sazonov, V. V.

[1] "On perfect measures." *Translations of Am. Math. Soc.* **48** (Ser 2), 229–254 (1965).

Scalora, F. S.

[1] "Abstract martingale convergence theorems." PJM **11**, 347–374 (1961).

Scheffer, C. L.

[1] "Limits of directed projective systems of probability spaces." ZW **13**, 60–80 (1969).

Schreiber, B. M., Sun, T.-C., and Bharucha-Reid, A. T.

[1] "Algebraic models for probability measures associated with stochastic processes." TAMS **158**, 93–105 (1971).

Schwartz, L.

[1] Radon Measures on Arbitrary Topological Spaces and Cylindrical Measures. Tata Institute, Bombay, 1973.

Segal, I. E.

[1] "Equivalence of measure spaces." *Am. J. Math.* **73,** 275–313 (1951).

Shepp, L. A.

[1] "Gaussian measures in function space." PJM **17**, 167–173 (1966).

[2] "Radon-Nikodým derivatives of Gaussian measures." AMS **37**, 321–354 (1966). Correction, AP **5**, 315–317 (1977).

Šidák, Z.

[1] "On relations between strict sense and wide sense conditional expectations." TV **2**, 283–288 (1957).

Sikorski, R.

[1] Boolean Algebras. (3d edition). Springer-Verlag, New York, 1969.

Sion, M.

[1] Introduction to the Methods of Real Analysis. Holt, Rinehart and Winston, New York, 1968.

Sudderth, W. D.

[1] "A 'Fatou equation' for randomly stopped variables." AMS **42**, 2143–2146 (1971).

Tjur, Tue

[1] Conditional Probability Distributions. Lecture Notes No. 2, Institute of Math. Statist., Univ. of Copenhagen, 1974.

Traynor, T.

[1] "An elementary proof of the lifting theorem." PJM **53**, 267–272 (1974).

Uhl, Jr., J. J.

[1] "Abstract martingales in Banach spaces." PAMS **28**, 191–194 (1971).

[2] "Martingales of vector valued set functions." PJM **30**, 533–548 (1969).

Varberg, D. E.

[1] "Gaussian measures and a theorem of T. S. Pitcher." PAMS **63**, 799–807 (1962).

Vasilach, S.

[1] "Direct limits in the categories of nonfinitary heterogeneous algebras." *Rend. Circo. Mat. Palermo* (2) **28**, 337–350 (1979).

Wright, J. D. M.

[1] "Stone algebra valued measures and integrals." *Proc. London Math. Soc.* **19** (Ser 3), 108–122 (1969).

[2] "Applications to averaging operators of the theory of Stone algebra valued modular measures." *Quart. J. Math.* **19** (Ser 2), 321–331 (1968).

[3] "A Radon-Nikodým theorem for Stone algebra valued measures." TAMS **139**, 75–94 (1969).

[4] "Martingale convergence theorems for sequences of Stone algebras." *Proc. Glasgow Math. Soc.* **10**, 77–83 (1969).

Wulbert, D. E.

[1] "A note on the characterization of conditional expectation operators." PJM **34**, 285–288 (1970).

Yosida, K.

[1] Functional Analysis. Springer-Verlag, Berlin, 1965.

Yosida, K., and Hewitt, E.

[1] "Finitely additive measures." TAMS **72**, 46–66 (1952).

Zaanen, A. C.

[1] Integration. North-Holland, Amsterdam, 1967.

Zygmund, A.

[1] Trigonometric Series. Cambridge Univ. Press, London, 1958.

Index

D

E

F

G

H

I

Probability and Mathematical Statistics

A Series of Monographs and Textbooks

Thomas Ferguson. Mathematical Statistics: A Decision Theoretic Approach. 1967

Howard Tucker. A Graduate Course in Probability. 1967

K. R. Parthasarathy. Probability Measures on Metric Spaces. 1967

P. Révész. The Laws of Large Numbers. 1968

H. P. McKean, Jr. Stochastic Integrals. 1969

B. V. Gnedenko, Yu. K. Belyayev, and A. D. Solovyev. Mathematical Methods of Reliability Theory. 1969

Demetrios A. Kappos. Probability Algebras and Stochastic Spaces. 1969

Ivan N. Pesin. Classical and Modern Integration Theories. 1970

S. Vajda. Probabilistic Programming. 1972

Sheldon M. Ross. Introduction to Probability Models. 1972

Robert B. Ash. Real Analysis and Probability. 1972

V. V. Fedorov. Theory of Optimal Experiments. 1972

K. V. Mardia. Statistics of Directional Data. 1972

H. Dym and H. P. McKean. Fourier Series and Integrals. 1972

Tatsuo Kawata. Fourier Analysis in Probability Theory. 1972

Fritz Oberhettinger. Fourier Transforms of Distributions and Their Inverses: A Collection of Tables. 1973

Paul Erdös and Joel Spencer. Probabilistic Methods in Combinatorics. 1973

K. Sarkadi and I. Vincze. Mathematical Methods of Statistical Quality Control. 1973

Michael R. Anderberg. Cluster Analysis for Applications. 1973

W. Hengartner and R. Theodorescu. Concentration Functions. 1973

Kai Lai Chung. A Course in Probability Theory, Second Edition. 1974

L. H. Koopmans. The Spectral Analysis of Time Series. 1974

L. E. Maistrov. Probability Theory: A Historical Sketch. 1974

William F. Stout. Almost Sure Convergence. 1974

E. J. McShane. Stochastic Calculus and Stochastic Models. 1974

Robert B. Ash and Melvin F. Gardner. Topics in Stochastic Processes. 1975

Avner Friedman, Stochastic Differential Equations and Applications, Volume 1, 1975; Volume 2. 1975

Roger Cuppens. Decomposition of Multivariate Probabilities. 1975

Eugene Lukacs. Stochastic Convergence, Second Edition. 1975

H. Dym and H. P. McKean. Gaussian Processes, Function Theory, and the Inverse Spectral Problem. 1976

N. C. Giri. Multivariate Statistical Inference. 1977

Lloyd Fisher and John McDonald. Fixed Effects Analysis of Variance. 1978

Sidney C. Port and Charles J. Stone. Brownian Motion and Classical Potential Theory. 1978

Konrad Jacobs. Measure and Integral. 1978

K. V. Mardia, J. T. Kent, and J. M. Biddy. Multivariate Analysis. 1979

Sri Gopal Mohanty. Lattice Path Counting and Applications. 1979

Y. L. Tong. Probability Inequalities in Multivariate Distributions. 1980

Michel Metivier and J. Pellaumail. Stochastic Integration. 1980

M. B. Priestly, Spectral Analysis and Time Series. 1980

Ishwar V. Basawa and B. L. S. Prakasa Rao, Statistical Inference for Stochastic Processes. 1980

M. Csörgö and P. Révész. Strong Approximations in Probability and Statistics. 1980

Sheldon Ross. Introduction to Probability Models, Second Edition. 1980

P. Hall and C. C. Heyde. Martingale Limit Theory and Its Application. 1980

Imre Csiszár and János Körner, Information Theory: Coding Theorems for Discrete Memoryless Systems. 1981

A. Hald. Statistical Theory of Sampling Inspection by Attributes. 1981

H. Bauer. Probability Theory and Elements of Measure Theory. 1981

M. M. Rao. Foundations of Stochastic Analysis. 1981

in preparation

Jean-Rene Barra. Mathematical Basis of Statistics (Translated and Edited by L. Herbach). 1981